Hydrogen Engines

Hydrogen Engines

Design, Performance Evaluation, Combustion Analysis, and Exhaust Emissions

L.M. Das

Former Professor
Indian Institute of Technology
Hauz Khas, New Delhi

WILEY

Registered Offices
John Wiley & Sons, Inc., 111 River Street, Hoboken, NJ 07030, USA
John Wiley & Sons Ltd, The Atrium, Southern Gate, Chichester, West Sussex, PO19 8SQ, UK

For details of our global editorial offices, customer services, and more information about Wiley products visit us at www.wiley.com.

The manufacturer's authorized representative according to the EU General Product Safety Regulation is Wiley-VCH GmbH, Boschstr. 12, 69469 Weinheim, Germany, e-mail: Product_Safety@wiley.com.

Wiley also publishes its books in a variety of electronic formats and by print-on-demand. Some content that appears in standard print versions of this book may not be available in other formats.

Library of Congress Cataloging-in-Publication Data has been applied for.

Hardback ISBN: 9781394245680
ePDF ISBN: 9781394245703
ePub ISBN: 9781394245697
oBook ISBN: 9781394245710

Cover Design: Wiley
Cover Image: © Andriy Onufriyenko/Getty Images

Set in 7 × 10 inches STIXTwoText by Lumina Datamatics
Printed and bound by CPI Group (UK) Ltd, Croydon, CR0 4YY

C9781394245680_210525

Dedication

To the memory of Professor T. N. Veziroglu, Founding President of the International Association of Hydrogen Energy and Founding Editor-in-Chief of the *International Journal of Hydrogen Energy*, whose profound influence for more than four decades from our first meeting in 1982 sparked my research passion and steered me into hydrogen energy. Sadly, while I was working on this book, he passed away, leaving behind a legacy that will inspire generations of hydrogen enthusiasts and researchers.

Contents

List of Figures and Tables

Chapter 1

Chapter 2

Chapter 3

Chapter 4

Chapter 5

Chapter 6

Chapter 7

Chapter 8

Chapter 9

Chapter 10

The Journey Begins

As we embark on this journey together, we venture into the fascinating world of nearly zero-emission hydrogen engines and vehicles. These pages offer insights into how a sustainable future can be shaped.

Our planet faces two urgent challenges—the rapid plummeting of fossil fuel reserves and escalating environmental degradation—both threatening humankind's survival. Amid this gloomy atmosphere, hydrogen stands out as a transformative force to reshape our current lifestyle. Many countries, in their plans to shift toward energy transition, have integrated hydrogen technology into their strategies. The Conference of Parties to the UNFCCC, more commonly known as COP, has been discussing hydrogen's potential substantively since COP21 (2015) in Paris. That was when hydrogen's role as part of clean energy solutions was recognized within the broader Paris Agreement framework. Since then, hydrogen has featured increasingly prominently at later COPs. At COP26, 32 countries and the European Union committed to making affordable renewable and low-carbon hydrogen globally available by 2030. COP27 further emphasized green hydrogen, highlighting its potential to achieve zero emissions and benefit local communities. The 4th Annual Hydrogen Transition Summit, held in Dubai during COP28 on 7 December 2023, united 500 global stakeholders to drive the hydrogen agenda, expedite clean energy transition, and address the climate crisis with a hydrogen-based economy. At COP29 in November 2024, the COP29 Hydrogen Declaration, highlighted hydrogen as a crucial enabler for decarbonization. "Global energy storage and grids pledge," one of eight areas of commitment at COP29, received a lot of attention to expand energy storage capacities and infrastructure for supporting hydrogen use.

The push for renewable hydrogen energy is driven by concerns over energy security and climate changes aimed at decarbonization of hard-to-abate sectors and net-zero emissions by 2050. Dwindling fossil fuel availability heavily impacts sustainability of internal combustion engines (ICEs), escalating environmental degradation. Fuel cell vehicles, EVs, BEVs, HEVs, and PHEVs are being explored as alternatives to IC engines in several energy sectors. Of course, hybrid electric vehicles (HEVs) and plug-in hybrids (PHEVs) use an internal combustion engine. The environmental impact of BEVs, including greenhouse gas emissions (GHG) and toxicity levels, depends largely on the sources of electricity for charging. Manufacturing batteries for large battery BEVs can contribute up to 200 kg CO_2 eq/kWh to lifetime GHG emissions. So despite clean energy systems supporting BEVs, their production process significantly affects human toxicity potential and freshwater eco-toxicity, mainly due to the metals used in battery production. BEVs can have a human toxicity potential three to five times higher than similarly sized IC engine vehicles. Relevantly, the European Commission has raised concerns about the 2035 deadline to phase out ICE vehicles due to the high costs of EVs and insufficient charging infrastructure. Despite progress in expanding public charging locations, economic and logistical challenges persist.

Given the century-old established technology of the internal combustion engine (ICE), significant efforts are made to ensure its survival by utilizing alternative fuels. CNG, LNG, propane, biofuels, ethanol, methanol, and DME are being explored globally. Uncertainty about the future is the most challenging factor in the shift from petroleum-based fuels to locally available resources. Technological advancements, including turbocharging and direct fuel injection, continue to improve engines' efficiency and reduce emissions while accommodating alternative fuels. As the market transitions, consumer familiarity with IC engine technology, along with evolving regulatory frameworks to meet stringent emission standards, ensure that IC engines retain their importance. Their extensively established global infrastructure—refueling stations, repair networks, and supply chains—continues to offer immense practical advantages. So it is believed that IC engines will remain in use, with the focus on changing the fuel rather than the engine itself. Current R&D on alternative fuels aims at modifying conventional IC engines to operate on specific fuels, providing data and guidelines to design fuel-specific engines. For environmental impacts, a cradle-to-grave assessment is more rational than only considering tailpipe emissions. Currently, over 90% of the global transport system relies on IC engines, while fuel cells and EVs make up less than 10%.

Fuel cell and hydrogen IC engines are neither conflicting nor competing technologies. Rather, they are complementary, aiming at a common objective through different routes. Implementation depends on various factors. Fuel cells require hydrogen with over 99.9% purity, while IC engines can handle some impurities. Green hydrogen production and use in IC engines makes the entire system environmentally sustainable throughout its life cycle. The hydrogen engine isn't a radically novel concept. It dates back almost 200 years with sporadic efforts primarily driven by research curiosity or dire need as in world wars. In recent years, many hydrogen engine test rigs have been constructed in several academic institutes and in the automotive and oil industries. Hydrogen-fueled vehicles have been built in academic institutes at UCLA USA, Mushashi Institute of Technology, Japan, and Indian Institute of Technology, Delhi. There was industrial collaboration and government funding in those efforts. Apart from developing fuel cell vehicles, a number of automotive companies such as BMW, Toyota, Ford, Tata Motors-Cummins, and several others have developed hydrogen-fueled vehicles built on internal combustion engines (H_2ICE vehicles).

This book is tailored for readers with a foundational understanding of internal combustion engines and fuels. It refrains from overemphasizing the advantages of hydrogen-fueled IC engines and instead offers a rational assessment by extensively covering technical aspects of hydrogen's combustion behavior and exhaust emission characteristics. I have discussed advanced applications such as optical techniques, laser diagnostics, simulation, and modeling applied to hydrogen engines.

A single lifetime is a mere ripple in the vast ocean of knowledge required to grasp even a fraction of a subject's profound depth. Any effort to distill even the littlest I know into a single volume is too ambitious. This work humbly aspires to present a broader perspective that hopefully sparks the reader's curiosity to drill deeper into more granular and finer details.

January 23, 2024
Windsor, Ontario
Canada

L.M. Das
Former Professor and Emeritus Professor
Indian Institute of Technology Delhi (IIT-D)
New Delhi 110016, India

About the Author

Dr. Das is a former Professor at the Indian Institute of Technology Delhi (IIT Delhi), India, where he later became Emeritus Professor. He also served as a Visiting Professor at the University of California, Riverside, during 2001–2002. With an academic career spanning over 45 years, Dr. Das has remained deeply committed to teaching and research, focusing primarily on developing low-emission engines and vehicles powered by alternative fuels. His work includes supervising numerous PhD theses and master's projects, as well as publishing extensively in leading international journals. He has supervised a number of PhD theses and master's degree projects in addition to publishing many papers in international journals in this area.

Dr. Das has undertaken prestigious international research assignments in the USA, France, the UK, and Portugal. He has been invited to deliver talks in a wide array of countries, including the USA, Canada, England, France, Germany, Japan, Switzerland, Australia, Belgium, the Netherlands, Thailand, Russia, Turkey, Singapore, the Philippines, Nepal, and Sri Lanka, earning global recognition for his expertise. As Principal Investigator, Dr. Das led several high-impact projects funded by esteemed organizations, such as India's Ministry of New and Renewable Energy (MNRE) and Ministry of Science and Technology; General Motors (USA); the European Union (Brussels); the United Nations Industrial Development Organization (UNIDO); and Shell (UK). In 2012, the Indian Institute of Technology (IIT), Delhi, under his stewardship was successful in launching the first hydrogen-powered three-wheeler in the world in collaboration with Mahindra and Mahindra (India) and Air Products (USA). This innovative project, which was sponsored by the UNIDO with subsequent field trials of the vehicle funded by MNRE, was a significant contribution to sustainable transportation.

Dr. Das has been honored with several prestigious awards throughout his illustrious career including winner at the Innovator's Competition for the DST-Lockheed Martin Indian Innovation Growth Programme (2012); the Outstanding Service Award from the International Association of Hydrogen Energy (2013) for his altruistic contributions to advancing the hydrogen economy; and the T.N. Veziroğlu Award from the Hydrogen Association of India (2013). He was also the recipient of the Biju Patnaik Award for Excellence in Science (2014), conferred by the Odisha Bigyan Academy, and the Lifetime Achievement Award from the International Society of Energy, Environment, and Sustainability (ISEES) (2015), among others. These reflect his unwavering dedication to promoting sustainable energy and environmental solutions. As a researcher and author, he has contributed numerous chapters to internationally published books. His book, *Hydrogen Energy: Production, Storage, Safety, and Application*, was published by Wiley in January 2024. He is currently working on his next book, *Alternative Fuels for Sustainable Transportation: Production, Utilization, and Emissions*, which will be published by Wiley and is expected to be released by early 2027.

Acknowledgments

If truth be told, writing acknowledgments for this book, in a way, seems more challenging than sculpting its contents. Acknowledging the many individuals who inspired and supported me at different stages of my journey could fill an entire chapter on its own. While it may be presumptuous of me to mention each of them by name, please be assured that this book is a reflection of the collective effort of all those who made it happen.

In the early 1980s, witnessing Delhi's billowing vehicular traffic and imagining the disastrous consequences of vehicle pollution, I scrambled to dig deeper into potential solutions. Realizing that imminent radical changes in the designs of well-established engines weren't techno-economically feasible, I shifted gears to focus on low-emission, sustainable fuels. Much like the pollution that hung low, blanketing Delhi roads, among myriad options hydrogen stood out as a low-hanging fruit. I saw that it could be the eventual solution as a fuel with its endless source supply and clean-burning features, with virtually no pollutants from vehicular exhausts. Ironically, the staggering magnitude of vehicular pollution fired in me the relentless zeal to literally drill into hydrogen engines. Over four decades of research, I met numerous souls in classrooms, laboratories, industries, and at national and international conferences. Each interaction—whether teaching, conducting experiments, visiting shop floors, presenting research papers, or reviewing research proposals—often introduced key technical insights that steered my research on a goal-oriented path.

I wish to express my heartfelt gratitude to my doctoral supervisor Prof. H. B. Mathur for passionately encouraging me to investigate the hydrogen engine—a topic far ahead of its time in many countries in the 1980s. My interactions, either face-to-face or in correspondence with esteemed international researchers on hydrogen engines such as Prof. Wm. D. Van Vorst, Prof. P. C. T. De Boer, Prof. Roger Sierens, Dr. Giuseppe Spazzafumo, Prof. S. Furuhama, Prof. Makato Ikegami Dr. W. J. McLean, Prof. M. R. Swain, Prof. G. A. Karim, Mr. Frank Lynch, Mr. J. J. Fagelson, Mr. C. A. Mac Carley, Prof. K. S. Varde, Prof. C. J. Winter, and Prof. Jim Wallace, have sustained my deep interest in delving deeper into the subject. Continuous support from the Ministry of New and Renewable Energy (MNRE), Government of India, through a number of sponsored projects since 1981 has provided the opportunity to conduct tests and derive practically applicable results. Subsequently, optimized backfire-free conditions from lab tests on vehicular engines could be successfully transferred to prototype vehicles and field-tested, thanks to the valuable support of projects sponsored by MNRE and United Nations Industrial Development Organization (UNIDO). Contributions by Mr. G. P. Singh, Mr. Attar Singh, along with the dedicated band of research scholars and lab staff during the experiments, were indeed invaluable.

My heartfelt gratitude goes to Dr. Pawan Goenka, Managing Director of Mahindra and Mahindra, for his visionary initiative in fostering collaboration between industry and academia to develop a prototype hydrogen-fueled vehicle. This project, along with the subsequent road tests, was made possible through the active involvement of Dr. Mathew Abraham and others from Mahindra and Mahindra, and Mr. Ravi Subramanian from Air Products, USA. I also thank Dr. Lymperopoulos Nikolaos and Mr. Federico Villatico from the International Centre for Hydrogen Energy Technologies (ICHET), a UNIDO center, for their unwavering support throughout all stages of the project.

I would be remiss to overlook my family's invaluable contributions and the support I have received. I extend my profound appreciation to Sudhansu Mohanty, my wife Bithika's younger brother, for his meticulous editing that vastly improved the text's readability. I also thank my son-in-law, Himanshu Bhankar, and my daughter, Ishani Das, for their technical input, refined graph designs, and precise formatting, all of which significantly enriched the quality of the book. My sincere thanks go to Dr. Divesh Bhatia for his input on the modeling and simulation chapter. I wish to extend my heartfelt appreciation to the Wiley executives with whom I have had the privilege of interacting on several occasions, beginning with submission of the initial book proposal to its development into the final form.

For the names missed whose encouragement and technical guidance have in one way or another helped this journey, I want to deeply thank them for more than words can convey on this acknowledgment page.

Last but not least, I wish to acknowledge the selfless nudging and unwavering support of my dear wife Bithika Das—alas, not with me today!—who always inspired me to be diligent to pass on every bit of knowledge to the reader, leaving not a shred of value unshared.

To my cherished readers, the ultimate judges, your feedback means the world to me. If my humble efforts stir your interest and appreciation, it would be immensely soul-satisfying and gratifying for me.

Abbreviations

ACIS	Advanced corona ignition system
ACO-BP	Ant colony optimization-back propagation
AFR	Air/fuel ratio
AHP	Analytic hierarchy process
AIT	Autoignition temperature
AMTA	Alberta Motor Transport Association
ANN	Artificial neural network
ARC	Active radical combustion
ASL	Apex seal leakage
ASTM	American Society for Testing and Materials
ATAC	Active thermo-atmosphere combustion
ATI	After-tail ignition
ATR	Autothermal reaction
BDC	Bottom dead center
BEV	Battery electric vehicles
BFL	Backfire limit
BLHSD	Bulk liquid hydrogen storage Dewar
BLIT	Backfire limit injection timing
BMEP	Brake mean effective pressure
BSDC	Brake specific diesel consumption
BSEC	Brake specific energy consumption
BSFC	Brake specific fuel consumption
BTDC	Before top dead center
BTE	Brake thermal efficiency
C_2H_2	Acetylene
C_2H_6	Ethane
C_3H_8	Propane
C_3HF_7	Heptafluoropropane
C_4H_{10}	Butane
CH_3OCH_3	Dimethyl ether
CA	Crank angle
CABARET	Compact accurately boundary-adjusting resolution technique
CAI	Controlled autoignition
CAPEX	Capital expenditures
CARS	Coherent anti-Stokes Raman spectroscopy

CCD	Charge coupled device
CCS	Carbon capture and storage
CCU	Carbon capture and utilization
CCV	Cycle-by-cycle variation
CDC	Colorless distributed combustion
CE-CERT	College of Engineering, Centre for Environmental Research and Technology
CEA	Chemical equilibrium with applications
CFC	Chlorofluorocarbons
CFD	Computational fluid dynamics
CFRP	Carbon fiber-reinforced plastic
CHP	Combined heat and power
CHR	Cumulative heat release
CI	Compression ignition
CICH	Compression ignited homogeneous charge combustion
CIHC	Compression ignited homogeneous charge
CMB	Compagnie Maritime Belge
CNG	Compressed natural gas
CO	Carbon monoxide
CO_2	Carbon dioxide
COG	Coke oven gas
COP	Coil-on-plug (ignition)
COPs	Conference of the parties
COV	Coefficient of variation
CPI	Cryogenic port injection
CPM	Coarse particles method
CR	Compression ratio
CVCB	Constant volume combustion bomb
CVI	Constant volume injection
CVS	Constant volume sampler
DA	Diesel alone
DDT	Deflagration-to-detonation transition
DH	Diesel and hydrogen
DHW	Diesel and hydrogen with water injection
DI	Direct injection
DICI	Direct injection compression ignition
DISI	Direct injection spark ignition
DiSOME	Diesel/dairy scum oil methyl ester
DKI	Dimensionless knock indicator
DLC	Diamond-like coating
DME	Dimethyl ether
DOC	Duration of combustion
DOE	Department of Energy
DR	Dilution rate
DRI	Direct reduced iron
DSOI	Diesel start of injection
DSP	Double spark plug
ECFM	Extended coherent flame model

ECU	Electronic control unit
EFV	Environmentally friendly vehicles
EGR	Exhaust gas recirculation
EnTR	Ensembles of trees
EOI	End of injection
EPC	Exergetic performance coefficient
EPO	Exhaust port opening
EV	Electric vehicles
EVC	Exhaust valve closing
FDS	Fire dynamics simulator
FDT	Flame development time
FEA	Finite element analysis
FFT	Fast Fourier transform
FRT	Flame rise time
FTP	Federal test procedure
GA-BP	Genetic algorithm-back propagation
GDI	Gasoline direct injection
GES	Green Energy Solutions
GGFES	Gaseous fuel energy share
GPR	Gaussian process regression
GRPE	Gasoline reciprocating piston engine
H2ICE	Hydrogen-fueled internal combustion engine
H_2O_2	Hydrogen peroxide
H_2S	Hydrogen sulfide
H_2SO_4	Sulfuric acid
HC	Hydrocarbon
HCCI	Homogeneous charge compression ignition
HCDC	Homogeneous charge compression ignition diesel combustion
HCNG	Hydrogen + compressed natural gas
HDI	Hydrogen direct injection
HENGE	Hydrogen-enriched natural gas engine
HES	Hydrogen energy share
HFCP	High-frequency component of (in-cylinder) pressure
HHV	Higher heating value
HI	Head ignition
HIA	Hydrogen injection angle
HICE	Hydrogen internal combustion engine
HIL	Hardware-in-the-loop
HIT	Hydrogen induction technique
HIT	Hydrogen injection timing
HLIA	High-pressure laser ignition apparatus
HNG	Hydrogen natural gas
HPDI	High-pressure direct injection
HRE	Hydrogen-fueled rotary engine
HRPE	Hydrogen reciprocating piston engine
HRR	Heat release rate
HRR_{max}	Maximum heat release rate

HRS	Hydrogen refueling station
HUCR	Highest useful compression ratio
HVO	Hydrotreated vegetable oil
HWRE	Hydrogen Wankel rotary engine Hythane Hydrogen + methane
IAA	Internationale Automobil-Ausstellung
IC	Internal combustion
ICCD	Intensified charge-coupled device
ICE	Internal combustion engine
ICEV	Internal combustion engine vehicle (vehicle built on internal combustion engine)
IDT	Ignition delay time
IFP	Institut Français du Pétrole
IIR	Infinite impulse response
IIT	Indian Institute of Technology
IITD	Indian Institute of Technology, Delhi
IMEP	Indicated mean effective pressure
IMPO	Integral of modulus of pressure oscillations
IMSC	In-cylinder mildly stratified charge
IPC	Intake port closing
IR	Infrared
IRENA	International Renewable Energy Agency
ISSC	In-cylinder severely stratified charge
ITE	Indicated thermal efficiency
ITU	Intermodal transport unit
IVO	Inlet valve opening
KD	Knock duration
KI	Knock intensity
KLER	Knock-limiting equivalence ratio
KLλ	Minimum λ limited by knock
KPI	Key performance indicator
L	Thermal flame thickness
λ	(Lambda) Excess factor
LANL	Los Alamos National Laboratory
LEL	Lower explosion limit
LES	Large eddy simulation
LFL	Lower flammability limit
LFS	Laminar flame speed
LH_2	Liquid hydrogen
LHC	Lean homogeneous combustion
LHV	Low heat value
LHV	Lower heating value
LHV	Lower heating value in CFD (computational fluid dynamics)
LIBS	Laser-induced breakdown spectroscopy
LIF	Laser-induced fluorescence
LII	Laser-induced ignition
LIPI	Laser-induced plasma ignition
LIRIAM	Late injection, rapid ignition and mixing

LNG	Liquefied natural gas
LOL	Lean operating limit
LPDI	Low-pressure direct injection
LPG	Liquefied petroleum gas
LR	Linear regression
LSC	Lean stratified charge
LTC	Low-temperature combustion
LTR	Low-temperature reaction
LWI	Low wobble index
M_a	Markstein number
MAP	Manifold absolute pressure
MAP	Mean average precision
MAPO	Maximum amplitude of pressure oscillations
MAS	Multi-agent system
MBT	Minimum for best torque
MESG	Maximum experimental safe gap
M_F	Mass of the fuel dose burned
MFB	Mass fraction burned
MFT	Mixture formation time
MHRR	Maximum heat release rate
MIE	Minimum ignition energy
ML	Machine learning
MLIT	Ministry of Land, Infrastructure, Transport and Tourism
MLS-SVR	Multi-output least squares-support vector regression
MN	Methane number
MNRE	Ministry of New and Renewable Energy
MON	Motor octane number
MPRR	Maximum pressure rise rate
MRI	Magnetic resonance imaging
MSE	Mean square error
M_T	Total mass of the fuel dose burned in the working cycle
MTPA	Metric tons per annum
MULDIC	Multiple staged diesel combustion
N_2O	Nitrous oxide
NCHR	Normalized cumulative heat release
NEDC	New European Driving Cycle
NH_3	Ammonia
NHRR	Net heat release rate
NIMEP	Net indicated mean effective pressure
NMHC	Non-methane hydrocarbons
NO	Nitric oxide
NO_2	Nitrogen dioxide
NOx	Oxides of nitrogen
NRG	Numerical reactive gas-dynamics (software package)
NRPD	Nanosecond repetitive pulse discharge
NSR	catalyst (NOx storage-reduction catalyst)
NTSEL	National Traffic Safety and Environment Laboratory

NVO	Negative valve overlap
NZE	Net zero emissions
OPEX	Operational expenditures
OPF	Outwardly propagating spherical flame
ORC	Organic ranking cycle
PAH	Polycyclic aromatic hydrocarbons
PCC	Plume ignition combustion
PCCI	Premixed charge compression ignition
PCIC	Premixed compression ignited combustion
PCSI	Pre-chamber spark ignition
PCV	Positive crankcase ventilation
PDE	Pulse detonation engine
PDF	Probability density function
PDI	Phase Doppler interferometer
PFI	Port fuel injection
PFI-HICE	Port fuel injection hydrogen engines
PG	Producer gas
PIV	Particle image velocimeter
PLIF	Planar laser-induced fluorescence
PM	Particulate matter
P_{max}	Maximum cylinder pressure
PMLS	Physical model with machine learning laminar flame speed post-ivc after the intake valve closure
PP	Peak pressure
PPD	Power per displacement
ppm	Parts per million
PPP	Peak pressure position
PRD	Pressure relief device
pre-ivc	Prior to intake valve closure
PREDIC	Premixed lean diesel combustion
PRR	Pressure release rate
PRR	Pressure rise rate
PTI	Polytropic index
PTOs	Power take-offs
PTV	Particle tracking velocimetry
ϕ	Equivalence ratio
Q_{ch}	Chemical energy released by combustion
RANS	Reynolds-averaged Navier–Stokes
R_c	Radius of curvature of the flame front
RCCI	Reactivity controlled compression ignition
RCEM	Rapid compression and expansion machine
RF	Radio frequency
RFSI	Radio frequency sustained plasma ignition
RFSP	Radio frequency spark plug
RHRR	Representative heat release rate
RMS	Root mean square
RNG	Renewable natural gas

RNG	Renormalization group
ROHR	Rate of heat release
RON	Research octane number
RPE	Reciprocating piston engine
RPM	Revolutions per minute
RRC	Reaction rate constant
RSM	Response surface methodology
RSM	Reynolds stress model
RTHR	Representative total heat release
SCC	Stratified charge combustion
SCI	Serial communication interface
SCR	Selective catalytic reduction
SEMA	Specialty Equipment Market Association
SEOI	Secondary end of injection
SI	Spark ignition
SIBS	Spark-induced breakdown spectroscopy
SIMF	Secondary injection mass fraction
SKI	Standard knock intensity
SMR	Steam methane reforming
SNSC	Spark plug mildly stratified charge
SO_2	Sulfur dioxide
SOI	Start of injection
SOM	Self-organizing maps
SRC	Spearman rank correlation
SSP	Single spark plug
SSSC	Spark plug severely stratified charge
SST	Shear stress transport
ST	Spark timing
SVM	Support vector machines
TAN	Total acid number
TBC	Thermal barrier coating
TBN	Total base number
TCPL	Tata Cummins Private Limited
TDC	Top dead center
TEA	Triethylamine
TEL	Tetraethyl lead
TFC	Tsuneishi Facilities & Craft
TFC	Turbulent flame speed closure
THC	Total hydrocarbon
TI	Tail ignition
TJI	Turbulent jet ignition
TKE	Turbulent kinetic energy
TKM	Thermo-kinetic model
Tkm	Tonne-kilometer
TMA	Trimethylamine
TMI	Timed manifold injection
TR	Regression tree

TTL	Transistor–transistor logic
TWC	Three-way catalyst
UBHC	Unburned hydrocarbons
UCLA	University of California, Los Angeles
UEL	Upper explosion limit
UFL	Upper flammability limit
UHC	Unburned hydrocarbon
UNFCC	United Nations Framework Convention on Climate Change
UNIDO	United Nations Industrial Development Organization
VGT	Variable geometry turbocharger
VOC	Volatile organic compounds
VVT	Variable valve timing
WFOME	Waste frying oil methyl ester
WLTC	Worldwide harmonized light duty test cycle
WOT	Wide-open throttle
WRC	World Rally Championship
WRE	Wankel rotary engine

About the Companion Website

This book is accompanied by a companion website that includes color versions of figures:

www.wiley.com/go/das/hydrogenengines

1

Hydrogen: A Promising Frontier for Energy-Environment Solutions

No force on earth can stop an idea whose time has come.

Victor Hugo

1.1 Introduction

The world, at present, is caught in the clutches of two formidable crises: the relentless depletion of fossil fuels and the rapid pace of environmental degradation. Geopolitical complexities lead to localized fuel crises, and geographic disparities lead to the exhaustion of our finite petroleum reservoir. At the same time, rapid industrial expansion has sparked an insatiable demand for energy globally. The growing world population and radical change in lifestyle patterns put further strain on energy resources. Shortages in the energy supply affect various industrial sectors, including transportation, power generation, agriculture, and chemical production. Addressing these challenges requires a rational comprehensive strategy that diversifies energy sources, invests in renewable energy infrastructure, and promotes energy efficiency. A wide range of cataclysmic events, both natural and anthropogenic, threaten the very fabric of our planet. A number of environmental problems—from floods, cyclones, heavy rain, deforestation, soil erosion, etc., to global warming, climate change, greenhouse gas emissions, acid rain, ozone layer depletion, smog, urban heat islands, and vehicular pollution—demand a paradigm shift toward renewable energy sources to ensure energy security and the reordering of nature and its sustainability.

1.2 Fossil Fuel Combustion: Energy Security and Environmental Impact

Consumption rates and projected depletion deadlines of fossil fuels vary among researchers based on their projected assumptions and analytical methodologies.

The life cycle of the world oil industry extends beyond 300 years (Figure 1.1). Oil production (in Gtoe or gigatons of oil equivalent) has taken an upward trend since 1986 [1]. Figure 1.2 shows global energy consumption patterns during the past two decades based on data by the International Energy Agency [2]. The transport sector, in which hydrogen is likely to play a significant role, shows a higher consumption level than other sectors.

Hydrogen Engines: Design, Performance Evaluation, Combustion Analysis, and Exhaust Emissions, First Edition.
Lalit Mohan Das.

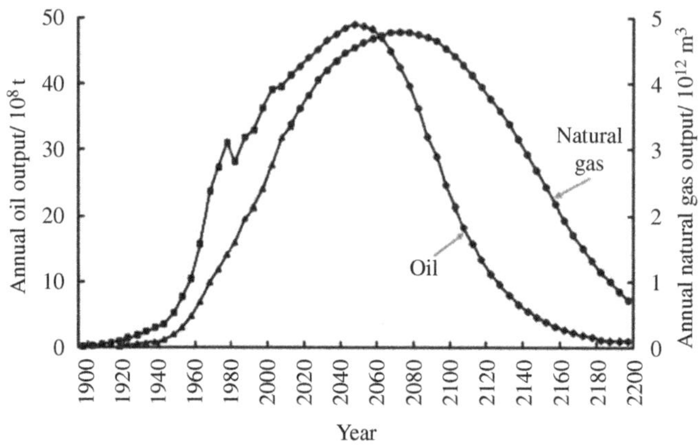

Figure 1.1 Projections of peak oil production [1] / with permission of Elsevier.

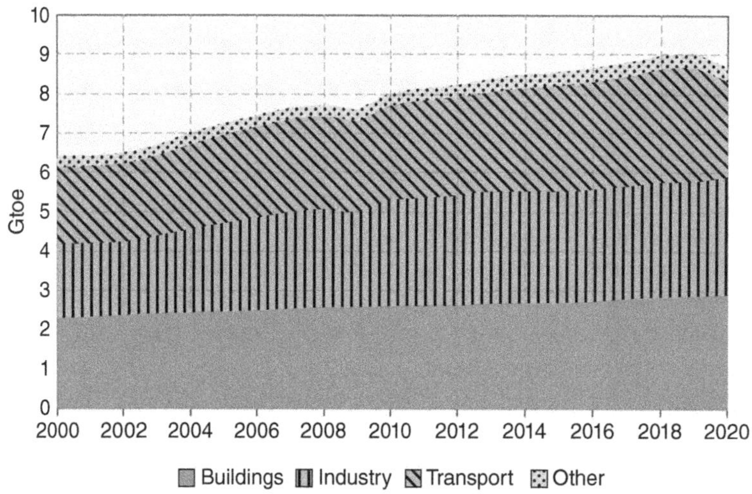

Figure 1.2 Pattern of global energy consumption [2] / with permission of Elsevier.

1.3 Alternative Fuels for Internal Combustion (IC) Engines

Reciprocating internal combustion engines (ICE), operating on liquid petroleum fuels, have long dominated land transportation. The extensive infrastructure developed around them over the past century is now under threat from dwindling petroleum resources. Scarcity of petroleum fuels spurs a rapid transition toward alternative fuels and electric vehicles (EVs) for sustainable transportation solutions. Hydrogen, compressed natural gas (CNG), biodiesel, ethanol, methanol, propane, green diesel or hydro-treated vegetable oil (HVO), hydrogen-added natural gas, ammonia, butanol, butane, liquefied petroleum gas (LPG), dimethyl ether, renewable natural gas (RNG), methane hydrates, diethyl ether, and P-series fuels present a wide range of alternatives to traditional petroleum-based fuels. Some of these fuels lack widespread recognition. Some others are still in early stages of development. However, a few have already found successful applications in engines and prototype vehicles. Vehicles operating solely on electric power are referred to as electric vehicles (EV) or battery electric vehicles (BEVs) in the automobile glossary. The transition to fully electric vehicles faces significant challenges. For instance, the mining of the metals required for batteries raises serious environmental and human toxicity issues. Overcoming these challenges demands substantial investment and innovation in battery recycling and sustainable mining practice. In some cases, urban air with particulates, oxides of nitrogen (NOx) and sulfur dioxide (SO_2), is worse for BEVs than for IC engine vehicles [3].

1.4 Pollution from Vehicles

Figure 1.3 depicts the major sources of air pollution due to a vehicle with an internal combustion engine (ICEV). In an IC engine, incomplete combustion often occurs, leading to formation of carbon monoxide (CO). The majority of carbon dioxide emissions typically occur outside the engine upon release into the atmosphere. In the context of overall environmental degradation, CO_2 is of great concern. It is a potent greenhouse gas responsible for global warming, ozone layer depletion, and other associated environmental problems. As shown in Figure 1.4, the atmospheric CO_2 level keeps increasing. In the graph, the monthly CO_2 concentrations from observations up to 2023 (black) are compared to the expected future projection of 1.5 °C global warming (purple). The figure also shows IPCC data and the Met Office forecast for 2024 [4].

There has been a steady uptick in CO_2 concentration in the atmosphere. Despite global lockdowns and reduced economic activity during the COVID-19 pandemic, there was no overall tangible impact on atmospheric levels of greenhouse gas emissions. Although emissions temporarily declined during the lockdown period, they bounced back rapidly afterward. On May 20, 2024, CO_2 emissions reached 426.28 ppm, according to NOAA Global Monitoring Laboratory (Mauna Loa Observatory).

Table 1.1 shows the pollutants coming out of the exhaust pipe of an automobile operated by an internal combustion engine and the characteristics of the pollution this causes in the atmosphere. Hydrogen combustion does not produce any of the pollutants listed in Table 1.1 except nitrogen oxides (NOx). There are several techniques to reduce NOx concentration and develop an almost zero-emission engine as described in Chapter 3.

Representative concentration pathways (RCPs) are different climate change scenarios with projected greenhouse gas concentrations. The four RCPs—RCP2.6, RCP4.5, RCP6, and RCP8.5—are labeled after a possible range of radioactive forcing values in year 2100 (2.6, 4.5, 6, and 8.5 W/m^2,

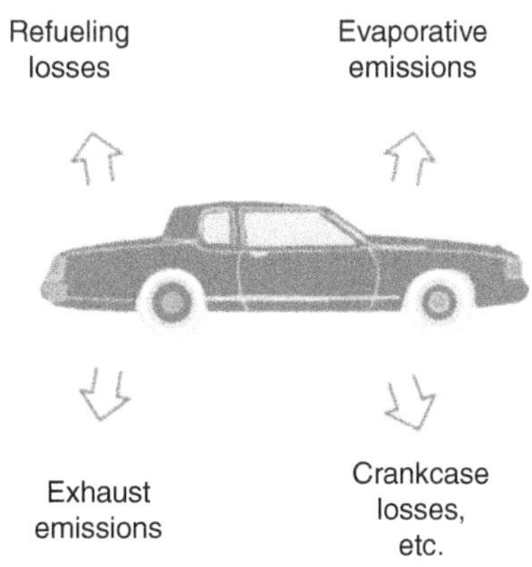

Figure 1.3 ICEV as a source of air pollution.

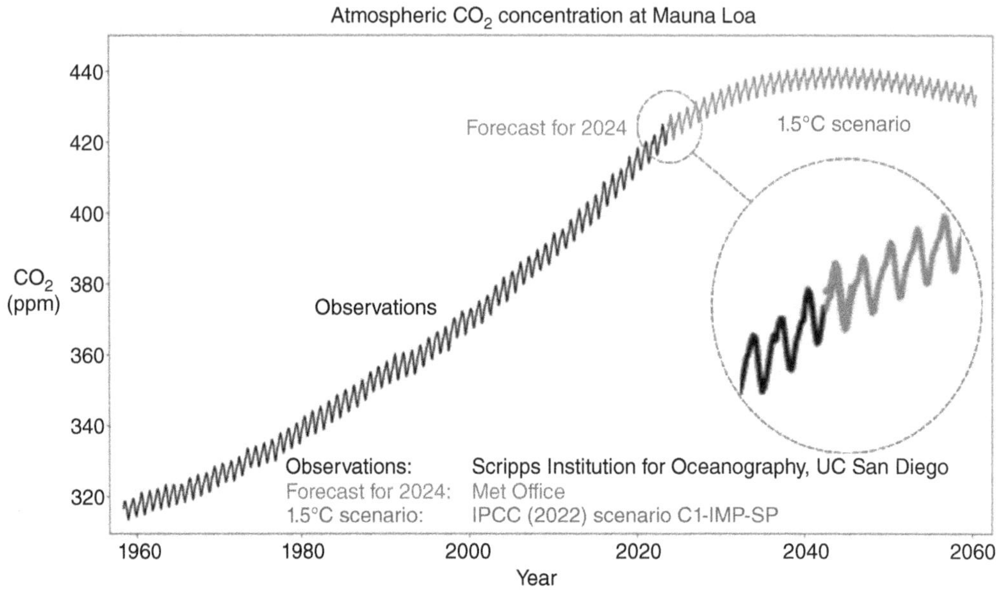

Figure 1.4 Increasing trend of atmospheric CO_2 concentration [4] / Earth.Org.

Table 1.1 Major pollutants from vehicles.

Pollutant	Effect on environment
Carbon monoxide (CO)	Reduces blood's ability to carry oxygen to organs and tissues, leading to headaches, dizziness, and nausea and impairs cognitive functions. Prolonged exposure to low levels of CO can cause chronic health problems, especially for individuals with cardiovascular diseases. CO contributes to the formation of ground-level ozone, a key component of smog.
Unburned hydrocarbons (UBHC) and volatile organic compounds (VOC)	UBHC contributes to the formation of ground-level ozone (smog), which can lead to respiratory issues and reduced visibility, cardiovascular problems, and can aggravate conditions of asthma and bronchitis. Volatile organic compounds (VOCs) are organic molecules, specifically hydrocarbons, that are volatile and evaporate quickly into the air. They react with nitrogen oxides in the air to form ground-level ozone and contribute to the formation of secondary particulate matter. VOCs can react with other pollutants in the atmosphere to form secondary pollutants such as formaldehyde and acrolein, which are toxic to both humans and the ecosystem.
Oxides of nitrogen (NOx)	Contributes to the formation of smog, acid rain, and eutrophication. Inhalation of NOx can aggravate respiratory conditions such as asthma, bronchitis, and emphysema. Contributes to the formation of fine particulate matter.

Pollutant	Effect on environment
Particulate matter (PM)	Can penetrate deep into the lungs, causing respiratory and cardiovascular problems. Contributes to reduced air quality, especially in urban areas, leading to visibility reduction and haze. Carries toxic substances and heavy metals, posing health risks when inhaled.
Sulfur dioxide SO_2	Reacts with water vapor, oxygen, and other chemicals in the atmosphere to form sulfuric acid (H_2SO_4), which contributes to the formation of acid rain. Acid rain can damage vegetation, soil, aquatic ecosystems, and infrastructure such as buildings and statues. Causes respiratory problems, particularly for individuals with pre-existing conditions like asthma or bronchitis.
CFC	Emitted from the coolant in vehicles when not properly disposed of. The CFCs lead to depletion of the ozone layer. Sometimes they also cause skin cancer.

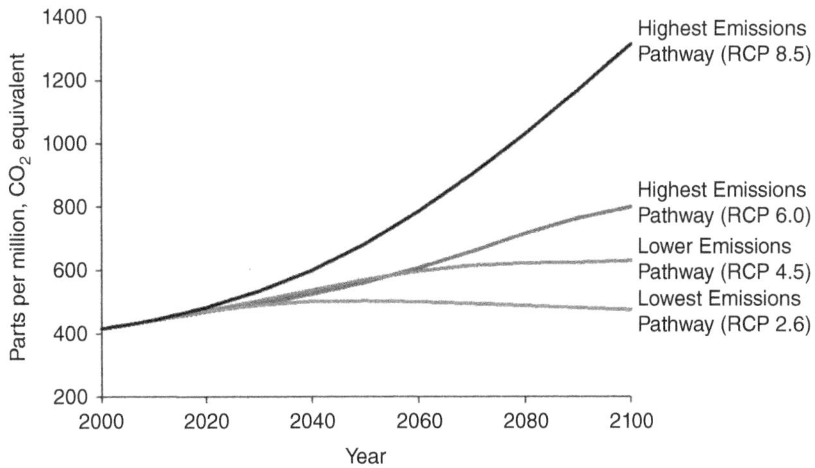

Figure 1.5a (a) Projected atmospheric greenhouse gas concentration. *Source:* IPCC / https://www.iamrenew.com/environment/crossing-carbon-threshold-at-415-ppm-what-does-it-mean/.

respectively). Figure 1.5a shows projected greenhouse gas concentrations for four different emissions pathways. The most stringent case (top pathway with RCP 8.5) assumes that greenhouse gas emissions will continue to rise throughout the current century. The mildest case (bottom pathway with RCP2.6) assumes emissions peaks between 2010 and 2020, after which it takes a declining trend. Figure 1.5b shows the projected mean global temperature. The vertical bars on the right indicate the likely ranges in temperature by the end of the century.

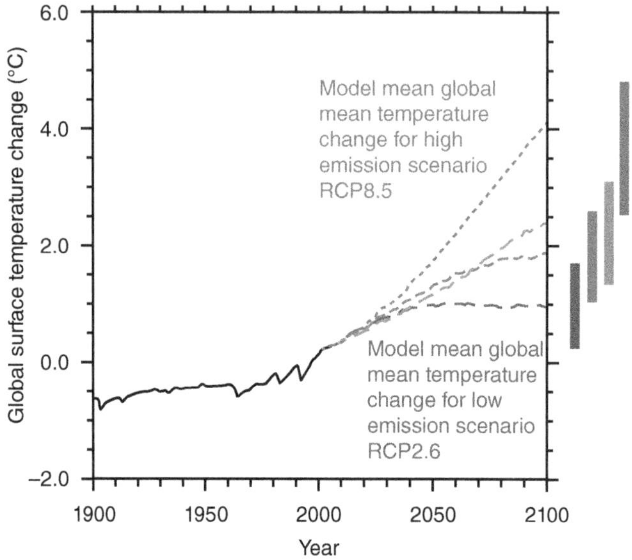

Figure 1.5b (b) Projected mean global temperature. *Source:* United States Environmental Protection Agency / https://climatechange.chicago.gov/climate-change-science/future-climate-change / Public domain. Please visit the companion website to view a color version of this image.

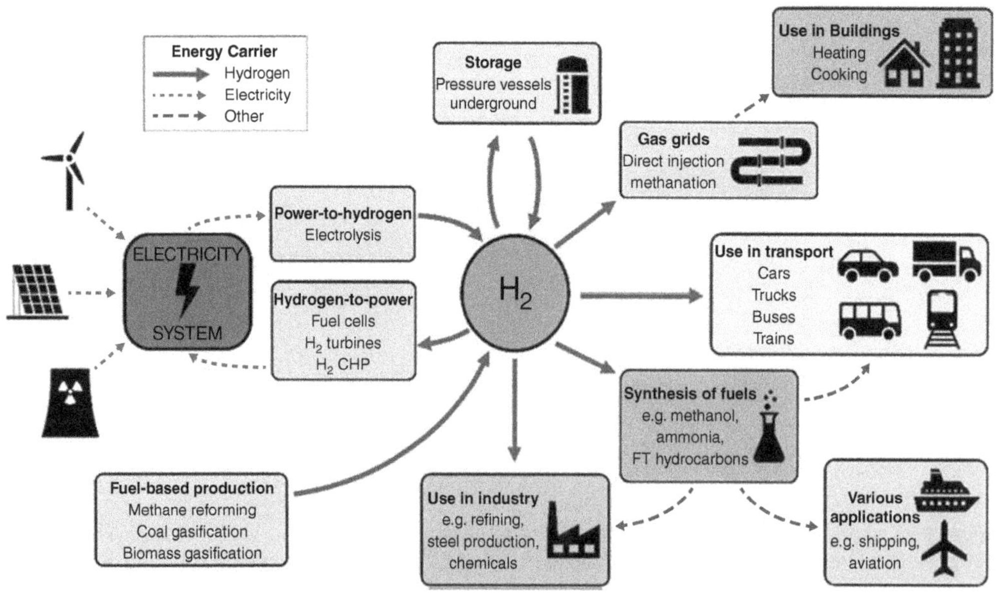

Figure 1.6 Hydrogen production and utilization pathways [5] / The Royal Society of Chemistry / CC BY 4.0.

1.5 Hydrogen Energy Pathways

Details of the production, storage, transport, safety, and utilization of hydrogen with life cycle analysis are important to have an overall picture of hydrogen's role in emerging energy-environment crises. There are several pathways for hydrogen production and use in an energy system. Figure 1.6 shows an overview of some of the major pathways.

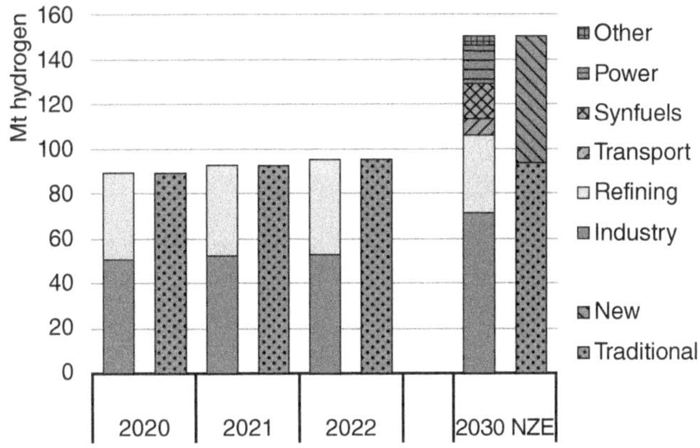

Figure 1.7 Hydrogen use by sector in net zero emissions by 2050 (2020–2030) [7] / International Energy Agency / CC BY 4.0.

1.6 Hydrogen in a Net Zero Emission (NZE) Scenario

Given its infinite source potential and clean-burning characteristics, hydrogen is emerging as one of the most sustainable environment-friendly alternative fuels. A global drive toward decarbonization and a stress on energy security have brought forward low-emission hydrogen to the leading edge of sustainable energy solutions. Chapter 2 gives details of properties of hydrogen and its suitability for application in engines.

Hydrogen's global use expands from less than 90 million Mt (metric tonnes) in 2020 to over 200 Mt by 2030. The proportion of low-carbon hydrogen in the global hydrogen market rises significantly from 10% in 2020 to 70% in 2030. This trend reflects the sustained efforts to decarbonize the hydrogen production process. At present, in several countries, hydrogen is blended with natural gas in gas networks to reduce carbon emissions and also to ensure hydrogen's early entry into the global energy infrastructure. Projections suggest that by 2030, the global standard blend will comprise 15% hydrogen in volume. This blending strategy would contribute to a reduction of approximately 6% in CO_2 emissions from gas consumption [6].

Figure 1.7 shows hydrogen use in different sectors during the span 2020 to 2030 considering the net zero emissions by 2050 (NZE) scenario. It is clear from the figure that although hydrogen use continues to increase, it still for now remains concentrated in traditional applications [7].

1.7 Hydrogen Color Codes and Carbon Footprint

It is extremely fascinating that the colorless gas hydrogen is often associated with a spectrum of colors. Depending upon the feedstock and mode of hydrogen production, hydrogen is categorized into several colors (Figure 1.8).

There is no universally accepted code for hydrogen. Orange hydrogen is not commonly mentioned in the literature. In some literature, even minute quantities of naturally occurring hydrogen are classified as "white" hydrogen. Pyrolysis of methane produces turquoise hydrogen. Turquoise hydrogen is a recent entrant into the color spectrum and is still in the pilot phase [8]. Universally identified colors of hydrogen based on production mechanism and emission character are grey, blue, and green hydrogen (Figure 1.9). Grey hydrogen is the most common form and is generated from natural gas, or methane, through the process of "steam reforming." Sometimes, depending on the coal, black

COLOR	DESCRIPTION: FEEDSTOCK
	Grey: natural gas reforming without CCUS
	Brown: brown coal (lignite) as feedstock
	Blue: natural gas reforming with CCUS
	Green: electrolysis powered through renewable electricity
	Pink: electrolysis powered through nuclear energy
	Turquoise: methane pyrolysis
	Yellow: electrolysis powered through electricity from solar
	Orange: electrolysis powered through electricity from wind

Figure 1.8 Hydrogen color spectrum. Please visit the companion website to view a color version of this image.

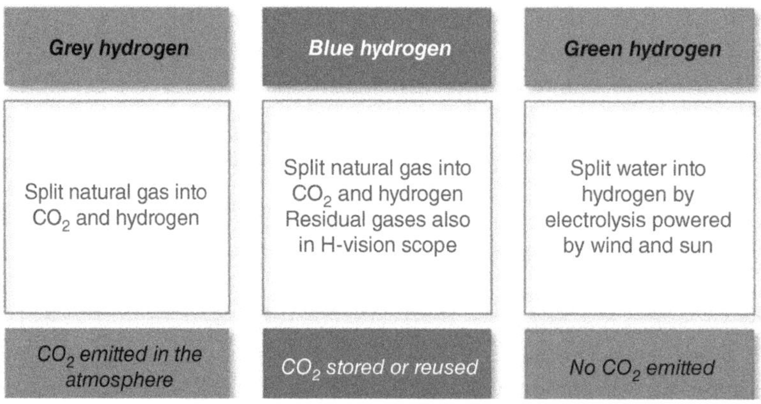

Figure 1.9 Hydrogen color codes: production and emission. Please visit the companion website to view a color version of this image.

(bituminous) and brown (lignite) codes are used. Grey hydrogen inherently releases carbon emissions during its production process and contributes to greenhouse gas emissions, impacting climate change. It has been estimated that the sum of emissions of carbon dioxide (75.6.0 g CO_2 per MJ) and unburned methane (77.4 g CO_{2eq} per MJ) to produce grey hydrogen is 153 g CO_{2eq} per MJ [9]. Traditionally blue hydrogen, because of its carbon-capture ability, is projected as a preferred option to grey hydrogen. However, recent scientific studies challenging climate benefits of blue hydrogen have put into question the true environmental impact of blue hydrogen. Analysis by Howarth et al. [9] discusses the life cycle of greenhouse gas emissions associated with blue hydrogen, taking into account both carbon dioxide emissions and unburned fugitive methane. This study shows that release of fugitive methane, in particular, contributes significantly to the overall emissions profile of blue hydrogen. The greenhouse gas footprint of blue hydrogen, even when carbon dioxide is captured from exhaust flue gases, can be as large as or even larger than that of natural gas.

Ongoing research might introduce novel feedstocks and production methods, changing hydrogen categorization on the basis of colors. However, evaluation of carbon intensity of major hydrogen production processes will continue to be more relevant with decarbonization as the goal. In today's era of environmental consciousness, the focus must shift decisively toward green hydrogen.

1.8 Green Hydrogen

Green hydrogen, produced through electrolysis powered by renewable energy sources such as wind, solar, and water, holds the promise of power without the destructive impact on ecosystems. Embracing green hydrogen is not just a choice but an absolute necessity for a cleaner, sustainable future. Related literature [7–12] discusses several aspects of green hydrogen, including methods of production, benefits, and challenges. The ultimate goal for green hydrogen use is to achieve zero emissions across all stages of fuel production, storage, and utilization. Bion and colleagues [13] have discussed the design of a nano catalyst for production of green hydrogen from bioethanol. In this method, raw feedstock is used without significantly degrading the catalyst. Composition of bioethanol from biomass, catalytic ethanol steam reforming, and the design of catalysts for hydrogen production from bioethanol feed have been discussed.

Despite rapid global research and development efforts to reduce costs and improve production methods, researchers and developers have not yet fully commercialized green hydrogen technology. Currently, worldwide hydrogen production surpasses 120 million metric tonnes, two-thirds of which are pure hydrogen and the rest in the form of a mixture of other gases [14]. The significance of green hydrogen has received global attention, and more than 120 countries have committed to achieving net zero emissions, with seven nations having already attained this milestone [15]. The rapid shift toward a sustainable future in today's global markets is emerging with hydrogen as crucial player, and it is estimated that hydrogen production (Figure 1.10) capacity of nearly 600 million metric tonnes per annum (MTPA) will be necessary to achieve carbon neutrality targets for 2050 [16].

Figure 1.11 gives an overview of the hydrogen energy ecosystem, including its sources, production methods, status, and diverse applications across various sectors [17]. Of all the routes shown in the figure, some are technologically mature, while others are still in the R&D stage. The present thrust is on maximizing sustainable green hydrogen production at a reasonable cost. The International Renewable Energy Association (IRENA), in one of its studies [18], discussed the importance of adopting a comprehensive approach to designing and operating systems for green

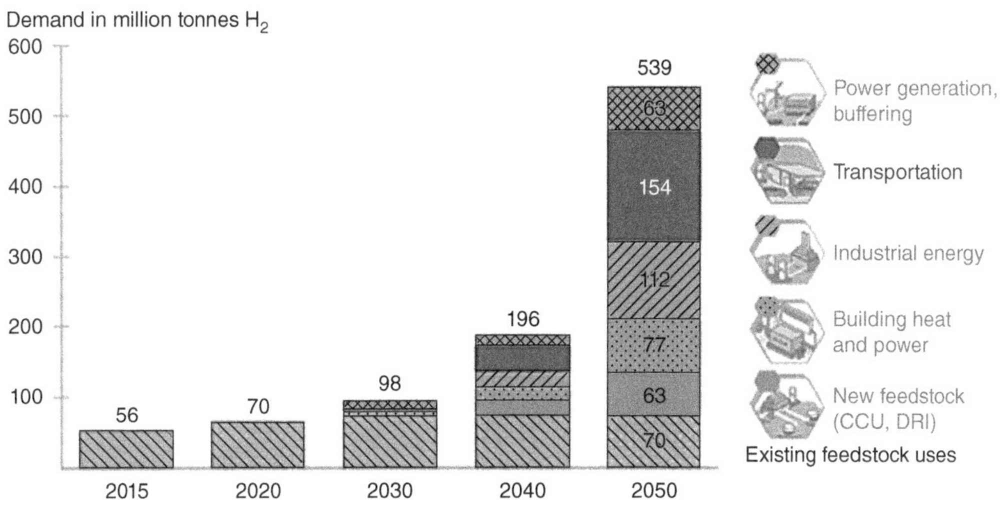

Figure 1.10 Growing demand of green hydrogen. Adapted from [16].

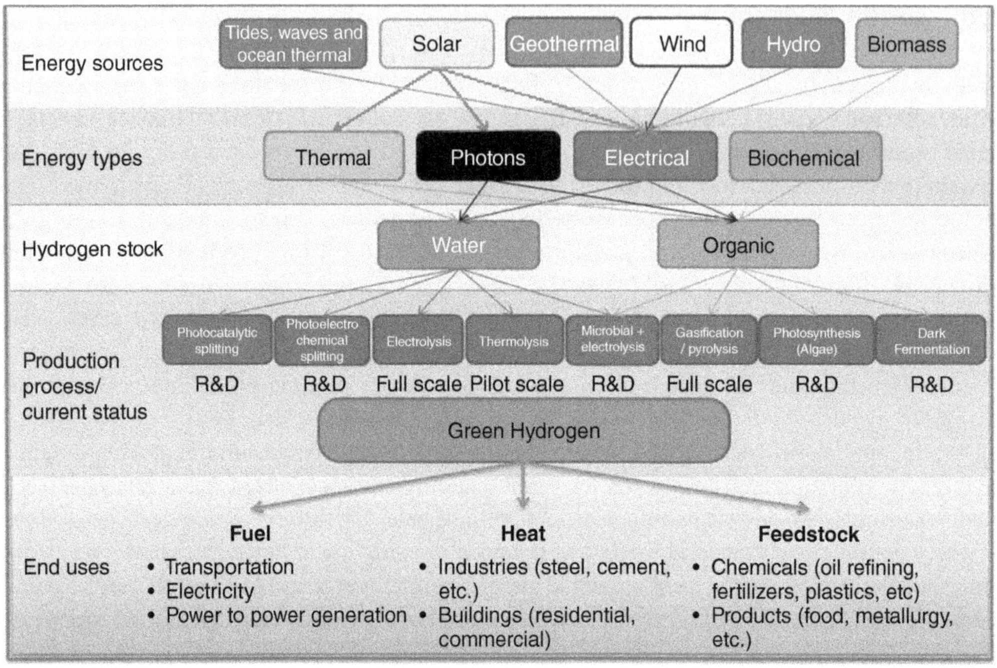

Figure 1.11 Hydrogen production and application of green hydrogen [17] / with permission of Elsevier.

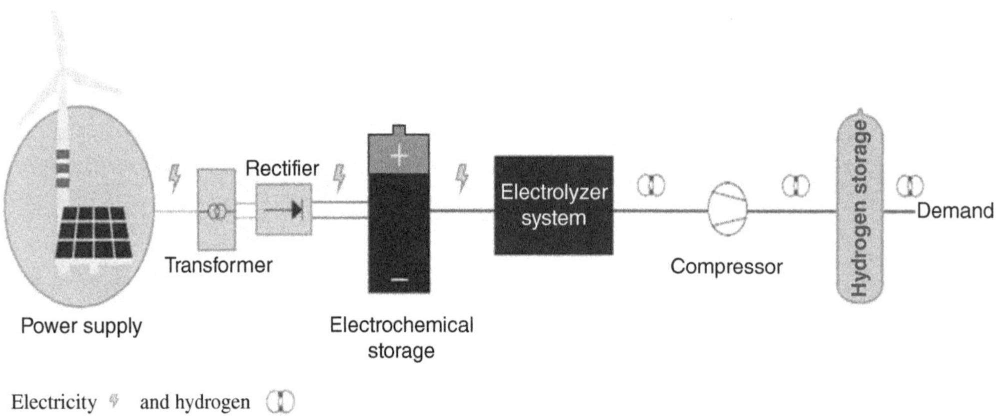

Figure 1.12 Green hydrogen production facility that includes electricity and hydrogen storage on-site [18] / IRENA / Public domain.

hydrogen production. Storage plays a crucial role in decoupling variable supply from hydrogen demand. Electrochemical storage manages short-term fluctuations before electrolysis, whereas hydrogen storage handles long-term fluctuations after electrolysis.

The combination of electrolyzer technology with electricity and hydrogen storage offers flexible supply-and-demand management (Figure 1.12).

1.9 Electrolysis Development: Energy Demand and Global Warming Potential

In their review paper, Yu et al. [19] discuss the recent developments in high-performance, earth-abundant, nonnoble electrocatalysts for hydrogen and oxygen evolution reactions, and overall water splitting. Work is being carried out to identify the most efficient catalysts that can withstand accelerated degradation under large current densities (500 mA/cm^2) over long periods. Currently, water electrolysis, despite its potential for zero-carbon hydrogen production, accounts for only 4% of global hydrogen production, mostly as a by-product of the chlor-alkali industry [20]. However, in a study in Jordan it has been proposed that if the hydrogen from chlor-alkali plants, which is often wasted, can be recovered, it could significantly contribute to on-site process heat [21]. Alkaline water electrolysis (AWE) is a mature technology and can be used for large-scale green hydrogen production. However, optimization of the process and development of affordable, active, and durable electrocatalysts are essential especially for the oxygen evolution reaction (OER). Atom economy is a crucial factor that needs to be considered for a sustainable catalyst design. Iron impurities in commercial KOH can activate some electrocatalysts but may cause deactivation in industrial applications due to contaminants. Therefore, the impact of these impurities at an AWE pilot plant needs to be investigated [22].

The journey of water electrolysis technology began in the 18th century; the first generation laid the groundwork, followed by a series of evolutions. Throughout the entire period of development in water electrolysis, several challenges have been overcome through technological innovations, resulting in substantial progress in each stage of application as shown in Figure 1.13, reproduced from a research paper by Shiva Kumar and Lim [23]. Materials play a critical role in global warming due to their extraction, processing, and usage. A study by Nuss and Eckelman [24] shows global warming potential and cumulative energy demand for critical materials used in electrolyzers (Figure 1.14).

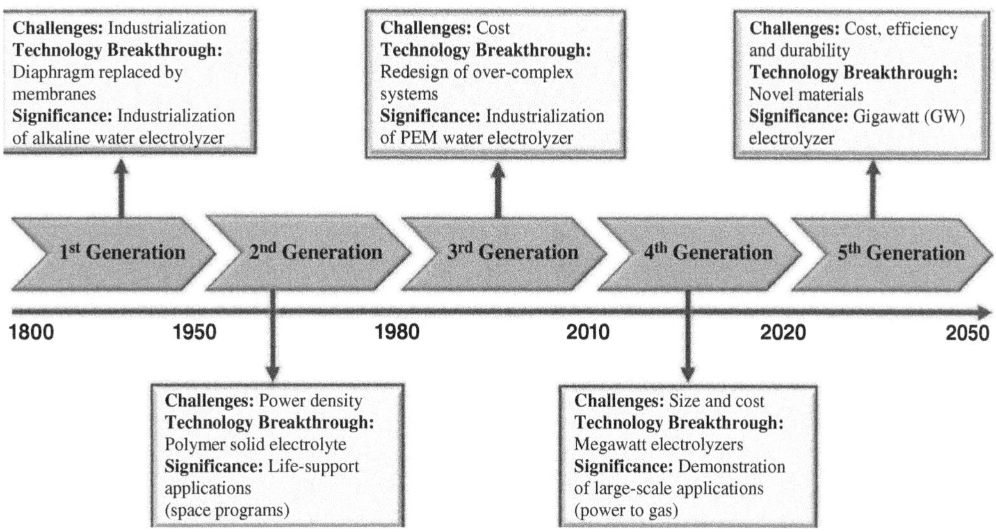

Figure 1.13 Generations of water electrolysis development [23] / with permission of Elsevier.

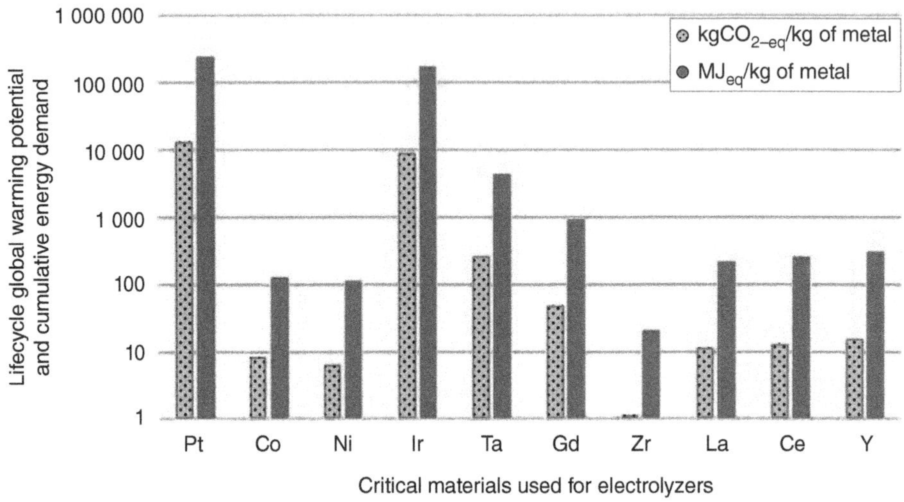

Figure 1.14 Global warming potential and cumulative energy demand for critical materials used in electrolyzers [24] / IRENA / Public domain.

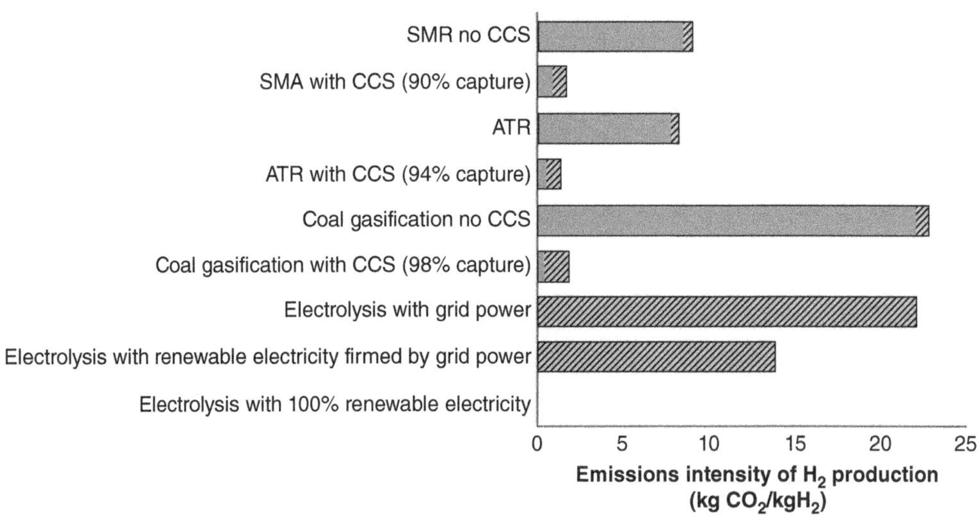

Figure 1.15 Emission intensity through hydrogen production processes [12] / S&P Global.

1.10 Emission Through Hydrogen Production by Different Routes

Steam methane reforming (SMR) with and without carbon capture and storage (CCS), auto thermal reaction (ATR), gasification, and electrolysis are now major routes of hydrogen production. However, there are emissions from different modes of hydrogen production except for electrolysis with 100% renewable electricity, which emits no emissions (Figure 1.15).

M. Wappler and a team of researchers investigated green hydrogen's potential to ensure a "no emission" condition through production [25]. They studied 25 hydrogen road maps from various countries and observed that over 700 GW of water electrolysis units were required to meet the net zero goal by 2030, and that the electrolyzer-manufacturing capacity must increase by 200% to

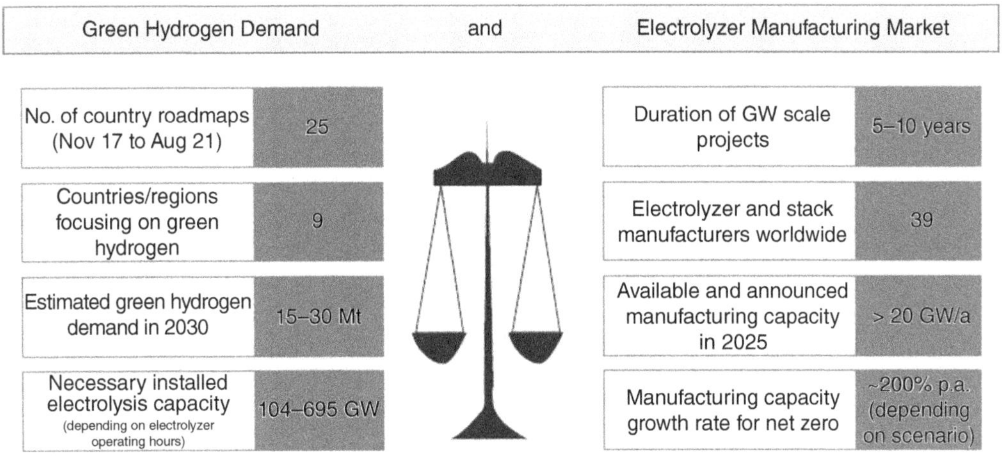

Figure 1.16 Building up the green hydrogen market [25] / with permission of Elsevier.

achieve this target. Figure 1.16 shows the current demand for hydrogen and the strategies to manufacture electrolyzers to build up the green hydrogen market.

1.11 Exhaust Emissions from Hydrogen Combustion in Engines

After hydrogen production, combustion in engines is the stage at which significant pollution could occur. Emissions during intermediate stages like storage, transport, and refueling are typically much lower in magnitude. Similarly, as discussed in Chapter 7, safety steps adopted for hydrogen engine drastically reduce emission problems due to leakage. None of the major pollutants of petroleum-fueled engines such as CO, HCs, SOx, smoke, particulate matter, and any other toxic metals or carcinogenic compounds are formed during hydrogen combustion. Oxides of nitrogen (NOx) are the single pollutant of concern. Chapter 3 discusses formation of oxides of nitrogen and their control techniques.

Concluding Remarks and Perspectives

It has been recognized that in the realm of energy, hydrogen plays a strategic role to achieve energy security, economic growth, and environmental sustainability. The hydrogen-operated internal combustion engine offers a promising solution to decarbonizing various sectors such as transportation, industry, and power generation. When produced using renewable energy sources such as wind, solar, or hydroelectric power through electrolysis, hydrogen production becomes entirely sustainable and carbon-neutral. Looking at the several production technologies from the point of view of merits, shortcomings, and technological readiness, the solid oxide electrolysis cell (SOEC) stands as a top performer [26]. Hydrogen combustion produces no noxious pollutants that are usually in the exhausts of petroleum-fueled engines. Green hydrogen from renewable sources has the potential to play a crucial role in achieving global climate targets and mitigating air pollution. Looking forward, the global hydrogen economy, at present, is on the brink of rapid expansion and integration into the existing energy system. Ongoing research and development efforts aim at improving the hydrogen storage and distribution infrastructure to ensure widespread implementation among several energy sectors. Policy initiatives and robust regulatory frameworks are crucial for an early and effective transition to a hydrogen-based economy. Governments and industry leaders have realized the strategic role of hydrogen toward the goal of achieving energy security, economic growth, and environmental sustainability. By effectively utilizing its unique properties and by overcoming its technological and infrastructure challenges, hydrogen can offer eventual freedom from energy constraints and environmental concerns worldwide.

References

1 Zou, C., Zhao, Q., Zhang, G. et al. (2016). Energy revolution: From a fossil energy era to a new energy era. *Natural Gas Industry B* 3 (1): 1–11.

2 González-Torres, M., Pérez-Lombard, L., Coronel, J.F. et al. (2022). A review on buildings energy information: trends, end-uses, fuels and drivers. *Energy Reports* 8: 626–637.

3 Kalghatgi, G. (2018). Is it really the end of internal combustion engines and petroleum in transport?. *Applied Energy* 225: 965–974.

4 Igini, M. (2024). *Atmospheric CO_2 jump in 2024 off track with trajectory needed to meet 1.5C goal, Met Office says* Earth.org.

5 Quarton, C.J., Tlili, O., Welder, L. et al. (2020). The curious case of the conflicting roles of hydrogen in global energy scenarios. *Sustainable Energy Fuels* 4: 80.

6 IEA (2021). Net zero by 2050. A roadmap for the global energy sector. www.iea.org.

7 IEA (2023). Global Hydrogen Review 2023. Clean Energy Ministerial Hydrogen Initiative. www.iea.org › reports › global-hydrogen-review-2023.

8 Philibert, C. (2020). *Methane splitting and turquoise ammonia. Ammonia Energy Association.* New York, http://www.ammoniaenergy.org/articles/methane-splitting-and-turquoise-ammonia.

9 Howarth, R.W. and Jacobson, M.Z. (2021). How green is blue hydrogen?. *Energy Science and Engineering* 9 (10): 1676–2687.

10 Liang, Q., Huang, Y., Guo, Y. et al. (2024). Efficient osmosis-powered production of green hydrogen. *Nature Sustainability* 7: 628–639. https://doi.org/10.1038/s41893-024-01317-7.

11 Maka, A.O. and Mehmood, M. April (2024). Green hydrogen energy production: Current status and potential. *Clean Energy* 8 (2): 1–7. https://doi.org/10.1093/ce/zkae012.

12 Naschert, C. September (2021). 29 Blue hydrogen under microscope over emissions, role as transition fuel. *S&P Global* https://www.spglobal.com/marketintelligence/en/news-insights/latest-news-headlines/blue-hydrogen-under-microscope-over-emissions-role-as-transition-fuel-66572882.

13 Bion, N., Duprez, D., and Epron, F. et al. (2011). Design of nanocatalysts for green hydrogen production from bioethanol. *ChemSusChem* 5 (1): 76–84. https://doi.org/10.1002/cssc.201100400.

14 IRENA (2019). *Hydrogen: A Renewable Energy Perspective*. Abu Dhabi: International Renewable Energy Agency.

15 WEF (World Economic Forum) (2020). Insight report: The net-zero challenge: Fast- forward to decisive climate action, *World Economic Forum* in collaboration with Boston Consulting Group. http://www3.weforum.org/docs/WEF_The_Net_Zero_Challenge.pd.

16 Hydrogen Council November (2017). Hydrogen, scaling up: A sustainable pathway for global energy transition. https://hydrogencouncil.com/en/study-hydrogen-scaling-up/.

17 Olabi, A.G., Abdelkareem, M.A., and Mahmoud, M.S. et al. (2023). Green hydrogen: Pathways, roadmap, and role in achieving sustainable development goals. *Process Safety and Environmental Protection* 177: 664–687.

18 IRENA (2020). Green hydrogen cost reduction: Scaling up electrolysers to meet the 1.5 °C climate goal. Abu Dhabi: International Renewable Energy Agency.

19 Yu, F., Yu, L., Mishra, I.K. et al. (2018). Recent developments in earth-abundant and non-noble electro catalysts for water electrolysis. *Materials Today Physics*: 121–138.

20 Hall, W., Spencer, T., Renjith, G. et al. et al. (2020). The potential role of hydrogen in India: A pathway for scaling-up low carbon hydrogen across the economy. In: *The Energy and Resources Institute (TERI)*. New Delhi, India, 1–136.

21 Khasawneh, H., Saidan, M.N., and Al-Addous, M. et al. (2019). Utilization of hydrogen as clean energy resource in chlor-alkali process. *Energy Exploration and Exploitation* 37 (3): 1053–1072. https://doi.org/10.1177/0144598719839767.

22 Tüysüz, H. (2024). Alkaline water electrolysis for green hydrogen productionChemical. *Research* 57 (4): 558–567.

23 Shiva Kumar, S. and Lim, H. (2022). An overview of water electrolysis technologies for green hydrogen production. *Energy Reports* 8: 13793–13813.

24 Nuss, P. and Eckelman, M.J. (2014). Life cycle assessment of metals: A scientific synthesis. *PLOS One* 9 (7): e101298, https://doi.org/1371/journal.pone.010298.

25 Wappler, M., Unguder, D., Lu, X. et al. (2022). Building the green hydrogen market—current state and outlook on green hydrogen demand and electrolyzer manufacturing. *International Journal of Hydrogen Energy* 47 (79): 33551–33357.

26 Zainal, B.S., Ker, P.J., and Mohamed, H. et al. January (2024). Recent advancement and assessment of green hydrogen production technologies. *Renewable and Sustainable Energy Reviews* 189. Part A, 113941.

2

Hydrogen's Properties and Fuel Induction in Engines

Sooner or later, we will have to recognise that the Earth has rights, too, to live without pollution. What mankind must know is that human beings cannot live without Mother Earth, but the planet can live without humans.

Evo Morales

2.1 Introduction

Today, hydrogen emerges as a pivotal energy carrier in situations when there is pressing need to transition away from fossil fuels due to fuel scarcity and environmental degradation. Figure 2.1 outlines the trajectory of a hydrogen economy, emphasizing its role from production source to transmission, storage, and utilization. The broad view in Figure 2.1 shows the major hydrogen production routes: steam methane reforming, gasification, and electrolysis. Hydrogen can be stored in solid, liquid, and gaseous modes. It can be moved by road, rail, and ship or through pipelines and used in engines and fuel cells. Detailed discussions of all these aspects are beyond the scope of this chapter and can be found in detail in another study [1].

This chapter discusses the technical features essential for assessing hydrogen's applicability to internal combustion engines. Produced from water and renewable energy sources, hydrogen is a gateway to a greener future. Upon combustion it gives only water vapor, ensuring zero carbon emissions. It has an impressively high energy-to-weight ratio, which makes it possible for a small volume of hydrogen gas to store a significant amount of energy. This feature can be useful for the land transport sector and in aerospace applications where weight is a critical factor. Hydrogen engines offer longer driving ranges than battery-electric vehicles. The fast refueling of hydrogen engines makes them suitable for long-haul transportation and heavy-duty vehicles.

2.2 Physical Properties

Hydrogen's properties vary significantly depending on the phase (solid, liquid, or gas) of matter and operating conditions. Depending upon temperature and pressure, most of these phases can change their state by freezing, boiling, melting, etc. A few properties of hydrogen are discussed here (Table 2.1) along with similar properties of gasoline (the traditional liquid fuel for ICE, internal combustion engines) and compressed natural gas (a gaseous fuel, now used in ICE vehicles in many countries). Hydrogen has a boiling point of 20.28 K and a melting point of

Hydrogen Engines: Design, Performance Evaluation, Combustion Analysis, and Exhaust Emissions, First Edition.
Lalit Mohan Das.

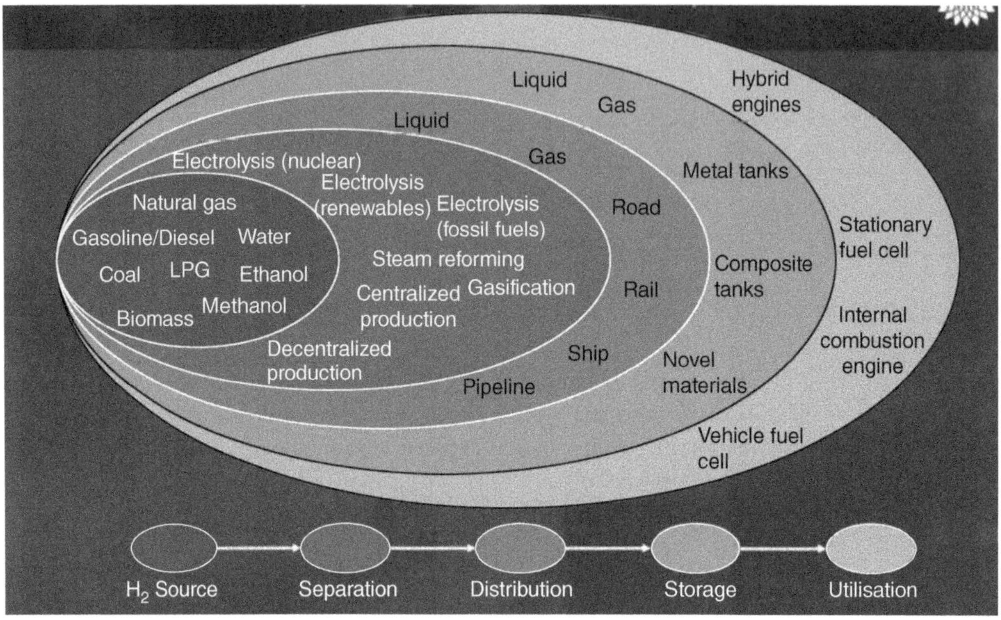

Figure 2.1 Hydrogen pathway: from production to utilization.

13.81 K. Increasing the pressure on a substance causes its boiling point to rise. For example, propane exists as a gas at atmospheric pressure, but if pressurized, it can become a liquid even at relatively moderate pressures. However, hydrogen's boiling point increases to a certain extent (up to about −240 °C) by applying pressure, specifically around 13 times atmospheric pressure. Beyond this point, adding more pressure does not raise hydrogen's boiling point higher. Pure hydrogen is an almost colorless, odorless, tasteless, nontoxic, noncorrosive, nonmetallic gas. Since it is colorless, hydrogen stream from a leak is almost invisible in daylight. Experimental studies sometimes add colorants and odorants to hydrogen so it can be detected. In many industrial and laboratory settings, hydrogen gas detectors or sensors are used to sense hydrogen concentration in the surrounding air and provide warning if it is at a hazardous level.

Schefer and colleagues [2] studied the emission bands in ultraviolet, visible, and infrared zones of the spectrum in hydrogen flames and observed that even though visible emissions from hydrogen flames are much weaker than in corresponding hydrocarbon flames, still they are sometimes visible in reduced light. Hydrogen itself is nontoxic, but if there is too much hydrogen in the air, it can act as an asphyxiant by displacing oxygen in the air. Oxygen levels below 19.5% can lead to faster breathing, tiredness, confusion, faulty judgment, depression, emotional instability, and various other problems affecting biological activity. If oxygen levels drop below 12%, a person might suddenly lose consciousness without warning. A small hydrogen leak in a closed space usually is not a big deal because hydrogen moves around quickly and spreads over the entire volume. In the event of a large leak, hydrogen diffuses and spreads fast to fill up the volume by replacing the air. The risk of asphyxiation in open areas is much lower because hydrogen rises and disperses quickly.

2.2.1 Density, Diffusivity, and Expansion Ratio

The density of a liquid (liquid density) or gas (vapor density) is determined by specific temperature and pressure, which greatly influence molecular packing density. This effect is more pronounced in gases. Table 2.2 shows the vapor and liquid density of hydrogen, methane, and

Table 2.1 Thermodynamic properties of hydrogen, methane, and gasoline (generally accepted values from the literature).

Properties	Hydrogen	Methane	Gasoline
Molecular weight	2.016	16.043	107
Heat of combustion (low) (kJ g^{-1})	119.93	50.02	44.5
Heat of combustion (high) (kJ g^{-1})	141.86	55.53	48
Specific heat (c_p) of NTP gas (J $g^{-1}K^{-1}$)	14.86	2.22	1.62
Viscosity of NTP gas (g $cm^{-1}s^{-1}$)	0.0000875	0.00011	0.00005
Specific heat ratio (γ) of NTP gas	1.383	1.308	1.05
Gas constant (R) (cm^2 atm g^{-1} K^{-1})	40.703	5.11477	0.77
Diffusion coefficient in NTP air (cm^2s^{-1})	0.61	0.16	0.005
Flammability limit volume in air (%)	4–75	4.3–15	1.4–7.6
Minimum ignition energy (mJ) at stoichiometry	0.02	0.28	0.24
Flame velocity at stoichiometry (m s^{-1})	1.85	0.38	0.37–0.43
Heat of combustion MJ/kg_{air}	3.37	2.9	2.83
Adiabatic flame temperature (K)	2480	2214	2580
Autoignition temperature (K)	858	723	550
Quenching distance (mm)	0.64	2.1 (methane)	Nearly 2
Percentage of thermal energy radiated from flame to surrounding (%)	17–25	23–32	30–42
Normalized flame emissivity (2000 K, 1 atm)	1	1.7	1.7
Lower heating value MJ/kg	119.7	45.8	44.79
Heat of combustion at stoichiometry (MJ/kg_{air})	3.37	2.9	2.83
Stoichiometric volume fraction (%)	29.53	9.48	Nearly 2 (vapor)
Stoichiometric fuel/air mass ratio	0.029	0.069	0.068

Table 2.2 Vapor and liquid density of hydrogen, methane, and gasoline.

Substance	Vapor density At 20 °C and 1 atm	Liquid density at normal boiling point and 1 atm
Hydrogen	0.08376 kg/m^3	70.8 kg/m^3
Methane	0.65 kg/m^3	422.8 kg/m^3
Gasoline	4.4 kg/m^3	700 kg/m^3

gasoline. Hydrogen's low density is one of its significant properties for practical applications because it can be compressed or liquefied. Hydrogen remains a gas above its boiling point of 20.28 K (−252.87 °C) and is buoyant from this point up to much higher temperatures, such as room temperature. This buoyancy indicates its gaseous state over a wide temperature range. Hydrogen acts as an insulator in both its gaseous and liquid states due to its low density and thermal conductivity that limits heat and electrical conduction. However, when a critical voltage is applied, it undergoes ionization and becomes a conductor of electricity, allowing flow of electrical current. Hydrogen's electrical properties under various conditions are used especially in plasma physics and certain technological applications.

From the data in Table 2.2, it is clear that because of hydrogen's low density, a large volume of hydrogen is required to provide an adequate driving range for a vehicular engine. The specific volume of hydrogen gas is 11.964 m^3/kg at 20 °C and 1 atm, and liquid hydrogen's specific volume is 0.0141 m^3/kg at −253 °C and 1 atm. The relative density of a gas is usually expressed as specific gravity with reference to air. Gaseous hydrogen, with a density of 0.0837 kg/m^3, has a specific gravity of 0.0696 and is thus approximately 7% of air density. Liquid hydrogen, with a density of 70.8 kg/m^3, has a specific gravity of 0.0708. Coincidentally, this is also approximately 7% of water's density. Hydrogen's low volume and density cause an efficiency drop when added to the intake of an engine. At the same time, low energy density has positive features to ensure rapid dispersal in the atmosphere. Figure 2.2 shows intake air losses for various combustible gases [3]. Hydrogen's diffusivity, like that of any gas, is a measure of how easily hydrogen molecules can spread or move through a medium, typically another gas or liquid. Hydrogen is a very small and lightweight molecule with a diffusion coefficient of 61 cm^2/s in air. It has the ability to diffuse through many metals, plastics, and elastomers that are airtight or impermeable to other gases. Hydrogen's buoyancy and diffusivity are crucial properties for safety and ventilation strategies in the event of a leak. In confined spaces, leaked hydrogen, when mixed with air, can even cause explosion due to buildup of flammable concentrations. A leak from the high-pressure hydrogen storage tank can create a flammable mixture inside the vehicle if the hydrogen concentration in the air reaches the lower explosive limit (LEL) of around 4%. In large, open areas, leaked hydrogen disperses rapidly and gets diluted due to its diffusivity. Liquid hydrogen, because of its very low boiling point, evaporates quickly enough and poses safety problems. Precautions need be taken to ensure that leaked liquid hydrogen does not transition into the gaseous phase, leading to formation of a flammable atmosphere within an enclosed space. In the case of hydrogen-operated vehicles, gaseous hydrogen is stored in high-pressure tanks, typically at pressures ranging from 350 to 700 bar. Selection of appropriate material and design criteria are important for these tanks to withstand hydrogen gas at such high pressures.

Expansion ratio is the ratio of the volume of a given amount of a substance in liquid form compared to the volume of the same amount of substance in gaseous form at room temperature and normal atmospheric pressure. The expansion ratio of hydrogen is 1:848 (Figure 2.3), which

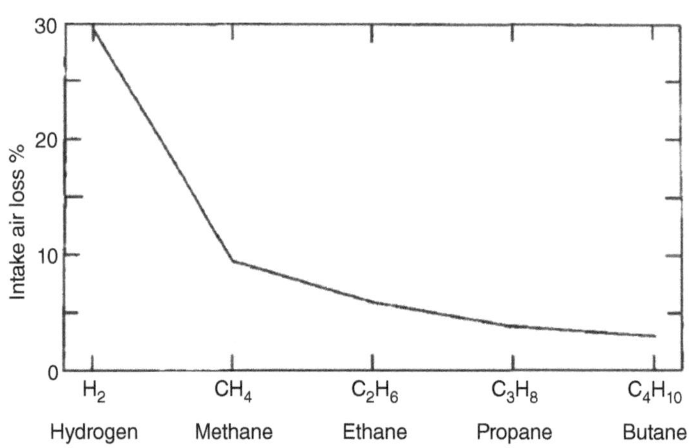

Figure 2.2 Intake air losses of hydrogen and other combustible gases [3].

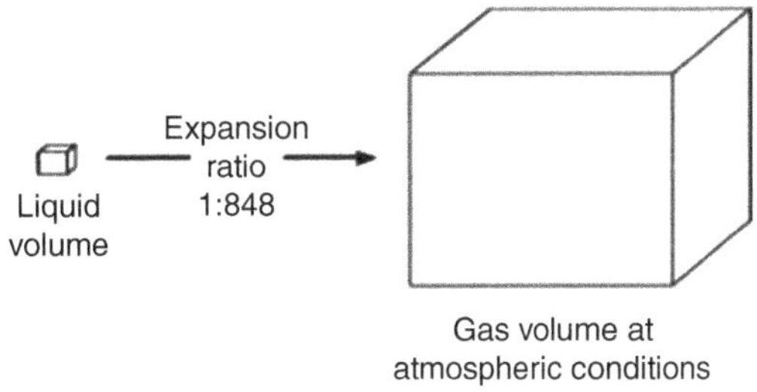

Figure 2.3 Expansion ratio for hydrogen (from liquid to gas).

indicates that hydrogen in the gaseous state occupies a volume that is 848 times the volume it occupies in the liquid state.

One water molecule (H_2O) contains two hydrogen (H) atoms. Water's (H_2O) molar mass is approximately 18.015 g/mol, while hydrogen's (H) molar mass is approximately 1.008 g/mol. Apart from being the lightest gas, hydrogen is also not very dense in liquid form. One cubic meter of water (with hydrogen and oxygen as constituents) contains approximately 112 kg of hydrogen. One cubic meter of liquid hydrogen contains approximately 71 kg of hydrogen.

2.2.2 Joule-Thomson Effect

The Joule-Thomson effect refers to temperature changes resulting from adiabatic and irreversible gas expansion during the flow of an actual gas caused by throttling. In engine applications, gas throttling is a vital phenomenon. High-pressure hydrogen pressure-reducing valves in hydrogen refueling stations and vehicles apply the Joule-Thomson effect through throttling to lower hydrogen pressure. Current research concentrates on several factors such as improving valve stability, minimizing noise, resolving issues on material durability, etc. to optimize performance.

2.3 Combustive Properties of Hydrogen

On Earth, elemental hydrogen is mostly bound to other elements, such as oxygen or carbon, forming compounds like water (H_2O) or hydrocarbons. Hydrogen is a predominant element in celestial bodies like the sun, where extreme temperatures lead to its ionization into a plasma state, consisting mainly of hydrogen ions (protons) and electrons. This highly ionized plasma state is not exclusive to the sun alone but applies to other stars also.

Hydrogen exhibits a broad range of chemical reactivity, interacting with various elements and compounds to form diverse products. When hydrogen reacts with metal ores like copper, lead, and mercury, it can reduce them to produce free metal. It can also convert salts like nitrites and nitrates into corresponding metals. Hydrogen forms hydrides with both metals and nonmetals and reacts with sulfur and gives various sulfur hydrides, hydrogen sulfide (H_2S) being the most common. In the presence of oxygen, hydrogen can generate hydrogen peroxide (H_2O_2). In organic chemistry, hydrogen reacts with a variety of compounds and oxidizers to form ethane (C_2H_6), butane (C_4H_{10}), and water vapor. Atomic hydrogen can act as a reducing agent and react with compounds such as metal chlorides and oxides including silver compounds at room temperature. One of the important well-known common reactions of hydrogen is its combination with oxygen to form water vapor. This reaction, which can occur through combustion or electrochemical means, is typically slow. However, it could be accelerated by an electric spark or the presence of an excellent catalyst like platinum.

2.3.1 Reactivity, Energy, and Energy Density

Chemical fuels are characterized by high reactivity. When chemical fuels undergo combustion, their molecules react with oxygen from air, and bonds are formed between fuel molecules and oxygen atoms. The final molecules formed after the reaction are in a lower energy state than the initial unreacted molecules. When molecules react and move to a lower energy state, a corresponding amount of energy is released in the form of heat, which performs useful work. Hydrogen's adiabatic flame rate is higher than that of gasoline, which improves its combustion stability. Hydrogen has a significantly greater diffusion coefficient than gasoline, which facilitates a homogeneous mix of air and fuel. Hydrogen's broad flammability makes it easier to run engines in lean mode than gasoline. However, because hydrogen has a higher auto ignition combustion temperature it is better suited to SI engines than to compression ignition (CI) engines. Ignition energy, flammability limit, burning speed, quenching distance, and explosion phenomena, such as deflagration, detonation, etc., are some of the typical properties and phenomena related to hydrogen combustion, which needs close monitoring. Appropriate application of some of these could provide ample benefit for smooth combustion without any abnormal combustion phenomena. Hydrogen reacts with metals having low electronegativity as well as with nonmetals having high electronegativity and can form ionic or covalent hydrides.

On the Pauling scale, hydrogen's electronegativity is 2.20, which indicates its moderate ability to attract electrons when it participates in a chemical bond. Energy density levels of different fuels correspond to equivalent energy released through combustion. Hydrogen has one of the highest energy density values per mass. Its energy density is between 120 and 142 MJ/kg. This means that for every 1 kg of hydrogen mass, it has an energy value of 120–142 MJ in contrast to the heat content of natural gas, which varies in the range 42–55 MJ/kg.

The energy content of a fuel is expressed on either a mass basis or a volume basis. It is meaningful to consider both perspectives when comparing different fuels. The energy content on a mass basis shows how much energy is contained in a given mass of fuel, while the energy content on a volume basis takes into account the volume occupied by that mass. Figure 2.4 shows that on a mass basis, hydrogen's energy is almost 3 times that of gasoline and 2.5 times that of methane.

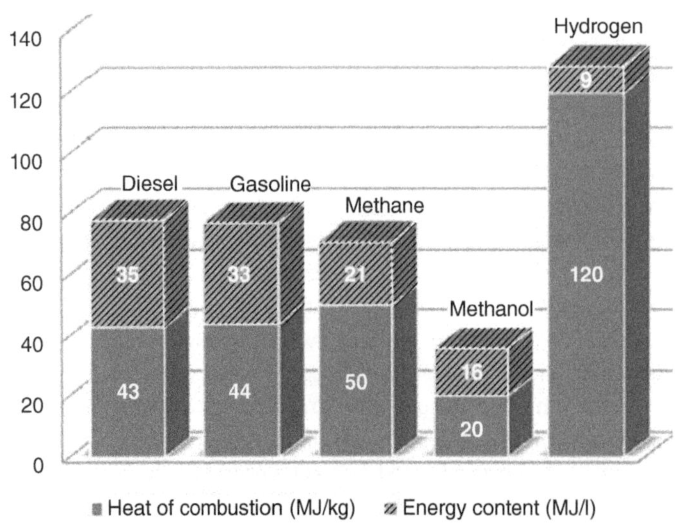

Figure 2.4 Hydrogen's energy on a mass and volume basis as compared to that of gasoline and methane.

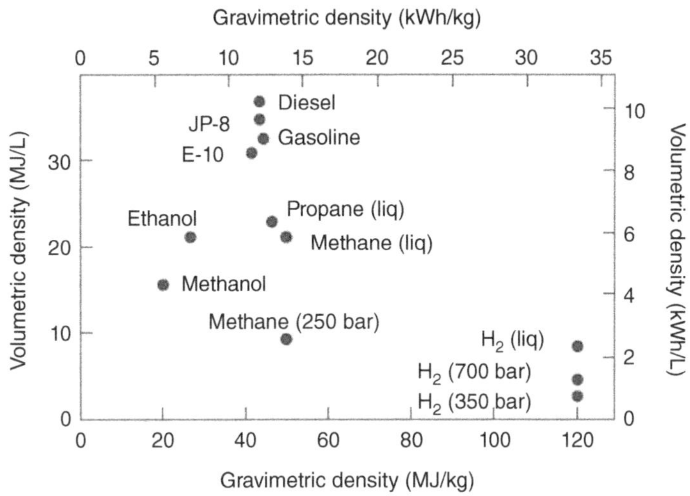

Figure 2.5 Volumetric energy density and gravimetric energy density of hydrogen and other fuels. *Source:* U.S. Department of Engergy / https://www.energy.gov/eere/fuelcells/hydrogen-storage / Public domain.

While hydrogen has a higher energy content per kilogram than gasoline and methane, its energy density on a volume basis is lower because hydrogen has a much lower density as a gas. Therefore, when discussing volume, practical storage and transportation need to be considered. Figure 2.5 shows the volumetric energy density and gravimetric energy density of hydrogen along with those of a host of other conventional and prospective alternative fuels [4]. Hydrogen requires much larger storage tanks or containers than gasoline or methane for the same amount of energy. This is a significant factor when assessing the feasibility of using hydrogen as a fuel in several applications.

2.3.2 Higher Heating Value (HHV) and Lower Heating Value (LHV)

The difference between the HHV and the LHV is the "heat of vaporization," also known as the enthalpy of vaporization, which is a thermodynamic property that represents the amount of heat energy required to change a given amount of a substance from a liquid to a vapor (gas) at a constant temperature and pressure. Table 2.3 shows the HHV and LHV of hydrogen along with those of conventional fuels and some prospective alternatives. A higher heating value indicates a more energy-dense fuel, meaning it releases more energy when burned. Because of the high energy content, energy of a hydrogen gas explosion is almost 2.5 times that of hydrocarbon fuels. On an equal mass basis, hydrogen gas explosions are more destructive. However, hydrogen fires tend to subside much more quickly than hydrocarbon fires due to the rapid release of energy. Depending on what form the fuel is stored in (as liquid or gas), the energy density is also affected. Consider the following situation: A 500 L diesel tank contains 400 kg of fuel. On an energy basis, this diesel tank is equivalent to 8000 L of hydrogen gas at 250 bar. This represents a 16-times volume increase. However, the weight of hydrogen is only 150 kg. This decrease in fuel weight is by a factor of about 2.8. Considering the same diesel tank, it is equivalent to 2100 L of liquid hydrogen. In such a case, there is a volume increase of 4.2 times. Besides gaseous and liquid hydrogen, the other option is in the form of metal hydride. In such a situation, to maintain the same hydrogen/diesel energy equivalence, approximately 4.5 kg of metal hydride is required instead of each kilogram of diesel. Therefore, in this case, a 500 L diesel tank with 400 kg of diesel replaces a hydride tank with 1725 kg of hydrogen [5].

Table 2.3 Heating values (HHV and LHV) of hydrogen.

Fuel	Higher heating value (at 25 °C and 1 atm)	Lower heating value (at 25 °C and 1 atm)
Hydrogen	141.86 kJ/kg	119.93 kJ/kg
Methane	55.53 kJ/kg	50.02 kJ/kg
Propane	50.36 kJ/kg	45.6 kJ/kg
Methanol	19.96 kJ/kg	18.05 kJ/kg
Gasoline	47.5 kJ/kg	44.5 kJ/kg
Diesel	44.8 kJ/kg	42.5 kJ/kg

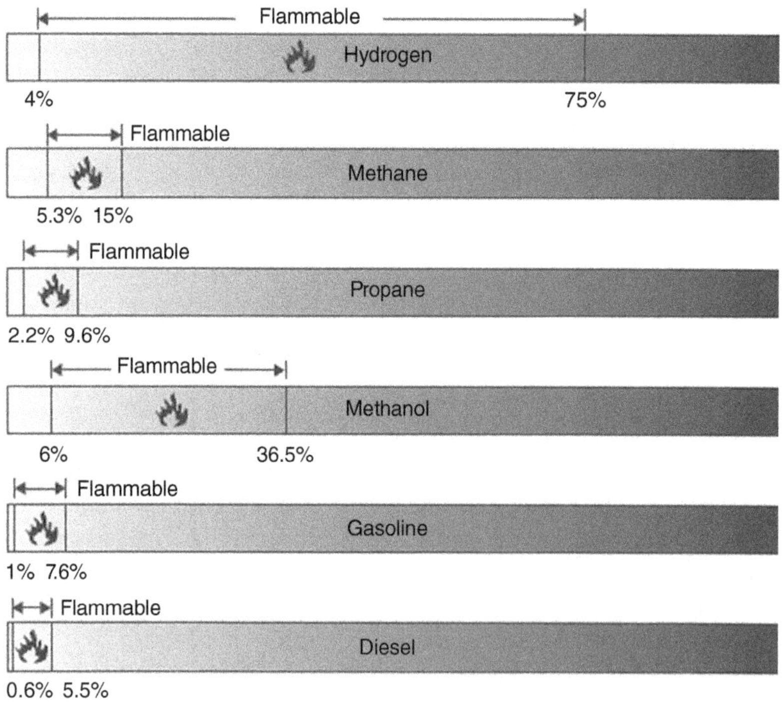

Figure 2.6 Flammability range of hydrogen and other fuels.

2.3.3 Flammability Limit

Hydrogen exhibits a wide flammability range, from 4% (lower flammability limit, LFL) to 75% (upper flammability limit, UFL) by volume. The LFL represents the minimum gas concentration that supports a self-propagating flame when mixed with air and ignited, while the UFL is the maximum concentration that supports combustion. Figure 2.6 shows the flammability range of hydrogen along with those of natural gas, propane, and gasoline. The percent of the combustion chamber occupied by hydrogen for a stoichiometric mixture is about 29.6%, which may be regarded as the theoretical optimal combustion condition.

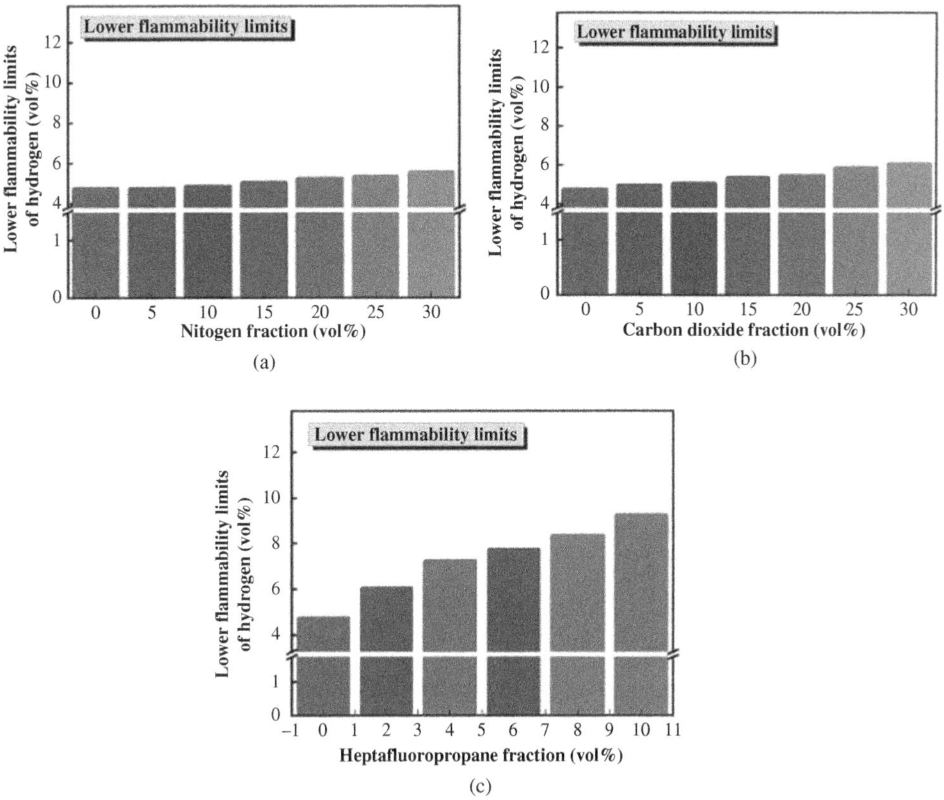

Figure 2.7 Effect of an inhibitor on the lower flammability of hydrogen [6] / with permission of Elsevier. Please visit the companion website to view a color version of this image.

Liang et al. [6] in a recent paper investigated the effect of inhibitors (carbon dioxide, nitrogen, and heptafluoropropane) on the lower flammability limit and explosive characteristics of hydrogen. Using chemical kinetics, the study determined the critical inhibitory concentrations that can effectively suppress undesirable combustion events. The results showed that increasing inhibitor quantity led to higher and lower flammability limits for hydrogen. The addition of 10% heptafluoropropane (C_3HF_7) increased the limit by 94.74%. Under lean burn conditions, carbon dioxide exhibited stronger inhibitory effects on maximum explosion pressure, adiabatic flame temperature, and heat release rate than nitrogen. Addition of heptafluoropropane initially increased these parameters, followed by a significant decrease. On the basis of this study, critical inhibitory concentrations of the three inhibitors for hydrogen with different volume fractions can be ranked in order from low to high as heptafluoropropane < carbon dioxide < nitrogen. The researchers in this study were able to identify the concentration of inhibitors to ensure complete suppression. A closer look at Figure 2.7 (a, b, c) shows that an increased quantity of inhibitors leads to a corresponding increase in the lower flammability of hydrogen. It is evident from these figures that the extent to which LFL increases, under the influence of C_3HF_7 (heptafluoropropane), is abnormally higher than the rise in LFL due to the addition of N_2 and CO^2 materials. The implication is that the material may have additional flammability or reactivity when exposed to higher oxygen concentrations.

The flammability limit widens with increasing temperatures as shown in Figure 2.8. The lower flammability limit (LFL) is of great practical relevance because in the event of leakage, this limit is reached first, and it might lead to backfire.

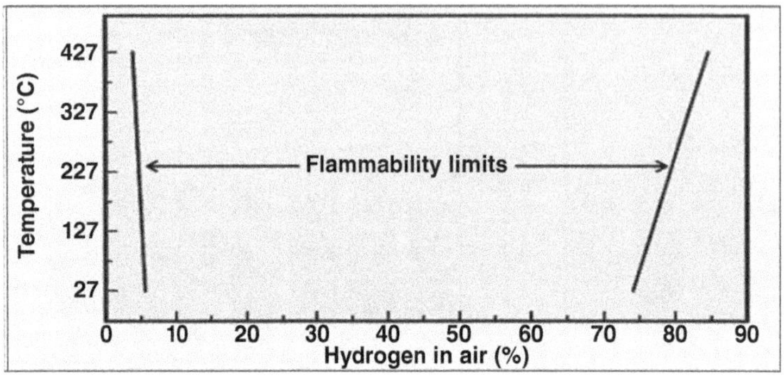

Figure 2.8 Variation of hydrogen's flammability limit with temperature.

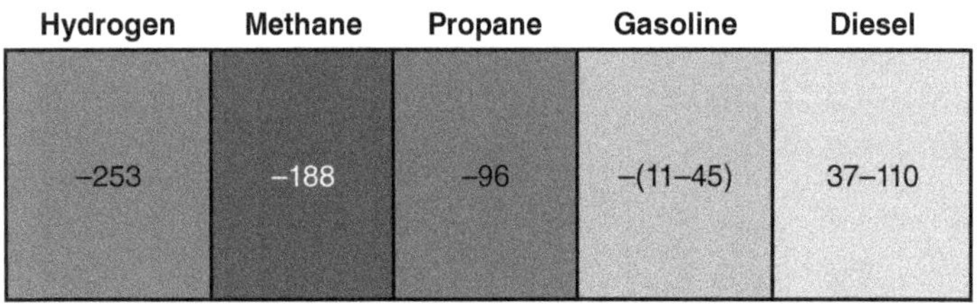

Figure 2.9 Flash point of hydrogen along with those of methane, propane, gasoline, and diesel. Please visit the companion website to view a color version of this image.

The wide flammability range of hydrogen makes it possible for use in an engine from ultra-lean equivalence ratio to just above the stoichiometric value. Lean hydrogen operation leads to improved fuel economy, better thermal efficiency, and reduced NOx emissions. However, there are reductions in power output due to a reduction in the volumetric heating value of the air–fuel mixture. These features of the hydrogen engine are discussed in detail in Chapter 3.

2.3.4 Flash Point

Flash point describes the characteristic property by which a fuel can produce enough vapor to form an ignitable mixture. Therefore, if fuel's temperature is below its flashpoint, it will not produce enough vapor to ignite when exposed to an open flame or another ignition source. Since a fuel burns only in its vapor or gaseous state, conventional fuels such as gasoline and diesel convert to a gaseous state to burn. Thus, the concept of a flash point is not as relevant for hydrogen gas as for liquid fuels. Figure 2.9 shows the flash point of hydrogen along with that of other fuels.

2.3.5 Minimum Ignition Energy (MIE)

Minimum ignition energy (MIE), which defines the minimum amount of energy required to initiate combustion under specific conditions, is a critical property while considering the ignition of fuel–air mixtures.

As evident in Figure 2.10, the minimum ignition energy of hydrogen is much lower than that of methane, propane, and gasoline. Thus, hydrogen is advantageous for use in spark ignition (SI) engines as it can be ignited with a very low-energy spark, without requiring sophisticated or high-energy ignition systems. On the contrary, hydrogen's low ignition energy increases the likelihood of backfire, as even relatively low-energy sources like hot surfaces or exhaust components can provide sufficient energy to ignite the hydrogen–air mixture.

Figure 2.10 Minimum ignition energy for hydrogen, methane, propane, and gasoline.

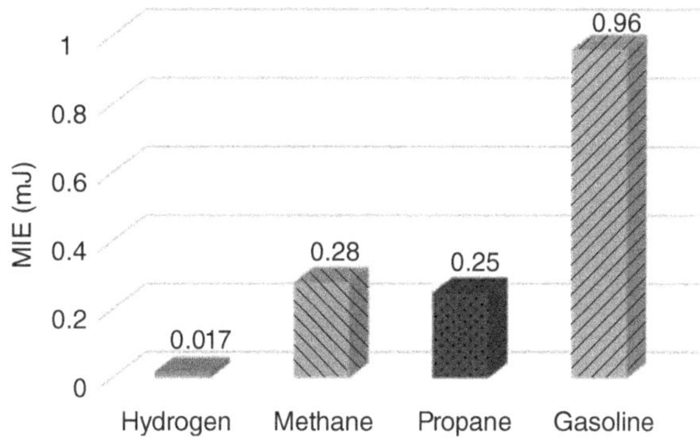

Figure 2.11 Autoignition temperature of hydrogen, methane, propane, and gasoline.

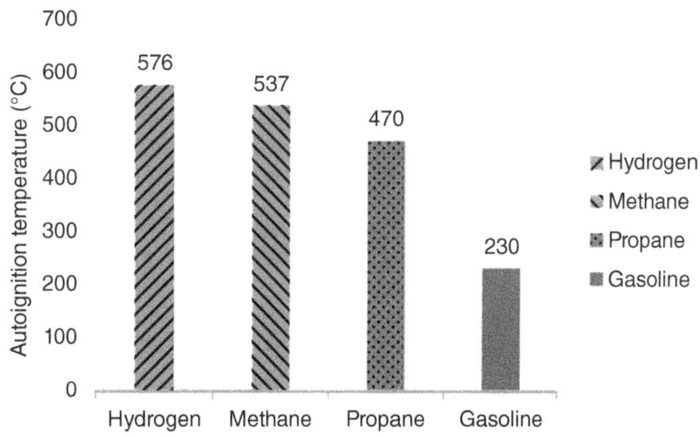

2.3.6 Autoignition Temperature

The autoignition temperature of fuel is defined as the minimum temperature to initiate self-sustained combustion in the absence of any ignition source. Hydrogen has an autoignition temperature of 576 °C, which is relatively higher than that of gasoline (Figure 2.11).

Autoignition temperature is a significant factor in determining what maximum compression ratio the engine can adopt for smooth combustion, because temperature rise during compression is closely linked to the compression ratio. Hydrogen's high autoignition temperature allows larger compression ratios in a spark ignition engine, which leads to improved thermal efficiency. At the same time, it is difficult to achieve the compression ignition of hydrogen in a diesel configuration without any additional ignition source, since the temperature required for ignition in this case is higher than that used for a conventional diesel engine.

2.3.7 Maximum Experimental Safe Gap (MESG) and Quenching Gap

The maximum experimental safe gap (MESG) is the maximum gap between two flat plates through which flame propagation can occur without causing explosion. The MESG for hydrogen is 0.08 mm. The quenching gap, on the other hand, is related to the ability to suppress ignition of a flammable mixture. It represents the smallest gap or diameter in a tube through which a flame cannot propagate. Thus, it implies the flame-extinguishing characteristics of a fuel. In the case of an internal combustion engine, it relates to the distance from the cylinder wall at which the flame extinguishes. Hydrogen has a much smaller quenching distance (0.064 cm), almost three times less than gasoline. This property makes it more difficult for hydrogen flames to extinguish. Hydrogen flames travel close to the cylinder wall, which increases the possibility of backfire because the flame from a hydrogen–air mixture can more easily pass through a nearly closed intake valve. Interestingly, the quenching gap of hydrogen, so vital for engine applications, has no specific relevance for use with fuel cells.

2.3.8 Octane Number

Knock has been an undesirable combustion phenomenon in a conventional spark ignition engine since its invention. It occurs when the air–fuel mixture ignites spontaneously before regular spark plug ignition. Knock happens due to heat build-up in the combustion chamber resulting in a temperature higher than the autoignition temperature of the fuel. Octane rating is a measure of a fuel's anti-knock characteristics. The octane rating is typically expressed as a number relative to a standard reference fuel, isooctane, which has an assigned octane rating of 100. Fuels with higher octane ratings are more resistant to engine knocks, while fuels with lower octane ratings are more prone to knocking. Hydrogen has an octane number above 130, which characterizes hydrogen's resistance to autoignition. Thus, hydrogen operates in lean mode without knocking.

2.3.9 Stoichiometric Composition

A stoichiometric mixture refers to a chemically correct air–fuel mixture ratio in which there is precisely enough air to completely burn the fuel in the combustion process. Thus, maximum combustion energy is released in this process. Let us find the stoichiometric combustion of hydrogen and oxygen.

$$2H_2 + O_2 \rightarrow 2H_2$$

Therefore, 2 moles of H_2 and 1 mole of O_2 are needed for complete combustion. In this analysis, air is the oxidizer instead of oxygen. Hence, the major constituent nitrogen in the air needs to be included in the calculations.

$$\begin{aligned}\text{Moles of } N_2 \text{ in air} &= \text{moles of } O_2 \times (79\%\ N_2 \text{ in air}/21\%\ O_2 \text{ in air})\\ &= 1 \text{ mole of } O_2 \times (79\%\ N_2 \text{ in air}/21\%\ O_2 \text{ in air})\\ &= 3.762 \text{ moles } N_2\end{aligned}$$

$$\begin{aligned}\text{Number of moles of air} &= \text{moles of } O_2 + \text{moles of } N_2\\ &= 1 + 3.762\\ &= 4.762 \text{ moles of air}\end{aligned}$$

$$\text{Weight of } O_2 = 1 \text{ mole of } O_2 \times 32\ \text{g/mole} = 32\ \text{g}$$

$$\text{Weight of } N_2 = 3.762 \text{ moles of } N_2 \times 28\ \text{g/mole} = 105.33\ \text{g}$$

$$\begin{aligned}\text{Weight of air} &= \text{Weight of } O_2 + \text{Weight of } N_2\\ &= 32\ \text{g} + 105.33\ \text{g} = 137.33\ \text{g}\end{aligned}$$

$$\text{Weight of } H_2 = 2 \text{ moles of } H_2 \times 2 \text{ g/mole} = 4 \text{ g}$$

Stoichiometric air/fuel (A/F) ratio for hydrogen and air is:

$$\begin{aligned}\text{A/F based on mass} : &= \text{mass of air/mass of fuel}\\ &= 137.33 \text{ g}/4 \text{ g} = 34.33{:}1\end{aligned}$$

$$\begin{aligned}\text{A/F based on volume} : &= \text{volume (moles) of air/volume (moles) of fuel}\\ &= 4.762/2\\ &= 2.4{:}1\end{aligned}$$

The percentage of the combustion chamber occupied by hydrogen for a stoichiometric mixture is

$$\begin{aligned}\%\ H_2 &= \text{volume (moles) of } H_2/\text{total volume}\\ &= \text{volume } H_2/(\text{volume air} + \text{volume of } H_2)\\ &= 2/(4.762 + 2) = 29.6\%\end{aligned}$$

Therefore, on a mass basis, the stoichiometric or chemically correct A/F ratio for complete combustion of hydrogen in air is about 34:1, which indicates that 34 kg of air is required for each kg of hydrogen to ensure complete combustion. This is much higher than the air/fuel ratio of 14.7:1, which is the corresponding stoichiometric composition needed for gasoline. Hydrogen, being in the gaseous state at ambient conditions, occupies more space in the combustion chamber than liquid gasoline. Therefore, air requires less space in the combustion chamber. Hydrogen displaces close to 30% of the combustion chamber, compared to barely 1 to 2% for gasoline. Figure 2.12 shows the relation between flame front velocity and equivalence ratio to achieve combustion stability in engine operation.

Drell and Belles [7] have recommended combustion properties for hydrogen–air mixtures under standard conditions, comprising a pressure of 1 atm and an initial temperature of approximately 25 °C. The data are for both stoichiometric mixtures and mixtures showing maximum or minimum values. In many experimental investigations, hydrogen engines have been operated from an ultra-clean mode to an equivalence ratio close to the stoichiometric mixture without backfire and with low NO_X emissions.

2.3.10 Flame Speed

Two terms, burning velocity and flame speed, represent different aspects of the combustion process. As a flame moves through a combustible medium, its reaction rate increases, whereas flame speed describes how rapidly a flame front moves through the same medium, which can be affected by various factors, including turbulence. In fact, it represents the velocity of the boundary between burned and unburned regions. Hydrogen's faster flame speed allows hydrogen engines to approach states closer to the ideal thermodynamic engine cycle. The flame speed is far greater than the burning speed because of expansion of combustion products, instabilities, and turbulent deformation of the flame.

$$S_u = S_b + U_d$$

where
S_u = flame speed,
S_b = burning speed, and
U_d = displacement velocity of the unburned mixture.

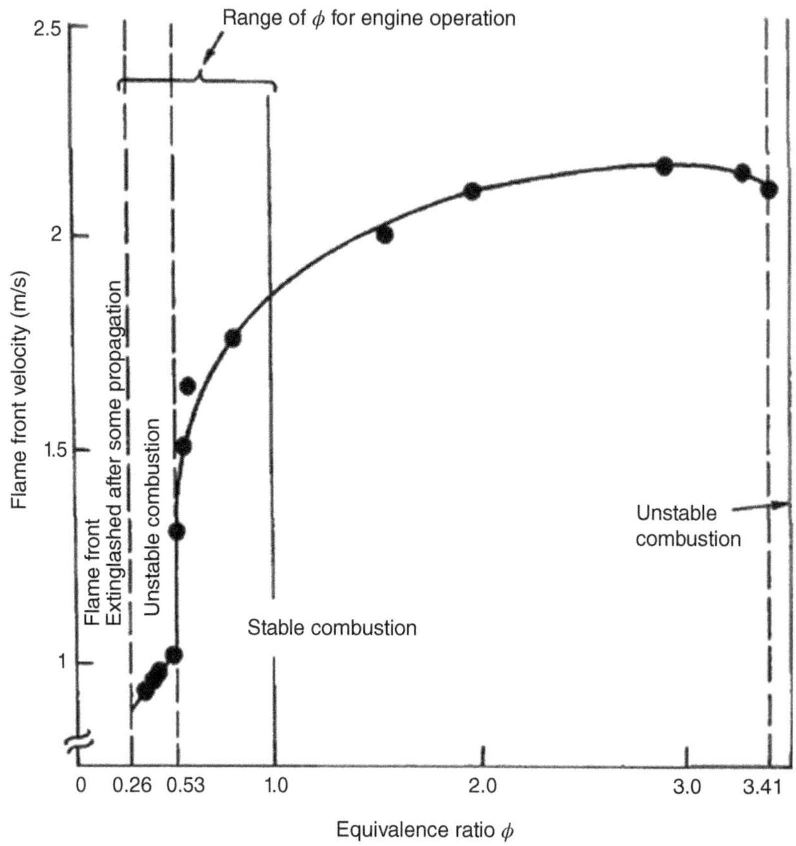

Figure 2.12 Flame front velocity of H_2–air mixture, p = 1 atm.

This relationship is a simplification that applies under specific assumptions (such as one-dimensional, steady state, laminar flame conditions), and it may not hold in more complex situations such as turbulent combustion. In realistic and varied combustion scenarios, the relationships between flame speed, burning speed, and displacement velocity can be more complicated. Hydrogen, upon combustion, releases a significant amount of heat. This heat causes the surrounding air to expand and rise, creating "heat ripples." These ripples are distortions in the air caused by varying temperatures, and they are visible. Hydrogen flames radiate significantly less infrared (IR) radiation (heat) with practically no visible radiation (light). Large hydrogen fires emit thermal radiation in the form of electromagnetic radiation produced by flame temperature. Thermal imaging cameras can detect infrared radiation, which is part of thermal radiation not otherwise visible to the human eye.

Compared to gasoline, hydrogen has a higher energy content per unit mass, and when it burns it releases a significant amount of energy. Even though hydrogen burns more vigorously, its combustion does not sustain itself for long. When liquid hydrogen is involved in a fire, it burns at 3 to 6 cm/min. In contrast, liquid methane burns more slowly at a rate of 0.3 to 1.2 cm/min; pools of liquid gasoline burn at a rate of 0.2 to 0.9 cm/min. This indicates rapid burning consistent with hydrogen's high reactivity. As discussed in Chapter 7 on safety, hydrogen fire is much safer than gasoline fire in many respects. Adiabatic flame temperature, along with the flame velocity, is a very

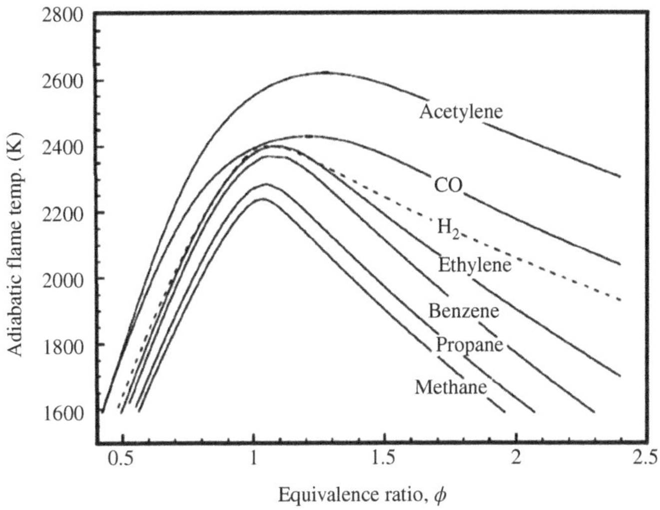

Figure 2.13 Adiabatic flame temperature of hydrogen as a function of the equivalence ratio [8].

crucial parameter for the combustion stability of engine operation. It also influences the thermal efficiency and exhaust emission characteristics of the engine.

2.3.11 Adiabatic Flame Temperature

In a combustion reaction, if no heat is lost in the process, the temperature achieved is referred to as the adiabatic flame temperature. Drell and Belles [7] calculated the adiabatic flame temperature for a pressure range from 0.01 to 100 atm and an initial temperature range from 0 to 1400 K for hydrogen–air mixtures. For hydrogen combustion in air, the adiabatic flame temperature is about 2127 °C or 2400 K at 1 atm. Figure 2.13 shows the pattern of variations of adiabatic flame temperature as a function of the equivalence ratio at STP conditions. It is clear from the figure that the maximum adiabatic flame occurs close to or a little above stoichiometric conditions. At a fixed equivalence ratio, the adiabatic flame temperature of a hydrogen/air premixed flame is higher than that of natural gas due to hydrogen's higher flame. Similarly, increasing hydrogen content in a methane/hydrogen mixture at a fixed equivalence ratio elevates the adiabatic flame temperature. This occurs because hydrogen burns more efficiently relative to methane, leading to increased heat release and a higher flame temperature.

2.4 Emissivity of the Hydrogen Flame

Hydrogen flames are typically nonluminous with poor heat transfer. The absence of carbon particles or soot in hydrogen flames results in reduced radiative heat transfer. Emissivity of hydrogen may be about 0.3 and, in some cases, it is even less than 0.1. Emissivity of hydrogen depends on various factors such as temperature, pressure, and the state of hydrogen (gas, liquid, or solid). Sometimes it differs significantly depending on the specific conditions and wavelengths of interest. At higher temperatures and pressures, such as those found in certain industrial processes or astrophysical contexts, the emissivity can vary but is always a lot less than that of other materials.

2.5 Hydrogen Embrittlement

While embrittlement itself is not strictly a property of hydrogen, hydrogen's ability to permeate metal structures and alter their mechanical properties is a key factor in hydrogen embrittlement. Therefore, hydrogen embrittlement can be considered as an outcome of specific hydrogen properties, such as its small atomic size and high diffusivity in metals. Embrittlement occurs when constant exposure of hydrogen to many materials leads to a situation in which hydrogen enters the cracks of these materials making them brittle. The magnitude of embrittlement depends on the amount of hydrogen accumulated over a period of time. Embrittlement effects can vary over a wide range: from simple leakage to catastrophic failure in both metal and nonmetallic components. The factors responsible for embrittlement include pressure, temperature, hydrogen concentration, purity of hydrogen, type of impurity, exposure time, heat treatment, etc. Embrittlement is often associated with industrial operations like pickling, where metal surfaces are treated with acids to remove scale and impurities. Inhibitor addition is a common strategy to mitigate hydrogen embrittlement. Inhibitors are some chemicals that are usually added to pickling solutions to minimize hydrogen gas generation. The inhibitor forms a protective layer on the metal surface and slows down corrosive reactions. Depending on the specific conditions and metals, various inhibitors such as organic compounds, surfactants, and film-forming additives are used.

Welding is another common process in which hydrogen embrittlement could be a concern. During welding, there is a possibility of hydrogen getting into the welded joint and leading to embrittlement of the metal. Hydrogen sources during welding include moisture, contaminants, and breakdown of moisture, particularly in high-strength steels and certain alloys, making them susceptible to hydrogen embrittlement.

2.6 Green Hydrogen and Hydrogen Use in Engines

Some discussion on hydrogen color codes and carbon footprints has already been made in Chapter 1. The References include related literature discussing several aspects of green hydrogen such as methods of production, benefits, and challenges [9–11].

Figure 2.14 shows the production routes and three major colors assigned to each route for hydrogen production. Rational evaluation of carbon intensity of the major hydrogen production routes is extremely significant when decarburization is the goal. Although hydrogen combustion itself is a clean-burning fuel, the production process can contribute to greenhouse gas emissions if fossil fuels are used. Grey hydrogen from fossil fuels releases carbon emissions during its production process, and this contributes to greenhouse gas emissions, which deteriorate climate conditions. While blue hydrogen mitigates some emissions through carbon capture and storage (CCS) technologies, it still relies on fossil fuels as its primary feedstock, resulting in a significant carbon footprint. Studies have shown that greenhouse gas emissions from blue hydrogen production are quite high, particularly due to the release of fugitive methane [12]. Thus, grey and blue hydrogen are not environmentally friendly. Green hydrogen, from renewable energy sources such as wind, solar, and water, cause no detrimental effects on ecosystems. As shown in Figure 2.15, green hydrogen, produced from a wide variety of renewables [13], is a zero-emission fuel from production to combustion and is the most desirable option from a life-cycle sustainability point of view. Green hydrogen in internal combustion engines marks a significant step toward sustainable transportation. A number of research initiatives are currently under way, and the global infrastructure for production and distribution of green hydrogen is being extensively developed.

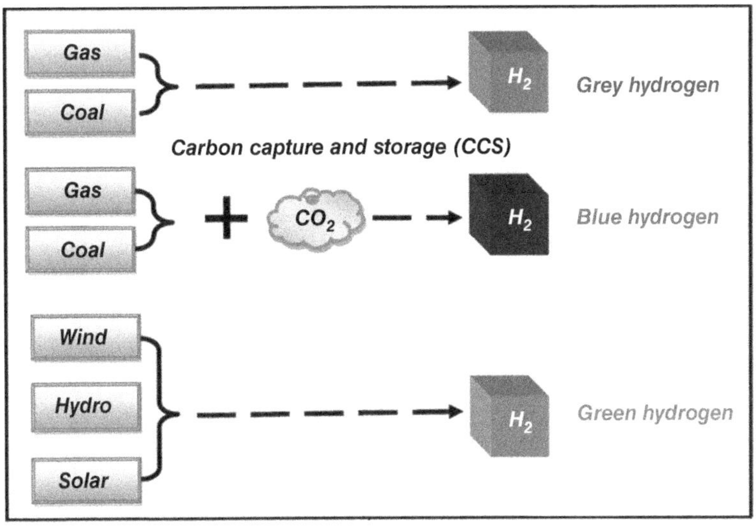

Figure 2.14 Color codes of hydrogen and feedstock. Please visit the companion website to view a color version of this image.

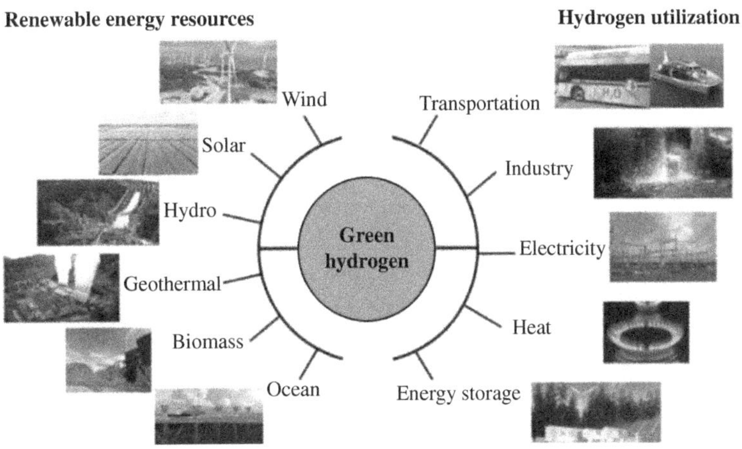

Figure 2.15 Green hydrogen from a wide spectrum of renewables [13] / Elsevier / CC BY 4.0.

2.7 Combustion Strategy

Figure 2.16 shows various combustion strategies, each having its own specific goals but with the common objective of contributing to improved efficiency, fuel consumption, and power, low emission, and stable combustion without knocking and backfire [14]. These strategies revolve around several variables such as fuel blending ratios, distribution, and compression ratios. The combustion process is influenced by the heat release rate, rate of combustion, duration of combustion (DOC), peak cylinder temperature, mixture formation time, ignition delay, and several other factors. While each strategy prioritizes its implementation route, they share the common goal of optimization. Appropriate monitoring of a single variable causes improvement, but integration of multiple variables gives much better results.

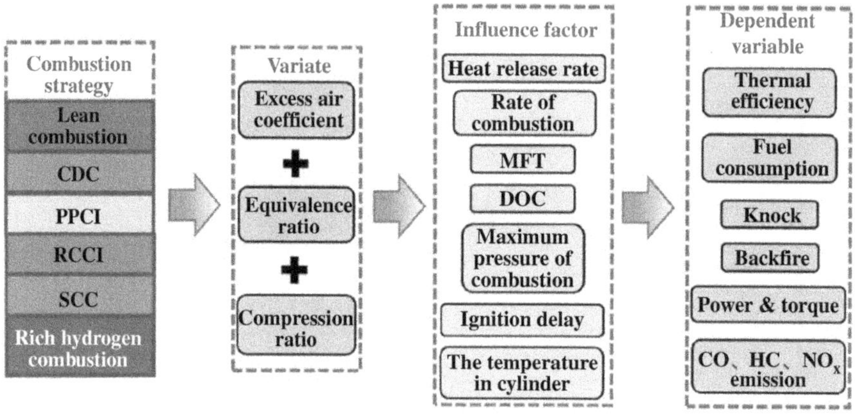

Figure 2.16 Different combustion strategies applied to the hydrogen engine [14] / with permission of Elsevier.

Modeling studies by Mustaf et al. [15] on the colorless distributed combustion (CDC) method were applied to gas turbines and discussed CDC control by regulating flame temperature regulation and reducing reaction rates. This technique proved effective in achieving ultra-low NOx emission and more uniform temperature distribution over a wide flame range. The CDC method effectively suppresses elevated temperatures and NOx emissions associated with hydrogen-rich conditions. It allows expansion of combustion range limitations, typical in hydrogen-rich environments.

The low-temperature combustion mode in a hydrogen engine uses several advanced compression ignition technologies such as homogeneous charge compression ignition (HCCI), premixed charge compression ignition (PCCI), and reactive controlled compression ignition (RCCI). Chapter 9 gives a detailed discussion of SCC, HCCI, PCCI, and RCCI. The compression ratio and equivalence ratio play very crucial roles in the combustion process. Combustion duration, maximum combustion, and maximum heat release rate are closely linked to engine knocking. Higher in-cylinder temperature, along with better fuel atomization, are significant for uniform premixing of the fuel–air mixture within the combustion chamber.

Injection systems with appropriate injection timing, injection pressure, and injection duration are crucial for smooth combustion and low emissions. Direct cylinder injection, an appropriately designed port injection system, or timed manifold injection (TMI) system, have proven successful by several researchers in avoiding backfire. All these methods demand that a precise fuel injection system needs a rapid injection valve with excellent repeatability, ample flow capacity, and high durability. In the case of early injection, the start of injection is often set after the intake valve closing to prevent displacement of fresh charge by low-density hydrogen. To enable direct hydrogen flow into the combustion chamber, the injection pressure has to be more than the cylinder pressure. Early injection allows the fuel adequate time to mix with air inside the combustion chamber and form a homogeneous mixture. On the other hand, the limited time available with late injection often leads to stratification of mixtures. Lee et al. [16] focused on analyzing the effects of mixture formation time (MFT) and ambient density on stratified charge combustion. They investigated the effects of mixture formation time (MFT) and ambient density on the hydrogen energy conversion process. Owing to the merits of employing hydrogen in lean burn, spark ignition engines, hydrogen stratified charge combustion (SCC) was tested in a constant-volume chamber. Depending on the degree of stratification, short and long MFTs had high and low burning velocities, respectively. Equivalence ratio, injection parameters, and the characteristics of hydrogen jet into the combustion chamber are key factors for SCC. The study also involved optical diagnostics to understand the behavior of the hydrogen flame. A well-pronounced, red-colored emission was visible in the rich combustion area of the hydrogen SCC.

Concluding Remarks and Perspectives

Hydrogen, with its distinctive physico-chemical properties, holds great promise as a key player in the transition toward a decarbonized energy landscape. Hydrogen demand is growing worldwide, in various sectors such as long-haul transport, power generation, and a number of industrial applications like steel and ammonia production. Several goal-oriented projects with a focus on technological innovations in hydrogen generation, storage, distribution, and safe utilization are being sponsored by governments and industrial organizations. Countries are collaborating on large-scale demonstration projects to ensure swift practical implementation.

Compared to grey or blue hydrogen, the use of green hydrogen is a crucial advantage in the automotive sector, where the goal is to reduce the overall carbon footprint of vehicles. Production of an adequate quantity of green hydrogen depends on deploying a large number of electrolyzers. Much research is being pursued to improve their performance and durability by integrating major contributions from solar, wind, and renewables. The International Energy Agency [17] has outlined key parameters, illustrating the trade-offs among efficiency, durability, and cost in electrolyzer technology (Figure 2.17).

Close and rational monitoring of parameters such as flammability limits, flame speed, minimum ignition energy (MIE), and autoignition temperature are essential to get the benefits of hydrogen

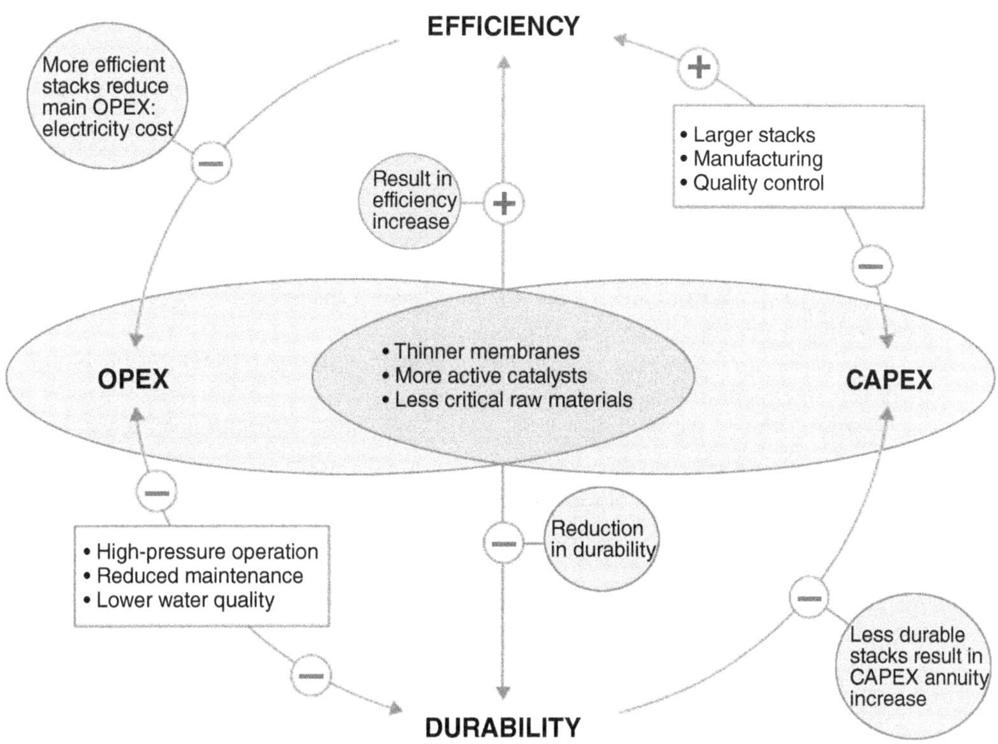

Figure 2.17 Trade-offs between efficiency, durability, and cost of electrolyzers [17] / IRENA / Public domain.

combustion in engines. Low ignition energy makes ignition easy with a low energy spark, and it also increases the possibility of backfire. High flame speed contributes to rapid and efficient combustion, while an uncontrolled flame speed can cause high-pressure rise, knocking, and higher temperatures leading to higher NOx levels. Engine control systems should be designed to ensure efficient and controlled combustion. Higher autoignition facilitates adopting a higher compression ratio and achieving improved efficiency. With the hydrogen engine, high compression ratios are achievable, but stable combustion under all operating conditions is sometimes challenging and requires sophisticated control strategies. Hydrogen has high energy density by weight, making it an attractive option for transportation applications, but achieving high energy density by volume, especially in storage, is difficult. Based on the distinctive combustion features, research and development efforts are focused on optimizing engine designs, developing advanced combustion control strategies, and identifying materials that can withstand hydrogen combustion.

References

1 Das, L.M. (2024). *Hydrogen Energy: Production, Safety, Storage and Applications*. Chichester: John Wiley.

2 Schefer, R.W., Kulatilaka, W.D., Patterson, B.D. et al. (2009). Visible emission of hydrogen flames. *Combustion and Flame* 156 (6): 1234–1241.

3 Binder, K. and Withaln, G. (1980). Mixture formation and combustion in interaction with hydrogen storage technology. *Hydrogen Energy Progress*. New York: Pergamon Press.

4 https://www.energy.gov/eere/fuelcells/hydrogen-storage.

5 Hydrogen fuel cell engines and related technologies (December 2001). Hydrogen properties, Module 1, College of the Desert, Palm Desert, California.

6 Liang, H., Wang, T., Luo, Z. et al. (2024). Investigation on the lower flammability limit and critical inhibition concentration of hydrogen under the influence of inhibitors. *Fuel* 356 (15): 129595.

7 Drell, I.L. and Belles, F.E. (1958). *Survey of Hydrogen Combustion Properties*. Technical Report 1383. National Advisory Committee on Aeronautics.

8 Law, C.K. (2006). *Combustion Physics*. Cambridge, UK: Cambridge University Press.

9 Hassan, Q., Algburi, S., Sameen, A.Z. et al. (2024). Green hydrogen: A pathway to a sustainable energy future. *International Journal of Hydrogen Energy* 50: Part B (2 January 2024): 310–333.

10 Stöckl, F., Schill, W.P., and Zerrahn, A. (2021). Optimal supply chains and power sector benefits of green hydrogen. *Scientific Reports* 11 (1): 92511.

11 IEA. (2019). *The Future of Hydrogen–Seizing Today's Opportunities*. Paris: IEA. https://www.iea.org/reports/the-future-of-hydrogen.

12 Howarth, R.W. and Jacobson, M.Z. (2021). How green is blue hydrogen? *Energy Science and Engineering* 9 (10): 1676–2687.

13 Agaton, C.B., Batac, K.I.T., and Reyes, Jr, E.M. (2022). Prospects and challenges for green hydrogen production and utilization in the Philippines. *International Journal of Hydrogen Energy* 47 (41): 17859–17870.

14 Sun, Z., Hong, J., Zhang, T. et al. (2023). Hydrogen engine operation strategies: Recent progress, industrialization challenges, and perspectives. *International Journal of Hydrogen Energy* 48 (1): 366–392.

15 Ilbas, M., Kumuk, O., and Karyeyen, S. (2022). Modelling of the gas-turbine colourless distributed combustion: an application to hydrogen enriched kerosene fuel. *International Journal of Hydrogen Energy* 47 (24): 12354–12364.

16 Lee, S., Kim, G., and Bae, C. (2021). Lean combustion of stratified hydrogen in a constant volume chamber. *Fuel* 301: 121045.

17 IRENA. (2021). *Making the Breakthrough: Green Hydrogen Policies and Technology Cost.* Abu Dhabi: International Renewable Energy Agency.

3

The Hydrogen Engine: Performance, Emission, and Combustion

I believe that water will one day be employed as fuel, that hydrogen and oxygen which constitute it, used singly or together will furnish an inexhaustible source of heat and light, of an intensity of which coal is not capable... water will be the coal of the future.

Jules Verne, The Mysterious Island (1874)

3.1 Introduction

Unpredictability in fossil fuel prices and finite fossil fuel reserves create uncertainty in the energy landscape. At the same time, improved economy, population explosion, and resource consumption contribute to substantial increases in carbonaceous greenhouse gas emissions, profoundly affecting the world climate and environment. An annual global carbon dioxide emission of approximately 33 billion tons and greenhouse gases have led to a plethora of problems like global warming, melting glaciers, rising sea levels, and frequent extreme weather events. The global land transportation sector is predominantly reliant on vehicles powered by century-old internal combustion (IC) engines. At the same time, the noxious pollutants emitted from the exhaust of vehicles have severely ravaged local air quality. Among a number of low-emission alternative fuels identified, hydrogen stands out as a unique candidate to provide an eventual solution to the energy-environment crises. It is not feasible, in the limited space here in this chapter, to cover an elaborate discussion on all the research and development activities on hydrogen engine technology for over a century. Historical perspectives [1] and several review papers on current research [2–9] shed light on hydrogen's role in sustainable energy systems. Production of hydrogen from renewable sources such as solar, wind, geothermal, and biomass represents a promising pathway toward energy independence and paving the way for a sustainable and eco-friendly future.

3.2 Hydrogen Engines: A Historical Journey of Nearly Two Centuries

Use of hydrogen in engines is not a radical concept. A closer examination of the literature reveals a long history that stretches back over 100 years. The evolution of perspectives on hydrogen engines has been marked by diverse phases, from experimental exploration as a laboratory curiosity to

Hydrogen Engines: Design, Performance Evaluation, Combustion Analysis, and Exhaust Emissions, First Edition.
Lalit Mohan Das.

deployment during world wars when supplies of conventional petroleum fuels fell short at the time of their urgent need. The 1970s witnessed a paradigm shift as emphasis on hydrogen intensified due to rapid fossil fuel depletion. Scanning the pages of history reveals accomplishments and challenges encountered by various researchers working on the use of hydrogen in internal combustion engines. It is not possible here to discuss extensively the intermittent, sporadic research on hydrogen engines carried out over more than a century. Discussion here is limited to some of the significant achievements by several investigators at different phases.

In 1820, Rev. W. Cecil made one of the earliest practical attempts at a hydrogen-operated engine. As documented in his paper titled "Hydrogen Gas as a Moving Power in Machinery" presented before the Cambridge Philosophical Society, Cecil observed, "The engine in which hydrogen gas is employed to produce moving force may be inferior in some respects to many engines at present employed, yet it will not be wholly useless if together with its own defects and it should be found to possess advantages also peculiar to itself" [10]. Such a statement shows that he had observed some unexplored potential as well as some problems with the hydrogen engine as compared to a "contemporary engine." His remarks suggest early awareness of a few special features of hydrogen fuel for engine operation. Hydrogen was investigated for automotive applications, and hydrogen-fueled vehicles were in operation in the 1930s [11]. Rudolf Erren was one of the pioneers in the development of hydrogen engines. Working initially in Germany, Erren conducted extensive tests on Otto and diesel engines, after modifying them to run on hydrogen. Unfortunately, records of his early research were lost during the Allied bombing of Berlin [11]. Later, Erren continued his work in England. In a paper presented at the Institute of Fuel in London [12], he claimed that brake thermal efficiency as high as 60% could be achieved in a hydrogen–oxygen engine with steam recirculation. K. H. Weil, who was the technical director of the Deutsche Erren Studiengesellschaft in Germany, observes in his paper that even though there were no ignition problems in Erren's engines, the patented system required special fuel injection and control mechanisms for hydrogen injection. As described by Weil [13], Erren and his collaborators did extensive work on the hydrogen engine, mostly in the decade before World War II, resulting in numerous patents. Engines modified by Erren were used in trucks, buses, and submarines [14].

Most of the earlier research focused on modifying carbureted spark ignition engines for hydrogen operation. Any modified system will exhibit reliability problems over time. Backfire issues were so severe that many researchers ceased their efforts, as they failed to identify the cause. Backfires, ranging from minor misfires to complete devastation of engines, stood as the Achilles' heel for hydrogen engine advancement for a long time. It is discussed later that direct cylinder injection, by design, avoids backfire. Almost eight decades ago, Etienne Oehmichen [15] conducted a comprehensive study of a single-cylinder hydrogen engine using direct cylinder injection early in the compression stroke. Even in contemporary research, direct cylinder injection continues to be a key area, since it is responsible for addressing the notorious challenge of backfire in hydrogen engines. His experiments covered a range of operating conditions, including speeds from 500 to 2000 rpm, compression ratios from 5.4 to 15.4, and equivalence ratios from 0.29 to 0.67, and were able to achieve a high thermal efficiency of 54% at an equivalence ratio of 0.4 and a compression ratio of 15.4. Oehmichen's research was trouble-free, with no instances of preignition or backfire during the tests. He did not conduct experiments beyond an equivalence ratio of 0.67 and therefore (perhaps) did not encounter backfire chiefly due to mixture strength. Most of the research results have demonstrated that the propensity for backfire increases significantly around or above the stoichiometric value.

Thus in the early phase, some researchers discontinued their efforts when confronted with backfire as they could not identify the cause. Some other researchers restricted their operating conditions (such as equivalence ratio and compression ratio, etc.) to avoid backfire. Even then, most studies with hydrogen engines showed improved performance compared to gasoline. However, in the absence of comprehensive long-term studies over a wide range of operating conditions, the overall understanding of hydrogen engines' response remained incomplete. The common conclusion from most of the investigations identified backfire as the primary obstacle to smooth combustion and improved performance. The next phase marked a period of research fueled by profound scientific curiosity. Certain research and development efforts for hydrogen engines were driven by strategic military applications, particularly during World War II. Captain R. O. King and his collaborators [16–18] deliberately created conditions that were suspected to cause preignition and knocking to identify the exact causes of these abnormal combustion phenomena. In some tests, they dismantled and cleaned the combustion chamber and piston crown before each test to find out the effect of combustion chamber deposits. These studies along with research by several subsequent investigations show that "hot spots" such as spark plugs, exhaust valves, or sharp projections or deposits in the combustion chamber can induce backfire. Sharp projections in combustion chambers could be due to casting imperfections, and deposits could be from particulate matter from pyrolysis of lubricating oil vapor or even inert dust in the intake air. In later years, the major thrust on hydrogen engine development targeted backfires and abnormal combustion while emphasizing performance optimization and reducing nitrogen oxides (NOx), the sole pollutant of concern in hydrogen engines.

3.3 Mixture Formation and Fuel Induction Techniques

As shown in Figure 2.12 in Chapter 2, the hydrogen engine has a window of equivalence ratio between 0.26 to a stoichiometric value where combustion stability occurs. Formation of the mixture at the appropriate point in the engine cycle with flow timing and supply pressure is extremely sensitive to backfire phenomenon. Continuous injection systems, such as those at the intake port, inlet manifold, or as pressurized carburetion are ruled out due to their inability to separate fuel and air intake streams to prevent backfire. Utilizing delayed fuel delivery through a timed injection technique ensures control over engine operation by monitoring the fuel–air ratio, as opposed to relying on changes in the intake manifold vacuum. This technique also minimizes pumping losses associated with traditional vacuum-based control and leading to increased thermal efficiency. Timed injection can be strategically implemented either through direct cylinder injection or injection at the intake port referred to as port fuel injection (PFI) or at a suitable point in the engine intake manifold called timed manifold injection (TMI).

Spark ignition engines have specific operational mixture limits influenced by various design considerations that determine engine efficiency and functionality. Figure 3.1 shows a strong agreement between calculated and experimental values when utilizing gaseous fuels such as methane, propane, and hydrogen [19]. The graph also shows hydrogen's potential for lean operation. Thermodynamic and heat transfer properties of hydrogen result in elevated compression temperatures, leading to improved engine efficiency during lean mixture operation. Hydrogen's fast burning rates and low minimum ignition energy levels make it suitable for high-speed and cold starting conditions for engines.

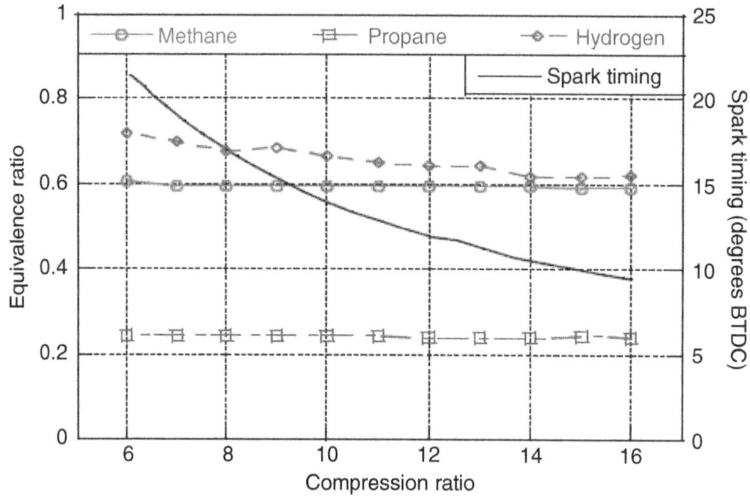

Figure 3.1 Variations of the predicted and experimental values on lean operational limits with changes in compression ratio for hydrogen, methane, and propane in a spark ignition (SI) engine at 900 rpm. Adapted from [19]. Please visit the companion website to view a color version of this image.

H. R. Ricardo, unable to control the phenomenon of "popping back into the carburetor" (backfire) even at a compression ratio as low as 3.8, thought hydrogen was impractical for most uses [20]. However, some degree of success was achieved by providing a quality-controlled mixture through a lean burn carburetor specifically developed for hydrogen operations [21]. The hydrogen induction technique (HIT) developed by Swain and Adt proved effective in overcoming backfire problems [22]. A multiyear DOE-supported project at the University of Miami produced very significant results from exhaustive tests on 19 hydrogen engine configurations. One of the objectives was to generate and document an experimental design database for hydrogen-fueled automotive engines [22–24]. Tests were conducted over a wide range of engine operating conditions, generating data for optimum spark timing, thermal efficiency, fuel consumption, power output, backfiring limit, and exhaust emissions. Quite often, flashback caused by hot spots was predictable and characterized. There was a period before hot spot–induced flashback where ignition occurred at increasingly earlier crank angles until it happened before the intake valve closed. Several studies show that a reduction in valve overlap does not reduce flashback tendencies. This contradicts the hypothesis that flashback is triggered by the interaction between the incoming fresh charge and the still-burning exhaust gases during the valve overlap period. The use of common valve overlaps does not support this assumption, which suggests that other factors may influence flashback. In some tests, oil in residual gas causes flashback. In a homogeneous charge engine, a judicious blending of throttling and mixture enrichment could lessen backfire tendency. Proper cylinder head passage design ensures backfire-free operation. The parallel induction system developed by Lynch [25] is an improved technique compared to traditional carburetion.

The schematic in Figure 3.2 shows the parallel induction system in a hydrogen engine where hydrogen and air enter the cylinder separately during the intake stroke. There, if ignition occurs during intake, the explosion cannot propagate through the intake manifold because it contains only air. This design reduces engine stalls and other damages common to the carbureted version. Figure 3.3 shows a parallel induction system in a hydrogen engine fitted with turbocharging and after cooling. The air is compressed in the turbocharger, cooled in the after-cooler, and mixed with

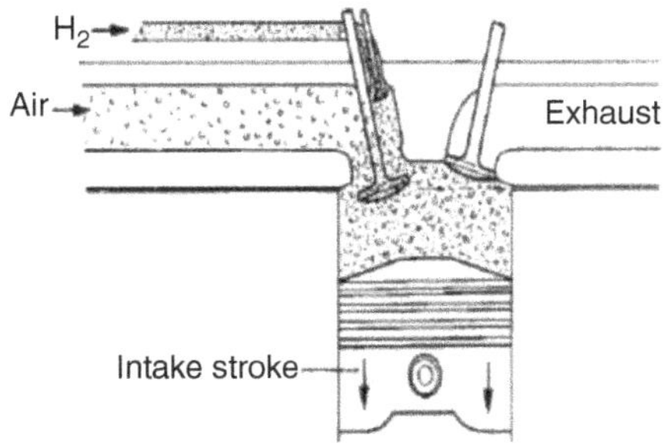

Figure 3.2 Parallel induction system [25] / with permission of Elsevier.

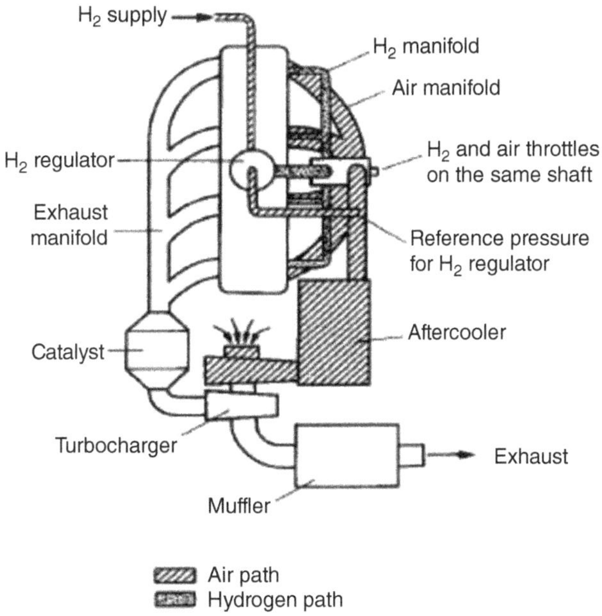

Figure 3.3 Intake and exhaust systems in a turbocharged, after-cooled, hydrogen engine using parallel induction [25] / with permission of Elsevier.

hydrogen. Air and hydrogen paths are separate until they reach the cylinders. The air throttle, similar to a conventional "butterfly" throttle, controls the flow of air through the manifold, past the intake valves, but it does not mix with hydrogen until reaching the cylinders. The hydrogen path has its own regulator, throttle, manifold, and intake valves. In this system, hydrogen valves can be operated synchronously with air intake valves. Parallel induction, which is conceptually closer to intake port injection, is an improvement over carburetion, but it needs major sophisticated controls for regular long-term reliable application.

Numerous studies have identified that an effective fuel induction technique not only reduces frequency and intensity of backfires but also eliminates them altogether. Risk of this abnormal combustion is highest with carburetion, lower with timed injection (port or manifold), and direct injection precludes backfire. Additionally, there is the possibility of stratification during direct injection. However, carburetion and timed injection, both external mixture formation techniques, operate at much lower injection pressures. In contrast, direct injection requires pressures between 5 and 250 bar depending on the mode of injection and timing. Thus, the injector has to be designed to encounter the severe thermal environment in the combustion chamber, more so if the in-cylinder injection is timed at the end of the compression stroke. Research and development activities on fuel injection started on a mild scale in the early phase of 1980 in several parts of the world and gained momentum. Many studies focused on fuel injection for various engine types including spark ignition (SI) and compression ignition (CI) engines, as well as two-stroke, four-stroke, single-cylinder, multi-cylinder, and vehicular engines.

Usually, a fuel suitable for spark ignition (SI) engines is not ideal for compression ignition (CI) engines, and vice versa. This is mainly due to fundamental differences between the combustion processes of these two types of engine. SI engines rely on a spark plug to initiate combustion and fuels with higher octane ratings, which resist spontaneous ignition. Combustion in CI engines occurs through the autoignition of the fuel due to the high temperature and pressure at the end of the compression stroke. Fuels that burn quickly and smoothly are suitable for SI engines whereas CI engines rely on a more controlled, slower combustion process. Therefore, fuels like hydrogen, which ignite too quickly, lead to knocking and increased pressure rise rates in CI engines. Even though most research and development activities focus on the existing spark ignition engine, modified for hydrogen operation, some modest efforts are directed at advanced engine technologies aimed at bridging the gap between the SI and CI engines. Technologies such as homogeneous charge compression ignition (HCCI) engines and multi-fuel and dual fuel engines often come with their own set of challenges and other complexities when hydrogen is used. Chapters 8 and 9 discuss the use of hydrogen in combination with other fuels and some augmented hydrogen engines and alternative configurations.

Hydrogen injection studies have been done on various test rigs with different injection strategies. Port injection involves introducing hydrogen into the intake ports, optimizing the fuel–air mixture before it enters the combustion chamber. Manifold injection distributes hydrogen throughout the intake manifold, while direct cylinder injection delivers fuel directly into the combustion chamber. Despite rigorous methodological diversity, the common objective remained aimed at achieving optimum performance, emissions, and smooth operation. SI engines focused on port injection [26–34], direct cylinder injection [35–43], and manifold injection [44–50]. Some researchers used hydrogen in compression ignition engines [51–66]. Most of the work used gaseous hydrogen while some used liquid hydrogen [67–71]. Exhaustive experimental studies, complemented by theoretical analyses and numerical investigations, have generated a wealth of technical information and guidelines for development of optimized test rigs with injection strategies. A modest effort has been made here to provide both a qualitative and realistic quantitative perspective, closely consistent with data reported in the literature.

3.4 Port Fuel Injection: Performance, Combustion, and Emission Features

Hydrogen port fuel injection (PFI) in spark ignition (SI) engines is the most extensively studied mode. Engines, operating at a lean equivalence ratio (ϕ) of less than or equal to 0.5, emit nitrogen oxide (NOx) below 100 ppm without additional after-treatment. Besides lean operation, exhaust gas recirculation (EGR) and a three-way catalytic converter contribute substantially to controlling

NOx emissions. In the P2000 Hydrogen Engine Design and Vehicle Development Program, researchers built and developed dedicated hydrogen spark ignition and tested with different compression ratios (10, 12.5, 14.5, and 15.3). Preignition, NOx emission, and combustion control were investigated using gaseous fuel injection [72]. Major activities of the system development involved oil control, design of the cooling system and cylinder head, improved ignition, the fuel supply system, and other hardware changes.

Port fuel injection (PFI) has features of excellent fuel distribution from cylinder to cylinder and chances of minimal fuel leakage in the intake manifold [73]. Lee and Kim [26] conducted tests with a solenoid-driven gas valve and hydrogen-controlled hydrogen by changing the duration of the solenoid-driving signal. Severe flashback was observed near the stoichiometric fuel–air equivalence ratio. This was due to a hot spot such as a lubricant deposit or spark plug rather than to the high temperature residual gas itself. To operate the engine at higher speeds and to determine the range of backfire-free operation, precise injection timing control should consider the solenoid's movement delay. In their study, Park and colleagues [27] performed tests at various engine speeds under wide-open throttle using a fuel injector originally designed for natural gas engines. The researchers adopted two distinct operation strategies: one aimed at limiting nitrogen oxide emissions and the other at maximizing torque output.

To address challenges concerning backfire, researchers have examined the effects of altering the spark plug's heat rating and adjusting the ignition energy. Despite the inherently low minimum ignition energy of hydrogen, the use of a cold-type spark plug has proved effective in suppressing backfires. However, a reduction in overall engine efficiency has been observed due to an increase in combustion temperature and consequent increase in heat loss. Figure 3.4 illustrates the variation in excess air ratio with different spark plug heat rating numbers and ignition energy across different engine speeds. The engine ran at full throttle for maximum torque. Increasing hydrogen flow reduced backfire, lowering intake airflow and excess air. Lowering ignition energy resulted in no significant difference in excess air ratio, and occurrence of backfire remained unaffected. The reduction in ignition energy was anticipated to prevent abnormal discharge during the intake

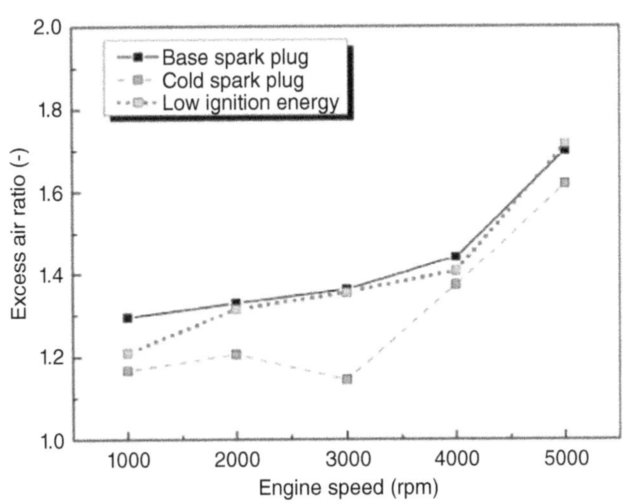

Figure 3.4 Excess air ratio with engine speed at wide-open throttle for each spark plug condition [27] / with permission of Elsevier.

stroke by maintaining a low air gap resistance between electrodes. This strategy was not effective at high loads. As reported, a decrease in ignition energy led to a slight increase in maximum torque at specific engine rpm conditions. It is evident in Figure 3.5 that the impact on thermal efficiency for different spark plugs at various ignition energies was minimal. As depicted in Figure 3.6, despite reducing ignition energy or changing the spark plug's heat rating, the combustion period did not change significantly. This is primarily due to initial flame development and flame propagation. In the context of the present work, "flame development period and flame propagation period are defined as the period from ignition to the 10% mass burned fraction and from the 10% mass burned fraction to 90% mass burn fraction, respectively" [27]. Gao et al. [28] reviewed the effects of high flame propagation velocity, low lean-burn limits, and low ignition energy as causes of backfire and discussed effective strategies to control them.

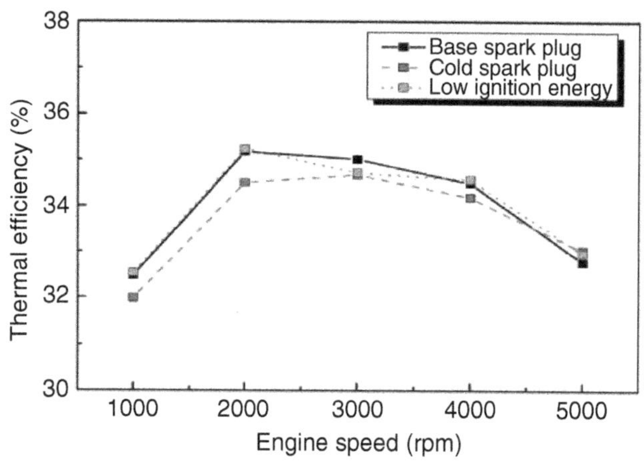

Figure 3.5 Thermal efficiencies at various engine speed at wide-open throttle for each spark plug condition [27] / with permission of Elsevier.

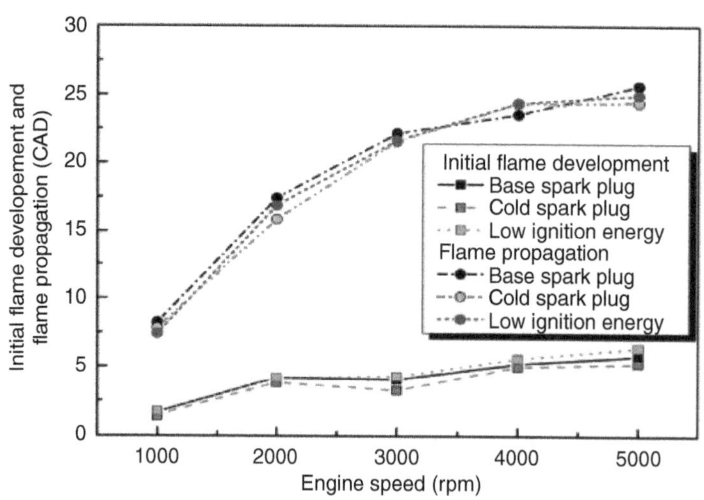

Figure 3.6 Variation in the initial flame development and flame propagation with engine speed at wide-open throttle for each spark plug condition [27] / with permission of Elsevier.

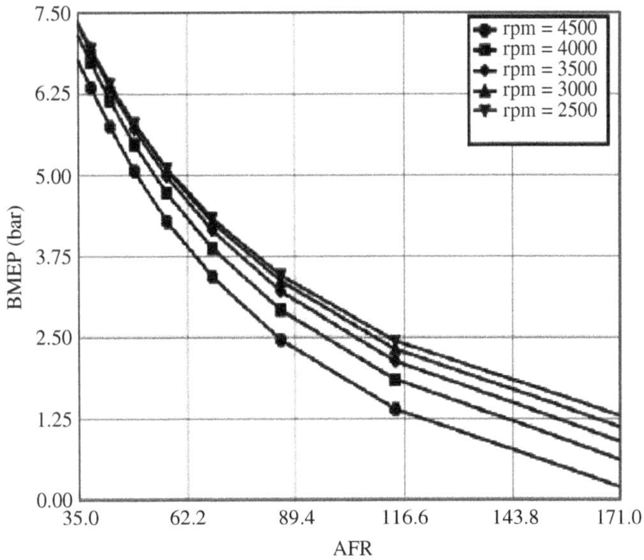

Figure 3.7 Variation of BMEP with air–fuel ratio [29] / Science Publications.

Bakar and colleagues performed some numerical studies with port injection [29] using a one-dimensional gas dynamic model to simulate flow and heat transfer. The experiment involved varying the air–fuel ratio from stoichiometric to lean limits and adjusting the speed between 2500 and 4500 rpm, while keeping the injector location fixed in the middle of the intake port. The results showed a decrease in BMEP and BTE with increasing the engine speed and the air–fuel ratio. Simultaneously, there was an increase in BSFC as both the speed and the air–fuel ratio increased. The cylinder temperature increased with higher engine speeds but decreased with increase in air–fuel ratio. The study identified volumetric efficiency with port injection, which affected overall engine performance. Figures 3.7 to 3.10 show the significant performance and related combustion parameters of the engine operation at different speeds within the ranges of 2500 to 4500 rpm. The air–fuel ratio (AFR) was varied from the stoichiometric limit (AFR = 34.33:1 based on mass, where the equivalence ratio $\phi = 1$) to an ultraclean operation (AFR = 171.65 based on $\phi = 0.2$). The trends in the variation of BMEP and thermal efficiency are evident in Figures 3.7 and 3.8, respectively.

Figure 3.9 indicates that BSFC tends to increase with a decrease in speed and an increase in AFR. However, within the speed range investigated in this work, the minimum required BSFC occurred within a specific range of AFR = 38.144 ($\phi = 0.9$) to 49.04 ($\phi = 0.7$). For an AFR of 171.65 ($\phi = 0.2$), a BSFC of 144.563 g/kWh was observed at 2500 rpm, in contrast to 1038.85 g/kWh at 4500 rpm. This significant increase is due to very lean operating conditions causing unstable combustion and also greater power loss due to a reduction in the volumetric heating value of the hydrogen–air mixture.

Figure 3.10 clearly demonstrates how an increase in AFR results in a significant decrease in maximum cylinder temperature, following a steep curve. Also, the relationship between maximum cylinder temperature and AFR is not affected by speed. Under stoichiometric operating conditions (AFR = 34.33), the maximum cylinder temperature was 2752.83 K and it dropped significantly to 1350 K at an AFR of 171.65 (ϕ). Sun and Liu studied a 2.0 L naturally aspirated gasoline engine modified for port fuel injection (PFI) using hydrogen fuel [30]. A series of dynamometer tests

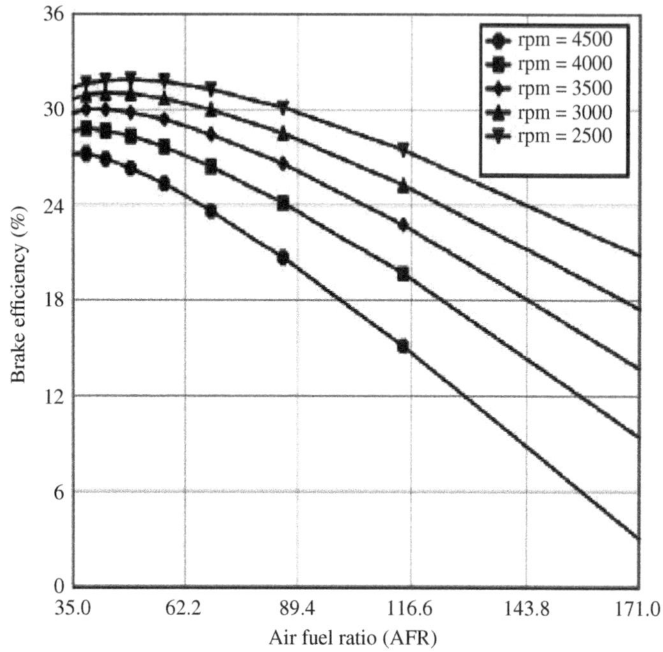

Figure 3.8 Variation of brake thermal efficiency with air–fuel ratio [29] / Science Publications.

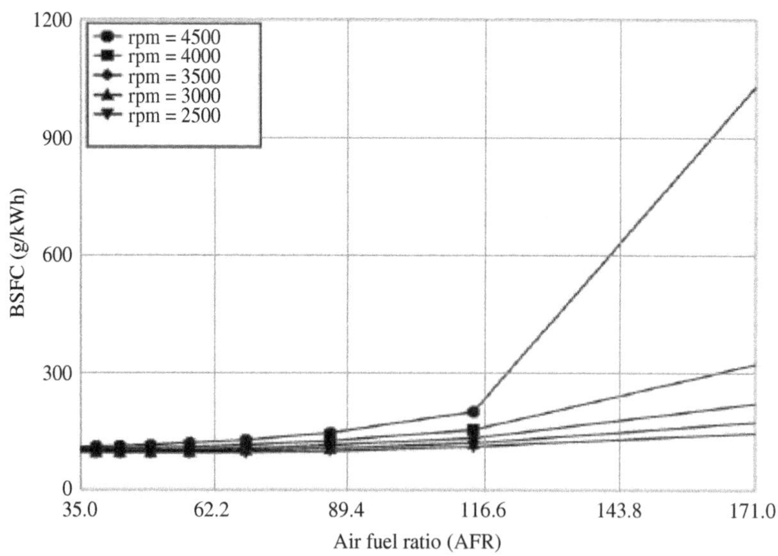

Figure 3.9 Variation of BSFC with air–fuel ratio [29] / Science Publications.

under various equivalence ratios showed that a naturally aspirated hydrogen engine exhibited 20–25% lower torque than a gasoline engine.

Significant results of this study showed that the hydrogen engine reached a peak indicated thermal efficiency of 45% at 5000 revolutions per minute (rpm) and a peak brake thermal efficiency of 35% at 2000 rpm. NOx emissions were minimal when the equivalence

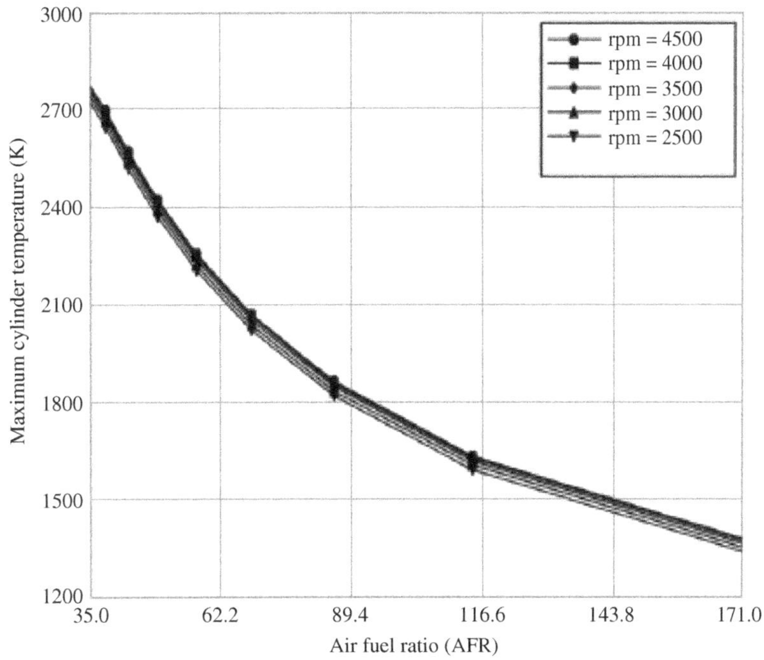

Figure 3.10 Variation of maximum cylinder temperature with air–fuel ratio [29] / Science Publications.

ratio was below 0.55 or above 1 with its peak level within the equivalence ratio range of 0.8 to 0.9.Combustion duration, defined as the period from 10 to 90% of the mixture burned, was influenced by laminar flame speed and turbulence intensity. Richer mixtures exhibited higher laminar flame speeds and shorter combustion durations. At higher engine speeds, intensified turbulence accelerated the mixing and reduced combustion duration, especially in lean mixtures. However, for equivalence ratios exceeding 0.6, the combustion duration was largely independent of engine speed. This is because, with increased mixture concentration, laminar flame speed rises rapidly, resulting in a balanced effect on the combustion duration between engine speed and laminar flame speed. In this study by Sun and Liu, after satisfactory bench calibration and testing, the engine was installed in a passenger car and evaluated on a chassis dynamometer. To control NOx emissions, the equivalence ratio was restricted to below 0.55. NOx emissions at the highway cycle test with corresponding vehicle speed and time showed that HICE met the Euro VI emission standard [30].

Watson et al. [33] had overcome backfire by delayed port admission of hydrogen at relatively low pressures. Backfire under high load condition was averted through water injection. Yang et al. [34] employed the analytic hierarchy process (AHP) entropy weight method to investigate the abnormal combustion index layer of port fuel injection (PFI) under various hydrogen injection angles and flow rates. Mixture uniformity coefficient, hot area temperature, maximum temperature rise rate, residual hydrogen in the intake port, and cylinder temperature at the exhaust end were considered for evaluating risk of abnormal combustion. Hot area temperature and maximum temperature rise rate initially decreased and then increased with rising hydrogen injection angle and flow rate. While higher injection angles and flow rates reduced cylinder temperature at the exhaust end, they also increased residual hydrogen in the intake port. Excessive or insufficient hydrogen injection angles and flow rates obstruct effective blending of hydrogen and increase the likelihood of abnormal combustion. Computational results indicate that at equivalence ratios of 0.5 and 0.67, a hydrogen injection angle of 45° and a flow rate of 4.96 kg/h ensured that all index layer parameters met the desired value levels.

3.5 Timed Manifold Injection: Injector Development

In an exhaustive series of experiments conducted at the Indian Institute of Technology, Delhi (IITD), four modes of fuel induction were exhaustively tested keeping in view long-term application [74, 75] for hydrogen engines. This comprehensive experimental investigation aimed at addressing the challenges of flashback and incomplete combustion. Injectors with different actuation mechanisms were developed and tested for both manifold and in-cylinder injection. Integrated control of the actuation mechanism and the flow capability was crucial to the development of a suitable leak-proof, reliable injector. Due to hydrogen's high flammability and undetectable flame, extreme precaution was taken in developing the injector's actuation mechanism and flow capability to ensure leak-proof injection throughout the engine cycle, preventing unscheduled hydrogen release. Suitable materials were selected for the injector considering hydrogen compatibility, corrosion resistance, and mechanical strength. Precision surface finish was achieved to ensure tight seals and prevent leakage. Proper alignment of moving and connecting parts was ensured for system integrity. Quite often, during continuous engine operation, the actuation mechanism of the injector fails due to misalignment of parts. Misalignment introduces a bending moment and can lead to actuator cracking during either manifold or in-cylinder injection. Taking into consideration all the above facts, three different actuation mechanisms such as hydraulically operated, cam-actuated and electronically controlled solenoid-actuated injection systems were developed at different phases and successfully tested for prolonged operation in research and small utility engines [76–78]. Figure 3.11 shows the three injectors installed in hydrogen engines. Figure 3.11a shows the hydraulically operated injector in which diesel oil was used as hydraulic fuel. A jerk from the diesel injector forced open the hydrogen injector. Diesel, after passing through the nozzle, was collected back. Figure 3.11b shows the cam-actuated injector, which used a lift rod moved by a cam, and the motion being transmitted through a specially designed linkage. An electronically controlled solenoid-actuated injector was mounted on a small utility engine in Figure 3.11c. This system offers adequate flexibility to control the injection timing and injection duration to provide the desired fuel quantity at the appropriate point during the engine cycle. Design of injectors with different specific physical design, actuation mechanism, and overall construction to meet the stringent requirements are relevant for smooth, sustained engine operation [78–82].

Flashback with different degrees of severity and frequency was encountered during tests with both carburetion and continuous manifold injection and so those methods were discarded. Direct cylinder injection was observed to be a robust mixture formation technique because it precludes backfire by definition, as the air–fuel mixture is formed within the cylinder after intake valve closure. Direct cylinder injection could be high-pressure direct cylinder injection (HPDI) or low-pressure direct cylinder injection (LPDI). A LPDI system injects hydrogen early in the compression stroke, allowing for improved mixing with air. On the other hand, a HPDI system operates with high pressures, enabling direct injection into the combustion chamber during the end of the compression stroke. During HPDI tests, it was a challenge for the injector to withstand the harsh thermal conditions of intense heat and pressure in the combustion chamber. Moreover, HPDI, scheduled to take place at the end of compression stroke in an almost stagnant condition of engine cylinder, does not generate the required level of in-cylinder turbulence necessary for

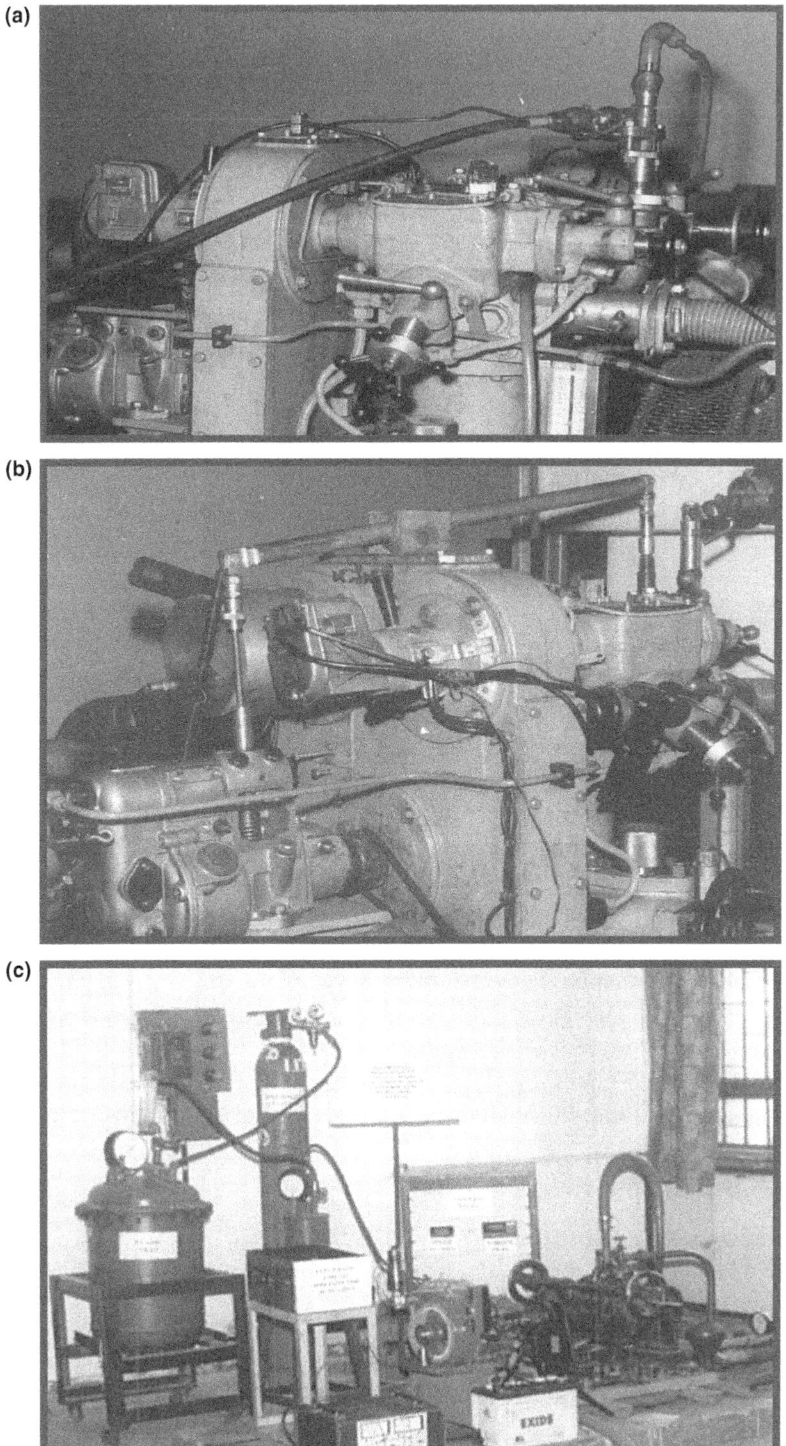

Figure 3.11 (a) Hydraulically operated injection system [76, 77, 81] / with permission of Elsevier. (b) Cam-actuated injection system [76, 77, 81] / with permission of Elsevier. (c) Solenoid-actuated electronically controlled injection system [78, 80] / with permission of Elsevier.

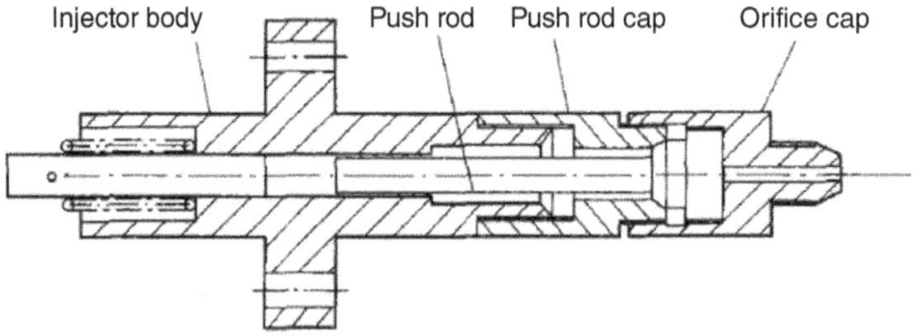

Figure 3.12 Hydrogen injector [76] / with permission of Elsevier.

adequate mixing. Formation of homogeneous charge was difficult [76, 77]. LPDI also sometimes showed problems with leaking injector tips. The extent of leakage in LPDI was considerably less compared to leakage during HPDI. Small leakage from LPDI could be resolved through careful material selection and heat treatment processes.

TMI provides a mechanism for delayed fuel delivery, cooling small thermal masses and residual exhaust gases, reducing the risk of hot spot-induced backfire. A TMI system worked effectively at a pressure a little above the atmospheric, demonstrating engine operation at lower pressure ranges and achieving optimal system performance and controlled combustion [76]. A TMI system strategically uses the time of fuel delivery, commencing sometime after initiation of the intake stroke, thus optimizing combustion efficiency. Since intake manifold, by design, does not contain combustible mixtures, the risk of backfire is almost ruled out. By ensuring that air is inducted before fuel delivery, the TMI system not only introduces a pre-cooling effect but also renders preignition sources ineffective. Moreover, it is possible to quench and dilute any residual combustion products near TDC, contributing to better combustion stability. A TMI-operated hydrogen engine optimally integrates the advantages of spark ignition and diesel engines. It demonstrates diesel-like quality governing and thermal efficiency while achieving the specific output characteristics of a spark ignition engine [76, 77].

Both TMI and LPDI are suitable to ensure backfire-free operation. A comprehensive comparative assessment was done between TMI and LPDI to determine the fueling technique for sustained engine tests. Figure 3.12 shows the hydrogen injector developed and used in both TMI and direct cylinder injection. As shown in Figures 3.13 and 3.14, timed manifold injection demonstrated better overall performance, making it the more pragmatic choice for hydrogen fueling [77].

3.5.1 Performance Characteristics

Exhaustive tests on a variable compression ratio engine with TMI show the performance characteristics as shown in Figures 3.15 through 3.18. A series of sensitivity tests were performed to identify minimum spark advance for best torque (MBT) conditions for the engine. To maintain consistency, all performance data are gathered under MBT ignition timing.

3.5.2 NOx Control: Lean Mixture, Exhaust as Recirculation

Combustion of hydrogen with air produces oxides of nitrogen (NOx):

$$H_2 + O_2 + N_2 = H_2O + N_2 + NOx$$

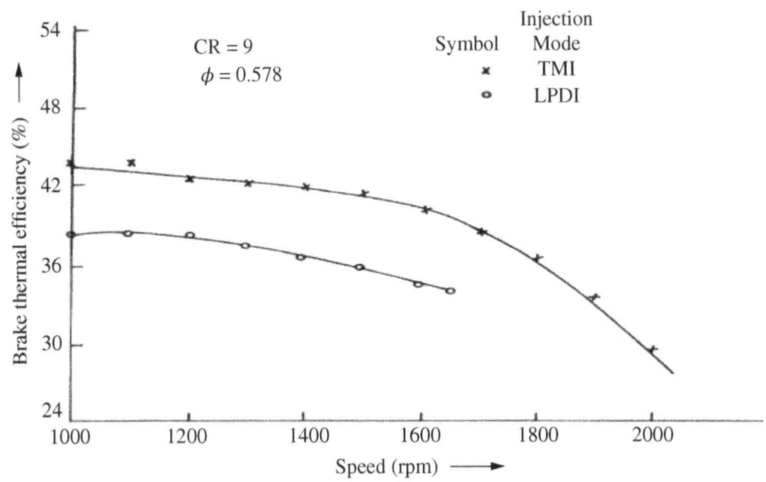

Figure 3.13 Brake thermal efficiency with respect to speed at a constant equivalence ratio for both TMI and LPDI operation [77] / with permission of Elsevier.

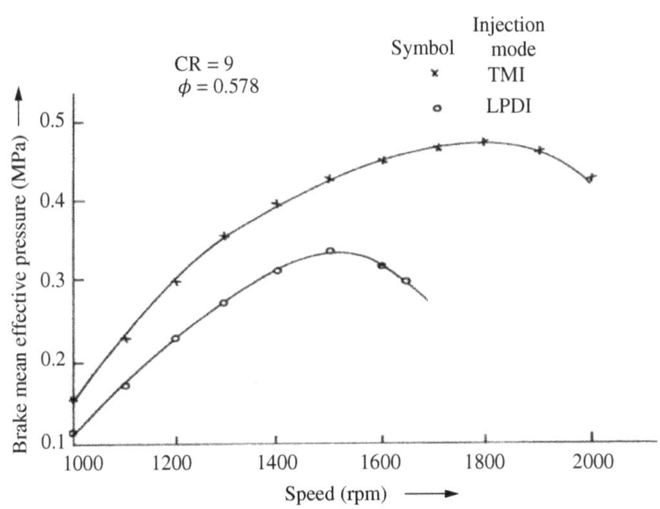

Figure 3.14 Brake mean effective pressure with respect to speed at a constant equivalence ratio for both TMI and LPDI operation [77] / with permission of Elsevier.

Considering the exhaust emissions, there are practically no hydrocarbons in hydrogen engine exhaust. Sometimes, minimal amounts of hydrocarbons are observed, mainly due to lubricating oil. Formation of nitrogen oxides (NOx) is of great concern, particularly with respect to the contributions of both flame-front and post-flame gases. Optimizing operating conditions for hydrogen engines is crucial to minimize NOx emissions. Formation of thermal NOx emerges as a complex endothermic process with substantial activation energy. Temperature emerges as a critical factor, and an increase in temperature results in higher levels of NOx formation. NOx emission levels are influenced by combustion temperature, duration of high-temperature combustion, and oxygen availability. Techniques such as engine operation at low equivalence ratio, controlling burn time, and introducing cooling measures like water addition or exhaust gas recirculation have been able to control NOx levels. Lean operation accompanied by slower chemical reactions weakens the kinetics of NOx formation [83, 84]. Moreover, catalytic converters and advanced control strategies,

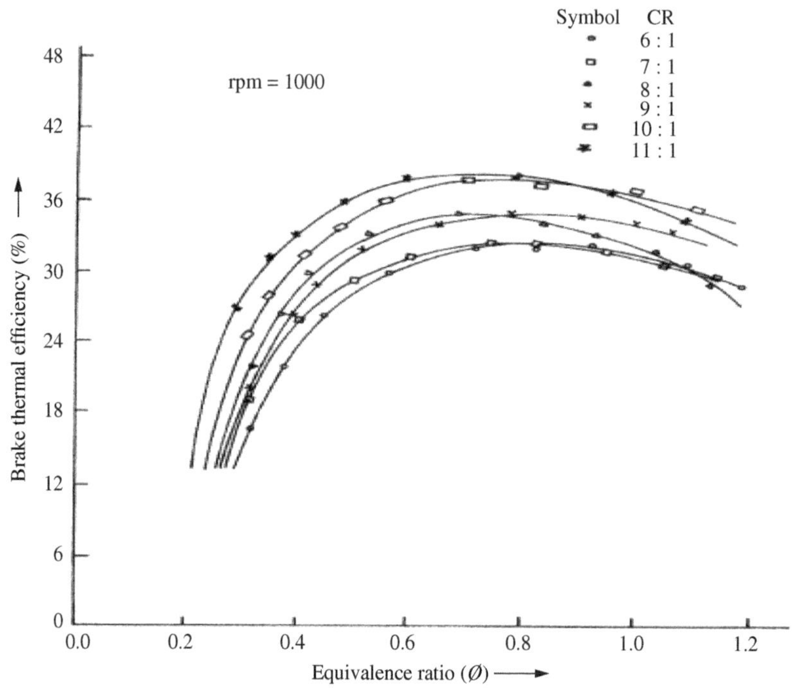

Figure 3.15 Variation of brake thermal efficiency with respect to the equivalence ratio at different compression ratios [76, 79].

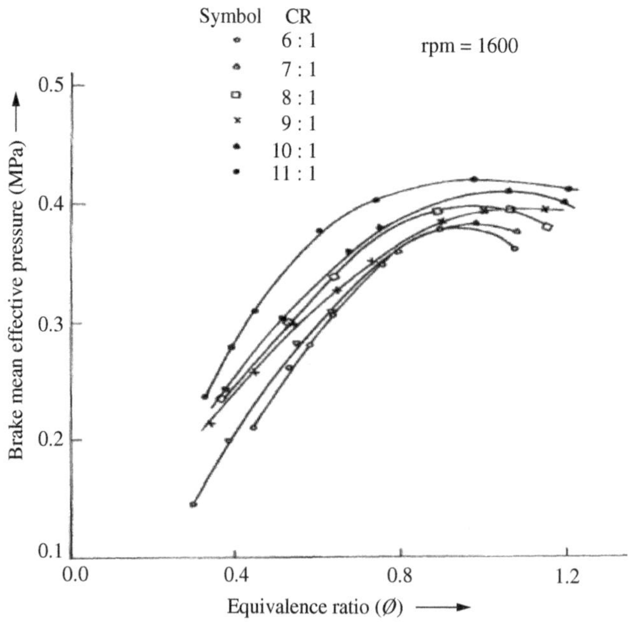

Figure 3.16 Variation of BMEP with respect to equivalence ratio at different compression ratios [76] / with permission of Elsevier.

water injection at the intake manifold [85–87], exhaust gas recirculation [88–91], and spark plug gap have proven effective for NOx control [92].

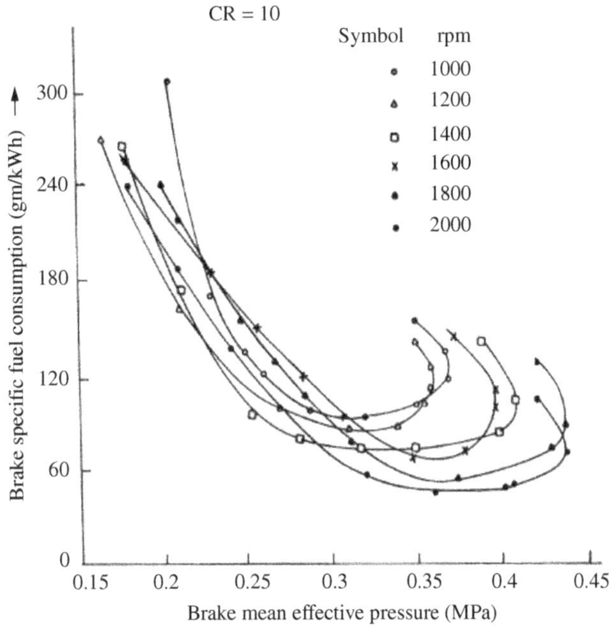

Figure 3.17 Relationship between BMEP and BSFC at different speeds [76] / with permission of Elsevier.

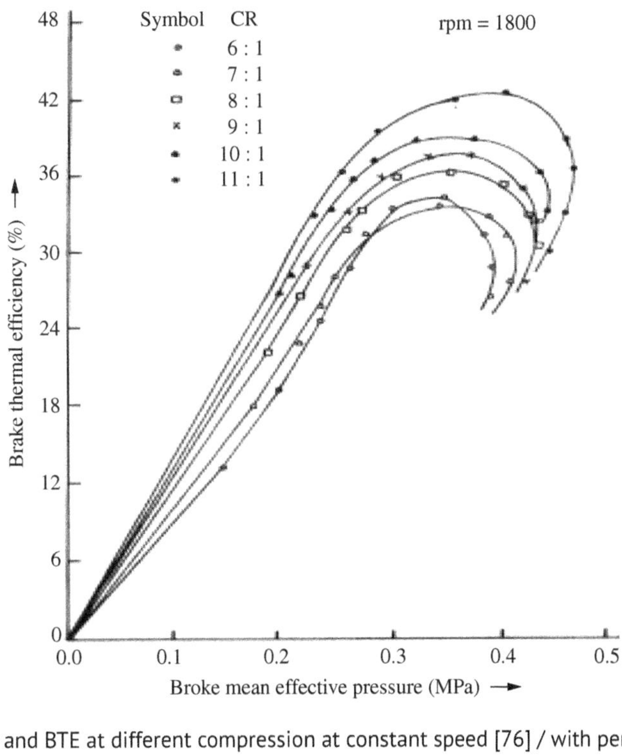

Figure 3.18 BMEP and BTE at different compression at constant speed [76] / with permission of Elsevier.

Increase in compression ratio is associated with higher temperatures, which results in rise in flame speed during combustion. This acceleration in flame speed reduces residence time. Thus, the situation leads to simultaneous increase in temperature and flame speed, coupled with decrease in

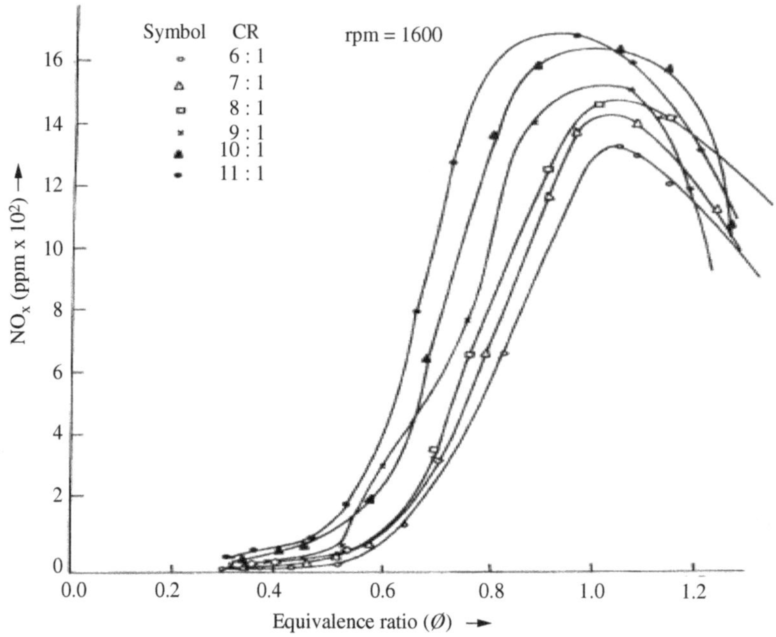

Figure 3.19 Variation of NOx emission with equivalence ratio [76] / with permission of Elsevier.

residence time. Between the two factors, increased temperature and decreased residence time, the impact of increased temperature predominates and has a more pronounced effect on NOx emissions compared to the influence of reduced residence time. Since hydrogen engine is capable of operating smoothly in lean mode, the simplest strategy to reduce NOx concentration is to run the engine in lean mode. As glaringly evident in Figure 3.19, there is hardly any NOx in the lower level of equivalence ratio.

3.5.3 Exhaust Gas Recirculation for NOx Control

Exhaust gas recirculation (EGR) is a mature technology with a proven method to curtail NOx emissions in conventional diesel and gasoline engines that contain higher concentrations of H_2O and CO_2, both possessing a high specific heat capacity. Recirculating a fraction of the exhaust gas containing these components back into the engine intake lowers the peak in-cylinder temperature, which in turn reduces NOx formation. The same strategy is applicable to hydrogen engines, where the exhaust is primarily composed of water vapor. As seen in Figure 3.20, NOx emission decreases as the level of EGR is increased, and this reduction is well pronounced at a 14% EGR rate. A NOx level with 18% EGR shows only a marginal benefit, suggesting that higher levels of EGR do not contribute to a proportional decrease in NOx emissions.

Figure 3.20 shows that at higher EGR levels, beyond 14%, the increased presence of inert gases with high specific heat capacities (such as water vapor and carbon dioxide) leads to a significant rise in heat transfer to the engine coolant.

While EGR contributes to reduce NOx emission by lowering the engine's peak temperature, it increases the overall heat loss. This increased heat loss quite often counters the benefits of lower peak temperature. As shown in Figure 3.21, when EGR is used, the engine's torque (the power available for doing work) decreases. Higher levels of EGR affects the combustion process. Thus, EGR implementation has to be considered as keeping NOx reduction and combustion stability.

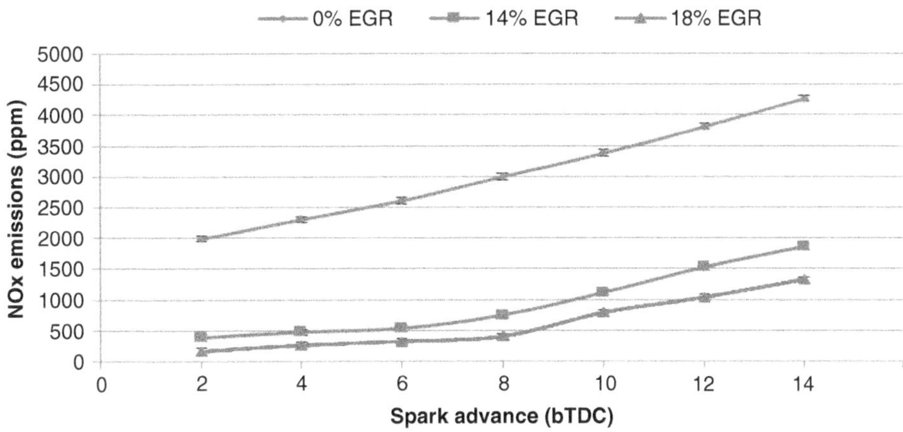

Figure 3.20 Effect of EGR on NOx emissions with varying spark timing (engine speed = 1200 rpm; Φ = 0.6 [93] / Indian Institute of Technology Delhi.

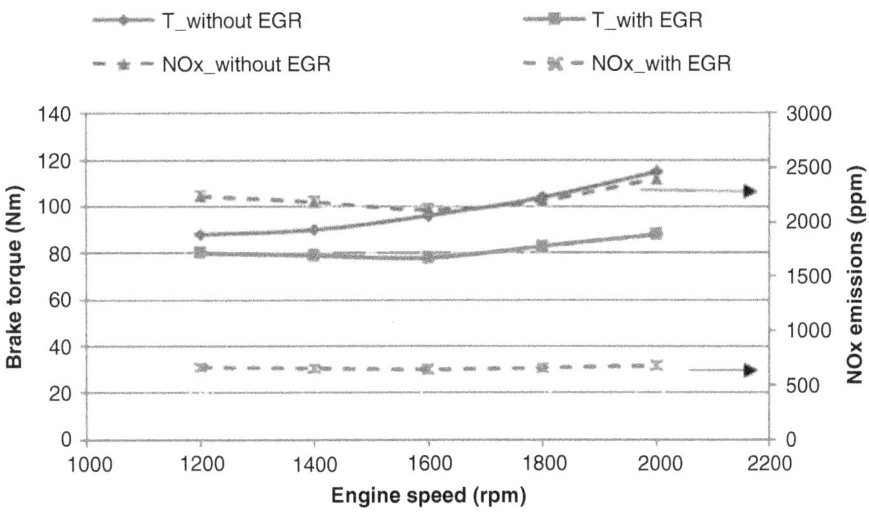

Figure 3.21 Brake torque and NO_x emissions with speed (12% EGR and at Φ = 0.6 [93] / Indian Institute of Technology Delhi.

Retarding the spark timing in spark-ignited engines and retarding the fuel injection timing in compression-ignition engines have proven effective in reducing NOx levels. The implications of these strategies extend beyond the fuel type, and tests on hydrogen engines showed increased NOx emissions by advancing spark timing. This trend is also reflected in Figure 3.20. In a gasoline vehicle, the concentration of carbon monoxide, unburned hydrocarbons, and nitrogen oxides in the exhaust can be achieved through a three-way catalytic converter. This device facilitates the oxidation of CO and hydrocarbons and causes reduction of NOx over the catalytic surface. Species such as hydrogen (H_2), carbon monoxide (CO), and hydrocarbons act as reductants, contributing to reduction of the NOx concentrations. The applicability of a similar approach could be extended to hydrogen (H_2) engines, where unburned hydrogen might be present in the exhaust. So, if the exhaust, containing both NOx and unburned hydrogen, is directed over a catalyst, a specific reaction (Reaction1) involving H_2 and NOx as reactants can effectively reduce the concentration of NOx. The reaction is expressed as follows:

$$H_2 + NO\ (or\ NO_2) \rightarrow\ H_2O + N_2 \qquad \text{(Reaction 1)}$$

Thus, the catalytic reduction not only aids in minimizing NOx emissions but also addresses safety concerns related to possible hydrogen leakage into the atmosphere.

Figure 3.22 shows the impact of a NOx reduction catalyst downstream of the engine when EGR is implemented. Addition of a NOx reduction catalyst along with EGR further decreases NOx concentrations. Direct measurement of exhaust composition using gas chromatography, illustrated in Figure 3.23, provides evidence for the presence of hydrogen in the exhaust.

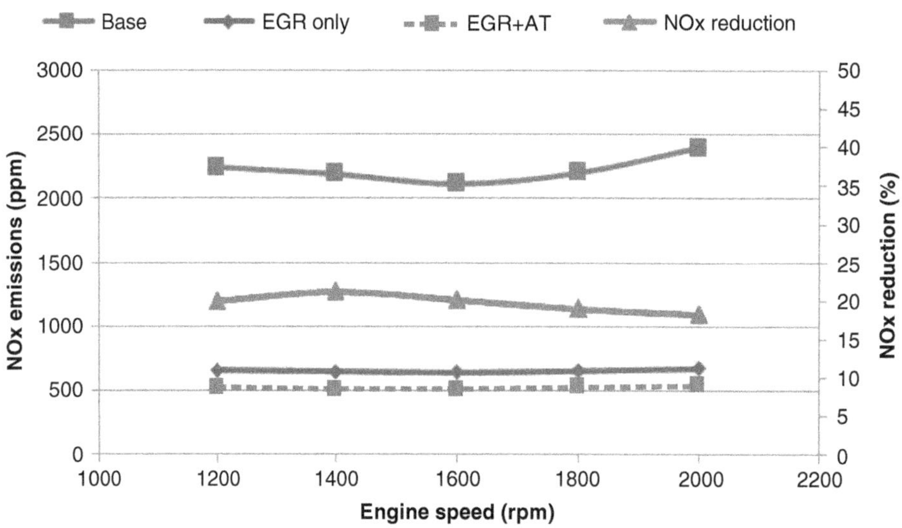

Figure 3.22 Variation of NO_x emissions with engine speed under the influence of EGR and catalytic reduction [93] / Indian Institute of Technology Delhi.

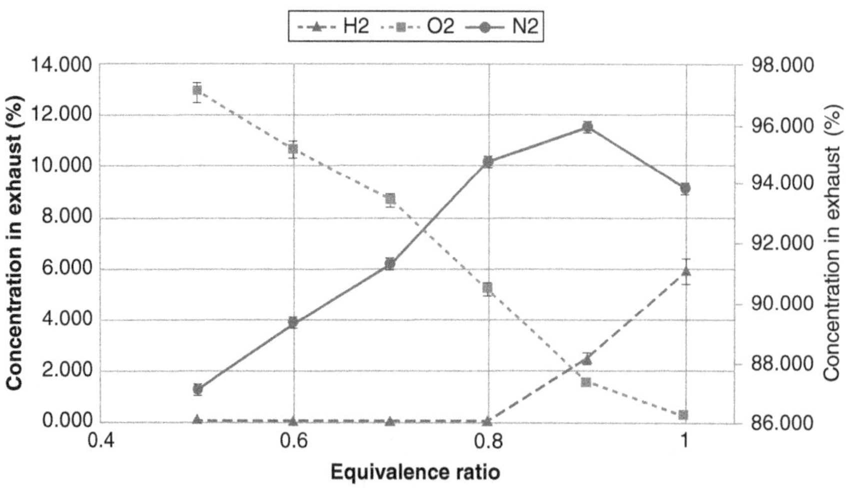

Figure 3.23 Variation of exhaust gas composition (unburned H_2, O_2, and N_2) with equivalence ratio [94] / with permission of Elsevier.

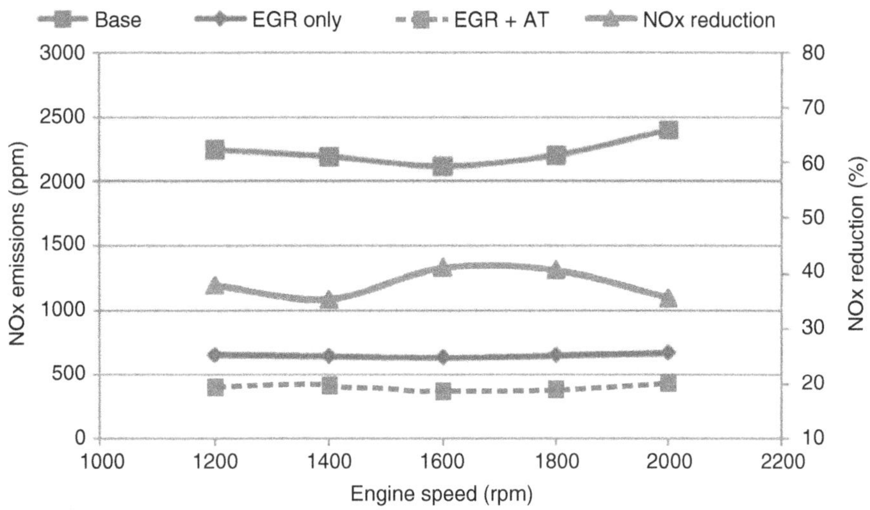

Figure 3.24 Variation in NO_x emissions using EGR and after-treatment (AT) catalyst with an external supply of hydrogen (equivalence ratio = 0.6). Adapted from [93].

In the automotive industry, urea-selective catalytic reduction (urea-SCR) technology is widely used to reduce NOx emissions. This approach involves introducing a urea solution into the exhaust pipe [95] where hydrolysis of urea generates ammonia (NH_3). NH_3 then acts as reductant for NOx, facilitating their conversion into harmless nitrogen and water. However, an alternative method for NOx reduction involves externally dosing additional hydrogen (H_2). It has already been discussed that unburned H_2 can effectively serve as a reductant for NOx (Figure 3.23).

Dosing of external H_2 as reductant offers a distinct advantage in hydrogen-fueled vehicles. Unlike urea dosing for diesel vehicles, which necessitates a separate urea tank and refilling, H_2 dosing can utilize the existing fuel tank in the vehicles. Figure 3.24 illustrates the impact of dosing hydrogen as an external reductant for NOx. Thus, higher H_2 availability leads to a more pronounced NOx reduction in the catalytic converter (indicated on the secondary Y axis). It is clear that in this case the NOx reduction is higher than in the case where hydrogen is not injected externally, as shown in Figure 3.22.

3.5.4 NOx Control Techniques at High Equivalence Ratio

It has been reported in the literature that maximum NOx formation in the smooth operation without backfire in a hydrogen engine occurs before the stoichiometric equivalence ratio, somewhere typically around 0.75 to 0.85 [95–97]. Operating at very low equivalence ratios reduces NOx emissions but also lowers the power output. Conversely, higher equivalence ratios achieve necessary torque levels with lower NOx concentrations compared to peak values within the 0.7–0.8 equivalence ratio range. Using a catalyst, it is possible to achieve higher conversion levels close to stoichiometric conditions. A typical commercial catalyst is effective in lowering NOx emissions under stoichiometric or even beyond stoichiometric conditions, but not under lean operation. Such a trend of low NOx conversion at an equivalence ratio of 0.6 is clearly reflected in the secondary Y axis of Figure 3.24 even after introducing external hydrogen as a reductant. NOx reduction is hindered under lean conditions due to competition between oxygen (O_2) and NO for hydrogen (H_2).

It is possible that hydrogen (H_2) could react with O_2 under lean conditions at the expense of NO, as shown in the equation.

$$H_2 + ½\ O_2 \rightarrow H_2O$$

In a NOx reduction, the catalyst is most effective in its reduced form, while active metal may be switched to an oxidized state when the engine runs lean. Operation at high equivalence ratios, while effective for NOx reduction, can lead to engine instability and abnormal combustion. Therefore, a substantial level of exhaust gas recirculation (EGR) becomes crucial for stable engine operation.

Figure 3.25 illustrates how different equivalence ratios affect NOx emissions. It also shows the range where engine-out emissions peak and where lower NOx levels are achieved at higher equivalence ratios. The effect of catalytic converters on reducing NOx emission levels was also studied. Catalytic converters are effective at bringing down NOx emission levels by using unburned and externally dosed H_2. As evident in Figure 3.26, the highest conversion was achieved at stoichiometric conditions.

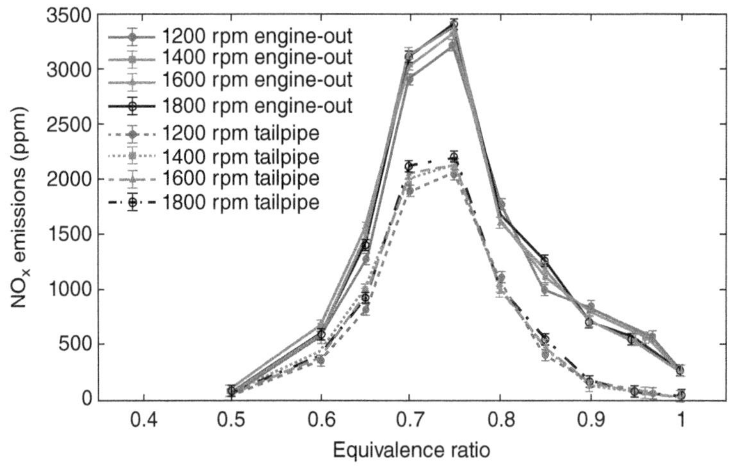

Figure 3.25 Variation of engine-out and tailpipe NOx emissions with equivalence ratio [94] / with permission of Elsevier. Please visit the companion website to view a color version of this image.

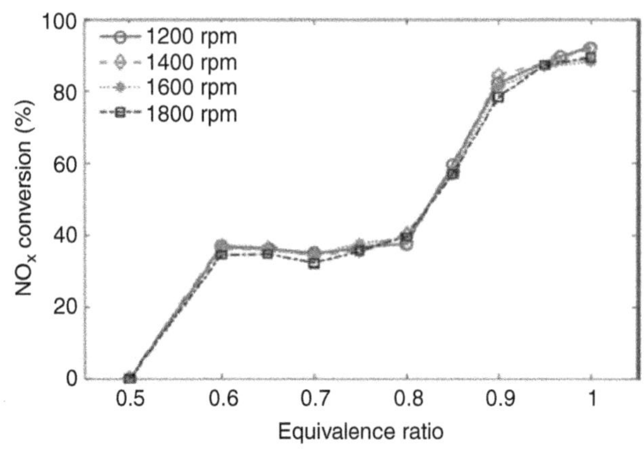

Figure 3.26 Variation in NOx conversion efficiency of the after-treatment device with equivalence ratio [94] / with permission of Elsevier. Please visit the companion website to view a color version of this image.

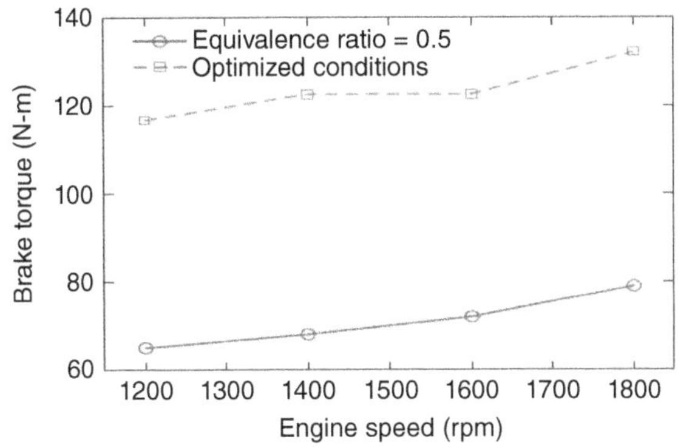

Figure 3.27 Comparison of brake torque generated under optimized and lean conditions at optimized spark timing [94] / with permission of Elsevier.

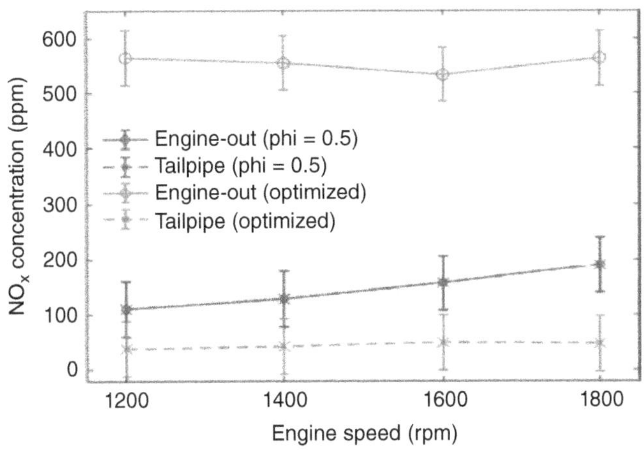

Figure 3.28 Comparison of NOx emissions under lean and optimized conditions [94] / with permission of Elsevier.

In these investigations, it was possible to achieve optimal NOx emissions control by operating the engine at a near-stoichiometric equivalence ratio. This condition resulted in remarkably low NOx formation, coupled with highest NOx conversion efficiency, leading to minimal tailpipe NOx emissions.

Operating at high equivalence ratios not only ensures low NOx emissions but also yields high torque output. This strategy of operating close to stoichiometric conditions with high EGR levels and complemented with a downstream NOx reduction catalyst and external hydrogen (H_2) dosing proved to be the optimum configuration. Figures 3.27 and 3.28 provide a comparative analysis of torque output and NOx emissions under optimized stoichiometric conditions and lean conditions.

3.5.5 Combustion Characteristics

The pressure crank angle in Figure 3.29 indicates smooth combustion [81]. Kulkarni and Thipse [98] carried out some studies on a hydrogen engine using a TMI system and observed smooth engine operation within equivalence ratios of 0.44 to 0.60. A spark advance strategy was used, and the spark angle was retarded to develop more engine power and optimize smooth engine operation without knock and backfire. At a constant speed of 1400 rpm, the peak cylinder pressure was

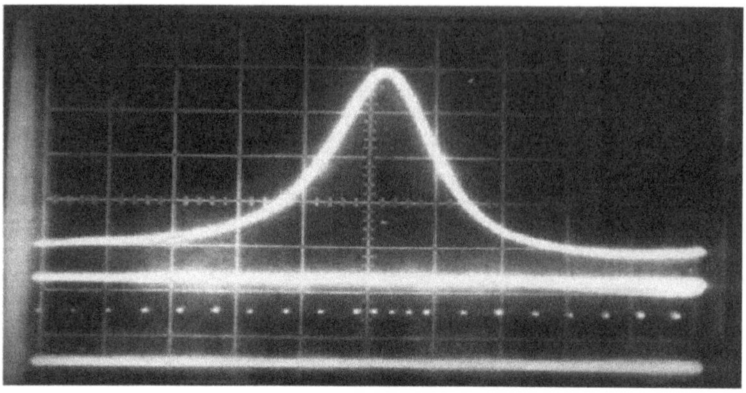

Figure 3.29 Pressure crank angle diagram [81] / with permission of Elsevier.

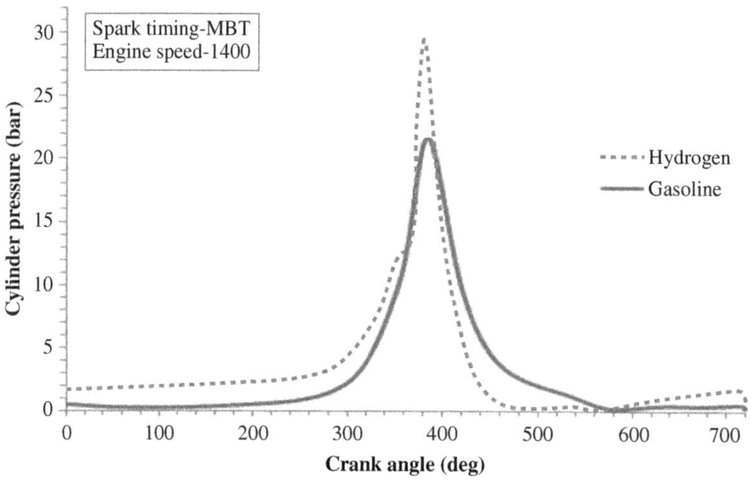

Figure 3.30 Peak cylinder pressure [98] / with permission of Elsevier.

1.36 times higher and the maximum net heat release rate was two times higher for hydrogen compared to gasoline (Figure 3.30).

Peak pressure for hydrogen reaches 29.44 bar at a crank angle of 375°, while gasoline peaks at 21.64 bar at a crank angle of 384°. This discrepancy arises due to hydrogen's rapid combustion, facilitated by its fast flame speed. The higher energy content per volume of hydrogen contributes to its faster combustion process. Figure 3.31 shows that maximum net heat release rate (NHRR) for hydrogen is 51.98 J/deg, while it was much lower for gasoline at 25.82 J/deg. Hydrogen's fast combustion kinetics resulted in a swift and intense heat release compared to the slower combustion of gasoline. Thus, faster burning speed and greater energy release per kilogram of hydrogen fuel contributed to the accelerated net heat release [98].

The researchers also studied the variation in MFB 90% between hydrogen and gasoline at different engine speeds [98]. Hydrogen consistently achieved the mass fraction burned (MFB) 90% earlier than gasoline, indicating its faster combustion characteristics. As engine speed increased, combustion durations decreased, and the difference in combustion timing between hydrogen and gasoline lessened due to faster flame speeds. Tests by Kumar and colleagues [99] in a TMI-operated

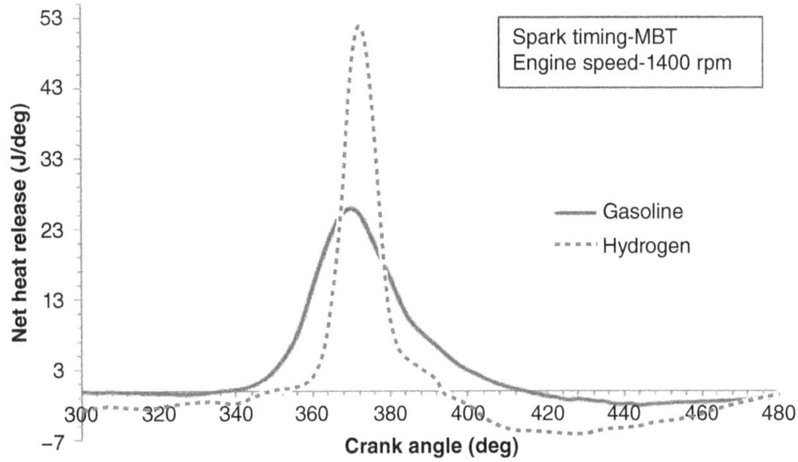

Figure 3.31 Net heat release rate [98] / with permission of Elsevier.

hydrogen engine showed that spark advance for knock-free operation was less than for gasoline. When controlling multiple operating parameters simultaneously, advancing the spark timing led to an increase in both peak pressure and pressure rise rate.

3.6 Direct Injection

The preceding paragraphs discussed the importance of a well-designed port and manifold injection system to mitigate the risk of backfire. However, in such cases, in the absence of a robust and technically well-designed injection system, backfire still remains a concern. In contrast, direct cylinder injection, by definition, precludes the possibility of backfire. The basic injection strategies in relation to engine positions occur at bottom dead center (BDC), intake valve closure (IVC), and combustion top dead center (TDC). Port fuel injection or manifold injection and direct cylinder injection, whether low pressure or high pressure, occurs at different positions for both low-load and high-load conditions. A fast-response injector allows subdivision of each injection event into multiple injections, which precisely controls the timing and duration of multiple injection. Thus, the combustion process can be fine-tuned to match the engine's operating conditions to improve performance, combustion, and emissions. Since direct injection involves injecting fuel directly into the combustion chamber, it induces higher in-cylinder turbulence, which leads to more efficient and rapid combustion. Increased turbulence causes better fuel stratification compared to using port fuel injection [100]. Requirements from injectors to withstand harsh thermal environments and the problem of injector leaks at the end of the compression stroke have raised challenges. Noise and vibration have been experienced by several researchers, particularly when running the engine at full load in the stoichiometric condition.

From the early stages of hydrogen engine research, dating back to the work by Erren in 1932 [12] and Oehmichen in 1942 [15], the inherent potential of direct cylinder injection for the hydrogen engine has been broadly recognized. Since almost last fifty years ago, some significant research results with definite guidelines have been generated intermittently from tests with direct cylinder injection. However, these efforts have been sporadic and were not consistently pursued. In the early 1970s Murray and Schoeppel [101–104] studied hydrogen injection in a small industrial single-cylinder engine. Their testing indicated that a relatively higher pressure of about 66 atm was necessary for the large engines with a higher output. The possibility of knocking was higher when a relatively large amount of hydrogen entered the cylinder prior to ignition by spark. The researchers did some studies on NOx emission and claimed that hydrogen engine was a technically feasible option, but safety related factors and reliability were essential for long-term industrial application. A study by Suzuki [105] has shown that low-pressure direct cylinder injection using an early mixture formation technique has been a reliable route to get rid of backfire, particularly at low load speed. Varde and Frame [36, 106] developed a system of electronically controlled fuel injection in which injection was designed to take place close to the intake valve when the valve was open. The investigators observed higher engine efficiency with injected hydrogen compared to carburetion. They conducted combustion-related studies, analyzing the rate of pressure rise and flame speed in relation to the equivalence ratio.

A well-designed series of research activities were carried out at the University of California, Los Angeles (UCLA) on the development of a hydrogen engine [45, 107]. Extensive experiments were conducted with both port injection and direct injection systems. Tests with port injection could be carried out with a less sophisticated injection valve. An electronically actuated injector for direct injection was developed to withstand the formidable conditions of combustion for precise control over the valve's opening and closing. The researchers observed that in the absence of adequate mixing, direct cylinder injection often resulted in heterogeneous charge formation and incomplete combustion. Researchers at Cornell University also carried out a series of experiments on the hydrogen engine [37, 108, 109]. Homan [37] in his extensive laboratory tests adopted the LIRIAM (late injection, rapid ignition and mixing) technique, based on hydrogen injection into the cylinder late in the compression stroke so that hydrogen mixed rapidly with air and initiated ignition. The rate of pressure rise was dependent upon the rate of hydrogen injection, which was monitored by varying injection pressure. In such an approach, since the combustion period coincided with the injection period, it was possible to control maximum pressure and rate of pressure rise. NOx emissions were extremely low for engine operation at an equivalence ratio less than 0.5.

Adequate research and development activities by different researchers show that it is possible to achieve diesel-like efficiency in an optimally designed hydrogen engine operating at high-pressure direct injection. Of course, when the fuel is injected late in the compression stroke, injection pressure has to be sufficiently high to overcome in-cylinder pressure. Moreover, higher injection pressure results in an increased fuel mass flow rate, which provides a higher energy input compared to low-pressure PFI systems. The flexibility of choosing early, mid, or late cycle direct injection timing can be adjusted to optimize the power density, efficiency, and NOx emission with stable combustion.

3.6.1 Injector for HPDI Operation

Many experimental and numerical studies have been done with a gaseous fuel like CNG [110–125], some of which discuss high-pressure direct injection. For hydrogen, direct cylinder injection stands out as the promising technique to eliminate backfire. However, development of a non-leaking injector operating at a high injection pressure with sufficient flow rate to ensure rapid mixing has been extremely challenging. Injectors designed for liquid hydrocarbon fuels or natural gas have not been successful at long-term durability and achieving high mass flow rates. Even though modeling studies provide very important guidelines to develop an effective injection system for the hydrogen engine, laboratory and field testing is very important. It is difficult to model thermal, chemical, and mechanical effects and their dynamic interaction. Hydrogen's low mass/energy density, high sonic velocity, and low viscosity are relevant for choice of appropriate material of the injector for overall tribology involving friction, wear, and lubrication of the injector. Low mass and energy density are important properties to design the appropriate injector geometry to ensure volumetric flow rates to achieve optimal combustion.

Welch has discussed the practical challenges faced in hydrogen injector development mainly from the material testing point of view [126]. The low viscosity of hydrogen decreases internal damping resistance experienced by the moving components. Such a situation intensifies the impact when moving parts reach their final positions, especially when the needle makes contact with the seat during closure of injector. This sometimes leads to part disassembly, material failures, wear, and needle bouncing. Application of dry lubricants or low-friction coatings on relevant surfaces improves overall lubrication and reduces friction between moving parts.

Due to its high diffusivity, hydrogen permeates a range of materials. This property is a critical design consideration for piezoelectric actuators employing epoxy materials. When subjected to pressure, hydrogen permeates the epoxy, and upon release of pressure, delamination of the epoxy material takes place leading to internal short circuit within the piezoelectric actuator. High sonic velocity and low viscosity dictate material choices and create challenges in maintaining proper lubrication. In this combustion mode, it is crucial to maintain sufficient air pressure to achieve sonic flow (where the fuel–air mixture reaches the speed of sound) and allow for an instantaneous mass flow rate regardless of combustion chamber pressure [126]. Achieving sonic flow is challenging, especially for engines operating under high peak cylinder pressure. Welch et al. discuss the J43Px series injector (Figure 3.32), which represents a shift to piezoelectric actuation from the magnetostrictive actuation developed earlier for natural gas injection [126].

The magnetostrictive effect uses materials that change dimensions in a magnetic field for mechanical motion and maintain precise nozzle adjustments in fuel injectors. In piezoelectric actuation, the material undergoes dimensional changes when subjected to an electric field. The actuators expand or contract to control the nozzle, allowing precise and rapid fuel delivery by varying pulse width. Friction and wear in metallic materials are influenced by several factors like material conditioning, thermal environment, lubrication, and surface oxide films. In hydrogen environments, wear-induced removal of oxide films leads to metal-to-metal contact, increasing friction and wear, especially in components made from a different composition of hardened tool steels. In some cases, common engineering steels have caused the problem of embrittlement. Therefore, material compatibility is essential to prevent leakage in high-pressure direct injection

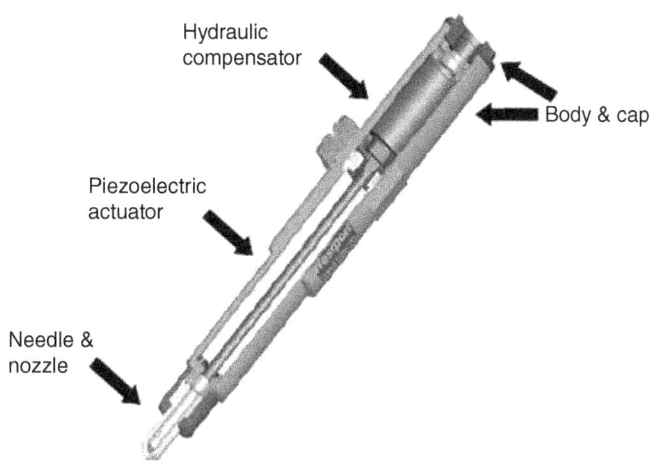

Figure 3.32 A J43Px injector [126] / SAE International.

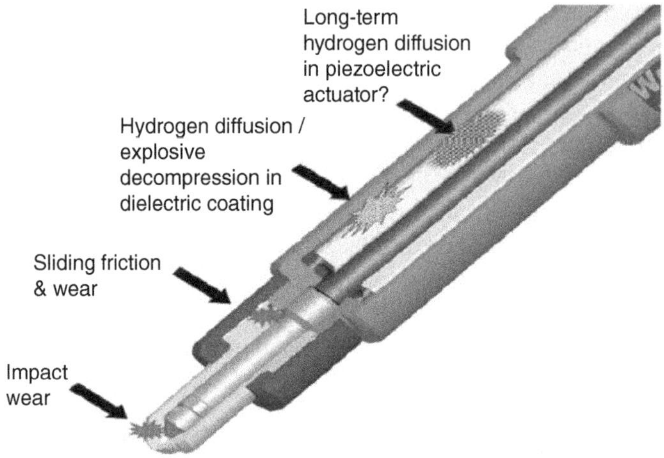

Figure 3.33 Material-related challenges in an injector leading to wear and failure [126] / SAE International.

systems. Figure 3.33 identifies material-related problems, which are responsible for wear and failure mechanisms [126] in the injector.

3.6.2 Spark-Activated Direct Injection

Engines with spark-assisted hydrogen direct injection (DI) have been investigated where injectors and spark plugs are placed in specific positions according to the geometry of the cylinder head that is modified for direct injection. Wimmer et al. [127] were able to get 40% ITE at low and medium loads in their tests on a modified single-cylinder automotive-sized DISI hydrogen engine. This efficiency was only marginally lower than that of a light-duty diesel engine. Their investigation showed that injection timing on the mixture homogeneity had a greater impact on engine performance and emissions than ignition timing. Retarding the start of injection (SOI) timing from 120° to less than 65° crank angle (CA) BTDC resulted in higher efficiency. This was due to a more stratified mixture, which caused faster initial flame development and improved combustion.

It was reported by Takagi et al. [128] that NOx emissions increased under some conditions, depending on the mixture concentration of the jet and the condition of its dispersion in the combustion chamber. Tests with high-pressure hydrogen injection showed that EGR was effective when "combined with stratified and diffusive combustion" [129]. The specific heat ratio was increased by dehumidifying EGR, and this led to improved efficiency and low NOx emission [129]. In a development project of hydrogen ICE for heavy-duty vehicles like trucks, researchers carried out tests on a multi-cylinder DISI hydrogen engine using JE05 mode transient emission and were able to achieve emission levels that were lower than specified by JAPAN 2009 regulation [130]. Using high-pressure direct injection in the ICE hydrogen truck development project by Kawamura et al. [131] and evolving a strategy of optimization of the EGR ratio, the injection timing, and the ignition timing, high thermal efficiency was achieved and NOx emission was reduced. Based on experimental results, NOx emissions and fuel economy were simulated for the heavy-duty vehicle with a six-cylinder DI hydrogen engine. Developing a simulation program and with the database from single-cylinder engine tests, an engine management strategy was made to produce high output power, energy efficiency comparable to a baseline diesel vehicle, and low NOx emissions of 0.5 g/kWh under the emissions test cycle [132].

3.6.3 Injection Strategies: Energy Proportion, Exergy Efficiency

There have been several investigations into various hydrogen injection strategies to optimize combustion efficiency and control emissions. A well-mixed charge with close-coupled injection and ignition strategies enables more fuel–air mixture stratification. These strategies include plume-head ignition (just after injection starts) and plume-tail ignition (just after injection ends). Takagi et al. [133] conducted a study on a direct injection hydrogen engine in which the researchers optimized the jet plume location relative to the combustion chamber wall, jet geometry, and injection timing. By selecting an injection angle for separating the jet from the chamber wall surface and by narrowing the nozzle hole diameter, they were able to increase the power output, improve the efficiency, and reduce the NOx emission level. The NOx emission level obtained with the plume ignition and combustion concept (PCC) was in the order of single-digit parts per million. Roy et al. [134] carried out an investigation on the structure of the jet in addition to other physical processes resulting from hydrogen gas injection. They used an Ar-ion laser beam as the source of light to visualize the hydrogen jet in the constant-volume chamber. They also carried out tests on an optical DISI engine to analyze early kernel growth and flame propagation and assess the impact of equivalence ratios and injection timings on combustion. With the objective of getting a deeper understanding of the combustion optimization in a DISI hydrogen engine, Roy et al. [135] further investigated how the hydrogen jet influenced fuel concentration distribution and the mixing process. Stratified combustion was developed by using a GDI injector with its jet directed toward the spark plug. They used spark-induced breakdown spectroscopy (SIBS) to measure the local fuel–air ratio at the ignition point under stratified charge conditions. During stratified charge operation, the jet injection relative to ignition timing was observed to affect the local mixture strength at the spark gap just after ignition. The equivalence ratio was found to be very rich and showed significant variation from one cycle to another. A high-speed camera synchronized with the spark was also installed at the appropriate location in the system. Images of jet-guided combustion relative to ignition timing were captured through the camera. Single-shot images at different times were useful in studying the development of the flame and its subsequent propagation.

A lean mixture near the cylinder walls minimizes heat losses [136]. Flow visualization techniques and simulation studies have thoroughly investigated fuel–air mixing in high-pressure jets. Planar laser-induced fluorescence and particle imaging provide significant information on flow concentration and flow field. Injecting a hydrogen jet toward the engine side wall sometimes generates a wall–jet vortex. Upon hitting the wall, the jet redirects and decelerates due to air entrainment. This process, influenced by factors such as injection angle, timing, injector nozzle number, and piston geometry, forms a more stratified mixture with reduced tumble flow in the combustion chamber. Molina et al. [38] studied both port fuel injection (PFI) and direct injection (DI) of hydrogen in a spark ignition engine designed for precise and controlled hydrogen release. Tests were conducted with two distinct outwardly opening poppet valve gas injection systems: a port fuel injection (PFI) system with a maximum injection pressure of 5 bar and a medium-pressure direct injection (DI) system with a maximum injection pressure of 30 bar. Both PFI and DI indicated similar trends in NOx concentration. Specific design of the injector proved crucial for mixture stratification. Even when the overall air-to-fuel ratio was same, design of the injector influenced fuel distribution and stratification in the combustion chamber, which often altered the local equivalence ratio. Moreover, distribution and stratification of fuel in the combustion chamber due to injector design also affected NOx formation during combustion.

Liang et al. [39] investigated the effects of single and split injection strategies (Figure 3.34) on the combustion and emissions of a hydrogen DISI engine at medium/low load.

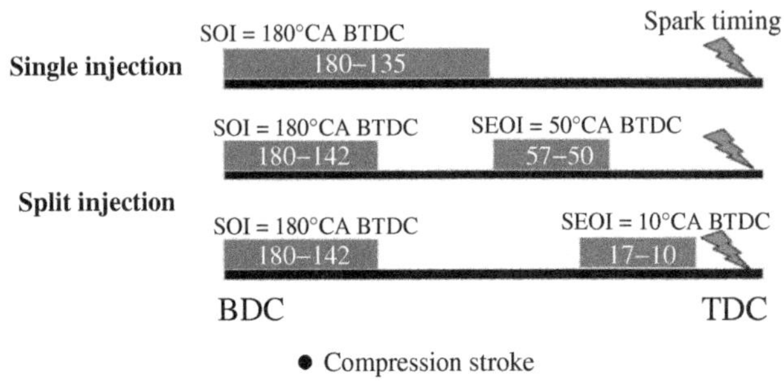

Figure 3.34 Injection strategies [39] / with permission of Elsevier.

Hydrogen injection into the cylinder before the intake valve was closed often led to preignition, especially after increased λ values, so the earliest hydrogen injection timing was set to 180° CA BTDC after intake valve closure. In the single injection strategy when SOI was delayed from 180° CA BTDC to 85° CA BTDC, brake thermal efficiency (BTE) initially increased and then decreased. NOx variation showed the opposite trend to BTE. The highest BTE of 36.1% was achieved at SOI of 100° CA BTDC, with NOx emissions below 0.1 g/(kWh). In the split injection strategy with delayed secondary end of injection (SEOI), BTE increased and then decreased, the optimal value of BTE being 35.6% at SEOI of 50° CA BTDC. It was also observed that if SEOI is delayed beyond 30° CA BTDC, it could raise NOx emission to over 1.5 g/(kWh).

Zhao et al. carried out both experimental investigation and modeling analysis on a 2.0 L turbocharged direct-injection hydrogen engine to achieve near-zero NOx emissions and enhanced performance through injection optimization [40]. Researchers developed a simulation model for hydrogen injection calibration and numerical simulation with local grid refinement. Implementation of an optimized piston ring and low viscosity, low ash lubrication oil controlled oil consumption rates and particle emissions. SOI from 200° CA BTDC to 120° CA BTDC at λ = 2.67 resulted in consistent NOx emissions below near-zero limits and an increase in brake thermal efficiency from 40.4 to 41.7%, due to reduced pumping and heat transfer losses. Optimal injection timing, particularly adjusting the end of injection (EOI) to around −80° CA, resulted in near-zero NOx emissions and a maximum brake thermal efficiency of 41.9%, a 1.5% increase under all conditions [40]. Early injection of high-speed hydrogen advances turbulence in the cylinder, leading to NOx emissions in regions with high turbulent kinetic energy. Delaying injection weakens cylinder turbulence but results in NOx generation in localized concentrated regions. There have been studies [41] on a phenomenological combustion model for an H_2-diesel engine based on experimental results. Yet there was significant variability in hydrogen content among cylinders with upstream injection. Thus, it is necessary to limit hydrogen use in all cylinders based on the highest concentration to prevent damage. However, it is possible to adjust diesel injection to balance cylinders with varying levels of hydrogen for uniform and optimized engine performance. In their study with numerical modeling Wu and colleagues [42] used gaseous hydrogen injected into the combustion chamber at 100 bar pressure through a centrally located single-hole injector. Two different in-cylinder flow patterns, characterized as low- and high-tumble configurations, were analyzed. Various numerical methods and model constants, including the turbulent Schmidt number, were evaluated. The study compared numerical results with experimental measurements of velocity and hydrogen concentration fields to assess the validity of modeling applied to a HPDI hydrogen engine. Zhang et al. [43] carried out a series of tests to analyze energy and exergy distribution characteristics of a turbocharged direct-injection hydrogen engine under various operating conditions of load and speed.

In Zhang et al.'s study, exhaust energy, coolant energy, intercooler energy and other energy losses were considered, and similarly, for exergy analysis, exhaust exergy, coolant exergy, and exergy destruction were calculated. Figure 3.35a shows that coolant energy proportion varies between 15.22% and 15.87% at engine speeds below 2500 rpm and further decreased to 12.86% at 3500 rpm. At the same time, it is evident in Figure 3.35b that at a constant speed of 2000 rpm, the coolant

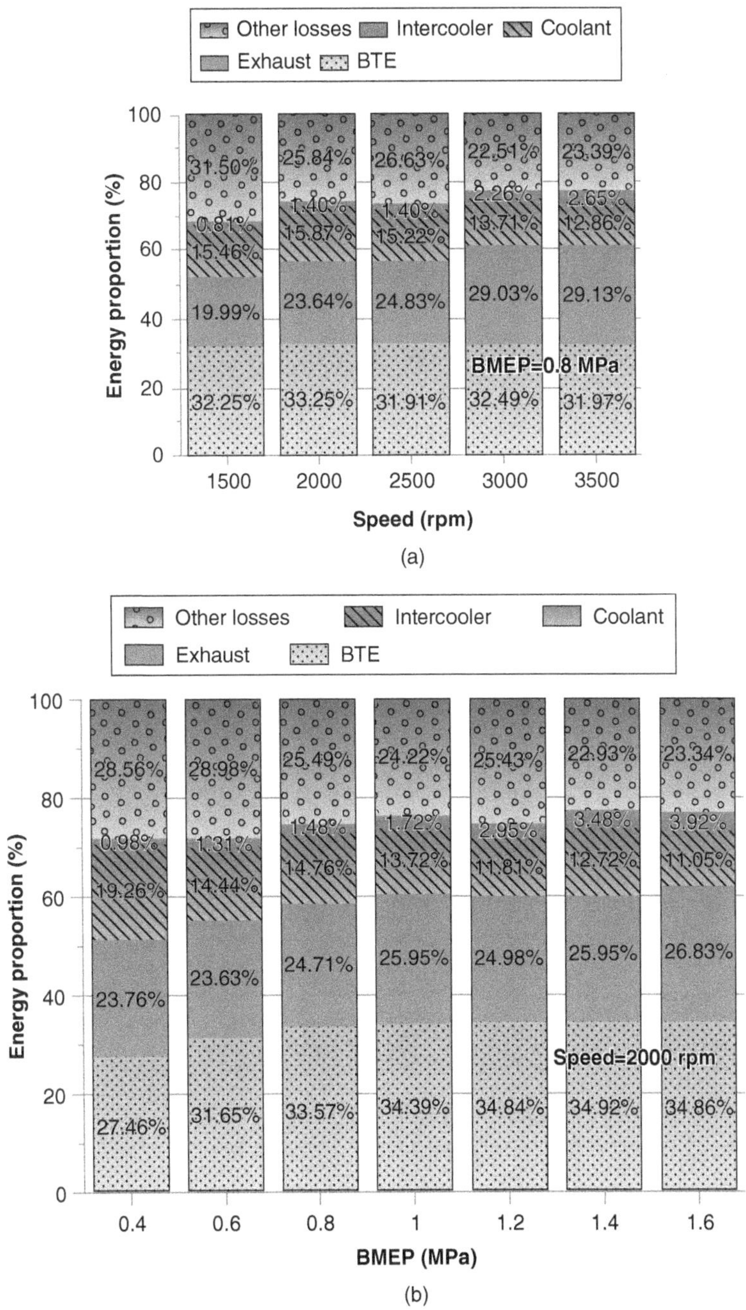

Figure 3.35 Effect of (a) speed and (b) load on engine energy proportions [43] / with Permission of Elsevier.

energy proportion varies with different loads, reaching 19.26% at a brake mean effective pressure (BMEP) of 0.4 MPa and decreasing to 11.05% at 1.6 MPa BMEP. The study indicates that effect of load on coolant energy was a more sensitive proportion compared to speed.

Figure 3.36a shows that the difference between indicated thermal efficiency (ITE) and brake thermal efficiency (BTE) decreases with increasing load at constant engine speed. Figure 3.36b shows that as load increases, the difference between exhaust and intake pressure narrows from 99 to 78 kPa, causing only slight improvement in BTE after BMEP exceeds 0.80 MPa. Thus, the study shows BTE to be more sensitive to load variations than engine speed.

The fact that coolant exergy efficiency peaks at 3.47% with a brake mean effective pressure (BMEP) of 0.4 MPa at 3500 rpm (Figure 3.37) indicates the influence of low-load operations where the coolant energy proportion is appreciably high. Despite coolant energy constituting around 15–20% of total energy, the available energy within the coolant remains below 5 kW. This study suggests that there are still challenges in recovering low-quality energy from coolant.

Figures 3.38a and 3.38b illustrate variations in combustion duration and exhaust temperature with respect to engine speed and load, respectively. As engine speed increases, combustion duration

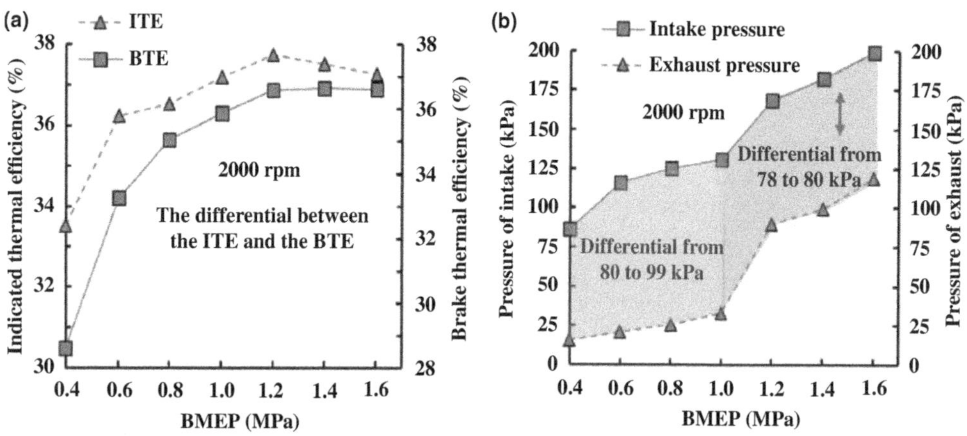

Figure 3.36 Effects of load on the (a) ITE/BTE and (b) intake/exhaust pressure [43] / with Permission of Elsevier.

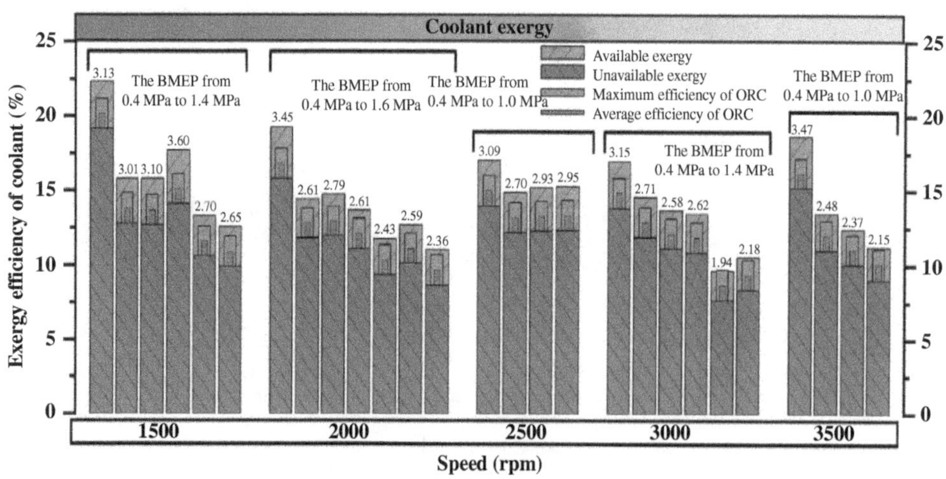

Figure 3.37 Variation of the exergy efficiency of coolant with speed and load [43] / with Permission of Elsevier. Please visit the companion website to view a color version of this image.

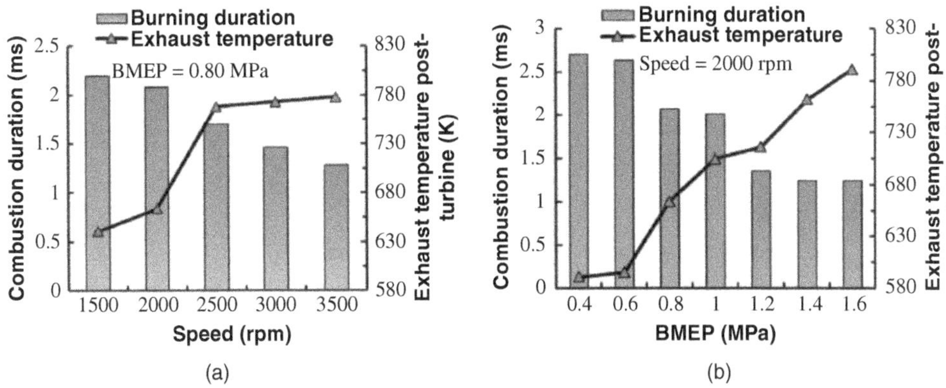

Figure 3.38 Combustion duration and exhaust temperature at various engine (a) speeds and (b) loads [43] / with Permission of Elsevier.

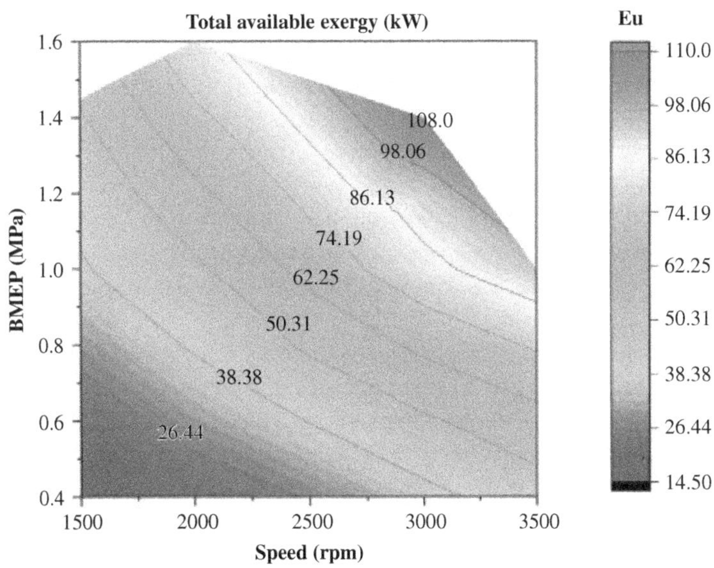

Figure 3.39 Variation of total exergy with speed and load [43] / with Permission of Elsevier. Please visit the companion website to view a color version of this image.

shortens, exhaust temperature rises significantly, and exhaust mass flow increases because the number of cycles per unit time is increased. These factors make exhaust energy more sensitive to engine speed changes compared to load variations. The total exergy map shows how fuel exergy converts to available energy under different operating conditions. Figure 3.39 shows that the highest available exergy is 108.0 kW at a BMEP of 1.4 MPa and 3000 rpm, while the lowest is 14.5 kW at a BMEP of 0.4 MPa and 1500 rpm.

3.6.4 Glow Plug-Assisted Direct Injection

Glow plugs are devices featuring electrically heated surfaces extending into the engine's combustion chamber (Figure 3.40). In conventional diesel engines, glow plugs facilitate cold starts by elevating the local charge temperature [35].

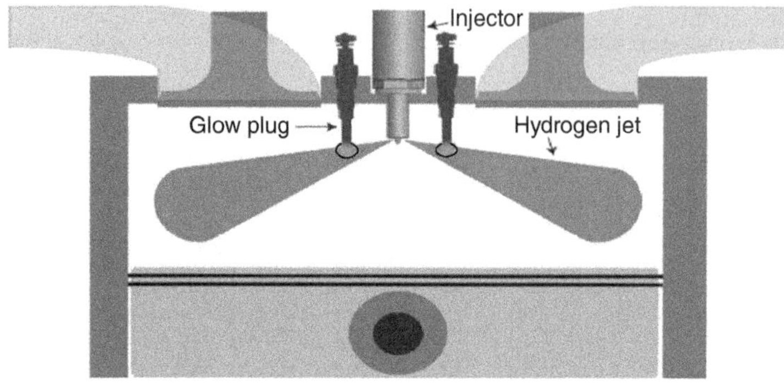

Figure 3.40 Glow plug-assisted ignition of hydrogen direct injection [35] / MDPI / CC BY 4.0.

Nevertheless, for hydrogen direct injection, glow plugs must operate continuously to ensure ignition in each engine cycle, requiring a surface temperature of 1200–1400 K. Tests by Homan et al. [137] on the stability of glow plug ignition in a hydrogen engine showed that glow plugs offer a more dependable ignition source for hydrogen compared to multi-strike spark plug systems. Glow plugs exhibit a short and stable delay between the start of injection and ignition. Spark plug ignition, on the other hand, exhibits fluctuating delays. In spite of such benefits, glow plug ignition has an approximately10% increase in specific fuel consumption relative to conventional diesel operation.

Welch and Wallace [138] converted a single-cylinder Lister ST-1 direct injection diesel engine to operate on hydrogen and studied its exhaust emissions. Hydrogen gas was injected at a high pressure of 10.3 MPa. The injection process was controlled through a hydraulically actuated injection valve that governs the timing and duration of hydrogen. A standard pencil type sheathed glow plug (Champion CH-78) was used. Tests showed significant improvements in power of approximately 40% at 1200 rpm and 30% at 1800 rpm. These improvement in power did not sustain for an extended period due to problems with overheating. There was significant reduction in NOx emissions. NOx emissions were sensitive to injection timing. Indicated efficiency of a hydrogen-fueled diesel engine was approximately 90% of that of the original diesel engine at moderate loads, but as full load approached, the efficiency of the diesel engine decreased and converged with the hydrogen-fueled results. This phenomenon is due to enhanced heat transfer resulting from the scrubbing of the boundary layer from the cylinder head surface by hydrogen jets. The research showed that it was possible to achieve smooth and reliable ignition with an ordinary commercial glow plug and not the platinum glow plugs used in other experimental set-ups [138]. Even though NOx emissions were relatively less compared to diesel operation, they still were significantly high, particularly under high loads. Glow plug ignition in hydrogen engine combustion has not received adequate attention. High surface temperatures have raised concerns about the durability of glow plugs for commercial applications. They are very rarely used today in the development of modern engines [139].

3.7 Hydrogen as a Fuel for the Compression Ignition Engine

Use of hydrogen in the diesel combustion process at high engine load conditions has resulted in a sharp increase in peak in-cylinder pressure and peak heat release rate [140]. Blending hydrogen with pure diesel fuel improves the brake thermal efficiency (BTE) of diesel engines [141–145]. The flame velocity of hydrogen is approximately nine times greater than that of pure diesel fuel.

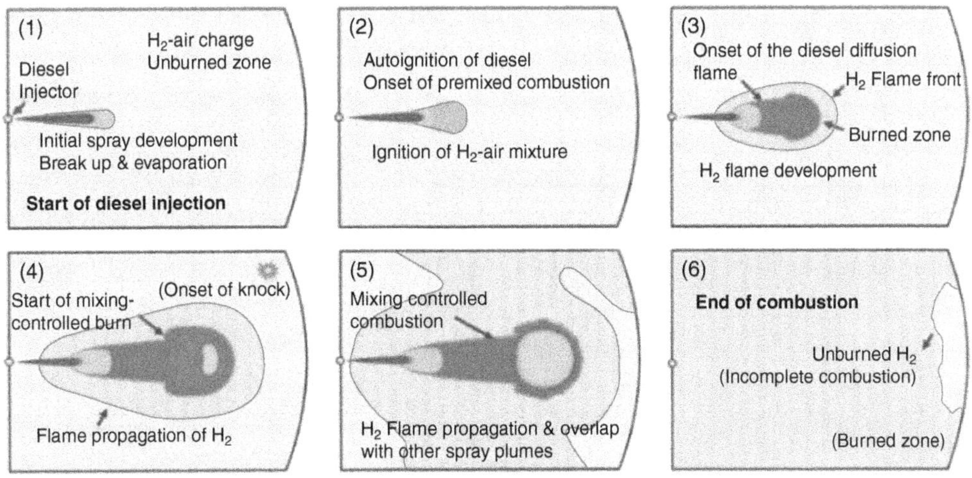

Figure 3.41 Sequence of combustion stages: diesel injection at near TDC [150] / IntechOpen. Please visit the companion website to view a color version of this image.

Thus the addition of hydrogen to neat diesel fuel results in a better combustion process owing to its elevated flame temperature and higher flame velocity [146–149]. Figure 3.41 shows the combustion process in a hydrogen-fueled diesel engine near the top dead center (TDC) [150].

3.7.1 Hydrogen–Diesel Dual Fuel Engine

Sandalci and Karagoz [151] experimented on a diesel engine in a hydrogen–diesel dual fuel mode in which hydrogen was injected into the intake manifold and pilot diesel through the usual diesel injector. The energy contents of the total fuel was 0% (neat diesel), 16%, 36%, and 46% hydrogen. Tests showed a significant decrease in smoke, carbon monoxide (CO), and carbon dioxide (CO_2) emissions, as illustrated in Figure 3.42. Total hydrocarbons showed a fluctuating pattern with increasing hydrogen fraction. At higher load conditions, the effect of hydrogen addition on hydrocarbon emissions is relatively much less. Specific NOx emissions at 36% and 46% hydrogen energy fraction increased dramatically by 58.8% and 159.7%, respectively (Figure 3.43).

A significant portion of the literature on the hydrogen–diesel dual fuel engine discusses the impact of the hydrogen energy share ranging from 10 to 40%. Strategic control of injection timing and duration in direct injection have a profound influence on combustion parameters such as the rate of pressure rise and maximum combustion pressure. Experimental studies by Gomes Antunes et al. [152] on direct injection CI hydrogen engines showed that hydrogen gave higher power-to-weight ratio with peak power being about 14% higher than neat diesel operation and higher efficiency of about 43%, while diesel engines gave only 28%. NOx emissions were around 20% lower. However, because of hydrogen's high self-ignition temperature, inlet air heating was necessary.

De Morais et al. [153] tested a hydrogen–diesel dual fuel engine with loads from 0 kW to 40 kW. Diesel was directly injected, and hydrogen was injected into the intake manifold at a 5%, 10%, 15%, and 20% energy basis. CO_2 emissions decreased, but efficiency and specific fuel consumption showed no significant change from baseline diesel operation. The engine used its original mechanical injection system. Use of an electronic injection system and optimization of injection timing might improve performance characteristics.

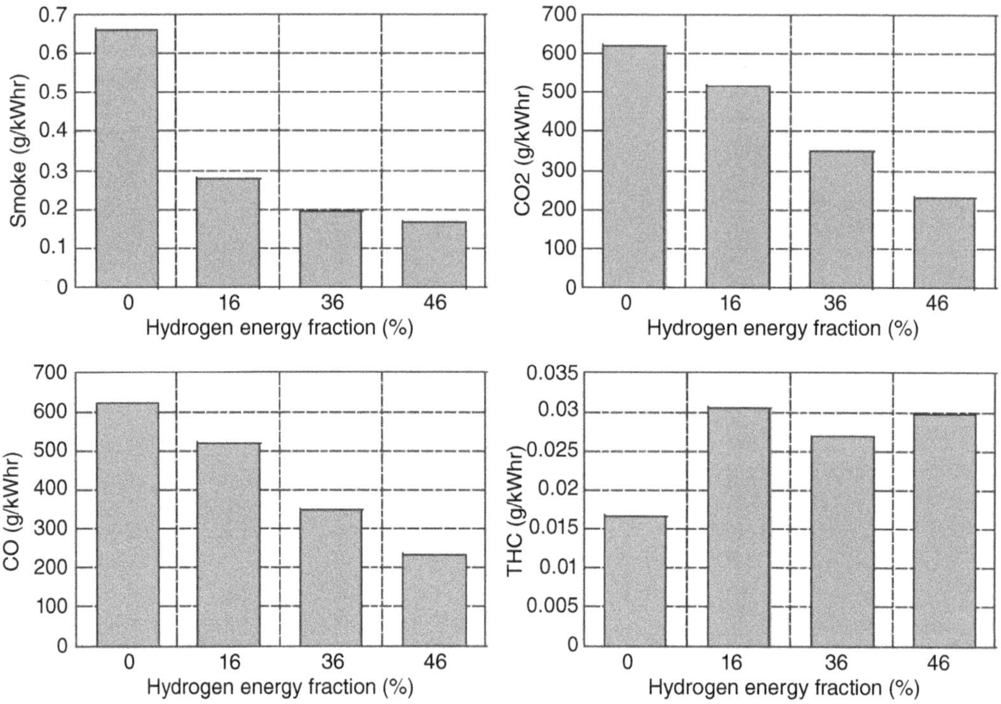

Figure 3.42 Effect of hydrogen energy share on smoke, CO_2, CO, and HC emissions [151] / with permission of Elsevier.

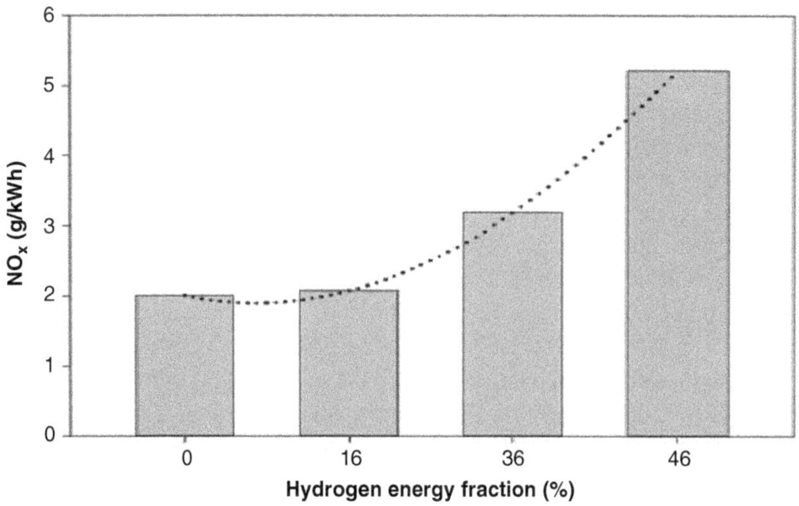

Figure 3.43 Increasing trend of NOx emission with hydrogen fraction increase [151] / with permission of Elsevier.

Deb et al. [154] tested a hydrogen–diesel dual fuel engine at 1500 rpm. Hydrogen was introduced into the intake manifold via an LPG-CNG injector and diesel was injected with conventional injector. Hydrogen energy contents tested in the blend were 11%, 17%, 30%, and 42%. Performance improved in brake specific energy consumption and brake thermal efficiency. CO, CO_2, and smoke emissions were lower than with pure diesel, though unburned hydrocarbons were slightly higher and NOx emissions much higher. Figure 3.44 shows heat release rates for 0%, 11%, 17%, 30%, and 42% hydrogen. Such a trend indicates a shift toward premixed combustion due to hydrogen's high flame speed. Peak heat release rates were 78.07 J/°CA (neat diesel), 75.70 J/° CA (11% H_2), 69.95 J/° CA (17% H_2), 60.68 J/° CA (30% H_2), and 88.04 J/° CA (42% H_2). The highest peak, at 42% hydrogen, was amplified by 12.77% compared to the no-hydrogen state.

In their study, Kose and Ciniviz used a four-cylinder, 3.908 L turbocharged CI engine with a 17:1 compression ratio [155]. Diesel was directly injected, and hydrogen was introduced into the inlet manifold at 2.5%, 5%, and 7.5% by volume. Results showed improvements in engine torque, power, and thermal efficiency, with reduced HC, CO, and O_2 emissions, but increased NOx emissions. Addition of hydrogen led to a decrease in O_2 emissions, with the lowest value recorded at 0.97% for 7.5% hydrogen addition at 1000 rpm compared to 3.7% for diesel injection. The highest O_2 of 9.3% with H2.5 addition was recorded with a hydrogen ratio at 2500 rpm, compared to about 11% for standard diesel injection (Figure 3.45).

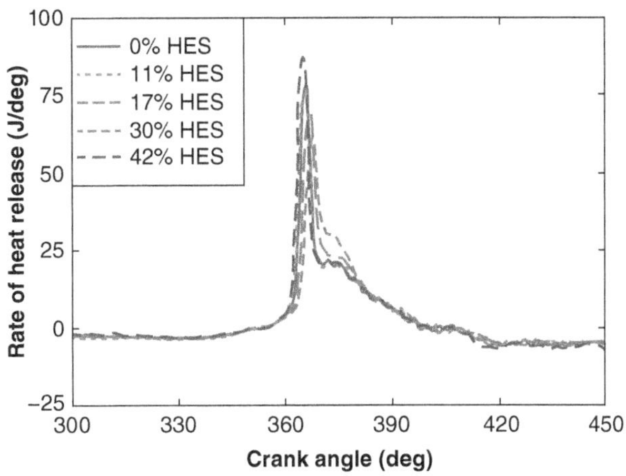

Figure 3.44 Variation of heat release rate with crank angle for hydrogen energy share [154] / with permission of Elsevier.

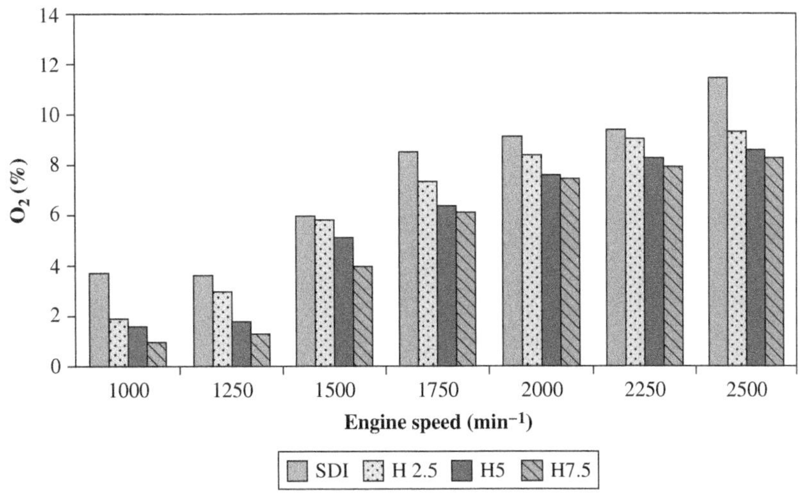

Figure 3.45 O_2 emissions for different values of hydrogen enrichment at different engine speeds [155] / with permission of Elsevier.

While hydrogen's autoignition temperature (576 °C) makes compression ignition difficult, it allows use of higher compression ratios [156–159]. Its high autoignition temperature and octane number increase resistance to knocking combustion compared to hydrocarbon fuels. The precise upper limits of hydrogen's octane number remain undefined [160]. Despite these advantages, hydrogen's minimum ignition energy poses risks of ignition from hot spots or residues, leading to uncontrolled combustion. Hydrogen's motor octane number (MON) is significantly lower than its research octane number (RON), with MON being a more accurate measure of knock resistance in hydrogen engines [161–165]. Gopal et al. [166] conducted tests to identify the optimal hydrogen proportion focusing on knock-limited power output and efficiency. They encountered challenges from the early onset of knocking well before the stoichiometric region.

Karim and his team were one of the early groups of researchers to investigate the use of hydrogen in compression ignition engines [167–169]. They studied the reaction courses and autoignition onset in homogeneous fuel–air mixtures with detailed chemical kinetic data. Computational predictions agreed reasonably with experimental data from a motorized autoigniting engine and a pneumatically driven compression apparatus [167]. In another study the researchers introduced hydrogen outside the engine cylinder, mixed it with air, compressed it, and ignited it with a small amount of liquid fuel at the end of compression. They compared the analytical results with experimental data and evaluated the influence of various operating parameters on the onset of autoignition in a motored engine without a deliberate ignition source [168].

Hydrogen–diesel dual fuel engines demonstrated stability within specific fuel mixture ranges and charge temperatures (Figure 3.46), which was limited by excessive pressure rise and knock on one side and by erratic performance and ignition failure on the other. A characteristic high-frequency pressure oscillation often marked the transition from non-knocking to knocking conditions [169]. Weigang and Lainfang [170] made some observations in their tests on a hydrogen–diesel blend. Introduction of hydrogen into the diesel engine enhanced the energy release rate during the initial combustion stage. While EGR brought down the NOx emissions level, volumetric efficiency significantly decreased, up to about 15% compared to a dual fuel hydrogen engine without EGR [171]. As with conventional diesel engines, hydrogen dual fuel engines with EGR emit increased particulate emissions and smoke. Kudryash et al. [172] compared a hydrogen–diesel engine with two different modes. The first system utilized a cylinder swirl chamber diesel engine with external mixture formation, incorporating hydrogen into diesel fuel. The second system featured a hydrogen-fueled diesel engine with internal mixture formation, utilizing glow plugs to initiate hydrogen ignition.

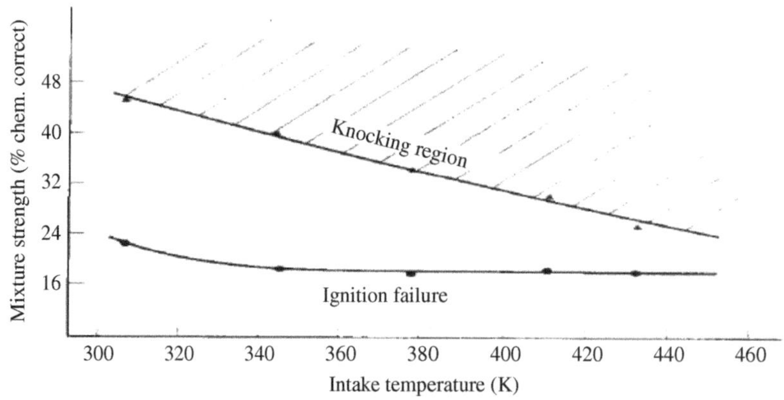

Figure 3.46 Variation of the knock limited mixture strength with charge temperature [169] / with permission of Elsevier.

The swirl chamber diesel engine offered improved conditions for mixture formation during the compression stroke, leading to a rapid rate of hydrogen combustion and reduction in combustion duration. It also demonstrated lower NOx emissions.

Ikegami et al. [51, 52, 173] investigated the feasibility of using hydrogen fuel in a compression ignition engine traditionally designed for diesel. Two experimental approaches were taken: one with an air-aspirated system [51] and another with an argon–oxygen mixture [52]. In the air-aspirated system, preliminary fueling significantly influenced ignition stability, suggesting that both pilot injection and fuel leakage from the injector facilitate hydrogen ignition leading to smooth engine operation. On analysis, it was found that the embers (hot remnants) from the previous firing stroke ignited the fuel. Even a little preliminary fuel keeps these embers hot until the main fuel is added. However, adding too much preliminary fuel can cause the fuel to ignite on its own, leading to uncontrolled autoignition [51]. In a subsequent study [52], a closed-cycle engine system that was simulated using a 21% oxygen mixture operated well without ignition aid.

Hydrogen can eliminate carbon emissions in compression ignition (CI) engines. Studies have been made on the ability of noble gases to eliminate oxides of nitrogen. Because of their high specific heat ratio, noble gases increase the in-cylinder temperature during compression. In their study, Taib et al. [173] optimized hydrogen combustion in an argon–oxygen atmosphere and investigated hydrogen combustion with various noble gases, using Converge CFD software and Yanmar NF19SK engine parameters [173]. Results were subsequently validated against experimental data generated in a rapid compression expansion machine (RCEM).

One basic design approach for hydrogen injectors by several researchers has been a hydraulically actuated system by a conventional diesel injection pump. The principle of operation is built on utilizing the pressure wave generated in the high-pressure pipe to open the injector and discharge hydrogen gas into the engine cylinder.

Researchers at Concordia University have also studied [53–56] hydrogen injection into compression ignition engines. The concept of fuel control system technology for automotive combustion engines by Krepec and co-workers [54] was built for liquid hydrogen system for on-board storage (Figure 3.47).

Liquid hydrogen from the cryogenic storage tank is pumped to the required pressure, then preheated to convert it to vapor at the desired pressure and temperature. It is then supplied to the engine through a metering system that regulates the flow based on engine characteristics. A microprocessor-based controller precisely adjusts hydrogen gas injection parameters such as timing, duration, and flow rate to match engine operational conditions [55]. Problems like injector needle seizure and gas leakage, encountered during tests, were solved through solenoid

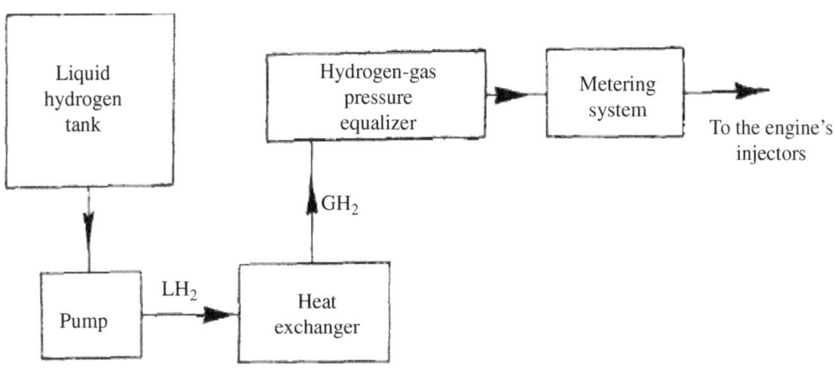

Figure 3.47 Block diagram of hydrogen supply system [54] / with permission of Elsevier.

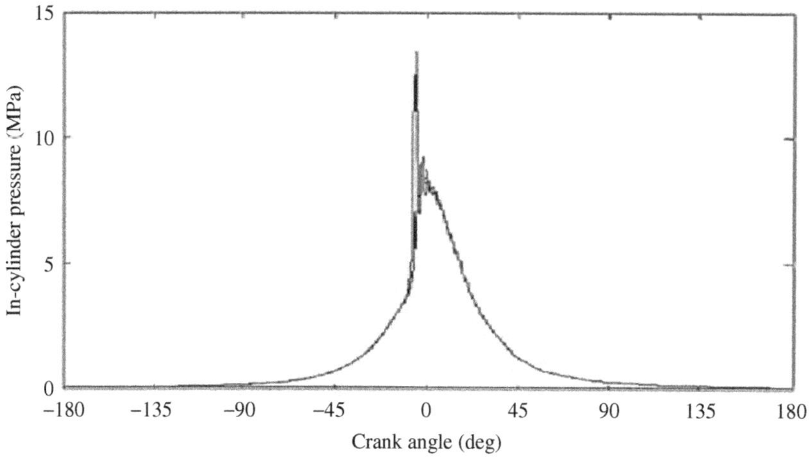

Figure 3.48 In-cylinder pressure versus crank angle for hydrogen combustion under the HCCI mode (CR: 17) for a single combustion event from a test [175] / with permission of Elsevier.

improvements and enhanced nozzle sealing. Tanabe et al. [174] studied the mixing dynamics of a transient hydrogen jet in a hydrogen injection engine using a fast-response concentration probe (<200 μs) to measure hydrogen gas concentration in quiescent air. The probe, similar to a hot-wire anemometer, transmitted electrical signals to a transient recorder, with data processed via microcomputer for average transient concentration. Hydrogen gas was injected through a single shot device or through the injection nozzle developed by Ikegami [51, 52], whose work was discussed earlier in the chapter.

Szwaja and Grab-Rogalinski [175] studied the combustion characteristics of pure hydrogen under HCCI mode and combustion of hydrogen with diesel in a CI engine. They varied the hydrogen fraction from 0 to 17% of the energy percentage and observed that a hydrogen concentration of 17% resulted in a stoichiometric hydrogen–diesel–air mixture, leading to combustion knock. The study examined hydrogen's effects on combustion knock intensity, mass fraction burned (MFB), and heat release rate. Unlike gasoline engines, where knock occurs late, hydrogen can show early knock, which intensifies with higher compression ratios [156, 175–177]. Hydrogen's self-ignition caused severe knock, whereas controlled ignition ensured gradual combustion. Figure 3.48 shows a hydrogen autoignition event in a CI engine producing a 6 MPa pin-shaped pressure pulse shortly after combustion starts.

Figure 3.49 shows 10% MFB (mass fraction burned), which indicates initiation of combustion. There was no discernible pattern in the distribution of this event across 120 consecutive combustion cycles. Such a random distribution suggests that the onset of combustion lacks clear regularity, and so the ignition process is inconsistent. A closer look at IMEP in Figure 3.50 strengthens earlier observations about irregular distribution of 10% MFB, which is mostly due to hot surface contributions. A comparative study on the high-frequency component of in-cylinder pressure (HFCP) is shown in Figure 3.51 for three different fuels. Figure 3.51a depicts the HFCP for a pure hydrogen engine while Figures 3.51b and 3.51c show the HFCP for a hydrogen–diesel mixture with an energy ratio of 25% and for pure diesel, respectively.

To analyze the HFCP, a band-pass (3–15 kHz) Butterworth digital filter was applied to the in-cylinder pressure data, utilizing Matlab signal processing tools. HCCI configuration showed that hydrogen combustion knock was characterized by an exceptionally high amplitude of the HFCP, reaching up to 4 MPa. This peak pressure (PP) for hydrogen combustion knock is substantially higher than the threshold considered harmful for gasoline engines, which is around 100 kPa.

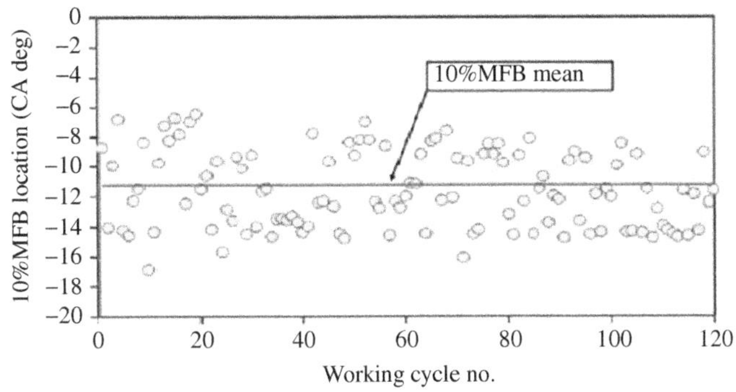

Figure 3.49 A 10% MFB location for consecutive engine working cycles [175] / with permission of Elsevier.

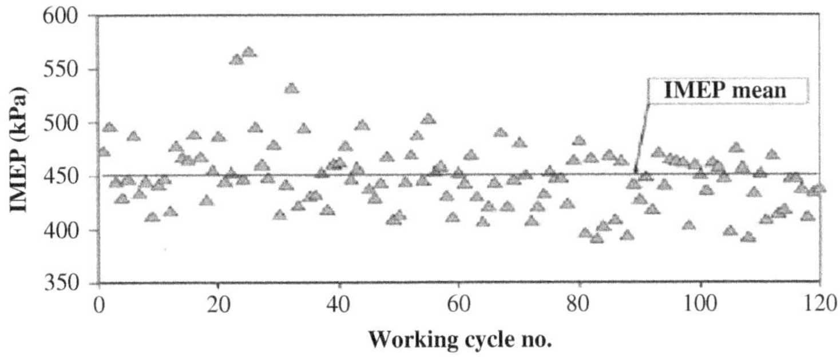

Figure 3.50 IMEP for consecutive engine working cycles [175] / with permission of Elsevier.

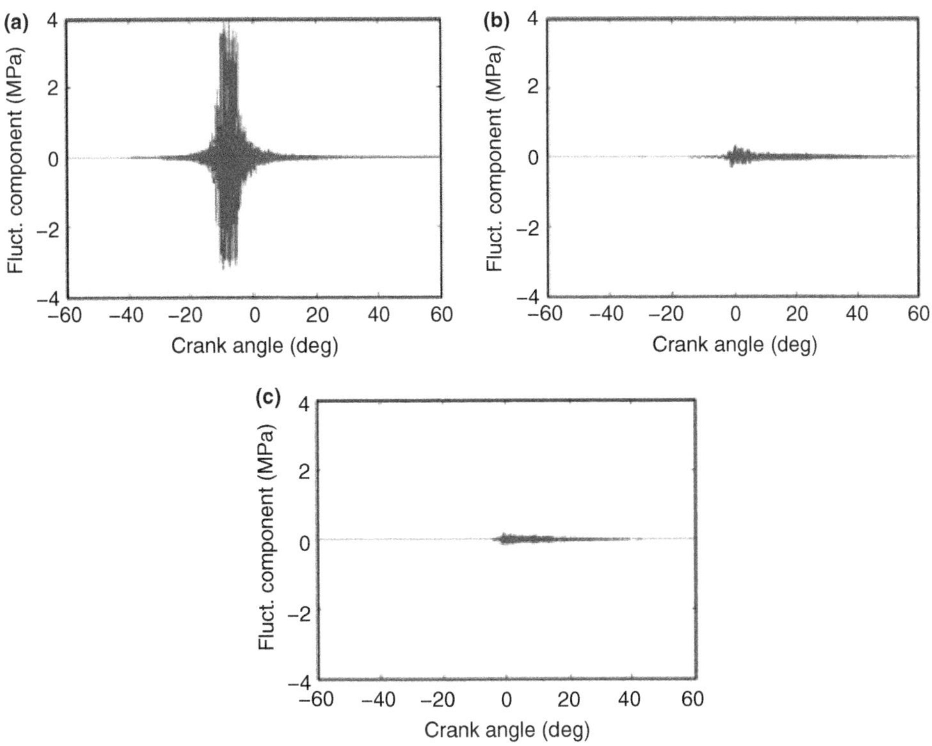

Figure 3.51 High-frequency component of in-cylinder pressure (HFCP) versus crank angle for (a) hydrogen fueled engine, (b) 25% hydrogen–75% diesel fueled engine, and (c) diesel engine [175] / with permission of Elsevier.

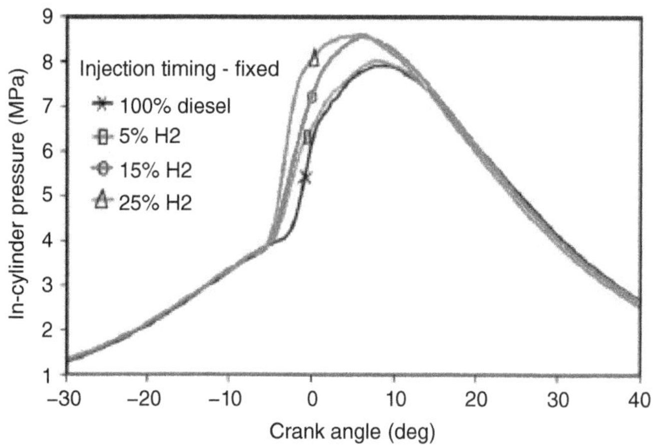

Figure 3.52 In-cylinder pressure traces for various hydrogen energy fractions in fuel charge [175] / with permission of Elsevier.

Figure 3.52 displays in-cylinder pressure plotted against crank angle for three different energy fractions of hydrogen and diesel to eliminate high-frequency components of in-cylinder pressure (HFCP). High-frequency components are characteristic of combustion knock, so removal of HFCP gives a cleaner picture of in-cylinder pressure.

A correct picture of the combustion process can emerge with a more precise determination of the MFB profile. In this analysis [175] the MFB is defined by

$$\mathrm{MFB} = M_F/M_T$$

where

M_F is mass of the fuel dose burned as a function of piston position measured by the crank angle, and

M_T is total mass of the fuel dose burned in the working cycle.

Mass of the fuel dose M_T, which is burned during the engine working cycle, can be calculated from $M_T = Q_{ch}/\mathrm{LHV}$, where Q_{ch} is chemical energy released by combustion and LHV stands for lower heating value of the fuel per unit of mass.

Combustion duration was calculated as the crank angle difference between 90% and 10% mass fraction burned (MFB). Despite varying total energy for fuels the analysis used normalized MFB (Figure 3.53) for standardizing the combustion progress [175]. The air–fuel ratio (lambda) varied during tests, affecting MFB curve shapes, but diesel injection timing was fixed.

Lilik et al. [178] investigated hydrogen-assisted diesel combustion on a 2.5 L, four cylinder, turbocharged, common rail, direct injection diesel engine. The substitution of hydrogen for diesel fuel was investigated over a range of energy proportions (0%, 2.5%, 5%, 7.5%, 10%, and 15%). Tests were conducted at four speed and load conditions: 1800 rpm at 25% and 75% of maximum output, and 3600 rpm at 25% and 75% of maximum output.

Injection timing adjustment resulted in substantial NOx emission reductions. It was observed that fumigation of hydrogen into the air intake of the diesel engine could replace up to 30% of the diesel fuel under high loads. This hydrogen substitution significantly altered the NO/NO_2 ratio. Computational fluid dynamics (CFD) models, specifically those considering turbulence–chemistry interactions using a transported probability density function (PDF) method, agreed with experimental observation of decreasing NO and increasing NO_2 levels with rising hydrogen amounts under certain operating conditions. The model emphasized the implication

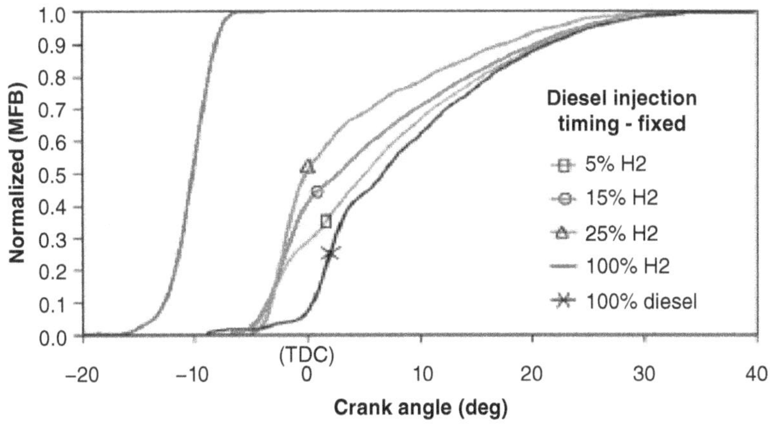

Figure 3.53 Mass fraction burned (MFB) vs. crank angle for various hydrogen energy percentages in fuel charge [175] / with permission of Elsevier.

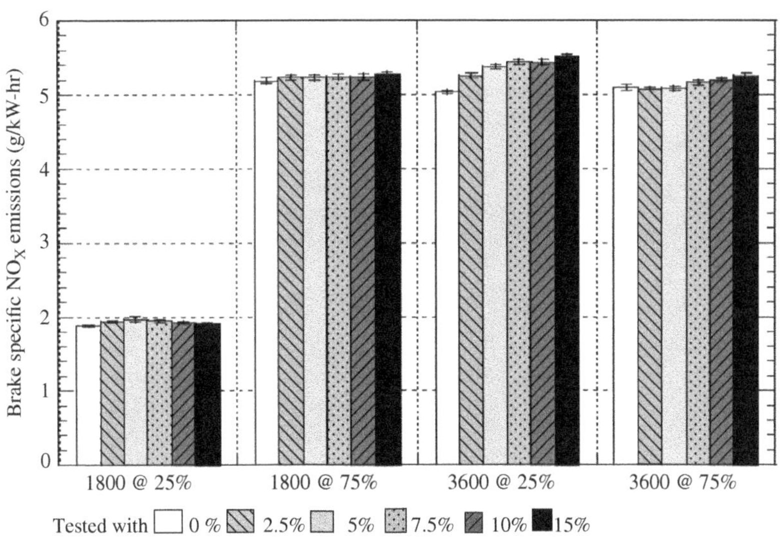

Figure 3.54 Brake specific NO_X emissions of the four modes [178] / with permission of Elsevier.

of turbulence–chemistry interactions using a transported PDF method, which had a stronger correlation with experimental results than the model ignoring turbulent fluctuations. CFD results confirmed that temperature changes alone were not sufficient to explain the observed reduction in NO and increase in NO_2 with increasing hydrogen. In high load modes, the increase caused by hydrogen is small, as seen in Figure 3.54.

Experimental investigations by Liew et al. put specific emphasis on the combustion aspect of heavy-duty hydrogen–diesel engines [179]. Introducing a substantial amount of hydrogen at 70% load significantly reduced combustion duration and increased the peak heat release rate due to rapid hydrogen combustion and ongoing diesel diffusion combustion. Hydrogen addition did not improve diesel combustion efficiency, and at lower hydrogen concentrations, combustion efficiency dropped–as evident in Figure 3.55, where at 70% load, 6% hydrogen addition reduced efficiency from over 99.5% to 97.9%.

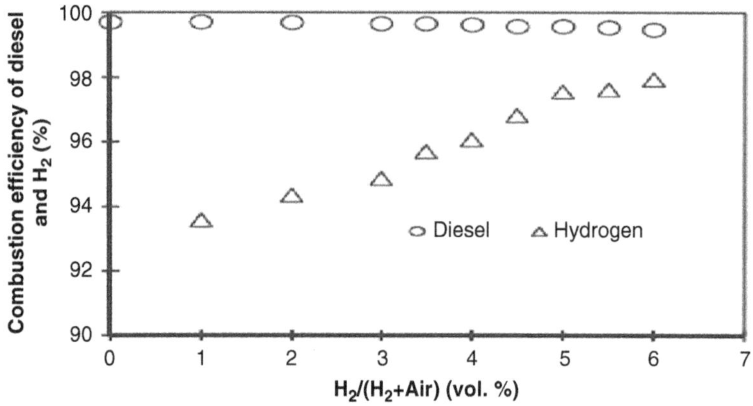

Figure 3.55 The effect of H_2 addition on combustion efficiency of diesel and H_2 at 70% load [179] / with permission of Elsevier.

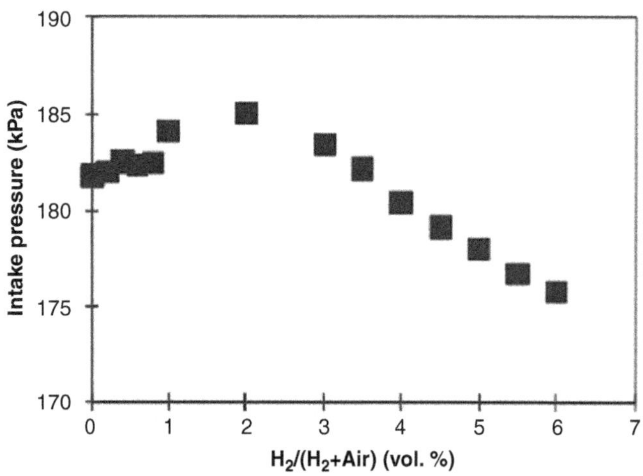

Figure 3.56 Effect of H_2 addition on the intake pressure at 70% load [179] / with permission of Elsevier.

Introducing a small amount of hydrogen increased intake pressure, with the highest pressure at 2% hydrogen. Beyond 3.5% hydrogen, intake pressure exhibited a rapid decreasing trend (Figure 3.56), due to alterations in exhaust temperature affecting the turbocharger and waste gate. The highest heat release rate occurred during the premixed combustion phase, with hydrogen marginally slowing down.

Figure 3.57a shows the effect of hydrogen addition on the heat release process. Introduction of hydrogen slightly retards the initiation of premixed combustion and reduces the peak heat release rate. The addition of 6% hydrogen delays the start of combustion) SOC by 0.8° CA. Hydrogen has a more pronounced effect on diffusion combustion, where it increases the heat release rate in the middle of this phase. An increase in the heat release rate occurs due to larger hydrogen addition because of the combined effects of diesel and rapid hydrogen combustion. This combined combustion involves multiple turbulent flames initiated by the diesel that propagate through the hydrogen–air mixture outside the diesel spray plume and accelerate the combustion process. Diffusion combustion, which includes both usual diesel diffusion and rapid flame development

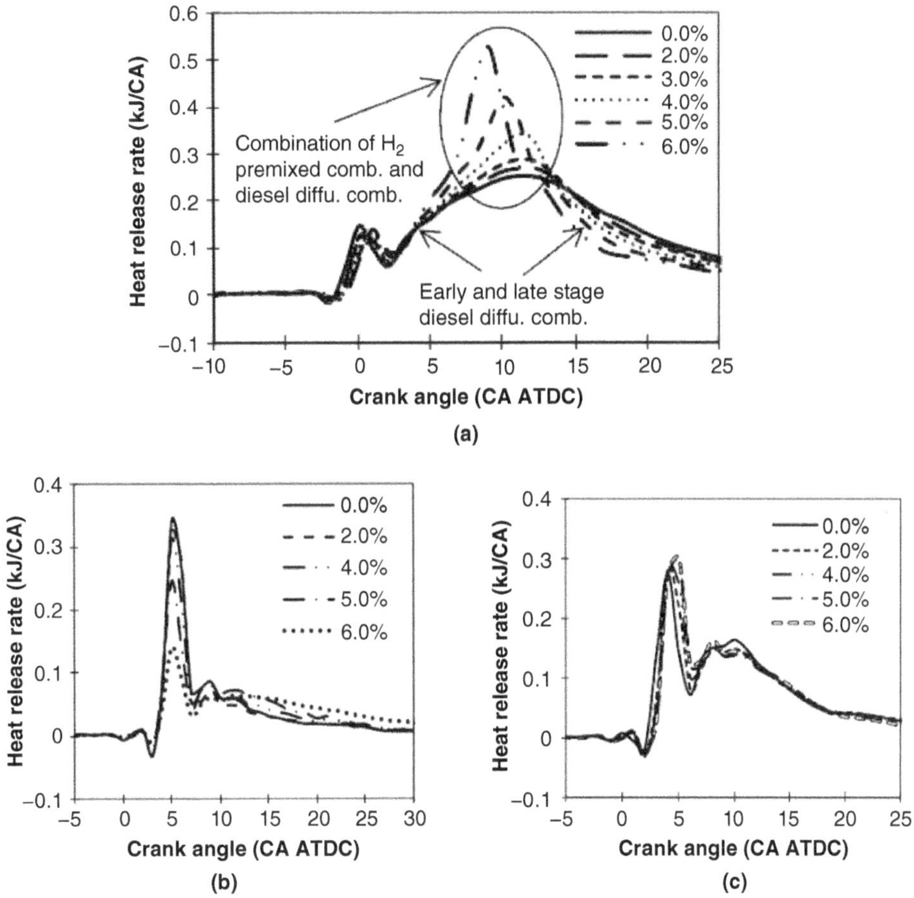

Figure 3.57 Effect of H_2 addition on the heat release process at: (a) 15% load, (b) 30% load, and (c) 70% load [179].

in the hydrogen–air mixture, has a similar effect of lean-burn spark ignition hydrogen engines but it is faster due to the multi-point ignition features of pilot diesel. Figure 3.57b shows the peak heat release rate occurring during the premixed combustion stage. Addition of hydrogen slightly retards but enhances this premixed combustion and improves the heat release process during the transition from premixed to diffusion combustion. As shown in Figure 3.57c, at 15% load, hydrogen addition significantly reduces the heat release rate during premixed combustion and slightly extends the diffusion combustion process. The result is a longer overall combustion duration and deterioration of engine performance. When operating at 15% load with over 5% hydrogen addition, the combustion duration significantly increases due to inhibited premixed combustion and long diffusion combustion.

Figure 3.58 presents the effect of hydrogen addition on combustion duration under different loads. A small amount of hydrogen had a negligible effect on combustion duration. The combustion process was improved leading to shorter combustion duration as shown at 70% and 30% load compared to 15% load. Drastic changes were visible in the combustion duration at 2% hydrogen and 70% load. Changes in the combustion duration were observed for 3% addition of hydrogen at 30% load and 5% addition at 15% load.

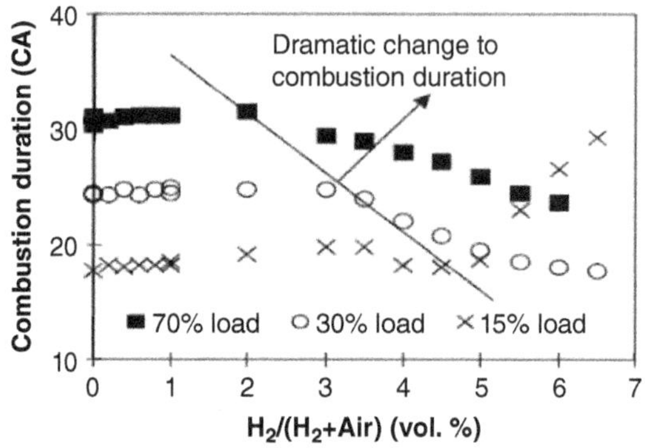

Figure 3.58 Effect of H_2 addition and engine load on combustion duration [179] / with permission of Elsevier.

3.8 Optimization of Dual Fuel Operation

Optimizing a hydrogen–diesel dual fuel engine from a performance, emission, and combustion point of view is very significant. Proper injection strategy with injection timing and duration is crucial for proper ignition by controlling in-cylinder pressure rise and NOx concentration. NOx is the only pollutant of concern in hydrogen combustion, and exhaust gas recirculation (EGR) has been a well-proven technique for reducing NOx. The compression ratio regulates the trade-off between efficiency and the risk of abnormal combustion such as knocking or preignition, ensuring improved power output, compliance with emission standards, and reliable engine performance.

3.8.1 Injection Timing

There have been many investigations on the impact of fuel injection strategies on performance, emissions, and knock behavior of hydrogen–diesel dual fuel engines [180–184]. In-cylinder pressure rise and NOx formation are reported to have been closely linked to injection timing and hydrogen energy-share ratio [181]. Tomita et al. [182] investigated a dual fuel engine combining hydrogen induction with air in the intake port and diesel oil injection into the cylinder. Mixing hydrogen with inlet air brings down HC, CO, and CO_2 emissions without smoke. Advanced injection timing lowers both smoke and NOx emissions with minimal HC, but sometimes with engine stability challenges. Roy et al. [180] studied engine performance and emissions of a supercharged engine using hydrogen and three other hydrogen-rich gaseous fuels, which included 13.7% H_2-content producer gas, 20% H_2-content producer gas, and 56.8% H_2-content coke oven gas (COG).Tests were conducted with diesel as a pilot fuel in a dual fuel mode where 90% or more of the energy was from primary fuels, with the rest from diesel. Tests were conducted at constant pilot injection pressure and quantity, varying fuel–air equivalence ratios and injection timings. Figure 3.59 shows the IMEP of different fuels at various equivalence ratios for different injection timings. In this series, injection timings were optimized for maximum engine power without knocking to be well within the safe pressure limits. It is seen in Figure 3.59 that IMEP increased with advanced injection timings at a constant equivalence ratio. Moreover, with the increased equivalence ratio, IMEP shows an increasing trend.

Saravanan and Nagarajan [183] studied the effects of hydrogen injection into the intake manifold of a diesel engine. By conducting tests, optimized injection timing was fixed to start at gas

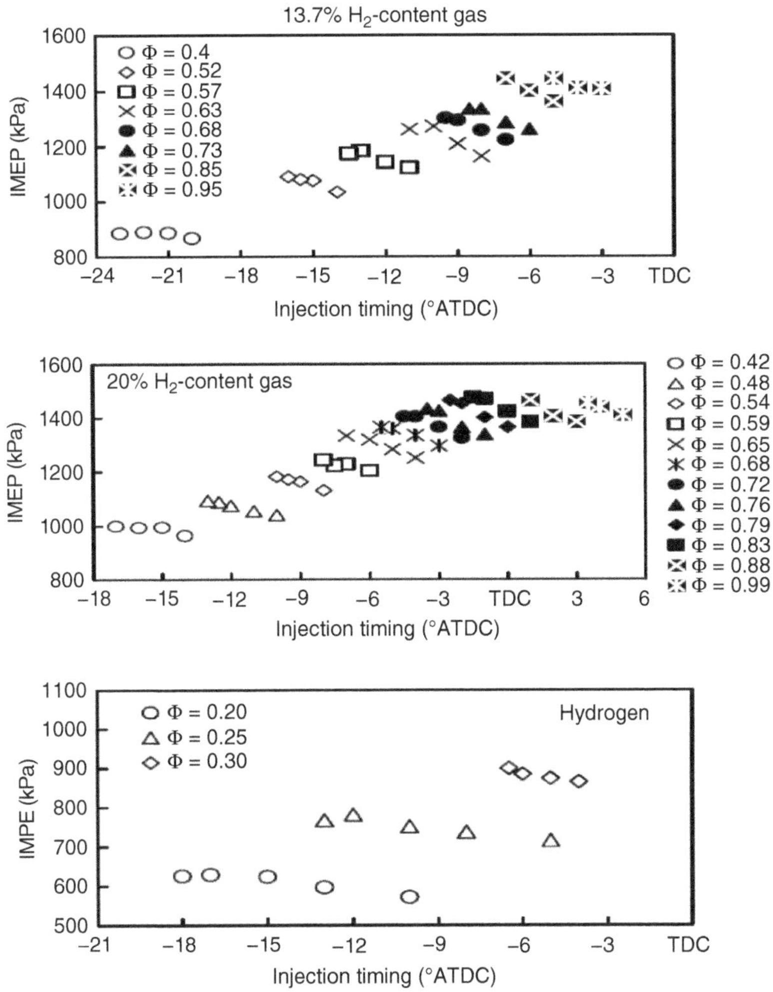

Figure 3.59 IMEP of different fuels at various equivalence ratios [180] / with permission of Elsevier.

exchange top dead center (GTDC), with an injection duration of 30° CA and a hydrogen flow rate of 7.5 lpm. Brake thermal efficiency increased by 9% compared to neat diesel operation. Full load smoke emission was reduced by four times while NOx emissions remained almost same as with diesel fueling except at no load conditions. CO_2 emissions decreased twofold at no load conditions due to higher hydrogen substitution. In their investigation on injection strategy on a single-cylinder, four-stroke diesel engine, Bayramo˘glu and Yilmaz [184] observed that increasing the hydrogen blending ratio and advancing the hydrogen injection timing contributes to substantial improvement in thermal efficiency and power. However, along with this the NOx emission level increases. Nguyen and Mikami [185] investigated noise generated in hydrogen–diesel dual fuel engines under different diesel fuel injection timings. Adding 10 vol% hydrogen to the intake air significantly reduced combustion noise when diesel fuel was injected late in the cycle. However, it increased cyclic variation compared to conditions without hydrogen. The maximum impact energy from combustion had a significant influence on the maximum engine noise power for each cycle. Specifically, combustion impact energy at a frequency of 830 Hz played a crucial role in contributing to overall engine noise during the early stage of the expansion stroke.

3.8.2 Exhaust Gas Recirculation

Exhaust gas recirculation (EGR), as in many other systems, also contributes to NOx reduction and knocking in hydrogen–diesel dual fuel engines. It is observed that a 20% EGR rate can achieve a 40% reduction in NOx emissions [186] in hydrogen–diesel dual mode operation, but EGR brings down the engine efficiency to dilution effects. Higher EGR rates cause further loss in efficiency and increase CO and HC emission due to reduced oxygen in the cylinder. EGR also adversely affects combustion efficiency and increases brake specific energy consumption (BSEC). Combining hydrogen and EGR have been useful in simultaneously reducing NOx and smoke emissions in single-cylinder DI engines [187]. In addition to thermal efficiency, many researchers including Selim [188] have experienced that combustion noise is affected when EGR is used in the dual fuel engine. The combustion noise decreased with increasing the engine speed for the dual fuel engine. At all engine speeds, upon increasing the EGR, the dual fuel engine showed a lower pressure rise rate. Bose et al. [189] carried out tests with a diesel engine with hydrogen flow rate at 0.15 kg/h. A light weight EGR cooler was used for NOx control. It was observed during tests that high rates of hydrogen led to uncontrolled combustion and a decrease in thermal efficiency. Performance and exhaust emission characteristics were evaluated with neat diesel operation, with diesel and 1.5 kg/h of hydrogen without EGR and also with 10% and 20% EGR, respectively, along with diesel and 1.5 kg/h of hydrogen.

3.8.3 Effects of Compression Ratio

Masood et al. [190] studied the impact of compression ratio on performance of a hydrogen–diesel dual fuel engine at various compression ratios ranging from 16.35:1 to 24.5:1. Higher compression ratios led to an increase in brake thermal efficiency (Figure 3.60) with higher hydrogen percentages, peaking at 100% hydrogen. CO, HC, and PM showed a consistent decreasing trend with higher hydrogen content at all compression ratios under full load conditions. However, with higher hydrogen rates, NOx emissions increased from 400 to 900 ppm with intake temperature rising from 65 to 85 °C.

It is evident in Figure 3.61 that a substantial reduction in particulate matter (PM) occurs as hydrogen substitution increases. This reduction is due to decreased diesel content in the fuel mixture. For instance, a compression ratio of 18.35 achieved a remarkable 92.7% reduction in PM with hydrogen substitution from 10 to 90%. Similarly, at a compression ratio of 22, there was an 82.6% reduction in PM under similar conditions.

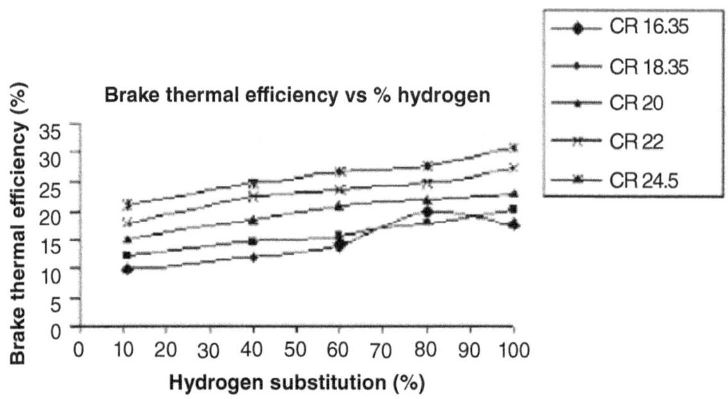

Figure 3.60 Effect of hydrogen substitution on brake thermal efficiency [190] / The American Society of Mechanical Engineers.

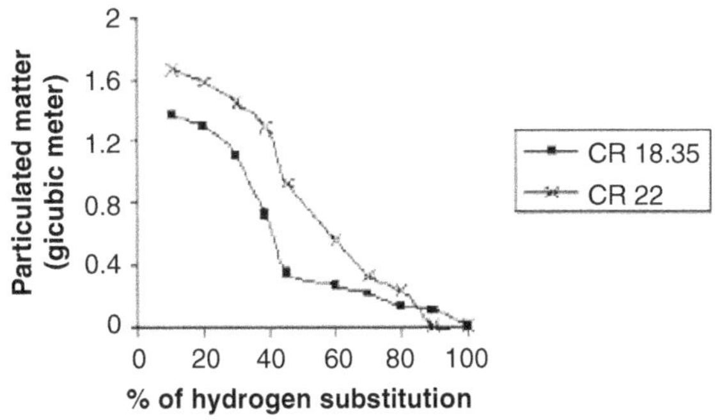

Figure 3.61 Effect of hydrogen substitution on particulate matter [190] / The American Society of Mechanical Engineers.

A study by Chaichan et al. [191] on a hydrogen–diesel dual fuel engine showed that with a 10% hydrogen volumetric fraction, the useful compression ratio increased to 18.1 from 17.7 for neat diesel. This increase persisted up to 40% hydrogen, with a maximum compression ratio of 19:1. However, brake power increased only at very lean equivalence ratios below 0.4. At higher equivalence ratios, retarded injection timing was needed to prevent knock, reducing brake power. High unburned hydrogen emissions at low loads remain a problem due to inefficient mixing of air and diesel. Such a situation leads to incomplete hydrogen combustion, as also observed by Suzuki et al. [192]. In this context, adopting low-temperature combustion (LTC) strategies can be used to enhance mixing and reduce unburned hydrogen emissions. LTC has emerged as a promising technique to improve efficiency and reduce soot, HC, CO, and CO_2 emissions over a wide range of operating conditions. Homogeneous charge compression ignition (HCCI) and premixed charge compression ignition (PCCI) have also been investigated as effective combustion strategies for hydrogen enrichment in diesel engines [192–195]. Detailed discussions on HCCI, PCCI, and RCCI are made in Chapter 9.

3.8.4 Water Induction for Knock Control

Hydrogen-assisted diesel engines face a significant challenge at hydrogen-rich conditions due to knocking. As discussed earlier, EGR, even though it extends knock-free limits, still increases harmful emissions and affects power output. Lambe and Watson [196] used up to 40% of water from exhaust gases as a coolant to suppress knock. Inducing 80% of exhaust water impacts ignition delay and burn time by cooling the mixture before compression, extending the delay period and increasing burn time. This is a crucial observation for water induction strategy to optimize the combustion process. Prabhukumar et al. [197] observed that water induction allowed for an increase in the hydrogen energy share ratio by up to 3.5% at high loads and 5% at medium loads before knocking occurred. At full load, water induction improved the knock-limited power output by up to 39%. However, brake thermal efficiency and maximum power output decreased due to the quenching effect of water, which caused gaseous fuel to escape during combustion and reduced the heat input from hydrogen. Despite these reductions, water addition improved brake thermal efficiency beyond the knock-limited power output of the hydrogen–diesel engine. The cooling effect of water increased ignition delay, slowed the combustion process, and reduced peak pressure and the maximum rate of pressure rise. The study identified an optimal water induction rate of 0.7×10^{-3} m^3/h for improved knock-limited power output. Water addition beyond this rate deteriorated lubricating oil quality.

Adnan et al. [198] studied water injection at various crank angles, ranging from 20° BTDC to 20° ATDC with injection durations of 20° CA and 40° CA. Early- and short-duration water injection increased gross indicated work and indicated thermal efficiency, while prolonged injection delayed combustion initiation and altered heat release characteristics. The study measured oxygen (O_2) and sulphur dioxide (SO_2) emissions during tests with diesel alone (DA), diesel and hydrogen (DH), and diesel and hydrogen with water injection (DHW). DA had the highest SO_2 emissions at higher speeds, while DH had the lowest, particularly at lower speeds. DHW operation decreased SO_2 emissions compared to DA. SO_2 emissions, in these tests, originated solely from the diesel pilot fuel. Minimal diesel injection of DHW resulted in lower SO_2 emissions. This reduction was because of lower mixing-controlled combustion temperatures and a leaner mixture in the cylinder charge. Moreover, the presence of oxygen during DHW operation caused oxidation of sulphur, forming SO_2, which subsequently combined with water to produce sulphuric acid (H_2SO_4). In DHW operation, oxygen concentration increased with speed up to 2000 rpm, after which it started to decrease. DH operation consistently generated higher oxygen emissions than DA operation over all ranges of speed. DHW operation showed higher O_2 emissions than both DA and DH operations. Lower O_2 concentrations in DA are due to reduced post-quench diffusion and oxidation rates. DHW demonstrated higher O_2 concentrations due to the dissociation of water droplets and CO_2 at high combustion temperatures. In lean mixture regions, DHW produced more O_2 from dissociation [199].

Chintala and Subramanian [200] identified that with a specific water consumption of 200 h/kWh and 20% hydrogen energy share, a 24% reduction in NOx emissions could be achieved, but a 5.7% decrease in energy efficiency occurred. Investigators adopted two strategies: retarding injection timing and water injection to achieve knock-free operation [201]. By applying retarded injection timing, the maximum energy share of hydrogen with knock-free operation could be raised from 18 to 24%, and with water injection it went up to 36%.

3.8.5 Charge Dilution

Tests have reported the effects of the charge dilution technique, using diluents on the performance and combustion characteristics of a hydrogen-fueled diesel engine with focus on knock [202–204]. A diluent, which is an inert fluid, does not get involved chemically in the combustion process and does not act as a fuel or oxidant. Instead, its main role is to absorb heat due to its heat capacity and control the temperature by acting as a heat sink. Addition of diluents improves knock-limited engine operation and increases the optimum hydrogen energy substitution percentage.

After carrying out exhaustive tests in the laboratory, it was observed that neat hydrogen induction up to 30 lpm ensured smooth combustion with improved performance. Experiments were carried out to assess what percentage of a particular diluent (helium, nitrogen, and water) corresponding to hydrogen to hydrogen flow rate was suitable for smooth combustion. An exhaustive series of tests conducted in the lab with various hydrogen induction rates (40, 50, and 60 lpm) along with 10%, 20%, and 30% of helium and nitrogen and various proportions of water are reported in research papers [203, 204]. Among various proportions, 10% helium by volume of hydrogen gave the highest thermal efficiency with smooth engine combustion over the entire range. Nitrogen, in 30% by volume of hydrogen, provided good performance with optimum full-load hydrogen energy substitution and higher knock-limited power output. The heat capacity of water is larger than those of helium and nitrogen. Since water displaces less volume of air for the same mass of diluents, it helps in a greater amount of hydrogen induction into the cylinder. Among various proportions of water tested, 2460 ppm was observed to be the optimal charge diluent for hydrogen–diesel dual fuel operation.

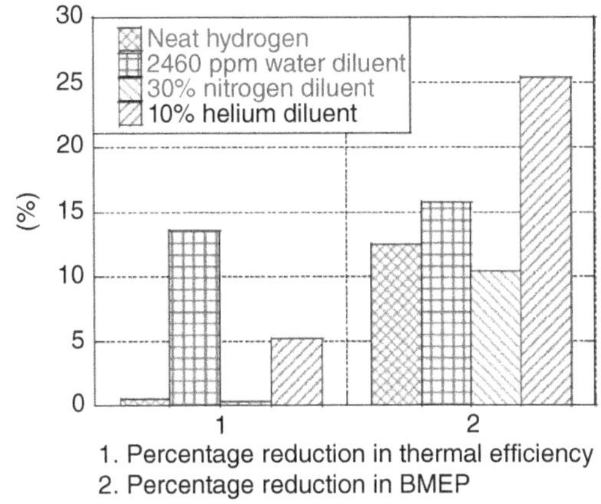

Figure 3.62 Performance characteristics through optimized diluent [204] / with permission of Elsevier.

Nitrogen was the best diluent for engine performance, while at emission levels water was much better than nitrogen. However, the choice between water and nitrogen depends on priorities and the application area. In general, as a diluent, water is preferred over nitrogen due to low cost, ample availability, and ease of storage and handling. Hydrogen–diesel engine systems with water, at 2460 ppm, deliver a balance between high hydrogen energy substitution, improved thermal efficiency, and knock-limited power output and can be of use in decentralized energy applications such as gensets or pump sets and other applications. Figure 3.62 gives a comparative picture of the three optimized diluents for a hydrogen–diesel engine.

Hydrogen and ammonia are primarily carbon-free fuels. Hydrogen exhibits exceptionally fast flame speed, leading to engine knock. In contrast, ammonia's low flame speed can lead to low heat release and cause ignition challenges. Ge and colleagues [205] carried out a computational study on premixed ammonia/hydrogen combustion in a spark ignition engine, with the focus on identifying the optimal blend that maximizes work output while avoiding knock. Identification of the optimal blend of ammonia was critical because ammonia reduced flame speed and flame temperature to eliminate knock and brought down NOx emission levels. At the same time, too much ammonia led to slow combustion and deteriorated engine performance. Subsequently, the engine in the ranges of low equivalence ratio was operated with pure hydrogen and as the engine approached a knock-limited condition, ammonia was introduced into the fuel mixture. This procedure proved effective in preventing engine knocks while simultaneously enhancing the energy content. However, the study had limitations with the current approach, particularly in terms of fuel injection. Port fuel injection of both hydrogen and ammonia is not the desirable option due to safety concerns related to hydrogen in the intake port and the potential for ammonia to slip from the intake system. Furthermore, the model did not account for crevice volume, leading to accumulation of unburned ammonia. Further research is needed to explore the merits of direct injection for both NH_3 and H_2, considering factors such as safety, ammonia slip, and crevice volume.

3.9 Liquid Hydrogen

Most of the research and development work on hydrogen engines or vehicle demonstrations is done with gaseous hydrogen as the fuel. It is easier to store and handle gaseous hydrogen than liquid hydrogen. Storing hydrogen in the liquid state requires extremely low temperatures (around −253 °C), making storage and transportation infrastructures more complex. Moreover, liquid hydrogen can experience boil-off losses and gradually evaporate over time, even in well-insulated containers, which can result in fuel loss if not properly managed. Gaseous hydrogen, on the other hand, does not have this issue. While both gaseous and liquid hydrogen have safety considerations, handling liquid hydrogen presents additional challenges due to its extremely low temperature. Sometimes, cold burns and other hazards may be higher when dealing with liquid hydrogen. Gaseous hydrogen is more flexible in terms of vehicle design and fueling systems. Fueling a vehicle with gaseous hydrogen is relatively simple now because a widespread infrastructure already exists for gaseous fuels such as compressed natural gas. It allows for a wider range of storage options and is generally more adaptable to different vehicle types. Perhaps the most important reason is that existing hydrogen fueling infrastructure supports gaseous hydrogen. The process of liquefying hydrogen involves a substantial amount of energy. Transporting and maintaining liquid hydrogen at extremely low temperatures also requires additional energy. Using gaseous hydrogen simplifies these processes and reduces energy losses associated with liquefaction.

Furuhama and his team [206–224] conducted extensive investigations into various hydrogen engine configurations, including two-stroke and four-stroke engines, liquid and gaseous hydrogen fuels, different ignition methods, direct injection, turbocharging, and backfire. Their research led to the development of a liquid hydrogen pump, a liquid hydrogen car, and medium-duty trucks. Documented over more than 15 years, their work significantly advanced hydrogen as a low-emission alternative fuel for internal combustion engines. In the course of engine development, serious challenges were encountered related to pressure rise rates and incomplete combustion during hydrogen injection at the end of the compression stroke, due to low hydrogen density and limited mixing time. The problem was solved by injecting hydrogen onto a hot surface near top dead center. This method, adopted in a turbocharged engine, improved performance and reduced noise and vibration. Their experiments with liquid hydrogen on the intake manifold increased the hydrogen engine's volumetric efficiency by 15% compared to a gasoline engine.

In an effort to reduce the amount of residual gas and prevent backfires, Furuhama and the team increased the compression ratio from 4.8:1 to 7.9:1, but could not succeed. Relocating the spark plug inside a small hole for better cooling extended operational limits to richer mixtures and eliminated preignition, but it led to an excessive rate of pressure rise. Steps to retard spark timing extended operational limits, increased output power, and reduced NOx formation, but could not completely suppress rapid pressure rise, and the NOx concentration remained high near stoichiometric mixtures. EGR up to 15% by weight was ineffective in lowering oxygen levels in combustion gases. Intake air throttling improved thermal efficiency and reduced NOx emissions under light loads but was less effective under heavy loads [215]. In another series of tests, Furuhama and team adopted the "combined combustion process" to control abnormal combustion. A pre-mixed hydrogen–air mixture ($\lambda = 4.0$) was introduced into the cylinder from the intake manifold, followed by hydrogen injection at 50 kg/cm^2 through seven holes before ignition (Figure 3.63). The study focused on injection initiation, duration, and spark timing. Optimized combined combustion yielded normal combustion and reduced NOx emissions at a stoichiometric ratio, improving power output due to augmented charge. A critical requirement was compressing hydrogen gas to at least 50 kg/cm^2 and ensuring high reliability and durability of injection equipment for precise control [215].

Apart from exhaustive tests in the laboratory restricted to engines, Furuhama and his team used liquid hydrogen for many development activities in prototype vehicles using liquid hydrogen [225–228]. In 1975, the team successfully ran a liquid hydrogen–powered car over 2800 km on public roads, demonstrating satisfactory acceleration, drivability, and better fuel economy compared to a gasoline car (Figure 3.64).

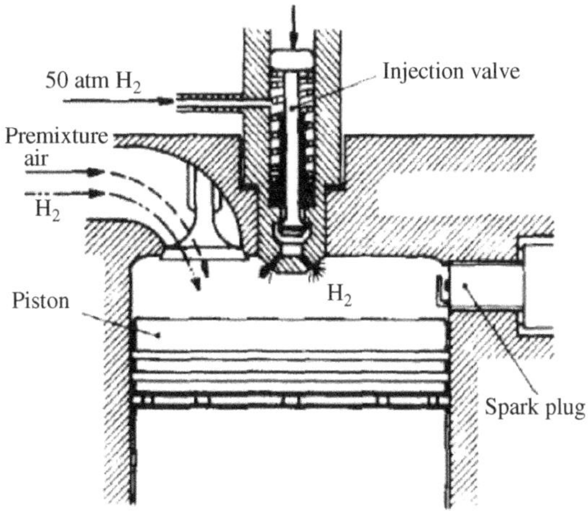

Figure 3.63 Engine using a combined combustion process [215] / with permission of Elsevier.

Figure 3.64 Liquid hydrogen car of Musashi Institute of Technology, which participated in the SEED Rally, 1975 [209] / with permission of Elsevier.

In June 1980, a hydrogen car with an LH_2 tank in the rear trunk offered a 400 km range. The two-stroke engine received −300 to −500 °C cold hydrogen from an LH_2 pump. Tests showed a 20–30% power increase over its gasoline counterpart, with no backfire, knocking, and minimal NOx emissions, and the car reached a maximum speed of 135 km/h [211]. A car was developed by Musashi Institute of Technology with an LH_2 tank, high-pressure LH_2 pump, hydrogen injector onto a hot surface, and an ignitor in a turbo-engine with a 12:1 compression ratio. Hino Motors Ltd. adapted the engine for a medium-duty truck with some modifications on the injector's gas tightness and endurance with better materials, size adjustments, finishing techniques, and lubricating oil [212]. Mixture formation was improved by optimizing the nozzle holes and injection timing with the hydraulic pump and valve. Tests showed that a small amount of hydrogen in the intake manifold reduced combustion noise and minimized LH_2 tank evaporation loss. A supplemental test with a constant volume combustion chamber by the Schlieren method showed that flame propagation from hot surface hydrogen jets was very different from that in diesel engines.

Apart from vibrations in the air column between pressure sensor and combustion chamber, there was vibration in the cylinder gas. The frequency of this vibration was dependent upon the size of the cylinder [213].

LH_2 on-board storage and handling technologies, such as using cold hydrogen gas from the LH_2 fuel tank to significantly increase the volumetric heating value of the cylinder charge, are important for LH_2 vehicle development. Peschka et al. [229–237] studied liquid hydrogen storage and refueling for road vehicles almost four decades ago as shown in Figure 3.65. As part of a project on the feasibility of liquid hydrogen as a vehicular fuel, a comprehensive investigation was carried out on a 1979 Buick Century sedan, equipped with a 3.8 L displacement turbocharged V6 engine and an automatic transmission [237].

A typical refueling arrangement is shown in Figure 3.65 where the unit 2 refueling station (U2RS) is connected to a bulk liquid hydrogen storage Dewar (BLHSD). The vehicle demonstrated fuel economy of 2.4 km/L of liquid hydrogen, equivalent to 8.9 km/L gasoline in energy. Impressively, hydrogen consumption resulted in a remarkable 22% reduction in energy requirements compared to gasoline for covering a given distance. Over a span of 17 months and 138 operational hours, the vehicle covered 3633 km on liquid hydrogen, establishing the extended operational feasibility of hydrogen as a fuel. The study involved testing two types of onboard liquid hydrogen storage tanks, one with capacity of 110 L and the other with 155 L. The unrefueled range of vehicles varied accordingly, reaching 274 km with the former and 362 km with the latter. Refueling, conducted at least 65 times, demonstrated efficient procedures with a 9 min refueling time.

During the constant speed tests the hydrogen flow rate data gathered during some field tests are shown in Figure 3.66. Liquid hydrogen losses during refueling were carefully measured. Despite successful achievements of many major objectives, inability to attain satisfactory vehicle acceleration capability and the persistence of engine backfiring were disappointing aspects that somewhat overshadowed the project's overall accomplishments.

Cryogenic hydrogen is attractive for use in internal combustion engines, as it increases the volumetric heating value of cylinder charge and avoids water injection. Research must focus on developing electronically actuated cryogenic fuel injection valves for port and direct injection. Investigating cryogenic internal fuel mixing for spark ignition and hydrogen–diesel engines is crucial for optimal configuration, and developing low-cost LH_2 tanks with extended storage time and high-pressure injection pumps separated from storage tanks is necessary for actual system

Figure 3.65 A typical refueling operation with a BLHSD [237] / with permission of Elsevier.

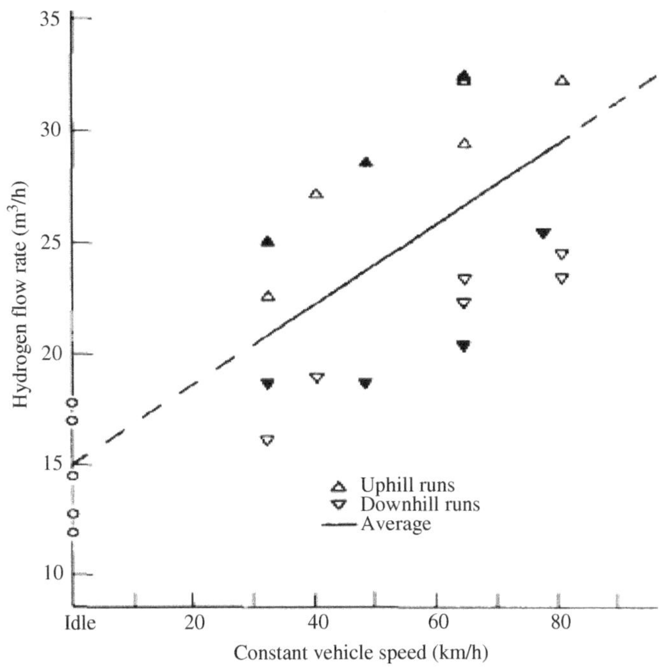

Figure 3.66 Hydrogen flow rate as a function of the constant vehicle speed. The solid triangles indicate data taken on the road near the Los Alamos airport [237] / with permission of Elsevier.

development. Research should focus on improving cryogenic components, including master solenoid valves for fuel shut-off and cryogenic solenoid valves for LH_2 tank refueling. Lubrication challenges in cryogenic technology are vital for optimizing injection systems. Cryogenic liquid injection systems face challenges such as poor lubrication and handling boiling cryogenic liquids with vapors. Developing dry lubrication materials for pump-piston systems, such as reinforced plastics, need to be investigated.

Concluding Remarks and Perspectives

In the realm of sustainable mobility, hydrogen engines shine as a beacon of hope, promising freedom from energy–environment crises. The application of complex strategies lead to improvements on many relevant parameters such as injection, ignition, combustion, and emission. With the capability to embrace benefits of both SI and CI engines, an optimized configuration of a hydrogen engine can adopt diesel-like quality governing, match thermal efficiency of a diesel engine, and develop the specific output of an SI engine. With thermal efficiency beyond 40% and nearly zero emission characteristics, these engines promise a paradigm shift to optimum performance and a transformative leap toward a clean future. An SI engine operates on neat hydrogen while a diesel engine runs in dual fuel mode. Hydrogen can be blended with compressed natural gas and liquefied natural gas, ethanol, dimethyl ether (DME), biodiesel, biogas, etc.

Extensive research and development activities carried out over a period of several decades have identified the fuel-specific design features of a hydrogen engine, but an optimized hydrogen engine needs some drastic alteration in the design of existing SI engines to get the best benefits from its use. Hydrogen engines are prone to backfires due to rapid combustion and low minimum ignition energy, which leads to decreased volumetric efficiency and damages intake and fuel injection systems.Improper intake valve timing, fuel injection timing, and high equivalence ratio need to be carefully monitored to avoid backfire. An appropriately designed port fuel injection (PFI) and timed manifold injection have been successful in avoiding backfire, but the possibility of backfire has not been entirely ruled out in these modes of fuel induction. Even minor design defects or alterations in implementation of PFI and TMI can at times lead to backfire of mild to moderate intensity. Direct injection completely eliminates backfire or flashback. Direct injection (DI) involves injecting hydrogen directly into the cylinder at elevated pressures, typically around 100 bar after intake valve closure. These injectors should have high injection pressure, reaching up to 250 bar, along with large fuel flow rate, fast response, remarkable repeatability, and precise control of the needle lift. A dedicated injector nozzle should adopt an optimized DI strategy, adjusting injection timing based on engine speed and load. The durability of injectors to withstand stringent thermal environments is vital. Hydraulically operated, cam-operated, solenoid-actuated, and piezo-actuated injectors have reduced needle seating impact and bouncing, improving injector durability. DI offers high flexibility for optimizing mixture formation. Injector geometry and mixture formation are critical to hydrogen mixture stratification around the spark plug and away from the cylinder walls.

Knock in hydrogen engines reduces power, increases fuel consumption, generates noise, and increases emissions. Interaction of flame and pressure waves during knock development needs to be critically analyzed to avoid knocking conditions. A comprehensive analysis of heat transfer is important to prevent loss in efficiency under various operating conditions. Hydrogen engines can run in an ultra-lean mode, which helps in drastically bringing down NOx emissions. The relative air/fuel ratio (λ) plays a significant role in high heat flux measured on the cylinder walls during combustion. Lean hydrogen–air mixtures produce lower heat transfer to the walls, resulting in lower efficiency loss. Safety features should be installed in both test cell and in-vehicle applications. Safety valves, flame arrestors, non-return valves, etc., should be put in appropriate locations in the engine test rig [238]. Adequate ventilation, hydrogen flame cameras, and detectors should be installed as safety measures to avoid threats from invisible hydrogen flames.

Efficiency of hydrogen engines could be further improved by using waste heat recovery from both coolant and exhaust systems. This approach can improve peak and cycle average fuel conversion efficiencies. H_2ICEs have the potential to achieve peak fuel conversion efficiencies exceeding 50% and average fuel conversion efficiency above 35%, especially when coupled with hybrid powertrains. These studies need to be amplified, and additional research and development efforts need to be done before widespread implementation. The difficulties of widespread application of hydrogen engines are challenges in infrastructure development, production methodologies, storage techniques, transportation logistics, establishment of refueling stations, and comprehensive safety considerations.

Phase-out of grey hydrogen is essential. Until an adequate quantity of green hydrogen is available, blue hydrogen may be used in this transition period. However, there is a critical need to enhance production of green hydrogen through electrolysis of water from renewable energy sources. Transport modes, such as road, railway, shipping, and pipelines, should be carefully considered on the basis of distance, quantity, and hydrogen form. Scarcity of hydrogen refueling stations (HRS) hinder its widespread application. Germany, France, Japan, the US, China, South Korea, and several other countries have launched plans to construct HRSs. Last but not least, there is a compelling need for the systematic development of safety standards for each stage such as production, storage, transportation, and application.

There are also advanced, upgraded, and augmented hydrogen engine configurations (discussed in Chapter 10). A lot of research and development is taking place in the area of integrating hydrogen with other fuels (discussed in Chapter 9). Similarly, the application of hydrogen in prototype vehicles such as trucks, buses, three-wheelers, cars, etc., for a sustainable and low-carbon future is discussed in Chapter 11. This chapter is limited to the discussion of regular spark and compression ignition engines. An effort has been made here to provide a comprehensive overview, yet much more still remains beyond the scope of the current chapter.

References

1 Das, L.M. (1990). Hydrogen engines: A view of the past and a look into the future. *International Journal of Hydrogen Energy* 15 (6): 425–443.

2 White, C.M., Steeper, R., and Lutz, A.E. (2006). The hydrogen-fueled internal combustion engine: A technical review. *International Journal of Hydrogen Energy* 31: 1292–1305.

3 Verhelst, S. and Wallner, T. (2009). Hydrogen-fueled internal combustion engines. *Progress in Energy and Combustion Science* 35: 490–527.

4 Stępień Z. (2024). Analysis of the prospects for hydrogen-fueled internal combustion engines. *Combustion Engines* 197 (2): 32–41. https://doi.org/10.19206/CE-1747.

5 Escher, W.J.D. (1983). Hydrogen as an automotive fuel: Worldwide update. *Symposium Papers of the Non-Petroleum Vehicular Fuels* 3: 143–180.

6 Dimitriou, P. and Tsujimura, T. (2017). A review of hydrogen as a compression ignition engine fuel. *International Journal Hydrogen Energy* 42: 24470–24486.

7 Wróbel, K., Wróbel, J., Tokarz, W. et al. (2022). A. Hydrogen internal combustion engine vehicles: a review. *Energies* 15: 8937. https://doi.org/10.3390/en15238937.

8 Stępień, Z.A. (2021). Comprehensive overview of hydrogen-fueled internal combustion engines: achievements and future challenges. *Energies* 14: 6504. https://doi.org/10.3390/en14206504.

9 Sun, Z., Hong, J., Zhang, T. et al. (2023). Hydrogen engine operation strategies: recent progress, industrialization challenges, and perspectives. *International Journal of Hydrogen Energy* 48: 366–392.

10 Cecil, W. (1822). On the application of hydrogen gas to produce a moving power in machinery, with a description of an engine which is moved by the pressure of atmosphere upon a vacuum caused by explosions of hydrogen gas. *Cambridge Philosophical Society* 1: 217.

11 Braun, H. (2008). The Phoenix Project: shifting to a solar hydrogen economy by 2020. *The Chemical Industry and Chemical Engineering Quarterly*, CI&CEQ 14 (2) 107–118.

12 Erren, R.A. and Campbell, W.H. (1933). Hydrogen: a commercial fuel for internal combustion engines and other purposes. *Journal of the Institute of Fuel* VI: 277–291.

13 Weil, K.H. (1972). The hydrogen I.C. engine–its origins and future in the emerging energy-transportation-environment system. *Proceedings of the 7th Inter-Society Energy Conversion Engineering Conference*. American Chemical Society, Paper 729212, 1355.

14 Bose, D. (2015). The hydrogen alternative. *International Letters of Chemistry, Physics, and Astronomy* 49: 15–26.

15 Oehmichen, M. (1942). *Wasserstoff als Motoreibmittel, Deutsche Kraft Fahart-Forschung Heft* 68. Berlin: VDI, Verlag GmbH.

16 King, R.O., Hayes, S.V., Allan A.B. et al. (1958). The hydrogen engine: combustion knock and the related flame velocity. *Transactions of the Engineering Institute of Canada* (EIC) 2(4): 143–148.

17 King, R.O. and Rand, M. (1955). The oxidation, decomposition, ignition and detonation of fuel vapours and gases–XXVII. The hydrogen engine. *Canadian Journal of Technology* 33: 445–469.

18 King, R.O., Wallace, W.A., and Mahapatra, B. (1957). The oxidation, ignition and detonation of fuel vapours and gases–V. The hydrogen engine and detonation of the end gas by the igniting effect of carbon nuclei formed by pyrolysis of lubricating oil vapour. *Canadian Journal of Technology* 34: 264–276.

19 Shrestha, B., Karim, GA., and Wierzba, I. (2000). Examination of operational limits in gas fueled spark ignition engines. SAE Technical Paper No. 2000-01-1944.

20 Ricardo, H.R. (1924). Further note on fuel research. *Proceedings of the Institute of Auto Engineers.* XVIII (Part 1): 327–341.

21 Bindon, J., Hind, J., Simmons, J. et al. (1985). The development of a lean-burning carburetor for a hydrogen-powered vehicle. *International Journal of Hydrogen Energy* 10: 297–304.

22 Swain, M.R. and Adt, R.R. (1972). The hydrogen-air fueled automobile. *Proceedings of the 7th Intersociety Energy Conversion Engineering Conference*, 1382.

23 Swain, M.R., Pappas John, M., Adt Jr Robert, R. et al. (1981). Hydrogen-fuelled automotive engine experimental testing to provide an initial design data base. SAE Technical Paper 810350.

24 Swain, M.R., Adt, Jr., RR., Pappas, JM. (May 1983). Experimental Hydrogen-Fueled Automotive Engine Design Data-Base Project Main Technical Report. Volume 2. US Department of Energy.

25 Lynch, F.E. (1983). Parallel induction: A simple fuel control method for hydrogen engines. *International Journal of Hydrogen Energy* 8: 721.

26 Lee, S.J., Yi, H.S., and Kim, E.S. (April 1995). Combustion characteristics of intake port injection type hydrogen fueled engine. *International Journal of Hydrogen Energy* 20: 317–322.

27 Park, C., Kim, Y., Oh, S. et al. (26 October 2021). Effect of the operation strategy and spark plug conditions on the torque output of a hydrogen port fuel injection engine. *International Journal of Hydrogen Energy* 46 (74): 37063–37070.

28 Gao, J., Wang, X., Song, P. et al. (2022). Review of the backfire occurrences and control strategies for port hydrogen injection internal combustion engines. *Fuel* 307: 121553.

29 Bakar, R.A., Mohammed, M.K., and Rahman, M.M. (2008). Numerical study on the performance characteristics of hydrogen fueled port injection internal combustion engine. *American Journal of Engineering and Applied Sciences* 1 (4): 338–346. ISSN 1941–7020.

30 Sun, D. and Liu, F. (2011). Research on the performance and emission of a Port Fuel injection hydrogen internal combustion engine. *2011 International Conference on Computer Distributed Control and Intelligent Environmental Monitoring*, Changsha, China 2: 963–966. https://doi.org/10.1109/CDCIEM.2011.523.

31 Saravanan, G., Nagarajan, C., and Dhanasekaran, K.M. (November 2007). Experimental investigation of hydrogen port fuel injection in DI diesel engine. *International Journal of Hydrogen Energy* 32 (16): 4071–4080.

32 Povel, R., Topler, J., Withalm, G. et al. (1984). Hydrogen drive in field testing. *Proceedings of the 5th World Hydrogen Energy Conference*, Toronto, Canada.

33 Watson, H.C., Milkins, E.E., Martin, W.R.B. et al. (1984). An Australian hydrogen car. *Proceedings of the 5th World Hydrogen Energy Conference*, Toronto, Canada.

34 Yang, Z., Wu, J., Yun, H. et al. (26 April 2022). Diagnosis and control of abnormal combustion of hydrogen internal combustion engine based on the hydrogen injection parameters. *International Journal of Hydrogen Energy* 47 (35): 15887–15895.

35 Yip, H.L., Srna, A., Yuen, A.C.Y. et al. (2019). A review of hydrogen direct injection for internal combustion engines: towards carbon-free combustion. *Applied Sciences* 9: 4842. https://doi.org/10.3390/app9224842.

36 Varde, K.S. and Frame G.A. (1984). Development of a high pressure hydrogen injection for S.I. engines and results of engine behavior. 5th World Hydrogen Energy Conference, Toronto, Canada.

37 Homan, H.S. (1978). An experimental study of reciprocating internal combustion engines operated on hydrogen. PhD thesis. Cornell University.

38 Molina, R., Novella, J., Gomez-Soriano, M. (15 March 2024). Olcina-Girona impact of medium-pressure direct injection in a spark-ignition engine fueled by hydrogen. *Fuel* 360: 130618.

39 Liang, Z., Xie, F., Lai, K. et al. (2024). Study of single and split injection strategies on combustion and emissions of hydrogen DISI engine. *International Journal of Hydrogen Energy* 49 (Part B): 1087–1099.

40 Zhao, F.C., Sun, B.G., Yuan, S. et al. (2 January 2024). Experimental and modeling investigations to improve the performance of the near-zero NOx emissions direct-injection hydrogen engine by injection optimization. *International Journal of Hydrogen Energy* 49 (Part B): 713–724.

41 Farzam, R. and McTaggart-Cowan, G. (2024). Hydrogen-diesel dual-fuel combustion sensitivity to fuel injection parameters in a multi-cylinder compression-ignition engine. *International Journal of Hydrogen Energy* 49 (Part B): 850–867.

42 Wu, B., Torelli, R., Pei, Y. (1 June 2023). Numerical modeling of hydrogen mixing in a direct-injection engine fueled with gaseous hydrogen. *Fuel* 341: 127725.

43 Zhang, S.W., Sun B.G., Lin, S.L. et al. (7 February 2024). Energy and exergy analysis for a turbocharged direct-injection hydrogen engine to achieve efficient and high-economy performances. *International Journal of Hydrogen Energy* 54: 601–661.

44 Das, L.M. (1987). Studies on timed manifold injection in hydrogen operated spark ignition engine: performance combustion and exhaust emission characteristics. PhD thesis. Indian Institute of Technology, Delhi.

45 MacCarley, C.A. and Van Vorst, W.D. (1980). Electronic fuel injection techniques for hydrogen-powered engines. *International Journal of Hydrogen Energy* 5: 179.

46 Liu, X.H., Liu, F.S., Zhou, L. et al. (July 2008). Backfire prediction in a manifold injection hydrogen internal combustion engine. *International Journal of Hydrogen Energy* 33 14: 3847–3855.

47 MacCarley, C.A. (January 1981). Study of factors influencing thermally induced backfiring in hydrogen fueled engines, and methods for backfire control. *Proceedings of the Intersociety Conference on Energy Conversion Engineering (United States)*, vol. 2.

48 Ganesh, R.H., Subramanian, M. et al. (2008). Hydrogen fueled spark ignition engine with electronically controlled manifold injection: an experimental study. *Renewable Energy* 33: 1324–1333.

49 Kumar, G.N., Subash, G.P., Das, L.M. (2010). Implications of combustion parameters on the performance of a hydrogen-fuelled research engine. 18th World Hydrogen Energy Conference, 2010-WHEC.

50 Das, L.M. (December 2007). The potential of hydrogen as an alternative fuel for the transport and power sector in India. *International Journal of Environmental Science* 64 (6): 749–759.

51 Ikegami, M., Miwa, K., Shioji, M. et al. (1980). A study on hydrogen-fuelled diesel combustion. *Bulletin of the JSME* 23 (181): 1187–1193.

52 Ikegami, M., Miwa, K., and Shioji, M. (1982). A study of hydrogen fuelled compression ignition engines. *International Journal of Hydrogen Energy* 7: 341–353.

53 Giannacopoulos, T., Krepec, T., and Lisio, C. (1986). Preliminary investigation on a microprocessor controlled gaseous hydrogen injector. *Computers in Engineering* 3: 233–238. New York: ASME.

54 Krepec, T., Tebelis, T., and Kwok, C. (1984). Fuel control systems for hydrogen-fuelled automotive combustion engines—a prognosis. *International Journal of Hydrogen Energy* 9: 109–114.

55 Krepec, T., Giannacopoulos, T., and Miele, D. (1987). New electronically controlled hydrogen gas injector development and testing. *International Journal of Hydrogen Energy* 12: 855–861.

56 Tebelis, T. and Krepec, T. (1985). A concept of electronically controlled hydrogen-gas injector for high-speed compression ignition engine. *Proceedings on Hydrogen from Renewable Energy*, Cape Canaveral, 397–405.

57 Shahpouri, S., Gordon, D., Shahbakhti, M.K. et al. (2024). Transient NOx emission modeling of a hydrogen-diesel engine using hybrid machine learning methods. *International Journal of Engine Research*. 25(12): 2249–2266. https://doi.org/10.1177/1468087424127289.

58 Welch, A.B., Wallace, J.S. (1990). Performance characteristics of a hydrogen-fueled diesel engine with ignition assist. SAE Technical Paper No. 902070.

59 Karagöz, Y., Sandalcl, T., Yüksek, L. et al. (2016). Effect of hydrogen-diesel dual-fuel usage on performance, emissions and diesel combustion in diesel engines. *Advances in Mechanical Engineering* 8: 1–13. https://doi.org/10.1177/ 1687814016664458.

60 Teoh, Y.H., How, H.G., Le, T.D. et al. (2023). A review on production and implementation of hydrogen as a green fuel in internal combustion engines. *Fuel* 333: 126525.

61 Debnath, B.K., Saha, U.K., and Sahoo, N. (2012). Effect of hydrogen-diesel quantity variation on brake thermal efficiency of a dual fuelled diesel engine. *Journal of Power Technology* 92: 55–67.

62 Lakshmanan, T. and Nagarajan, G. (2010). Experimental investigation of timed manifold injection of acetylene in direct injection diesel engine in dual fuel mode. *Energy* 35: 3172–8. https://doi.org/10.1016/j.energy.2010.03.055.

63 Patro, T.N., Das, L.M., and Mathur, H.B. (1994). Development of a small portable hydrogen fuelled diesel genset. Hydrogen Energy Progress X: *Proceedings of the 10th World Energy Conference*, FL, USA.

64 Siva Prasad, K., Tharun, K., and Krishnababu, M. (2024). Hydrogen energy with diesel engine. *International Research Journal of Modernization in Engineering Technology and Science* 6: 11.

65 Sipos, G., Bukovácz, K., Istvánkó, K. et al. (2024). Hydrogen engine conversion aspects. *Engineering Proceedings* 79 (6): 1–6. https://doi.org/10.3390/engproc2024079006.

66 Liu, X., Srna, A., Yip, H.L. et al. (1 January 2021). Performance and emissions of hydrogen-diesel dual direct injection (H2DDI) in a single-cylinder compression-ignition engine. *International Journal of Hydrogen Energy* 46 (1): 1302–1314.

67 Furuhama, S. and Fukuma, T. (1986). High output power hydrogen engine with high pressure fuel injection, hot surface ignition and turbocharging. *International Journal of Hydrogen Energy* 11 (6): 399–407.

68 Kobayashi, Y., Furuhama, S., Iida, M. et al. (1980). LH: car with a two-stroke direct injection engine and LH 2 pump. *Hydrogen Energy Progress.*

69 Bondarenko, V.L., Ilyinskaya, D.N., Kazakova, A.A. et al. (2022). Application of liquid hydrogen in hydrogen engines and fuel cells. *Chemical and Petroleum Engineering* 57: 1033–1037. https://doi.org/10.1007/s10556-022-01042-y.

70 Wetzel, F.-J. (23–28 June 1996). Handling of liquid hydrogen at filling stations. *Proceedings of the 11th World Hydrogen Energy Conference*, 1123–1134.

71 Furuhama, S., Kobayashi, Y., and Iida, M. (1981). A LH 2 engine fuel system on board cold GH2 injection into two-stroke engine with LH2 pump. ASME Paper No. 81-HT-81.

72 Tang, X., Kabat, D.M., Natkin, R.J. et al. (2002). Ford P2000 hydrogen engine dynamometer development. SAE Technical Paper No. 2002-01-0242.

73 Stockhausen, W.F., Natkin, R.J., Kabat, D.M. et al. (2002). Ford P2000 hydrogen engine design and vehicle development program. SAE Technical Paper No. 2002-01-0240.

74 Das, L.M. (2016). Evolution of hydrogen engine technologies. In: *Combustion and Propulsion: New Perspectives* 1–30. Athena Academia.

75 Das, L.M. (1990). Fuel induction technique for a hydrogen operated engine. *International Journal of Hydrogen Energy* 15 (11): 833–842.

76 Mathur, H.B. and Das, L.M. (1991). Performance characteristics of hydrogen-fuelled S.I. engine using timed manifold injection. *International Journal of Hydrogen Energy* 16 (2): 115–127.

77 Das, L.M. (2002). Hydrogen engine: research and development (R&D) programmes in Indian Institute of Technology (IIT), Delhi. *International Journal of Hydrogen Energy* 27: 953–965.

78 Das, L.M., Gulati, R., and Gupta, P.K. (2000). A comparative evaluation of the performance characteristics of spark ignition engine using hydrogen & compressed natural gas as alternative fuels. *International Journal of Hydrogen Energy* 25 (8).

79 Mathur, H.B. and Das, L.M. (1987). Development of hydrogen injection system for total hydrogen operation of spark ignition engine. 8th Miami International Conference on Alternative Energy Sources, Florida, USA.

80 Das, L.M., Gulati, R., and Gupta, P.K. (2000). Performance characteristics of a hydrogen-fuelled spark ignition engine using electronically controlled solenoid-actuated injection system. *International Journal of Hydrogen Energy* 25 (6).

81 Das, L.M. (2016). Chapter 7: Hydrogen-fueled internal combustion engines. *Compendium of Hydrogen Energy*, Volume 3: *Hydrogen Energy Conversion*, pp. 177–217. Elsevier.

82 Jilakara, S., Vaithianathan, J., Natarajan, S. et al. (2015). An experimental study of turbocharged hydrogen fuelled internal combustion engine. *SAE International Journal of Engines* 8 (1): 314–325. https://doi.org/10.4271/2015-26-0051.

83 Das L.M. and Mathur, R. (1993). Exhaust gas recirculation for NOx control in a multi-cylinder hydrogen-supplemented S.I. engine, *International Journal of Hydrogen Energy* 18 (12): 1013–1018.

84 Das, L.M. (1991). Exhaust emission characterization of hydrogen-operated engine system: nature of pollutants and their control techniques. *International Journal of Hydrogen Energy* 16 (11): 765–775.

85 Lanzafame, R. (1999). Water injection effects in a single cylinder CFR engine. SAE Technical Paper 1999-01-0568. https://doi.org/10.4271/1999-01-0568.

86 Subramanian, V., Mallikarjuna, J.M., and Ramesh, A. (2007). Effect of water injection and spark timing on the nitric oxide emission and combustion parameters of a hydrogen fuelled spark ignition engine. *International Journal of Hydrogen Energy* 32: 1159–1173.

87 Boretti A. (2011). Stoichiometric H2ICEs with water injection. *International Journal of Hydrogen Energy* 36: 4469–4473.

88 Heffel, J.W. (2003). NOx emission reduction in a hydrogen fueled internal combustion engine at 3000 rpm using exhaust gas recirculation. *International Journal of Hydrogen Energy* 28: 1285–1292.

89 Heffel, J.W. 2003. NOx Emission and performance data for a hydrogen fueled internal combustion engine at 1500 rpm using exhaust gas recirculation. *International Journal of Hydrogen Energy* 28 (8): 901–908.

90 Yao, H., Sun, B., Tian, H. et al. (2014). A study of hydrogen internal combustion engine EGR system. SAE Technical Paper 2014-01-1071. https://doi.org/10.4271/2014-01-1071.

91 Chaichan, M.T. (March 2016). EGR effects on hydrogen engines performance and emissions. *International Journal of Scientific & Engineering Research* 7 (3): 80–90. ISSN 2229-5518IJSER © 2016 http://www.ijser.org.

92 Ceper, B.A. (2012). Experimental investigation of the effect of spark plug gap on a hydrogen fueled SI engine. *International Journal of Hydrogen Energy* 37: 17310.

93 Jayakrishnan, K. (2018). Development of a hydrogen fuelled multi-cylinder vehicular engine with H2-SCR after treatment system. PhD thesis. Indian Institute of Technology, Delhi, India.

94 Krishnanunni, J., Bhankar, H., Chugh, D. et al. (2020). Strategies for stable operation of a H2–ICE at high equivalence ratios with use of unburnt and externally injected hydrogen for reducing NOx emissions. *International Journal of Hydrogen Energy* 45: 22125–22137.

95 Koebel, M., Elsener, M., and Kleemann, M. (2000). Urea-SCR: a promising technique to reduce NOx emissions from automotive diesel engines. *Catalysis Today* 59: 335–345.

96 Al-Baghdadi, M.A.R.S. (2004). Effect of compression ratio, equivalence ratio and engine speed on the performance and emission characteristics of a spark ignition engine using hydrogen as a fuel. *Renewable Energy* 29: 2245–2260.

97 Subramanian, V., Mallikarjuna, J.M., and Ramesh, A. (2007). Intake charge dilution effects on control of nitric oxide emission in a hydrogen fueled SI engine. *International Journal of Hydrogen Energy* 32: 2043–2056.

98 Navale, S.J., Kulkarni, R.R., Thipse, S.S. (2017). An experimental study on performance, emission and combustion parameters of hydrogen fuelled spark ignition engine with the timed manifold injection system. *International Journal of Hydrogen Energy* 42: 8299–8309.

99 Das, L.M. (2010). Implications of combustion parameters on the performance of a hydrogen-fuelled research engine. 18th World Hydrogen Energy Conference 2010-WHEC.

100 Thawko, A., Persy, S.A., Eyal, A. et al. (2020). Effects of fuel injection method on energy efficiency and combustion characteristics of SI engine fed with a hydrogen-rich reformate. SAE Technical Papers (2020), 1–12. http://dx.doi.org/10.4271/2020-01-2082.

101 Murray, R.G. and Schoeppel, R.J. (1971). Emission and performance characteristics of an air-breathing hydrogen-fuelled internal combustion engine. *Proceedings of the 6th Inter-society Energy Conversion Engineering Conference* (IECEC), Paper No. 719009.

102 Murray, R.G., Schoeppel, R.J., and Gray, C.I. (1972). The hydrogen engine in perspective. *Proceedings of the 7th IECEC*. San Diego, California, Paper 729216.

103 Schoeppel, R.J. (1971). Design criteria for hydrogen burning engines. Final Report, Environmental Protection Agency Contract No. EHS 70–103.

104 Murray, R.G. and Schoeppel, R.J. (01 February 1970). A reliable solution to the environmental problem: the hydrogen engine. Technical Paper 700608. ISSN 0148-7191, e-ISSN 2688 3627. https://doi.org/10.4271/700608.

105 Suzuki, K., Uchiyama, Y., and Hama, J. (1980). Research of hydrogen fueled spark ignition engine. *Proceedings of the 3rd World Hydrogen Energy Conference*, Tokyo, Japan.

106 Varde, K.S. and Frame, G.A. (1982). A study of combustion and engine performance using electronic hydrogen fuel injection. Paper presented at the World Hydrogen Energy Conference IV, Pasadena, California.

107 MacCarley, C.A. (1978). Electronic fuel injection techniques for hydrogen fuelled I.C. engines. MS thesis in engineering. University of California, Los Angeles.

108 Homan, H.S., De Boer, P.C.T., and McLean, W.J. (1983). The effect of fuel injection on NOx emissions and undesirable combustion for hydrogen-fuelled piston engines. *International Journal of Hydrogen Energy* 8: 131–146.

109 De Boer, P.C.T., McLean, W.J., and Homan, H.S. (1976). Performance and emissions of hydrogen-fueled internal combustion engines. *International Journal of Hydrogen Energy* 1: 153–172.

110 Li, F.B., Wang, Z., Wang, Y.F. et al. (2019). High-efficiency and clean combustion natural gas engines for vehicles. *Automotive Innovation* 2 (10): 284–304.

111 Pan, K. and Wallace, J. (2022). Soot and combustion models for direct-injection natural gas engines. *International Journal of Engine Research* 23 (1): 150–166.

112 Muhssen, H.S., Masuri, S.U., Sahari, B.B. et al. (2021). Design improvement of compressed natural gas (CNG)-air mixer for diesel dual-fuel engines using computational fluid dynamics. *Energy* 216: 118957.

113 Kim, S., Park, C., Jang, H. et al. (2021). Effect of boosting on a performance and emissions in a port fuel injection natural gas engine with variable intake and exhaust valve timing. *Energy Reports* 7: 4941–4950.

114 Li, M.H., Liu, G.F., Liu, X.R. et al. (2019). Performance of a direct-injection natural gas engine with multiple injection strategies. *Energy* 189: 116363.

115 Li, M.H., Wu, H.M., Liu, X.R. et al. (2021). Numerical investigations on pilot ignited high pressure direct injection natural gas engines: a review. *Renewable and Sustainable Energy Reviews* 150: 111390.

116 Huang, S., Xiong, S.T., and Zhao, Z.G. (2020). Discussion about the Cr-Mo steel materials using in CNG cylinders. *Materials Science Forum* 1001: 235–239.

117 Kim, Y.S., Park, K.S., and Kim, T.O. (2012). Suggestion for safety improvement of compressed natural gas vehicle. *Journal of the Korean Institute of Gas* 16 (4): 1–7.

118 Wei, Z.N., Li, M.H., Li, S. et al. (2021). Development of natural gas chemical kinetic mechanisms and application in engines: a review. *ACS Omega* 6 (37): 23643–23653.

119 Pan, K. and Wallace, J.S. (2019). A low temperature natural gas reaction mechanism for compression ignition engine application. *Combustion and Flame* 202: 334–346.

120 Han, Y. and Young, P. (2020). Natural gas engine model for speed and air-fuel control. *International Journal of Modelling, Identification and Control* 36 (2): 104–115.

121 Thipse, S.S., Dsouza, A., Sonawane, S.B. et al. (2017). Development of multi cylinder turbocharged natural gas engine for heavy duty application. *SAE International Journal of Engines* 10 (1): 27–38.

122 Zeng, Q. (2014). Study on the modification and performance of compressed natural gas engine. *Applied Mechanics and Materials* 568/570: 1690–1693.

123 Alanen, J., Simonen, P., Saarikoski, S. et al. (2017). Comparison of primary and secondary particle formation from natural gas engine exhaust and of their volatility characteristics. *Atmospheric Chemistry and Physics* 17 (14): 8739–8755.

124 Lehtoranta, K., Murtonen, T., Vesala, H. et al. (2017). Natural gas engine emission reduction by catalysts. *Emission Control Science and Technology* 3 (2): 142–152.

125 Zheng, J., Zhou, R., Zhan, R. et al. (2022). Combustion and emission characteristics of natural gas engine with partial catalytic oxidation of the fuel. *Fuel* 312: 122796.

126 Welch, A., Mumford, D., Munshi, S. et al. (2008). Challenges in developing hydrogen direct injection technology for internal combustion engines. SAE Technical Paper No. 2008-01-2379.

127 Wimmer, A., Wallner, T., Ringler, J. et al. (2005). H2-direct injection—a highly promising combustion concept. SAE Technical Paper No. 2005-01-0108.

128 Takagi, Y., Mori, H., Mihara, Y. et al. (2017). Improvement of thermal efficiency and reduction of NOx emissions by burning a controlled jet plume in high-pressure direct-injection hydrogen engines. *International Journal of Hydrogen Energy* 42: 26114–26122.

129 Tanno, S., Ito, Y., Michikawauchi, R. et al. (2010). High-efficiency and low-NOx hydrogen combustion by high pressure direct injection. *SAE International Journal of Engines* 3: 259–268.

130 Kawamura, A., Sato, Y., Naganuma, K. et al. (2010). Development project of a multi-cylinder DISI hydrogen ICE system for heavy duty vehicles. SAE Technical Paper No. 2010-01-2175.

131 Kawamura, A., Yanai, T., Sato, Y. et al. (2009). Summary and progress of the hydrogen ICE truck development project. *SAE International Journal of Commecial Vehicles* 2: 110–117.

132 Naganuma, K., Honda, T., Yamane, K. et al. (2010). Efficiency and emissions-optimized operating strategy of a high-pressure direct injection hydrogen engine for heavy-duty trucks. *SAE International Journal of Engines* 2: 132–140.

133 Takagi, Y., Oikawa, M., Sato, R. et al. (2019). Near-zero emissions with high thermal efficiency realized by optimizing jet plume location relative to combustion chamber wall, jet geometry and injection timing in a direct-injection hydrogen engine. *International Journal of Hydrogen Energy* 44: 9456–9465.

134 Roy, M.K., Kawahara, N., Tomita, E. et al. (2011). High-pressure hydrogen jet and combustion characteristics in a direct-injection hydrogen engine. *SAE International Journal of Fuels and Lubricants* 5: 1414–1425.

135 Roy, M.K., Kawahara, N., Tomita, E. et al. (2013). Jet-guided combustion characteristics and local fuel concentration measurements in a hydrogen direct-injection spark-ignition engine. *Proceedings of the Combustion Institute* 34: 2977–2984.

136 Verhelst, S., Demuynck, J., Sierens, R. et al. (2013). *Update on the Progress of Hydrogen-Fueled Internal Combustion Engines*, pp. 381–400. Amsterdam: Elsevier.

137 Homan, H.S., Reynolds, R.K., De Boer, P.C.T. et al. (1979). Hydrogen-fueled diesel engine without timed ignition. *International Journal of Hydrogen Energy* 4: 315–325.

138 Welch, A.B. and Wallace, J.S. (1990). Performance characteristics of a hydrogen-fueled diesel engine with ignition assist. SAE Paper 902070. SAE International, Warrendale, PA, USA.

139 Wei, L. and Geng, P. (2016). A review on natural gas/diesel dual fuel combustion, emissions and performance. *Fuel Processing Technology* 142: 264–278.

140 Saravanan, N., Nagarajan, G., Sanjay, G. et al. (2008). Combustion analysis on a DI diesel engine with hydrogen in dual fuel mode. *Fuel* 87: 3591–3599.

141 Naber, J.D. and Siebers, D.L. (1998). Hydrogen combustion under diesel engine conditions. *International Journal of Hydrogen Energy* 23: 363–371.

142 Putrasari, Y., Praptijanto, A., Nur, A. et al. (2018). Thermal efficiency and emission characteristics of a diesel-hydrogen dual fuel CI engine at various loads condition. *Journal of Mechatronics, Electrical Power, and Vehicular Technology* 9: 49. https://doi.org/10.14203/j.mev.2018.v9.49-56.

143 Koten, H. (2018). Hydrogen effects on the diesel engine performance and emissions. *International Journal of Hydrogen Energy* 43: 10511–10519. https://doi.org/10.1016/j. ijhydene.2018.04.146.

144 Hoang, A.T. and Pham, V.V. (2020). A study on a solution to reduce emissions by using hydrogen as an alternative fuel for a diesel engine integrated exhaust gas recirculation. *AIP Conference Proceedings* 2235: 20035. AIP Publishing.

145 Qin, Z., Yang, Z., Jia, C. et al. (2020). Experimental study on combustion characteristics of diesel–hydrogen dual-fuel engine. *Journal of Thermal Analysis and Calorimetry* 1–9. https://doi.org/10.1007/s10973-019-09147-y.

146 Kumar, R.S., Loganathan, M., and Gunasekaran, E.J. (2015). Performance, emission and combustion characteristics of CI engine fuelled with diesel and hydrogen. *Frontiers in Energy* 9: 486–494. https://doi.org/10.1007/s11708-015-0368-4.

147 Nguyen, V.N., Ganesan, N., et al. (2024). Eco-friendly perspective of hydrogen fuel addition to diesel engine: an inclusive review of low-temperature combustion concepts, *International Journal of Hydrogen Energy*. In press.

148 Hamdan, M.O., Selim, M.Y.E., Al-Omari, S.A. et al. (2015). Hydrogen supplement co-combustion with diesel in compression ignition engine. *Renewable Energy* 82: 54–60. https://doi.org/10.1016/j.renene.2014.08.019.

149 Zhou, J.H., Cheung, C.S., and Leung, C.W. (2014). Combustion, performance, regulated and unregulated emissions of a diesel engine with hydrogen addition. *Applied Energy* 126: 1–12. https://doi.org/10.1016/j.apenergy.2014.03.08.

150 Monemian, E. and Cairns, A. (2020). Hydrogen fumigation on HD diesel engine: an experimental and numerical study. *Diesel and Gasoline Engines*. IntechOpen. https://doi.org/10.5772/intechopen.8942.

151 Sandalcı, T. and Karagoz, Y. (2014). Experimental investigation of the combustion characteristics, emissions and performance hydrogen port fuel injection in a diesel engine. *International Journal of Hydrogen Energy* 39 (32): 18480–18489.

152 Gomes Antunes, J.M., Mikalsen, R., and Roskilly, A.P. (2009). An experimental study of a direct injection compression ignition hydrogen engine. *International Journal of Hydrogen Energy* 34: 6516–6522.

153 De Morais, A.M., Mendes Justino, M.A., Valente, O.S. et al. (2013). Hydrogen impacts on performance and CO2 emissions from a diesel power generator. *International Journal of Hydrogen Energy* 38: 6857–6864.

154 Deb, M., Sastry, G.R.K., Bose, P.K. et al. (2015). An experimental study on combustion, performance and emission analysis of a single cylinder, 4-stroke DI-diesel engine using hydrogen in dual fuel mode of operation. *International Journal of Hydrogen Energy* 40 (27): 8586–8598.

155 Kose, H. and Ciniviz, M. (2013). An experimental investigation of effect on diesel engine performance and exhaust emissions of addition at dual fuel mode of hydrogen. *Fuel Processing Technology* 114: 26–34.

156 Karim Ghazi, A. (May 2003). Hydrogen as a spark ignition engine fuel. *International Journal of Hydrogen Energy* 28 (5): 569–577.

157 Ono, R., Nifuku, M., Fujiwara, S. et al. (2007). Minimum ignition energy of hydrogen-air mixture: effects of humidity and spark duration. *Journal of Electrostatics* 65: 87–93.

158 Gandhi, R.D. (2015). Use of hydrogen in internal combustion engine. *International Journal of Engineering and Management Science* (IJEMS) 3: 207–216.

159 Ciniviz, M. and Kose, H. (2012). Hydrogen use in internal combustion engine. *International Journal of Automotive Engineering and Technologies* 18: 1–15.

160 Korn, T. (April 2020). The most efficient way for CO2 reduction: the new generation of hydrogen internal combustion engines. *Proceedings of the 41st International Vienna Motor Symposium*, Vienna, Austria, 22–24.

161 Bradley, D.M., Lawes, K., Liu, S. et al. (2006). Laminar burning velocities of lean hydrogen–air mixtures at pressures up to 1.0 MPa. *Combustion and Flame* 149: 162–172.

162 Luo, Q. and Sun, B. (2016). Inducing factors and frequency of combustion knock in hydrogen internal combustion engines. *International Journal of Hydrogen Energy* 41: 16296–16305.

163 Verhelst, S. and Sierens, R. (2001). Aspects concerning the optimization of a hydrogen-fueled engine. *International Journal of Hydrogen Energy* 26: 981–985.

164 Negurescu, N., Pana, C., and Cernat, A. (2012). Aspects of using hydrogen in SI engine. *University Politehnica Bucharest Science Bulletin Series D* 74: 11–20.

165 Abdelghaffar,W.A. (2010). Spark ignition engine fueled by hydrogen: comparative analysis. *European Journal of Scientific Research and Reviews* 44: 13–28.

166 Gopal, G., Rao, P.S., Gopalakrishnan, K.V. et al. (1982). Use of hydrogen in dual fuel engine. *International Journal of Hydrogen Energy* 7: 267–272.

167 Karim, G.A. and Watson, H.C. (1971). Experimental and computational considerations of the compression ignition of hydrogenous fuel oxidant mixtures. SAE Technical Paper No. 710133.

168 Karim, G.A. (1976). Hydrogen as a fuel in compression ignition engines. *Archiwum Termodynamiki i Spalania* 7 (1): 89–100.

169 Karim, G.A. (1980). A review of combustion processes in the dual fuel engine — the gas diesel engine. *Progress in Energy and Combustion Science* 6: 277–285.

170 Weigang, W. and Lainfang, Z. (1986). The research on internal combustion engine with the mixed fuel of diesel and hydrogen. *Papers Presented at the International Symposium on Hydrogen Systems*, 7–11 May 1985, Beijing, China, 83–94.

171 Kumar, N. (2015). Hydrogen use in internal combustion engine. *International Journal of Advanced Culture Technology* 2: 87–99.

172 Kudryash, A.P., Marakhovsky, V.P., and Kaidalov, A.A. (1988). Peculiarities of hydrogen-fueled diesel engine performance. *Proceedings of the 7th World Hydrogen Energy Conference, Moscow*. Oxford: Pergamon Press.

173 Taib, N.M., Mansor, M.R.A., and Mahmood, W.M.F.W. (2021). Combustion characteristics of hydrogen in a noble gas compression ignition engine. *Energy Reports* 7: 200–218.

174 Tanabe, H., Ohnishi, M., Fuimoto, H. et al. (1982). Experimental study of the transient hydrogen jet using a fast response probe. *International Journal of Hydrogen Energy* 7 (12): 967–976.

175 Szwaja, S. and Grab-Rogalinski, K. (2009). Hydrogen combustion in a compression ignition diesel engine. *International Journal of Hydrogen Energy* 34: 4413–4421.

176 Szwaja, S., Bhandary, K., and Naber, J.D. (2007). Comparisons of hydrogen and gasoline combustion knock in a spark ignition engine. *International Journal of Hydrogen Energy* 32 (18): 5076–5087.

177 Naber, J.D. and Szwaja, S. (2007). Statistical approach to characterize combustion knock in the hydrogen fuelled SI engine. *Journal of KONES Powertrain Transport* 14 (3): 443–450.

178 Lilik, G.K., Zhang, H., Herreros, J.M. et al. (2010). Hydrogen assisted diesel combustion. *International Journal of Hydrogen Energy* 35: 4382–4398.

179 Liew, C., Li, H., Nuszkowski, J. et al. (2010). An experimental investigation of the combustion process of a heavy-duty diesel engine enriched with H2. *International Journal of Hydrogen Energy* 35: 11357–11365.

180 Roy, M.M., Tomita, E., Kawahara, N. et al. (2011). Comparison of performance and emissions of a supercharged dual-fuel engine fueled by hydrogen and hydrogen-containing gaseous fuels. *International Journal of Hydrogen Energy* 36 (12): 7339–7352.

181 Miyamoto, T., Hasegawa, H., Mikami, M. et al. (2011). Effect of hydrogen addition to intake gas on combustion and exhaust emission characteristics of a diesel engine. *International Journal of Hydrogen Energy* 36 (20): 13138–13149.

182 Tomita, E., Kawahara, N., Piao, Z. et al. (2001). Hydrogen combustion and exhaust emissions ignited with diesel oil in a dual fuel engine. SAE Technical Paper 2011-01-3503: 1–8.

183 Saravanan, N. and Nagarajan, G. (June 2008). An experimental investigation on a diesel engine with hydrogen fuel injection in intake manifold. *SAE International Journal of Fuels and Lubricants* 23 (724): 776–790.

184 Bayramoğlu, K. and Yılmaz, S. (2021). Emission and performance estimation in hydrogen injection strategies on diesel engines. *International Journal of Hydrogen Energy* 46: 29732–29744.

185 Nguyen, T.A. and Mikami, M. (2013). Effect of hydrogen addition to intake air on combustion noise from a diesel engine. *International Journal of Hydrogen Energy* 38 (10): 4153–4162.

186 Singh Yadav, V., Soni, S.L., and Sharma, D. (2012). Performance and emission studies of direct injection C.I. engine in dual fuel mode (hydrogen-diesel) with EGR. *International Journal of Hydrogen Energy* 37: 3807–3817.

187 Wu, H.W. and Wu, Z.Y. (2012). Investigation on combustion characteristics and emissions of diesel/hydrogen mixtures by using energy-share method in a diesel engine. *Applied Thermal Engineering* 42: 154–162.

188 Selim, M.Y. (2003). Effect of exhaust gas recirculation on some combustion characteristics of dual fuel engine. *Energy Conversion and Management* 44: 707–721.

189 Bose, P.K. and Maji, D. (2009). An experimental investigation on engine performance and emissions of a single cylinder diesel engine using hydrogen as inducted fuel and diesel as injected fuel with exhaust gas recirculation. *International Journal of Hydrogen Energy* 34 (11): 4847–4854.

190 Masood, M., Mehdi, S.N., and Ram Reddy P. (2007). Experimental investigations on a hydrogen-diesel dual fuel engine at different compression ratios. *Journal of Engineering for Gas Turbines and Power* 129 (2): 572–578.

191 Chaichan, M.T. and Al-Zubaidi, D.S.M. (November 2014). A practical study of using hydrogen in dual-fuel compression ignition engine IPASJ. *International Journal of Mechanical Engineering* (IIJME) 2 (11): 1–9.

192 Suzuki, Y., Tsujimura, T., and Mita T. (2015). The performance of multicylinder hydrogen/diesel dual fuel engine. *SAE International Journal of Engines* 8 (5). 2015-2024-2458.

193 Rameez P.V. and Mohamed, I.M. (2024) Maximizing hydrogen utilization in CI engines: an investigation of dual fuel and RCCI combustion approaches under low and mid-load conditions in a medium duty automotive engine. *Energy Conversion and Management* 321: 119100.

194 Guo, H., Hosseini, V., Neill, W.S. et al. (2011). An experimental study on the effect of hydrogen enrichment on diesel fueled HCCI combustion. *International Journal of Hydrogen Energy* 36 (21): 13820-13830.

195 Park, H., Kim, J., and Bae, C. (2015). Effects of hydrogen ratio and EGR on combustion and emissions in a hydrogen/diesel dual-fuel PCCI engine. SAE Technical Paper 2015-1815. https://doi.org/10.4271/2015-01-1815.

196 Lambe, S.M. and Watson, H.C. (1992). Low polluting, energy efficient C.I. hydrogen engine. *International Journal of Hydrogen Energy* 17 (7): 513–525.

197 Prabhukumar, G.P., Swaminathan, S., Nagalingam, B. et al. (1987). Water induction studies in a hydrogen diesel dual-fuel engine. *International Journal of Hydrogen Energy* 12 (3): 177–186.

198 Adnan, R., Masjuki, H.H., and Mahlia, T.M.I. (2012). Performance and emission analysis of hydrogen fueled compression ignition engine with variable water injection timing. *Energy* 43 (1): 416–426.

199 Westbrook, C.K., Adamczyk, A.A., and Lavoie, G.A. (1981). A number study of laminar wall quenching. *Combustion and Flame* 40: 81–91.

200 Chintala, V. and Subramanian, K.A. (2014). Hydrogen energy share improvement along with NOx (oxides of nitrogen) emission reduction in a hydrogen dual-fuel compression ignition engine using water injection. *Energy Conversion and Management* 83: 249–259.

201 Chintala, V. and Subramanian, K.A. (2015). An effort to enhance hydrogen energy share in a compression ignition engine under dual-fuel mode using low temperature combustion strategies. *Applied Energy* 146: 174–183.

202 Mathur, H.B., Das, L.M., and Patro, T.N. (1992). Hydrogen fuel utilization in CI engine powered end utility systems. *International Journal of Hydrogen Energy* 17 (5): 369–374.

203 Mathur, H.B., Das, L.M., and Patro, T.N. (1992). Effects of charge diluents on the emission characteristics of a hydrogen fueled diesel engine. *International Journal of Hydrogen Energy* 17 (8): 635–642.

204 Mathur, H.B., Das, L.M., and Patro, T.N. (1993). Hydrogen-fuelled diesel engine: performance improvement through charge dilution techniques. *International Journal of Hydrogen Energy* 18 (5): 421–431.

205 Ge, H., Bakir, A.H., and Zhao, P. (2023). Knock mitigation and power enhancement of hydrogen spark-ignition engine through ammonia blending. *Machines* 11: 651. https://doi.org/10.3390/machines11060651.

206 Furuhama, S. and Kobayashi, Y. (1982). Development of a hot-surface-ignition hydrogen injection two-stroke engine. *Proceedings of the 4th World Hydrogen Energy Conference* 3: 100. Pasadena, California.

207 Furuhama, S. (1983). State of the art and future trends in hydrogen-fuelled engine. *International Journal of Vehicle Design* 4: (4): 359–385.

208 Furuhama, S. and Kobayashi, Y. (1982) A liquid hydrogen car with a two-stroke direct injection engine and LH2-pump. *International Journal of Hydrogen Energy* 7 (10): 809–820.

209 Furuhama, S., Hiruma, M. and Enomoto, Y. (1978). Development of a liquid hydrogen car. *International Journal of Hydrogen Energy* 3: 61.

210 Furuhama, S. and Azuma, H. (1978). Hydrogen injection two stroke spark ignition engine. *Proceedings of the 2nd World Hydrogen Energy Conference* 4: 1851.

211 Furuhama, S., Kobayashi, Y. and Iida, M. (1981). LH Sub 2 engine fuel system on board-cold GH Sub 2 injection into two stroke engine with LH Sub2 Pump. Corpus ID: 108052481.

212 Takiguchi, M., Furuhama, S., Suzuki, T. et al. (1987). Combustion improvement of liquid hydrogen fueled engine for medium-duty trucks. SAE Technical Paper 870535. https://doi.org/10.4271/870535.

213 Fukuma, T., Fujita, T., Pichainarong, P. et al. (1986). Hydrogen combustion study in direct injection hot surface ignition engine. SAE Technical Paper No. 861579.

214 Fukuma, T. and Furuhama, S. (1987). Development of 4-stroke hydrogen injection hot surface ignition engines for vehicles. *JSAE Transaction* 34: 94. https://doi.org/10.4271/870535.1981.

215 Furuhama, S., Yamane, K., and Yamaguchi, I. (1977). Combustion improvement in a hydrogen fueled engine. *International Journal of Hydrogen Energy* 2 (3): 329–340.

216 Furuhama, S. and Azuma, H. (1978). Hydrogen injection two-stroke spark ignition engine. *Proceedings of the 2nd World Higher Education Conference* 4: 1851.

217 Koyanagi, K., Hiruma, M., and Furuhama, S. (1994). Study on occurring mechanism of backfire in hydrogen engines. *Proceedings of the 72nd JSME Annual Meeting* 3: 341.

218 Koyanagi, K., Hiruma, M., Yamane, K. et al. (1993). Effect of hydrogen jet on mixture formation in a high-pressure injection hydrogen fueled engine with spark ignition. SAE Technical Paper No. 931811.

219 Kobayashi, H., Matsushita, T., and Furuhama, S. (1989). Studies on ignition methods for hydrogen engines with injection right before TDC. *Transactions of the Japan Society of Mechanical Engineers B* 55: 895–903.

220 Furuhama, S., Koyanagi, K., and Tomisawa, N. (1991). The power system of a computer controlled hydrogen car—GH2 injection and spark ignition engine with LH2 tank and pump. *IMechE, C* 430/028: 179.

221 Furuhama, S., Fukuma, T., and Kashima, T. (1984). LH2 fuel supply system to hot surface ignition turbocharged engine. *Cryogenic Processes and Equipment*, ASME Book No. G00282: 105.

222 Furuhama, S. and Kobayashi, K. (1984). Development of a hot-surface-ignition hydrogen injection two-stroke engine. *International Journal of Hydrogen Energy* 9 (3): 205–213.

223 Furuhama, S., Fukuma, T., and Kobayashi, Y. (1982). Combustion improvement in hot surface ignition hydrogen diesel engine. *Proceedings of the ASME-JSME Thermal Engineering Conference*, 261.

224 Furuhama, S. (1 January 1997). Problems of forecasting the future of advanced engines and engine characteristics of the hydrogen injection with LH2 tank and pump. *ASME Journal of Engineering for Gas Turbines and Power* 119 (1): 227–242. https://doi.org/10.1115/1.2815554.

225 Furuhama, S. and Kobayashi, Y. (1982). Hydrogen cars with LH2 tank, LH2-pump and cold GH2-injection two-stroke engine. SAE Technical Paper No. 820349.

226 Furuhama, S. (1977). Performance and exhaust emission of the LH2 hydrogen engine. *JSME Preprint*, No. 770–779: 61.

227 Furuhama, S. (1975). Drived the whole distance by LH2 car in the SEED rally. *Internal Combustion Engine* 14–17: 7373.

228 Kobayashi, Y., Furuhama, S., Iida, M. et al. (1980). LH2 car with a two-stroke direct injection engine and LH 2 pump. *Hydrogen Energy Progress*.

229 Stewart, W.F. (1980). Liquid hydrogen as an automotive fuel. In: *Advances in Cryogenic Engineering* (ed. K.D. Timmerhaus and H.A. Snyder). Boston: Springer. https://doi.org/10.1007/978-1-4613-9856-1_92.

230 Peschka, W. and Carpetis, C. (1976). A system consideration of the cryogenic storage tank for liquid-hydrogen fueled vehicles and the resulting tank concept for a passenger car. In: *Proceedings of the 1st World Hydrogen Energy Conference* (ed. T.N. Veziroglu), II: 28–43 to 28–64. Coral Gables, Florida: University of Miami.

231 Peschka, W. (1980). Operating characteristics of a LH2-fueled car and of a semiautomatic LH2 refueling station. In: *Hydrogen Energy Progress* (eds. T.N. Veziroglu, K. Fueki, and T. Ohta) 3: 1561–1576. New York: Pergamon Press.

232 Peschka, W. and Carpetis, C. (1980). Cryogenic hydrogen storage and refueling for automobiles. *International Journal of Hydrogen Energy* 5: 619–625.

233 Peschka, W., Edeskuty, F.J., and Stewart, W.F. (1980). Liquid hydrogen storage and refueling for automotive applications. *Proceedings of the 3rd Miami International Conference on Alternative Energy Sources*.

234 Peschka, W. (2–5 August 1981). Liquid hydrogen for automative vehicles — experimental results. ASME paper 81-HT-83. 20th Joint ASME/AIChE National Heat Transfer Conference, Milwaukee, Wisconsin.

235 Stewart, W.F., Edeskuty, F.J., Williamson, K.D. et al. (1975). Operating experience with a liquid-hydrogen fueled vehicle. In: *Advances in Cryogenic Engineering* (ed. K.D. Timmerhaus), 20: 82–89. New York: Plenum Press.

236 Peschka, W. (1984). Liquid hydrogen as a vehicular fuel: a challenge for cryogenic engineering. *International Journal of Hydrogen Energy* 9 (6): 515–523.

237 Stewart W.F. (1984). Operating experience with a liquid–hydrogen fuelled Buick and refueling system. *International Journal of Hydrogen Energy* 9 (6): 525–538.

238 Das, L.M. (1991). Safety aspects of hydrogen-fuelled engine system development. *International Journal of Hydrogen Energy* 16 (9): 619–624.

4

Undesirable Combustion Phenomena

We are made wise not by the recollection of our past, but by the responsibility for our future.
George Bernard Shaw

4.1 Introduction

Hydrogen, while being a valuable and clean energy source, possesses some typical combustion characteristics that can make it potentially dangerous unless judiciously handled. Proper understanding of hydrogen combustion is crucial for its safe utilization in engines. The impact of initial conditions on macroscopic parameters such as explosion limit, explosion pressure, flame propagation speeds, etc., needs to be considered from a safety and explosion point of view. The reaction kinetics of hydrogen combustion identifies first, second, and third explosion limits. Proper understanding of the lower explosive limit (LEL) and upper explosive limit (UEL) is useful in understanding the mechanistic aspects of hydrogen explosion limits. Hydrogen presents a greater explosion risk compared to ammonia, methane, and propane because of its greater range of explosion limits. A delicate balance between hydrogen and oxygen concentrations provides guidelines to maintain concentrations within safe limits.

Causes of undesirable combustion phenomena such as preignition, backfire, and knocking and their control techniques are extremely important for practical system development. Knowledge of the influence of injection pressure, injection duration, injection angle, ignition timing, cyclic pressure variation, etc. is relevant to ensure smooth combustion.

4.2 Hydrogen–Oxygen Reaction Mechanism

The hydrogen–oxygen reaction mechanism, which broadly signifies hydrogen combustion, is essentially a fundamental chemical process that involves a rapid and exothermic reaction between hydrogen (H_2) and oxygen (O_2) molecules. It has been extensively studied and applied to vital areas including land transport in vehicles, rocket application, electricity generation, and several industrial sectors.

Hydrogen and oxygen do not enter into direct reaction in the absence of a catalyst. At room temperature and standard pressure, the kinetic energy of hydrogen and oxygen molecules is not sufficient to overcome this relatively high activation energy. A catalyst facilitates the breaking of

Hydrogen Engines: Design, Performance Evaluation, Combustion Analysis, and Exhaust Emissions, First Edition.
Lalit Mohan Das.

the strong bonds in the hydrogen and oxygen molecules thereby reducing the energy required for reaction to occur. Hydrogen and oxygen molecules become activated when exposed to light through photochemical processes such as photo dissociation, leading to the formation of reactive oxygen atoms and radicals. In the context of photochemistry, sensitizers enhance sensitivity of photochemical reactions and make them more responsive to the effects of light. In a catalytic reaction, called a branching chain reaction, each cycle of the chain results in a net increase in the number of chain carriers, which are the radicals or atoms. A typical example of a hydrogen–oxygen branching chain explosion is as follows [1]:

$H_2 + O_2 \rightarrow 2OH$ (Initiating)

$H + O_2 \rightarrow OH + O$ (Branching)

$O + H_2 \rightarrow OH + H$ (Branching)

$OH + H_2 \rightarrow H_2O + H$ (Propagating)

(Branched chain)

$H + H + M \rightarrow H_2 + M$ (Terminating)

Figure 4.1 shows a typical explosion-limit diagram for a chain branching reaction of a hydrogen–oxygen system [2]. The figure provides significant information about the character of different types of preignition phenomena and under what circumstances preignition can occur. As evident in the graph, preignition can occur during the intake stroke specified by the second explosion limit of 50 mm of Hg. This type of preignition usually occurs due to heat from hot particles or the wall. During preignition at the intake stroke, the temperature level remains equal to or greater than the second explosion limit corresponding to a particular pressure. In fact, to ensure conditions leading to preignition, charge must remain at the temperature level for a minimum period specified by the induction time.

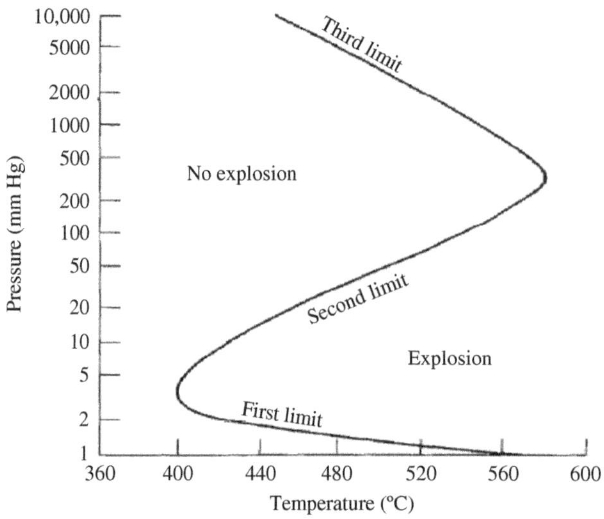

Figure 4.1 Explosion limits of a hydrogen–oxygen system [2].

4.3 Flammability Range and Explosion Limit

Abnormal combustion in hydrogen engines can be classified into three categories. These manifest their symptoms as abnormally high pressure rise, preignition in the combustion chamber leading to subsequent backfire into the intake manifold, and occasional backfire in a very lean hydrogen–air mixture or during idling conditions [3]. Figure 4.2 shows the flammability limit of a mixture of several fuels. This concentration range of flammable gas or vapor is also referred to as the explosive part of the cloud. The term cloud is used to convey that these flammable gases or vapors are dispersed and spread out, forming a cloud-like concentration rather than being concentrated at a particular location.

Hydrogen has a wide flammability range of 4 to 75% by volume. Hydrogen–oxygen systems' flammability characteristics are different in different zones as shown in Figure 4.3. Mixtures below 4% and above 75% are either too lean or too rich to be flammable.

Fire and explosion are two different outcomes of combustion. Fire occurs when fuel oxidizes, releasing heat and light. It occurs at a relatively slower rate, its intensity and size are visible, and it usually emits smoke. Explosion, instead, occurs due to a very rapid combustion process resulting in a sudden and violent instantaneous release of large amounts of energy in a confined space.

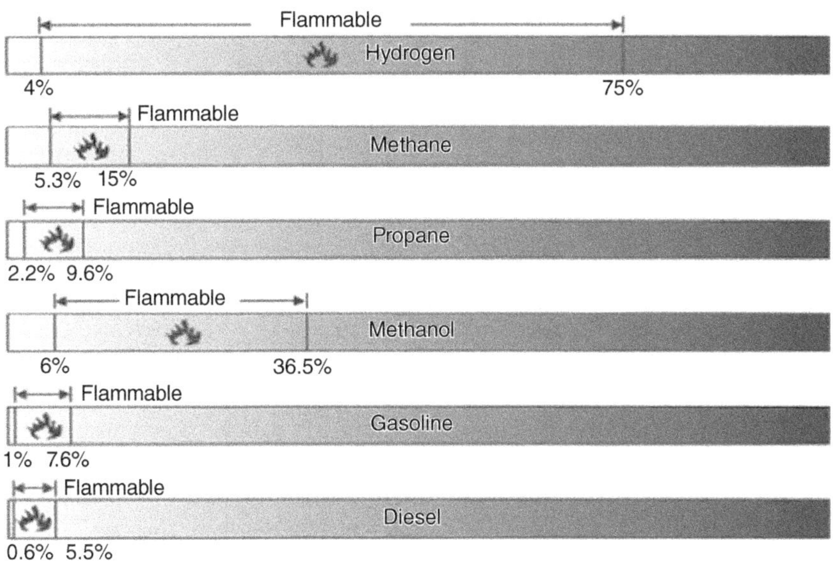

Figure 4.2 Flammability range showing the lower flammability limit (LFL) and upper flammability limit (UFL) of different fuels.

Figure 4.3 Too lean and too rich nonexplosive mixtures of hydrogen.

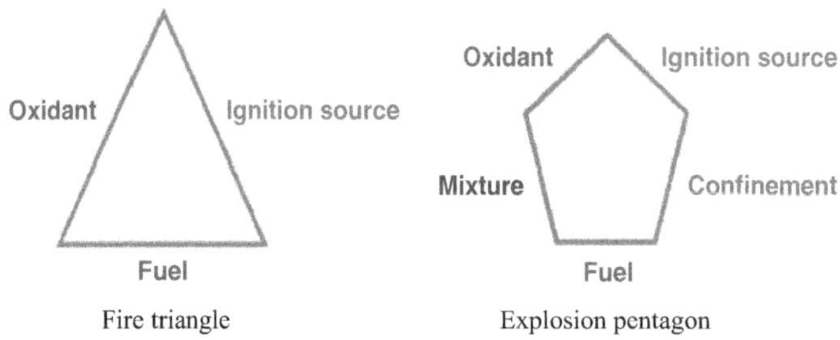

Figure 4.4 Fire triangle and the explosion pentagon.

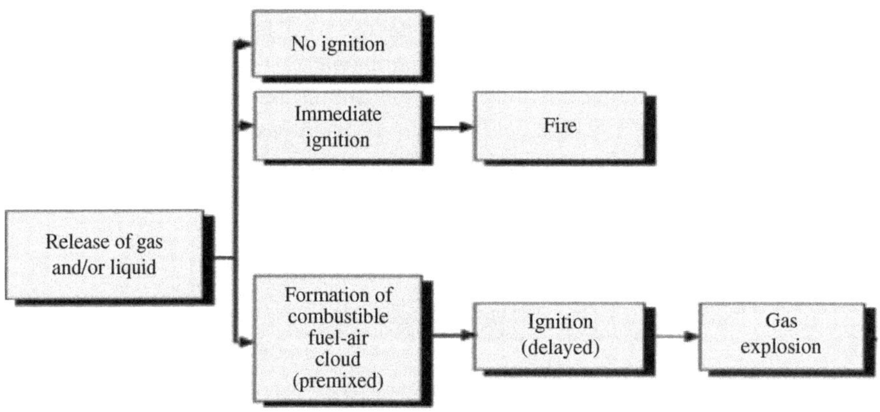

Figure 4.5 Possible sequence of events leading to fire or explosion.

Rapid expansion of gases generates a shockwave that travels outward from the explosion site and can cause immense damage. Explosions are associated with a number of factors such as the concentration of the fuel–air mixture, an ignition source, some sort of confinement such as an enclosed space, and a chemical reaction that releases energy extremely rapidly. Figure 4.4 shows the well-known fire triangle and explosion pentagon, indicating parameters responsible for these phenomena. The combustion process, represented by the fire triangle, shows three factors needed to cause combustion: ignition, oxidant, and fuel. The explosion pentagon completes the basic fire triangle concept and includes two additional factors such as mixture (of fuel and oxygen) and confinement of the mixture. Confinement of the combustible mixture, whether partially or fully enclosed, leads to development of overpressure that causes a rapid transition from a fast-burning flame to an explosion. Figure 4.5 shows the possible sequence of events that could lead to fire or explosion when a combustible gas is released.

The flammability limits of hydrogen vary with temperature due to changes in reaction kinetics, molecular mobility, and hydrogen's volatility. With increasing temperatures, hydrogen molecules move faster. This increases their collision frequency and accelerates rapid reactions. Figure 4.6 shows the effect of temperature on flammability limits.

Besides temperature, other factors such as pressure, impurities, and confinement also have an influence on the flammability limit. Figure 4.7 shows the effect of pressure on hydrogen's flammability limit.

During the reaction between hydrogen and oxygen, many intermediates like radicals, ions, and several reactive species are formed. Combustion reactions involve a series of chain reactions. Reactive intermediates initiate a series of reactions by generating other reactive species. Chain

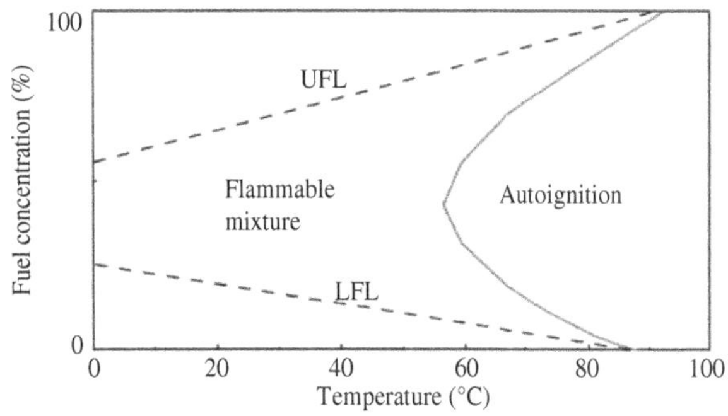

Figure 4.6 Effect of temperature on the flammability limits of hydrogen.

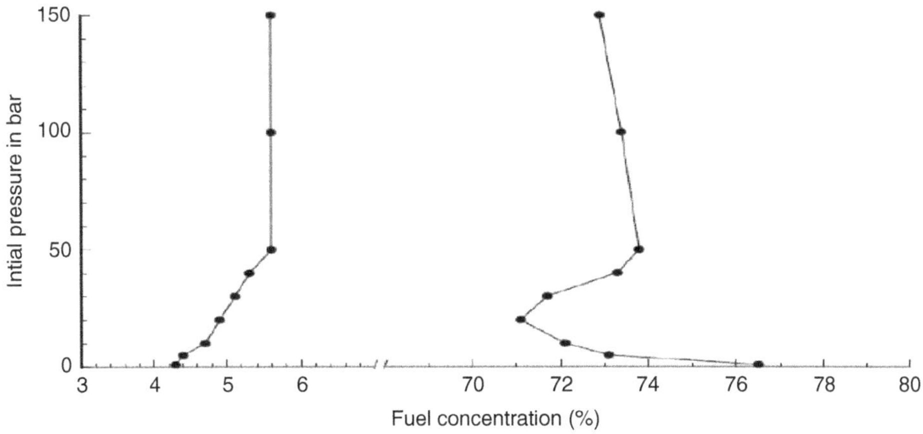

Figure 4.7 Effect of pressure on the flammability limits of hydrogen.

reactions also create additional pathways and species. Composition of the product depends on mixture, temperature, and pressure.

4.4 Flame Propagation

Flame propagation and acceleration mechanisms are fundamental aspects of hydrogen combustion. Several factors such as low molecular weight, high diffusivity, wide flammability limits, high flame speed, lean mixture, and high reactivity influence the characteristics of flame propagation. Besides rapid flame propagation, these factors contribute to fast combustion and flame stability, reducing the likelihood of flame quenching and improving combustion consistency. The high reactivity of hydrogen leads to a thin reaction zone where combustion reactions occur rapidly, and it influences the flame thickness and heat release rate. In hydrogen combustion, turbulence creates stretching of the flame due to velocity gradients, and these stretched zones enhance the combustion rate. Sometimes, the flame instability due to turbulence creates wrinkles in the flame surfaces. These wrinkles increase the surface area and allow for a greater heat release rate leading to an improved combustion process. In combustion theory and modeling, the flamelet regime is significant due to its thin, localized flame structures representing the localized combustion within the overall reaction zone. Analysis of the complex combustion process becomes simpler by modeling it as a collection of thin laminar flamelets, each propagating at a local stretched laminar flame speed.

4.5 Laminar Burning Velocity

Laminar burning velocity refers to the speed at which a laminar (smooth and nonturbulent) flame front advances through an air–fuel mixture under ideal conditions. Laminar burning velocity data provide important information for optimizing fuel formulations and designing combustion chambers. Turbulence in the engine results from high-pressure variations and fluid motion. Thus, turbulent burning velocity takes into account complex and turbulent flow patterns within the combustion chamber. Even though a higher turbulent burning velocity leads to better mixing of fuel and air, interaction between turbulence and the flame front is intricate and could affect combustion stability and flame structure. In the context of an internal combustion engine, achieving an optimal balance between laminar and turbulent burning velocities is essential for combustion strategies, fuel injection patterns, and combustion chamber designs. Implications of the laminar burning velocity on hydrogen–air mixtures have been reported since the 19th century.

An excellent survey of some very important combustion-related data is provided by Drell and Belles [4], such as burning velocity, flame temperature, quenching distance, flame stability explosion limit, and spontaneous ignition. Data on calculated adiabatic flame temperature are provided over a wide range of temperatures (0 to 1400 K) and pressure (0.01 to 100 atm). Most of the tests have been done to validate the reaction mechanism, and test data have been generated to ensure the boundary condition of safety. Several investigators have measured or computed the laminar burning velocity of hydrogen–air mixtures using several systems and techniques such as laser-Doppler anemometry, phase Doppler interferometer (PDI), schlieren photography, particle-tracking velocimetry (PTV), etc. on engine combustion and emission analysis [5–9]. Laminar burning velocities and their dependence on flame stretch have been determined. These velocities influence combustion duration, heat release rate, and pressure rise during the power stroke. Key features of flame propagation and acceleration include laminar flames, flame instabilities, turbulence-dominated flame propagation, deflagration-to-detonation transition (DDT), and detonation. Combustion in hydrogen engines involves a complex interaction between laminar and turbulent velocities, both crucial to determining combustion efficiency.

There are limited data on hydrogen's laminar burning velocity under engine conditions. Although most studies do not involve engines, they still provide valuable insights into laminar burning velocity, flammability limits, the effects of hydrocarbon substitution on hydrogen–air flames, and burning velocities in a closed vessel, laminar burning velocities of hydrogen–air mixtures from closed vessel gas explosions, etc. This information is crucial for analyzing engine combustion. Edmondson et al. [10] made efforts to measure the burning velocity of hydrogen air flames. Laminar conical flame stabilization was attained in a large, water-cooled nozzle burner. The "cone half angle" formed by the diverging sides of the flame in a combustion process was measured by the schlieren optical technique, and particle-tracking techniques were applied to measure the unburned gas velocity. By stabilizing high-burning-velocity laminar flames on such burners, burning velocity can be accurately measured. Andrews and Bradley [11] used the double kernel method, conducted experiments, and measured the burning velocities of methane–air mixtures, but they faced challenges while measuring higher burning velocities with hydrogen–air mixtures. In their study, the burning velocity of hydrogen–air mixtures was also determined using flame speed and thickness measurements. A modified procedure was developed to correct for the effects of finite flame thickness, assuming a linear temperature profile through the flame front. This approach provided a fairly accurate picture on burning velocities in a hydrogen–air mixture. Liu and MacFarlane [12] conducted experiments to measure the burning velocities of hydrogen–air and hydrogen–air–steam mixtures, taking into account the temperature and composition of the unburned gases. They employed laser-Doppler anemometry and schlieren photography with a constant-velocity nozzle burner for accurate measurements. Use of a two-cyclone in-series particle generator, designed to provide an appropriate particle-seeding rate for the laser-Doppler anemometer, showed excellent results. The study also introduced new data on the burning velocity as it varies with the temperature of the unburned gas. They derived a correlation equation from the observed burning velocities, covering a range of hydrogen concentration (18–65 vol%), steam concentration (0–15 vol%), and temperature (23–250 °C) that is useful to understand the burning behavior under different conditions for hydrogen–air–steam mixtures.

Milton, in his investigations on laminar burning velocities on stoichiometric hydrogen and hydrogen–hydrocarbon gas mixtures [13], conducted measurements of laminar burning velocities for various gases, including hydrogen, acetylene, propane, and methane, using a spherical combustion bomb. His study also simulated combustion products, all at stoichiometric ratios with air. The results for pure gases were quite in agreement with experimental results from combustion bombs and reasonably matched theoretical calculations. However, when hydrogen was mixed with other gases, such as methane–hydrogen and propane–hydrogen, peculiar patterns were observed in burning velocity curves. These mixtures exhibited a peak in burning velocity shortly after flame initiation, followed by a reduction and then a second increase in burning velocity. Over the entire range of values, burning velocity was lower than predicted by proportional averaging. Acetylene–hydrogen mixtures showed burning velocities moderately close to calculated values but still slower, indicating dominance of slower combusting molecules. No well-defined relationship between burning velocities and physical parameters like H-atom fraction or H/C ratio was evident.

The Markstein number (M_a) is defined as the ratio of flame thermal thickness to flame curvature. Mathematically, it can be expressed as:

$$M_a = \frac{L}{R_c}$$

where

L is the thermal flame thickness, and

R_c is the radius of curvature of the flame front.

The Markstein length is defined as the product of the thermal flame thickness (L) and the Markstein number (M_a). The Markstein number determines whether the flame is stabilizing or destabilizing with respect to curvature; the Markstein length gives a quantitative measure of the magnitude of this effect. In the context of premixed flames, the Markstein number plays a significant role. It characterizes the sensitivity of the burning velocity of a flame to curvature variations in the flame front. Large values of the Markstein number are not limited to fuels with large mass diffusivities, like hydrogen, relative to mass diffusivities of other major species. This observation shows significant effects of stretch on strongly turbulent flames in both hydrogen–oxygen–nitrogen and hydrogen–carbon–air mixtures. Large values of M_a do not necessarily require an exceptionally large mass diffusivity of one reactant, like hydrogen in the case of $H_2—O_2—N_2$ flames [14].

Law and Kwon [15] studied hydrogen–air flames and partial substitution with hydrocarbons (specifically methane, ethylene, and propane). Their investigation focused on both experimental and computational analyses. The study showed that substituting hydrogen with C1 to C3 hydrocarbons (methane, ethylene, and propane) resulted in reduced flame temperatures and burning velocities in hydrogen–air flames. This reduction in flame temperature holds potential benefits in lowering the level of oxides of nitrogen (NOx). Sometimes a slower burning rate promotes NOx production. Significantly, the study highlighted that propane substitution was useful in moderating the perturbations arising from variations in the diffusion rates and thermal characteristics of different species within the flame. It also smoothed hydrodynamic instabilities influencing the flame's structure and behavior. Wierzba and Kilchyk [16] conducted experiments to determine the flammability limits of various hydrogen–carbon monoxide mixtures across a broad range of compositions and initial temperatures up to 300 °C and at atmospheric pressure upward during flame propagation in stainless steel. The lean flammability limits adhered to Le Chatelier's Rule and indicated predictable behavior in response to temperature changes. However, for rich mixtures, there was a deviation from the values calculated using the same rule. To investigate potential preignition reactions and their impact on flammability, the mixtures were subjected to elevated temperatures for varying durations from 5 to 30 min before spark ignition. Waiting time duration significantly affected flammability limits, particularly at higher temperatures and in rich mixtures. Dahoe [17] studied the laminar burning velocity of hydrogen–air mixtures by initiating deflagration in a 169 ml cylindrical vessel under initial conditions of 1 bar and 293 K. The study covered a range of equivalence ratios from 0.5 to 3.0. Pressure variations over time were measured and the data were interpreted in an integral balance model. The data generated from this analysis on laminar burning velocities agreed with those reported by previous researchers and computed values from detailed kinetics models. However, for equivalence ratios above 0.75, advanced methods considering flame stretch effects did not show a significant advantage over the methodology adopted. At lower equivalence ratios, the laminar burning velocity was slightly higher than advanced methods but was deemed a conservative estimate for fire and explosion safety purposes. The study also noted pressure oscillations during deflagration explained by the velocity of the burned mixture induced by the expansion of combusting fluid layers near the vessel wall. Using both stable and unstable preferential diffusion conditions, Kwon, Tseng, and Faeth [18] examined the effects of positive flame stretch on laminar burning velocities and transitions to unstable flames. Measurements for $H_2{-}O_2{-}N_2$ mixtures at flame stretch values were done up to 7600 per second and for flame stretch values up to 900 per second for $C_3H_8{-}O_2{-}N_2$ mixtures. The laminar burning velocities demonstrated linear variation with increasing Karlovitz numbers, either decreasing or increasing under stable or unstable preferential-diffusion conditions. This resulted in Markstein numbers that varied with the fuel-equivalence ratio, but neutral preferential-diffusion conditions shifted toward the unstable side of the maximum laminar burning velocity. Such an observation shows that the relationship between diffusion rates and flame stability is quite complex.

Hongchao et al. [19] studied numerical simulations analyzing differential diffusion on the development of the hydrogen flame kernel in an engine. A spark ignition engine's performance depends on the development of flame kernels, because the small flame produced by the spark plug grows into flame kernels, which ignite the rest of the mixture in the combustion chamber. Shape, size, and location of the flame kernel are extremely crucial to engine performance. Differential diffusion is a phenomenon that occurs in small flame kernels due to interaction between the flame front and local flow field. Differential diffusion caused by large curvature of small kernels significantly affects the performance. In a lean hydrogen–air flame, where the proportion of hydrogen is lower than air, the flame tends to be thin and hot. High differential diffusion in these lean hydrogen air flames causes fluctuations in the flame front leading to "thermos-diffusive" or "diffusion-flame instability," which manifests as visible flame flickering, wrinkling, or even extinction of flames. In a diffusion flame, when the fuel and oxidizer mix, the process of mixing and reactions brings diffusion of different chemical species. Sometimes, the unequal diffusivities of different species lead to variations in their concentrations, which result in temperature fluctuation across the flame front. The character of the flame instability can be referred to as a "fuel-diffusion-dominated" flame or "oxidizer-diffusion-dominated" flame depending upon whether the fuel has higher diffusivity than the oxidizer or the oxidizer has higher diffusivity than the fuel. Such variations affect the local variation in reaction rates and so also the rate of heat release. Monitoring the flame instability to ensure optimized combustion for improved efficiency and reduced emissions is an important factor. There are several conditions of instability in a laminar flame related to hydrogen combustion. The intrinsically unstable laminar flame after ignition is affected in several ways through diffusion, thermal expansion, buoyancy, and heat loss processes. Depending upon the situation, several forms of instabilities such as Darrieus–Landau instability, Rayleigh–Taylor instability, and thermal-diffusive instabilities affect combustion. All these instabilities are relevant to the behavior and stability of flames. Darrieus–Landau instability appearing from thermal expansion at the flame front plays the primary role, whereas in the case of nonpremixed flames the main role is played by thermo-diffusive instabilities, and Darrieus–Landau instability does not have any significant effect on the flame. Darrieus–Landau instability is characteristically a hydrodynamic instability, caused due to the difference in the density of the two fluids at the interface.

In a combustion process, as the flame propagates, fluctuations in density occur at the interface between the burned and unburned gases. Such fluctuations in density and the temperature result in the formation of perturbations or wrinkles on the flame front, which can grow in size and lead to turbulent structure. This might affect the shape of the flame, the burning rate, and the combustion efficiency. There is another effect caused by the lower density of the burned gases as compared to that of the unburned gases. This is referred to as Rayleigh–Taylor instability, which occurs in flames, particularly in the presence of gravity or density gradients. This instability arises, particularly in case of an upward propagating flame, when a less-dense fluid is present beneath a fluid of higher density. Because of this gravity-driven motion, the flame front often gets disrupted, causing it to become corrugated or folded, and sometimes finger-like structures are developed at the interface between the two fluids. Thus, even though density difference is the cause for both the Darrieus–Landau instability and Rayleigh–Taylor instability, the basic difference between the two lies in the mechanism that drives these two types of instabilities. Flame instabilities can also be caused by unequal diffusivities, which arise due to the interplay between heat release and diffusion processes in the flame. As the flame front advances, heat is released through combustion reactions, creating a temperature gradient across the flame. In that case, the flame temperature has a strong influence on the flame propagation rate. Diffusion, on the other hand, tends to smooth out the temperature variations. In the case of high heat release rate and with reactants having different diffusivities, the thermal gradient becomes steep. Subsequently, the situation leads to amplification of temperature fluctuations and the onset of thermal-diffusive instabilities. These instabilities often affect the shape and structure of the flame front, leading to changes in burning rates and flame characteristics.

Three diffusivities have more relevance to combustion process. These are (i) thermal diffusivity of the unburned mixture, (ii) mass diffusivity of deficient reactants, and (iii) mass diffusivity of excess reactants. In chemical reactions, deficient reactants refer to reactants consumed in smaller quantities than those required for a stoichiometrically balanced reaction. Rates of chemical reaction depend on the relative concentrations of the reacting species. Reactant present in lower quantities vis-à-vis its stoichiometric coefficient in the balanced chemical equation is referred as "deficient reactant." Similarly, in chemical reactions, excess reactant refers to reactant present in greater quantities than is required for a stoichiometrically balanced reaction. In other words, excess reactant is reactant not entirely consumed during reaction because there is more of it than is needed to react with other reactants. In a lean flame the excess reactant is oxygen, and in a rich flame the excess reactant is fuel.

Ilbas et al. [20] investigated the laminar burning velocity of hydrogen–air and several blends of hydrogen-methane–air. In this study, laminar flame velocities of hydrogen–air and hydrogen–methane–air were measured at different ranges of equivalence ratio (0.8–3.2). The researchers observed that burning velocity increased by adding more hydrogen to the existing hydrogen–methane–air mixture. The schlieren photographic technique was adopted to measure flame speed and burning velocities at room temperature under atmospheric pressure. Flame speeds of progressive flame growth of hydrogen, methane, and several blends from the point of ignition until the end of explosion were analyzed. Figure 4.8 shows flame speeds of 100% CH_4, 70% CH_4 and 30% H_2, 30% CH_4 and 70% H_2, and 100% H_2. All these data were generated at stoichiometric combustion conditions. The tests suggest that 30% hydrogen added to 70% methane is a feasible alternative to existing systems. Figure 4.9 shows flame speeds of hydrogen, methane, and a hydrogen–methane blend (50% each) at different equivalence ratios. Figure 4.10 shows varying flame speeds for different percentages of hydrogen in methane under stoichiometric conditions. Figure 4.11 shows burning velocities for hydrogen, methane, and a hydrogen–methane blend at different equivalence ratios [20], while burning velocities for different percentages of hydrogen are shown in Figure 4.12.

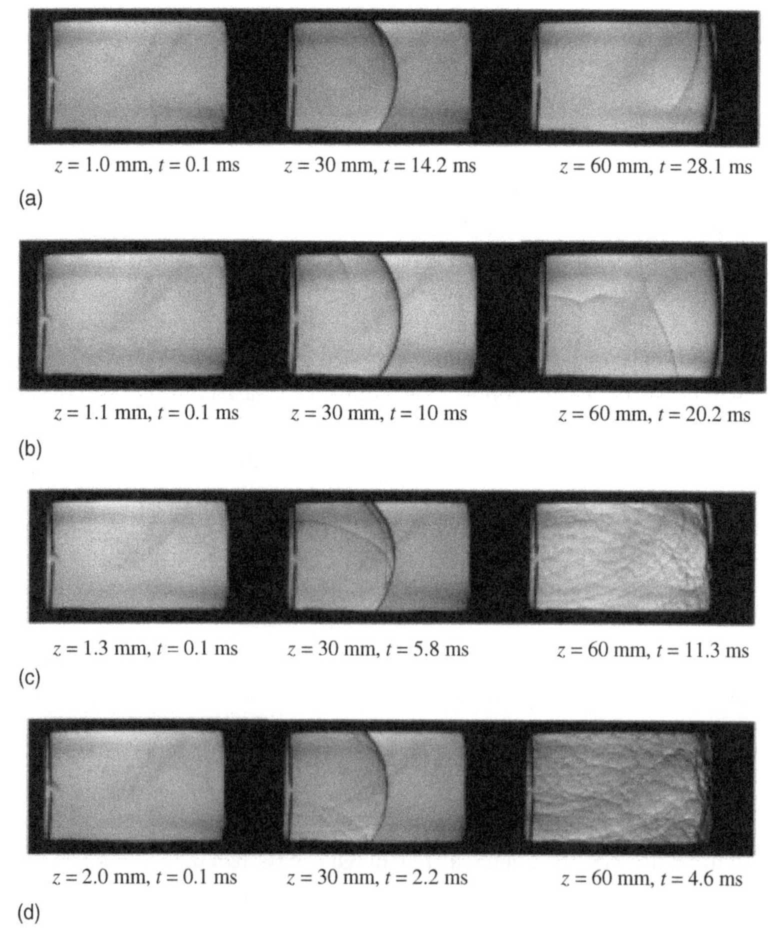

Figure 4.8 Schlieren photographs for methane, hydrogen, and blends [20] / with permission of Elsevier: (a) 100% CH_4, (b) 30% H_2 + 70% CH_4, (c) 70% H_2 + 30% CH4, and (d) 100% H_2.

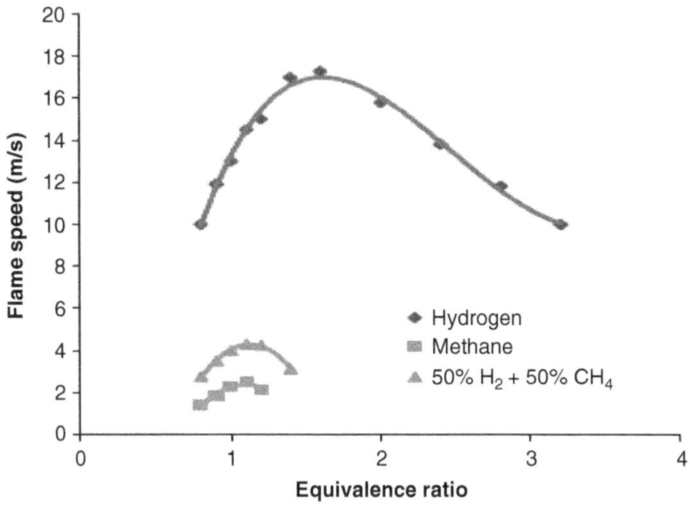

Figure 4.9 Flame speeds of hydrogen, methane, and the blend at different equivalence ratios [20].

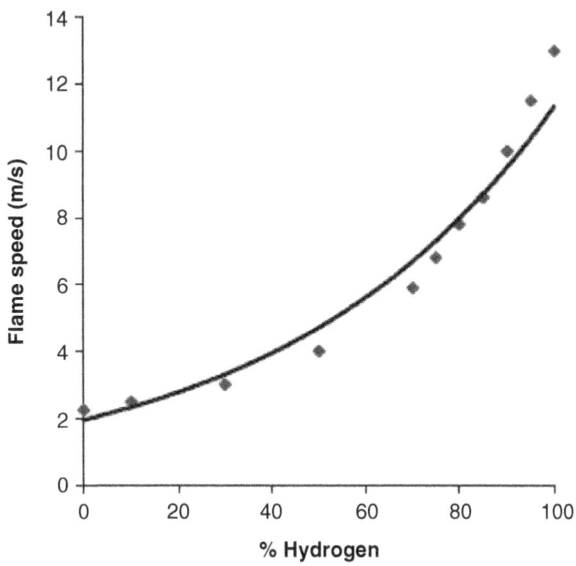

Figure 4.10 Flame speed for different percentages of hydrogen in methane under stoichiometric conditions [20].

As far as pure (100%) hydrogen is concerned, Ilbas et al. observed that maximum burning velocity is about 3.2 m/s at an equivalence ratio of about 1.8. On the other hand, the minimum burning velocity of methane is about 0.25 m/s for 100% CH_4 at an equivalence ration of 0.8. The graphs show that pure hydrogen combustion can occur over a very wide range of equivalence ratio (from 0.8 to 3.2) while pure methane combustion is restricted to a very narrow range of equivalence ratio (0.8–1.2). Thus, it is evident that increasing the hydrogen content in the mixture led to a drastic increase in burning velocity. Burning velocity of pure hydrogen combustion was 2.2 m/s, 2.6 m/s, 3.2 m/s, and 2.1 m/s for ϕ values of 0.8, 1.0, 1.8, and 3.2, respectively.

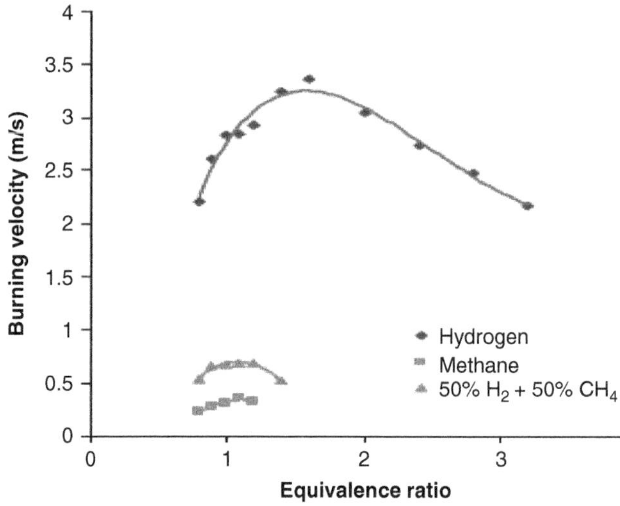

Figure 4.11 Burning velocities for hydrogen, methane, and a hydrogen–methane blend at various equivalence ratios [20] / with permission of Elsevier.

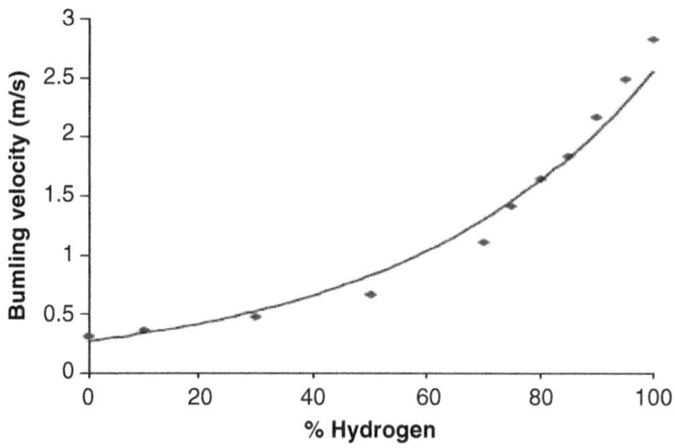

Figure 4.12 Burning velocities for different percentages of hydrogen [20] / with permission of Elsevier.

Liu et al. [21] proposed a laminar flame speed correlation of hydrogen–methanol blends for computational fluid dynamics (CFD) simulation. This analysis simultaneously considered several related parameters in various ranges such as hydrogen volume fractions (0–10%), equivalence ratios (0.6–1.5), unburned gas temperatures (400–2600 K), pressures (1–50 bar), and residual gas mass fractions (0–20%). The correlation developed was tested through an extended coherent flame model (ECFM) and found valid by CFD simulation. Moreover, calculated data of heat release rate, mean in-cylinder pressure, and various engine combustion parameters were found to be in close agreement with the experimental results.

Unlike hydrogen-enriched natural gas and hydrogen–methanol blends, ammonia has not been extensively studied primarily because of the uncertain chemistry of ammonia–hydrogen in the stringent engine environment under high pressure and high temperature. Yan et al. found that while high temperature contributes to an increased laminar flame speed (LFS) of ammonia, hydrogen and their mixtures, on the other hand, have an "inhibitory effect" [22]. The team of researchers also observed that even though hydrogen addition improved the efficiency and emission, "the combustion kinetics of ammonia–hydrogen mixtures were still dominated by ammonia with hydrogen addition levels below 60%" [23]. The study suggests that at least 20% of hydrogen needs to be added to ammonia to ensure engine operation under various operating conditions, even though this would not enable ammonia to achieve the rapid burning possible with gasoline or methane.

4.6 Preferential Diffusion and Turbulent Burning Velocity

Preferential diffusion defines the phenomenon in which different species in a mixture of gases have different diffusion rates depending upon the molecular weight, molecular size and shape, and the character of the medium through which diffusion occurs. In view of its low molecular weight and high diffusivity, hydrogen diffuses faster than other species and forms a thin reaction zone near the flame front. This preferential diffusion of hydrogen causes rapid mixing with the surrounding air. It leads to locally rich mixtures near the flame front, which result in faster reaction rates and flame propagation. Quite often, higher levels of NOx are formed because of this high temperature. In certain situations, it is possible for diffused hydrogen to ignite the flame at a location away from the ignition source. Hydrogen sometimes diffuses rapidly and forms a preheated zone ahead of the main flame. This preheated zone may be farther away from the ignition source. Both the shape and stability of the flame are governed by hydrogen's diffusivity. Wrinkling and stretching, created due to the effect of turbulence on the flame in turbulent combustion, increases the flame's surface area. As a result, more fuel and air come in contact and lead to improved combustion.

Laminar flow is a smooth flow with controlled combustion, whereas turbulent flow represents a complex flow characterized by chaotic, irregular fluid. Turbulent burning velocity depends on numerous factors, including flow conditions, geometry, and flame history. The Reynolds number indicates a transition to turbulent flow influencing burning velocity. Turbulent flows are inherently unsteady, leading to variations in flame shape, location, and propagation speed. Turbulent velocity disrupts the boundary layer, which improves mixing of fuel and oxidizer. Flame speed of turbulent flames is always higher than laminar flame speed. The geometry of the combustion chamber influences turbulence levels. The interaction of the flame with the walls can influence flame stability and propagation. Different geometries lead to variations in combustion behavior, thereby affecting the turbulent burning velocity. Since turbulence increases the surface area of the flame front by stretching and wrinkling, this improves the rate of combustion. It also stabilizes the flame front, and creates conditions of consistent combustion under varying operating conditions. At the same time, balancing turbulence levels is crucial for optimal combustion. Sometimes intense turbulence can lead to flame instabilities, such as oscillations or blow-off, and can also quench the flame. Turbulent mixing also sometimes creates pockets of unburned mixture, leading to combustion instability.

Wu et al. [24] studied the mean and fluctuating reaction progress variable and streamwise velocities for turbulent premixed hydrogen–air flames burning at relatively high Reynolds numbers up to 1800. The analysis was based on streamwise RMS (root mean square) velocity fluctuations and integral length scales. The integral length scale is often used to quantify the size of the energy containing eddies in the turbulent flow. Tests were carried out over a wide range of equivalence ratio between 0.3 and 3.6. Fully developed turbulent flow at the burner exit had a Reynolds number within 7000 to 40,000. The dominant effects of diffusive-thermal (preferential-diffusion) phenomena were evident for both stable and unstable conditions even at high Reynolds number.

Effective turbulent burning velocities for unstable conditions were twice as large as burning velocities for stable conditions.

Kido et al. [25] carried out some experiments to evaluate the turbulent burning velocity of lean hydrogen mixtures. The study put emphasis on local burning velocity. Several mixtures were prepared for tests with nearly same laminar burning velocity and with different equivalence ratios within the lean zone of $\phi = 03$ to 0.9. Measured turbulent burning velocity, at the same turbulent intensity, showed a large increasing trend as the value of ϕ decreased until it reached a value of 0.5. There was no such large increase for values of ϕ lower than 0.5. The researchers had taken a critical look at this trend and made an analysis by the estimated mean local burning velocity considering the preferential diffusion, tomograms of turbulent flames, and Markstein number. Chen and Im [26] made significant observations on preferential diffusion–unsteady stretch interactions in the turbulent premixed hydrogen–air flames. This study was based on ambient conditions using direct numerical simulations with the relevant chemical kinetics and transport phenomena. They investigated preferential diffusion–unsteady stretch interactions by varying the fuel mass equivalence ratio over a wide range, from lean to fuel-rich conditions. In the analysis they found that the rate of diffusion had a strong relationship with curvature and strain rate. This was distinct for moderate turbulence intensities and non-equidiffusive mixtures where different components do not diffuse at the same rate. Varying eddies, vortices, and fluctuations in turbulent conditions caused these situations. This study also showed that diffusive thermal effects had a significant effect on flamelets. Flamelet speed could be substantially faster for unstable mixtures and much slower for a stable mixture as compared to a laminar (nonturbulent) flame. The neutral stability point was observed close to an equivalence ratio of 0.6. This condition refers to the transition from being stable in a laminar to becoming unstable and undergoing turbulence. Near the stoichiometric condition, the correlation between the flame speed and the curvature depends upon a number of factors. Connection between flame speed and flame curvature (shape) changes depending upon the amount of heat released and the amount of fuel used. Peak heat release rate and peak fuel consumption do not always happen at the same time or at the same location within the flame. This separation or physical segregations of peak heat release and peak fuel consumption contributes to the difference in how flame speed correlates with curvature. Sometimes, tiny particles in the flame such as H atom or H_2 molecules moving around the flames make the analysis more complicated. The relationship between flame propagation and strain rate was nearly linear over a wide range of mixture strengths. In moderate turbulence, the effect of strain on the Markstein number matched with both simulations and experiments for lean fuel conditions. As turbulence increased and the rate of change in flame travel altered, the flame's response to stretching decreased, leading to a lower Markstein number. This reduced sensitivity of the flame to being stretched or compressed in turbulent conditions is supported by both experiments and theoretical analysis [26]. Investigations using direct numerical simulations on the interaction of twin premixed hydrogen–air flames showed how flame stretch affected the overall burning rate [27]. By varying the equivalence ratio from rich–rich to lean–lean interactions, it was found that preferential diffusion caused species concentration variations. Rich premixed flames intensified in regions of negative strain or curvature due to excess fuel, while lean flames were weakened or extinguished. The burning rate was influenced by surface density variation, which differs for rich and lean mixtures: rich flames exhibited rapid burning initially, whereas lean flames showed increased burning rates after an initial decrease in surface density.

Heywood and Vilchis [28] carried out tests with propane and hydrogen on a specialized spark ignition engine equipped with facilities to generate schlieren photographs of the flame. Hydrogen flame developed much faster from inception, and the cycle-to-cycle variation for hydrogen was substantially lower than that for propane. Thickness of the hydrogen turbulent flame was about one-third to one-quarter of the propane flame. The analysis using the wrinkled laminar flame model shows that hydrogen flames burn much faster and have finer wrinkles compared to propane flames. Specifically, hydrogen's burning time is one-sixth that of propane and wrinkles on hydrogen flames are 60%, the size of propane flames [28]. Melgar et al. [29] studied turbulent premixed combustion in an engine using natural gas, hydrogen, and some intermediate blends of the two. Engine speed and equivalence ratio were two crucial parameters considered in the analysis. They adopted a "two-zone diagnosis thermodynamic model combined with a geometric model using a spherical flame front hypothesis" for calculating combustion speed. An exhaustive analysis was done under 48 operating conditions. Cycle averaged turbulent combustion speed for each flame front radius was calculated by processing the data from pressure records of 200 cycles at each test point. The researchers considered two relevant factors, (i) the ratio between turbulence intensity and laminar combustion speed and (ii) the ratio between the integral spatial scale and the thickness of the laminar flame front increased by instabilities. Position of the flame front was crucial for developing expressions to evaluate the combustion speed. This study developed a unified correlation, applicable across various engine speeds, air–fuel ratios, and flame front positions, that can accurately estimate the heat release rate in an engine using quasi-dimensional predictive models [29]. Turbulent jet ignition (TJI) is ideal for hydrogen engines due to its ultra-lean operation. It reduces fuel consumption and enhances combustion efficiency. Pielecha et al. [30] conducted a numerical analysis on a heavy-duty engine at 1500 rpm and 10 bar IMEP, featuring a prechamber with seven holes. Using AVL FIRE software, they analyzed lean mixture operation with a constant global excess air coefficient. The study used coherent flame and extended Zeldovich models for combustion and NOx emission analysis. Significant data such as excess air coefficient variations, peak combustion pressures, interchamber mass flow, temperature changes, and heat release rates were evaluated. This information helped determine the necessary prechamber fuel dose for optimal TJI system efficiency and low NOx emissions.

4.7 Undesirable Combustion Phenomena in a Hydrogen Engine

Preignition, knock, and backfire are three well-documented abnormal combustion phenomena experienced in different phases of a hydrogen engine. In the intricate domain of hydrogen combustion, quite often the influence of one of these phenomena triggers another, disrupting the entire combustion process. The critical parameters responsible for these problems include laminar burning velocity, valve overlap period, valve timing, injection strategy with injection timing, injection pressure and duration, equivalence ratio, compression ratio, ignition timing, low combustion mixtures, misfires, hot spots in the cylinder, and abnormal spark plug discharges, etc. Interestingly, the distinctive properties of hydrogen such as wide flammability limit, low minimum ignition energy and high flame speed, which improve the performance and exhaust emission characteristics of the engine, are found to be the cause of a series of these problems if inappropriately used.

Preignition occurs during the compression stroke prior to spark timing at the start of combustion. On the other hand, backfire leads to the combustion of hydrogen–air charge in the intake manifold, whereas engine knock develops due to autoignition of remaining end gas with high-pressure oscillation and typical pinging noise. Preignition and knock are two such phenomena, which occur in the absence of any spark from the spark plug. In the literature, some researchers have liberally interchanged the two terms primarily because of mildly distinguishable features between the two; few others have clearly discussed the distinction between the two depending upon their "controlling phenomena" [31–34].

4.7.1 Preignition

Premature ignition or preignition, by definition, refers to the situation in which the ignition of the mixture takes place prior to the spark discharge. Premature ignition near the fuel intake valve can cause the flame to travel back into the induction system and lead to backfire. It often leads to inefficient, erratic engine operation and power loss. Usually, spark plugs and other hotspots such as the valve surface lead to such a situation chiefly due to hydrogen's low minimum ignition energy. Preignition causes greater chemical heat release rate, and the situation develops with a rapid rate of pressure rise and higher peak cylinder pressure, acoustic oscillations, and higher heat rejection, which results in higher in-cylinder surface temperature. In some cases higher cylinder temperature before ignition forms hotspots, which lead to preignition and backfire. In certain situation, the high temperature even results in engine failure, unless appropriate preventive steps are taken in time. Strategies such as delayed hydrogen injection and exhaust gas recirculation have proven effective.

Figure 4.13 shows the relationship between the equivalence ratio and the minimum ignition energy of hydrogen, methane, and heptane [35]. The common trend with all three fuels shows that minimum ignition energy takes a downward trend and decreases as equivalence ratio increases from lean to stoichiometric conditions. For hydrogen, minimum ignition energy occurs around stoichiometric conditions. Mixture strength, reactivity, and flame propagation contribute to preignition.

Research results by several investigators have clearly shown that the phenomenon of preignition is engine-specific. Preignition limits depend upon the design features of the engine, fuel used, and its operating characteristics. Several operating conditions such as compression ratio, speed, and increased temperature affect preignition. Preignition puts a practical limit on the maximum power output from a hydrogen engine. Sadiq Al-Baghdadi [31] developed a quasi-one-dimensional model for a hydrogen SI engine simulating the compression, combustion, and expansion processes of engine cycles. He studied the performance, emission, and combustion of a hydrogen engine at varying operating conditions. Figure 4.14 depicts the preignition limit of an engine at 1500 rpm with optimal spark timing for the best torque over compression ratios and equivalence ratios. The operational limit of preignition comes down with increase in compression ratio because of the temperature increase inside the engine cylinder. Figure 4.15 indicates the effect of engine speed on the operational limits of preignition. It takes into account the validity of the model for optimum spark timing for best torque and highest compression ratio (HUCR) to avoid abnormal combustion.

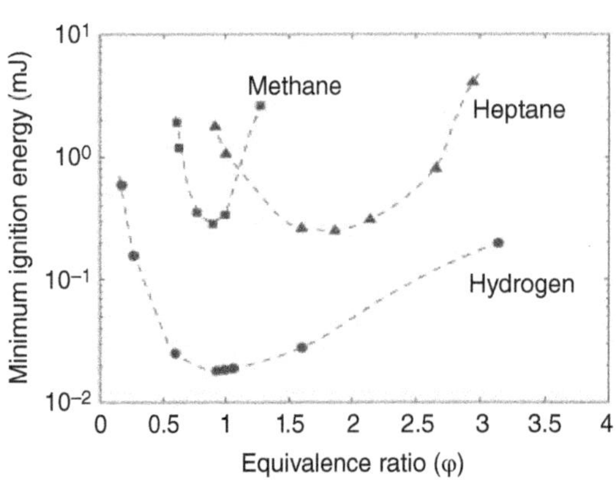

Figure 4.13 Minimum ignition energies of hydrogen–air, methane–air and heptane–air mixture with respect to the equivalence ratio at atmospheric pressure [35] / with permission of Elsevier.

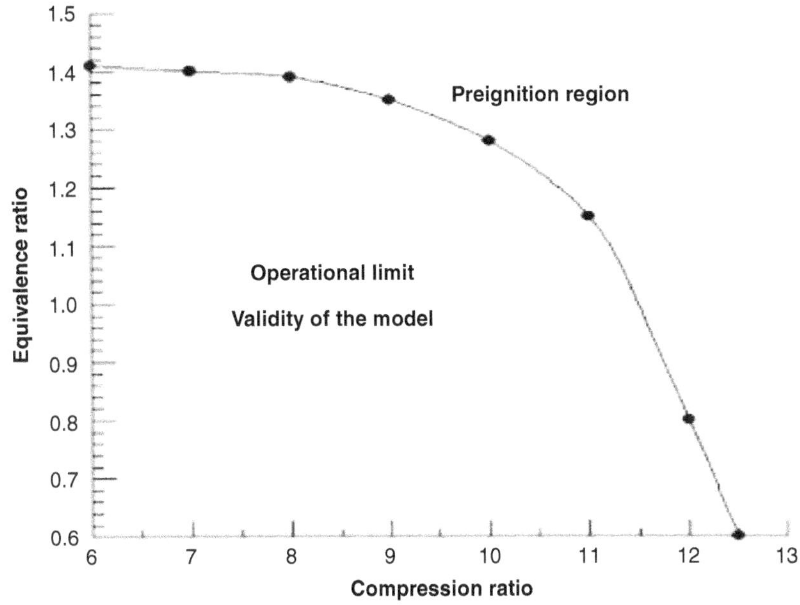

Figure 4.14 Variation of the compression ratio with equivalence ratio within operational limits for preignition [31] / with permission of Elsevier.

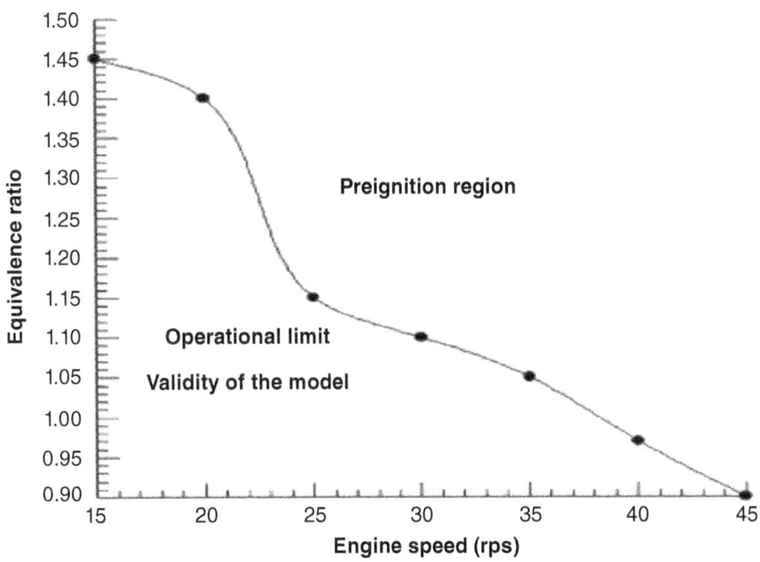

Figure 4.15 Variation of operational limits of preignition with engine speed [31] / with permission of Elsevier.

As engine speed increases, turbulence inside the cylinder intensifies, thus accelerating the flame speed. As a result, the rate of mass burning becomes faster and time to attain complete combustion is reduced. This sequence of events leads to higher peak pressure and temperature, which increases the tendency of unstable combustion. As in case of preignition, operating limits of knocking are engine-specific, and so it is necessary to know the operating conditions of the engine to analyze the knocking profile. There is another difference between experimental and analytical studies. If knocking behavior is shown on experiment data, the onset of knocking is mostly observed from the high-frequency pressure oscillations and characteristic knocking sound. On the other hand, such observations are not possible in the case of theoretical analysis with modeling.

Sometimes combustion occurring in the crevice volume between the piston and cylinder also cause preignition. It has been a complicated task to work out common and uniform control strategies to combat preignition problems. Elimination of hot spots and protrusions, use of spark plugs with narrow gap settings, use of a sodium-filled exhaust valve, an ignition system with low residual charge, well-designed crankcase ventilation, variable valve timing for effective scavenging of exhaust residuals, optimized engine cooling passages, and an appropriate fuel injection system have proven effective to avoid preignition in hydrogen engines.

4.7.2 Knock Fundamentals: Autoignition, Octane Rating, Methane Number

Engine operating parameters such as engine speed, equivalence ratio, intake temperature and intake pressure, and engine design parameters such as compression ratio, combustion chamber design, and bore-to-stroke ratio are crucial and relevant to knock. Among several parameters, temperature affects knocking the most. The intake air temperature, engine coolant temperature, residual gas fraction, specific heat ratio, and mixture humidity contribute to the temperature effect. Higher intake air temperature contributes significantly to raise the maximum temperature of the unburned gas. The temperature of the engine coolant affects the temperature of the intake port, cylinder head, cylinder wall, and piston. In fact, this plays a crucial role in heat transfer from and to the mixture. Burned gases left over from the previous cycle are always at higher temperatures and contribute to increasing the temperature of the air, fuel, and the residual gas mixture in the cylinder.

Some specific properties of hydrogen need to be carefully monitored from the point of view of knock. Hydrogen's high flame speed is conducive to rapid combustion. Higher autoignition temperature and fast combustion velocity reduce the propensity of knock. Hydrogen, with a high-octane rating in the range of 130 to 140, offers much higher resistance to knock. Poursadegh et al. [36] carried out both experimental and numerical analysis to study autoignition behavior and the octane rating of hydrogen to evaluate its knocking characteristics. After some initial RON tests, they found that hydrogen was more knock-resistant than gasoline under conditions of similar energy to the premixture. ASTM octane standards are "ill-poised" [36]. When assessing hydrogen-rich fuel blends, the methane number method is questioned because it assumes hydrogen is prone to knock. Hydrogen's high flame speeds near stoichiometric conditions cause rapid pressure rise, which can be misidentified as autoignition. This misinterpretation is possible even with modern in-cylinder pressure measurements, as done with hydrogen's RON measured as per ASTM standards. Malenshek and Olsen [37] adopted a knock rating technique using the methane number of wood gas, coal gas, reformed natural gas, digester gas, and landfill gas. They simulated the compositions with a gas blending system and also tested in an engine. Tests done three times over each gas showed significant variations in the methane number between runs. Data from one of the runs, selected arbitrarily, is shown in Figure 4.16.

Combustion chemistry closely links up the three major abnormal combustion events. Preignition often causes backfire, and backfire leads to knock. Thus knocking is characteristically a preignition phenomenon, if preignition is broadly defined as an event in which a mixture ignites prior to spark discharge. However, knock is also an autoignition phenomenon of the mixture ahead of the flame front that originates from the spark.

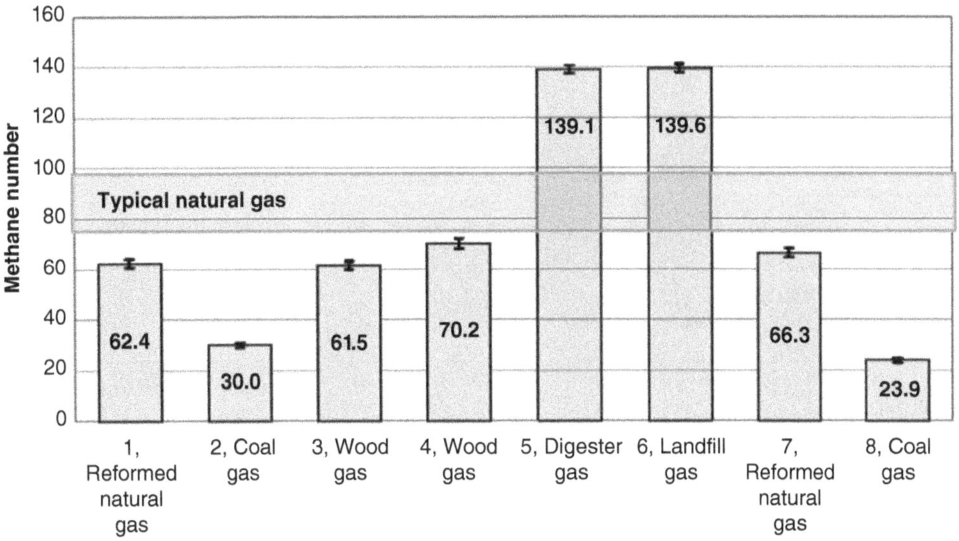

Figure 4.16 Methane number of several gaseous fuels [37] / with permission of Elsevier.

4.7.3 Parameters That Influence Knock

There have been several conflicting observations on the precise location and character of knocks. Some studies suggest that higher flame temperatures near the front trigger autoignition, while others suggest autoignition in the end gas away from the flame front. Some visual observations and simulation analyses have identified formation of autoignition kernels and rapid propagation leading to knocks. Most of these studies have detected knocks in hydrogen engines in modified gasoline engines. For a clearer picture, more specific studies need to be carried out with hydrogen. Some investigators have used optical engines for more precise identification of knocking conditions. Earlier studies focused on low-speed engines (400–900 rpm), whereas now engines with higher speeds are investigated. Kawahara and Tomita [38] used high-speed color and monochrome cameras to study knock and observed intense oscillations due to pressure waves as glaring symptoms of knocking caused by autoignition of the end gas. Occurrence of autoignition and high-pressure waves disrupts the thermal boundary layer near the cylinder wall and piston head. This leads to combustion of lubricant oil grease with a strong visible luminous flame. This autoignition and the associated pressure waves with heightened temperatures may damage the cylinder wall and piston head due to engine knocking [38]. Accurate temperature measurement is vital to identify knocking under turbulent conditions. Nakada et al. [39] used the technique of coherent anti-Stokes Raman spectroscopy (CARS) to investigate knocking in a spark ignition engine in a high range of compression ratios. The CARS technique can be useful for the study of knock in hydrogen engines because of its ability to measure temperature in combustion. By using CARS, accurate data of the unburned gas temperature at different locations in the combustion chamber can be obtained that provide a clear picture of the temperature pattern distribution of unburned gas at autoignition.

Li et al. [40] conducted a numerical analysis to study the spontaneous combustion of the end mixture along with the interaction between flame and pressure wave during knocking. A 3D geometrical model for HDI operation in a modified gasoline engine showed that knock in a hydrogen engine does not necessarily occur due to spontaneous combustion of the end mixture. Propagation speed of the pressure wave, which was much faster than that of the flame, caused a

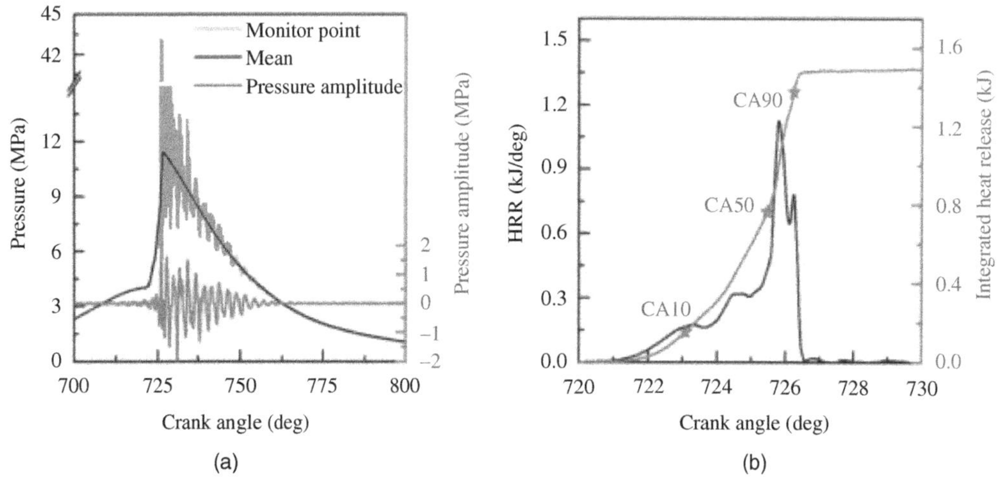

Figure 4.17 (a) Monitor points and mean pressure and (b) HRR and integrated heat release CA10, CA50, and CA90 represent, respectively, the crank angle when the 10%, 50%, and 90% of heat is released [40] / with permission of Elsevier. Please visit the companion website to view a color version of this image.

rise in temperature and pressure in the unburned zone in flame front. As a result, the reaction rate increased significantly, and the flame propagation speed accelerated. This interaction between the pressure wave and flame leads to developing knocks in the hydrogen engine. Figure 4.17a shows that under knocking conditions, local pressure underwent severe oscillations and peaks exceeding 42 MPa. This was a threat to the engine even if the mean pressure was around 12 MPa. The highest heat release rate (HRR) exceeds 1.0 kJ/deg with a narrow crank angle window from CA50 to CA90 of just 0.8° CA (Figure 4.17b). Moreover, the rise in heat release rate occurs almost at the same time with the start of pressure fluctuations. This coincidence suggests a close interaction between pressure waves and the flame during knocking.

Li and Karim have shown in their theoretical and experimental analysis that knocking may occur in a hydrogen engine in both lean and rich conditions. Hydrogen engines can exhibit knock accompanied by varying flame propagation rates over a wide range of mixture strength, from lean to above-stoichiometric [34].

Figure 4.18 shows that in the lean zone, the highest knock-free equivalence ratio takes up a decreasing trend as the compression ratio increases and the knock limit is lowered as temperature increases. A hydrogen engine exhibits knock along with varying flame propagation rates from lean mixtures to a mixture with equivalence ratios higher than stoichiometric compositions.

This analytical study shows the effect of some of the significant operating parameters such as compression ratio, intake temperature, and spark timing on knock-limiting equivalence ratios. Figures 4.19 and 4.20 show that the zone of knocking widened with increasing compression ratio while operational ranges got narrower [34].

Both compression ratio and intake temperature demonstrated a well-pronounced effect on onset of knock (Figure 4.21). The study also showed predicted data from the analysis against variation in knock-limiting equivalence ratio (KLER) with respect to changes in the compression ratio for an unthrottled engine under conditions in which the spark timing (ST) is further retarded (Figure 4.22). By retarding the spark timing, the combustion process and conditions leading to knock were delayed [34].

Identification and control of knock for smooth combustion was a topic of qualitative SI engine research. Of late, there is a surge of interest to identify parameters and quantify combustion knock

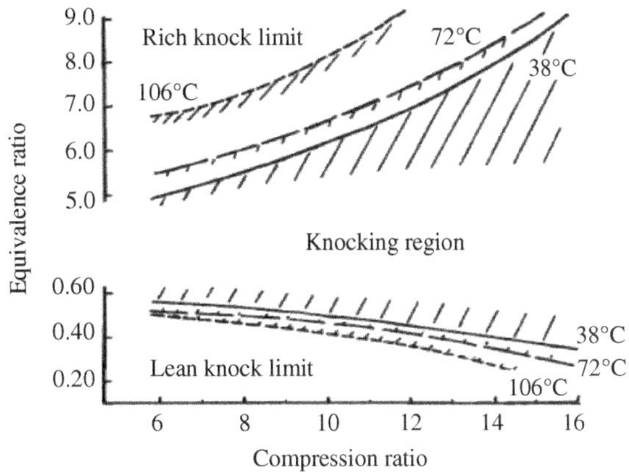

Figure 4.18 Rich and lean knock limit of unthrottled hydrogen engine at three temperatures and constant speed [34] / with permission of Elsevier.

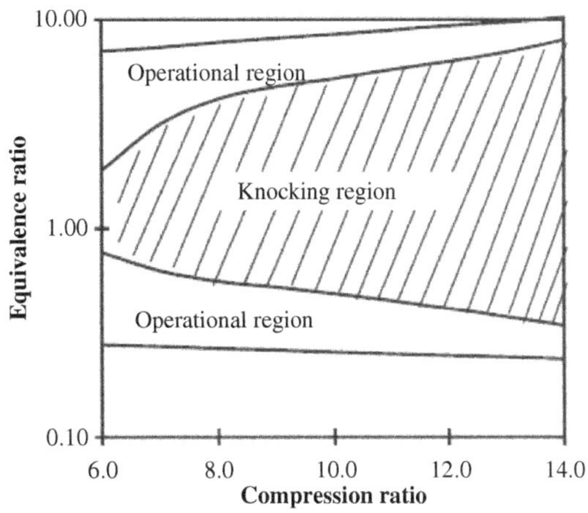

Figure 4.19 Predicted variation in the knocking and operational regions with CR in an unthrottled engine at 311 K and 900 rpm [34] / with permission of Elsevier.

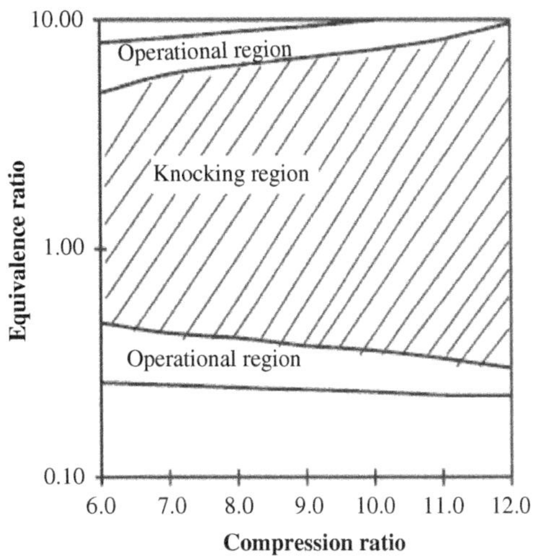

Figure 4.20 Predicted variation in the knocking and operational regions with CR in an unthrottled engine at 345 K and 900 rpm [34] / with permission of Elsevier.

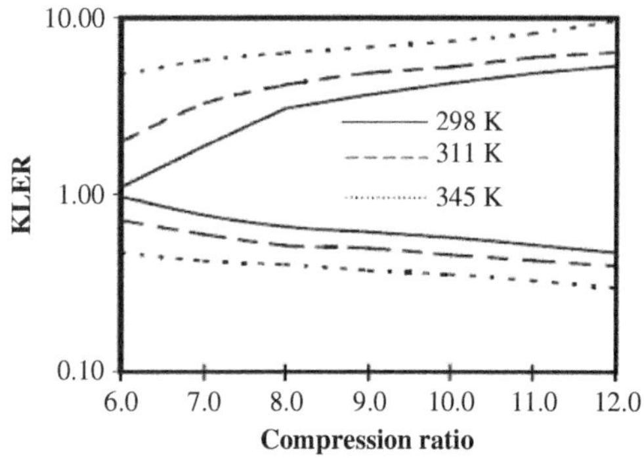

Figure 4.21 Variation of predicted KLER with compression ratio under unthrottled operation in a CFR engine at 900 rpm [34] / with permission of Elsevier.

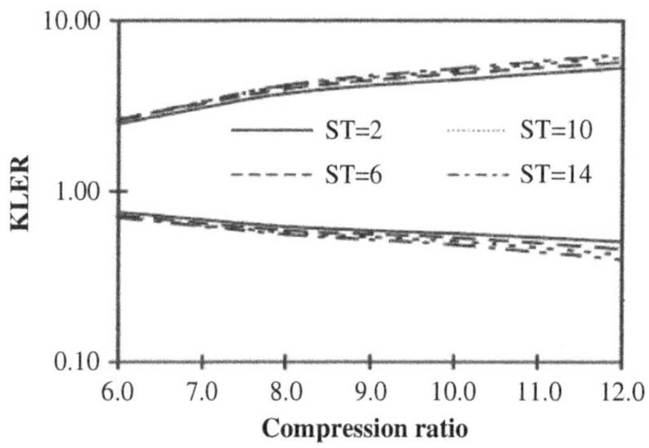

Figure 4.22 Predicted variation of KLER with CR for a range of spark timing in an unthrottled engine at 900 rpm and $T_{in} = 311$ K [34] / with permission of Elsevier.

in modern spark ignition engines [41]. Swarts et al. [42] analyzed knock intensity from the perspective of the complex dynamics of the combustion process. They observed in this study that the mass fraction burned at knock point also decreased with decreasing compression ratio, reaching as low as 30%. The relationship between metrics such as knock rating (octane), knock onset, and knock intensity still remains unclear, especially concerning how different fuels behave under varied conditions. Thus, as far as hydrogen fuel is concerned, it is important to understand octane rating, knock onset, and knock intensity in a diverse range of engine conditions.

Manzoor et al. [43] used the Reynolds averaged Navier–Stokes (RANS) framework to investigate the response of hydrogen combustion to spark timing variations. They conducted tests at a fixed compression ratio with four spark timings using a CFR engine. The commercial software CONVERGE facilitated 3D RANS simulations to predict pressure trace and autoignition timing. They found that initially, temperature stratification occurred when spark timing preceded the piston reaching TDC because early ignition caused parts of the air–fuel mixture to burn sooner than others. This led to formation of initial hot spots, which prompted autoignition and pressure wave propagation. Subsequently, such a situation induced formation of a secondary hot spot with significant pressure oscillations exceeding 100 kPa. Interestingly, it was found that a simple zero-dimensional model for end-gas autoignition could show similar trends as obtained through a more complex three-dimensional model and give a fairly correct estimate of when knocking would occur. Ford and the University of California, Riverside, collaborated in hydrogen engine development for automotive applications under the P2000 hydrogen-fueled internal combustion engine (H2ICE) vehicle program. The project put emphasis on investigating the problems of preignition and backfire with controlled combustion [44]. Dynamometer testing was carried out on hydrogen engines in both institutions with varying compression ratios (10, 12.5, 14.5, and 15.3:1). Three operation strategies were evaluated: constant fuel–air ratio, unthrottled induction, and a combination of both; and crucial mapping data were generated. Ford introduced the H2V, a hybrid electric vehicle (HEV) integrating a supercharged hydrogen IC engine [45] using knowledge from the P2000 hydrogen engine project [46]. Ford's Scientific Research Lab carried out research and development programs to create safe systems with improved combustion strategies. Natkin et al. [47] experimented on a Ford 6.8 L hydrogen IC engine for the E-450 shuttle van, demonstrating smooth knock-free operation at an equivalence ratio of 0.55 over a speed range of 1000 to 4000 rpm.

A study by Stephen et al. [48] showed the relationship of torque with spark timing for a hydrogen engine operating at 1500 rpm in various ranges of equivalence ratios (Figure 4.23). It showed zones of knocking and backfiring. Pressure traces of a heavy knocking cycle and a relatively lighter one are shown in Figure 4.24, referenced against a motored cycle.

Diéguez et al. [49] used in-cylinder pressure, block engine vibration, and acoustic measurements to detect knocking in a four-cylinder port fuel injection hydrogen engine over a speed range

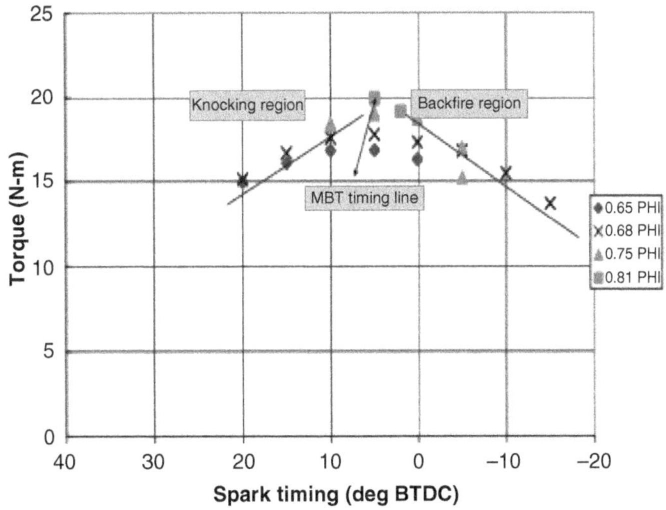

Figure 4.23 Knock and backfire regions as functions of spark timing and equivalence ratio at 1500 rpm [48] / The American Society of Mechanical Engineers.

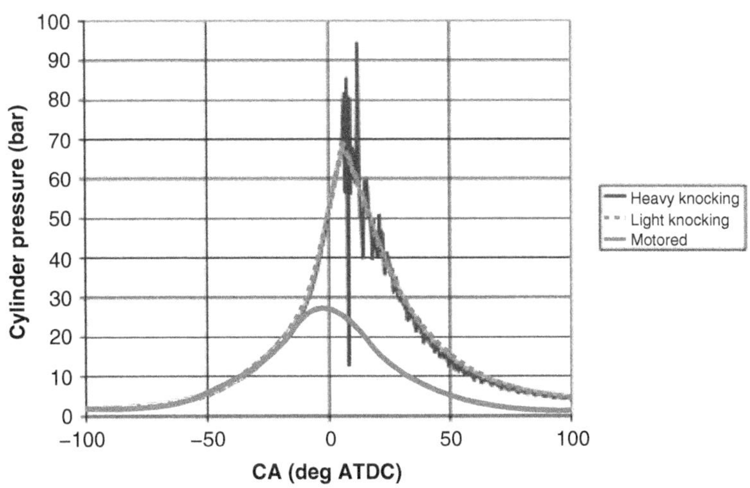

Figure 4.24 Cylinder pressure trace with knock, along with motored pressure trace at 3000 rpm [48] / The American Society of Mechanical Engineers.

of 1000 to 5000 rpm. By analyzing high-frequency components in in-cylinder and engine block acceleration signals using Fourier transform analysis, they developed a sensitive method to detect knocking cycles. Ambient noise was measured with a microphone, and spectral analysis with third octave bands characterized the noise. Intense octave bands found at higher frequencies (8–20 kHz) for knock and between 250 Hz and 4 kHz for backfire. This shows that knock generates distinctive frequency components observable in ambient noise.

Brecq et al. [50] took into account the effect of equivalence ratio, spark advance, volumetric efficiency, and width of the computational window on maximum amplitude of pressure oscillations (MAPO) and integral of modulus of pressure oscillations (IMPO) and evolved a dimensionless knock indicator (DKI):

$$\mathrm{DKI} = \mathrm{IMPO}/(\mathrm{MAPO}\ \times\ \mathrm{W})$$

where MAPO is maximum amplitude of pressure oscillations (bar), which occurs due to knock. IMPO stands for integral of modulus of pressure oscillations representing energy contained in the high frequency oscillations of the cylinder pressure signal due to knock. W is for width of computational window (CA). In this analysis, the pressure signal was processed in a band-pass filter (4–20 kHz) and then rectified in a rectifier to give IMPO and MAPO (Figure 4.25). Both IMPO and MAPO were calculated on a 600° CA window starting with spark timing. MAPO and IMPO values were obtained in each cycle and average values of 400 consecutive cycles were considered for analysis ϕ. Figure 4.26 shows DKI as a function of spark advance at an equivalence ratio of 0.7. Data for different volumetric efficiency are indicated in the graph.

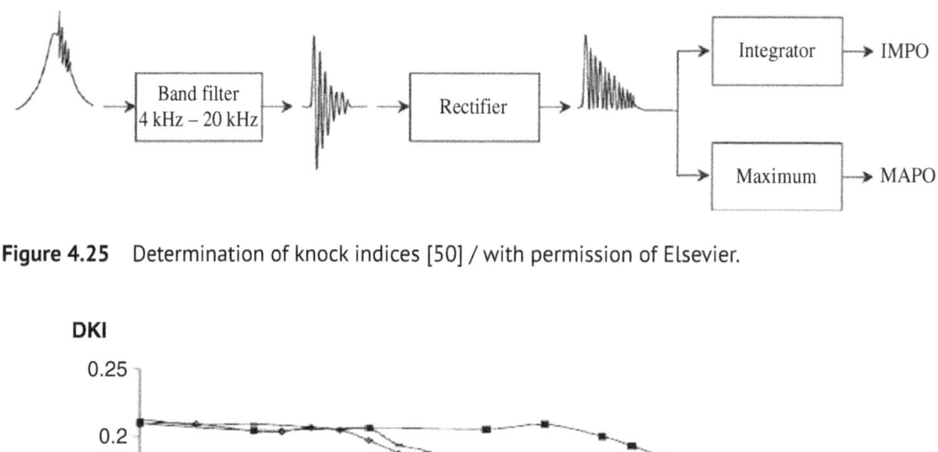

Figure 4.25 Determination of knock indices [50] / with permission of Elsevier.

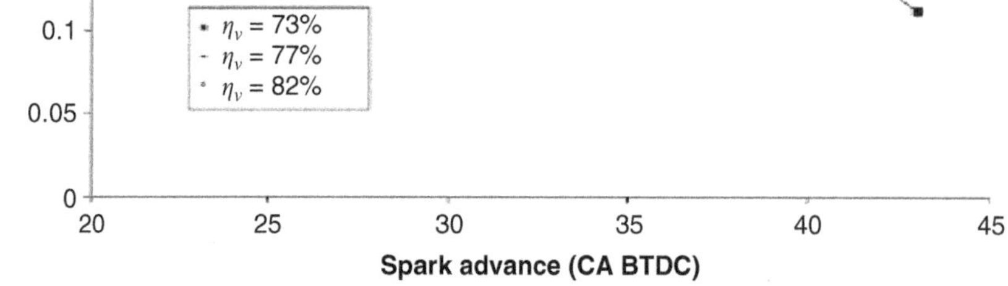

Figure 4.26 Variation of DKI with spark advance at $\phi = 0.7$ [50] / with permission of Elsevier.

Figures 4.27 through 4.29 show the pressure oscillations with no knock, mild knock, and severe knock, respectively.

The dimensionless indicator calculated in this study involved natural gas and adjunction of propane and carbon dioxide into the standard natural gas. Pressure traces in Figures 4.26 through 4.29 provide a qualitative picture of knock behavior that can be useful in the research pursuit of detecting knock in hydrogen engines using this new indicator.

Lai et al. [51] investigated knock intensity of a turbocharged hydrogen engine using direct cylinder injection and observed that adjusting the start of injection (SOI) slightly later than intake

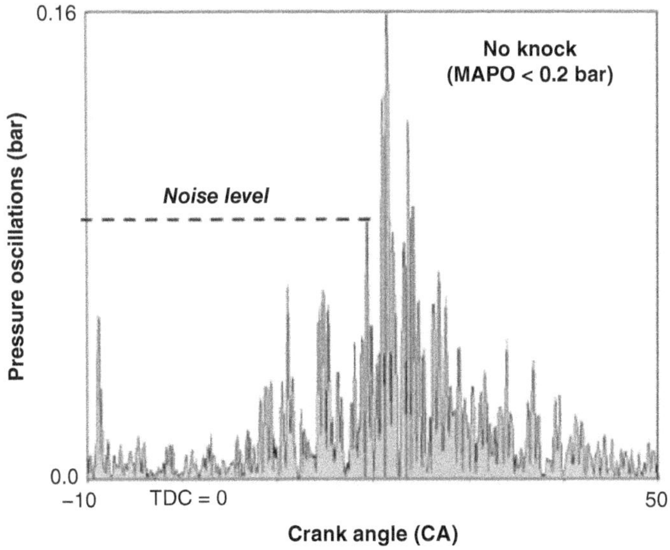

Figure 4.27 Pressure traces without knock [50] / with permission of Elsevier.

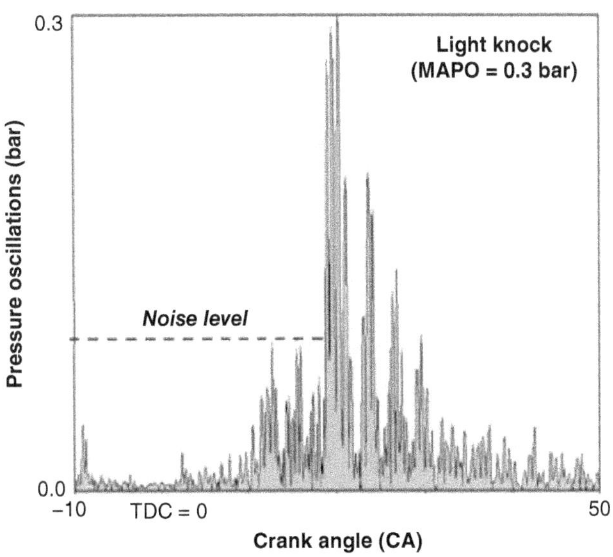

Figure 4.28 Pressure traces with mild knock [50] / with permission of Elsevier.

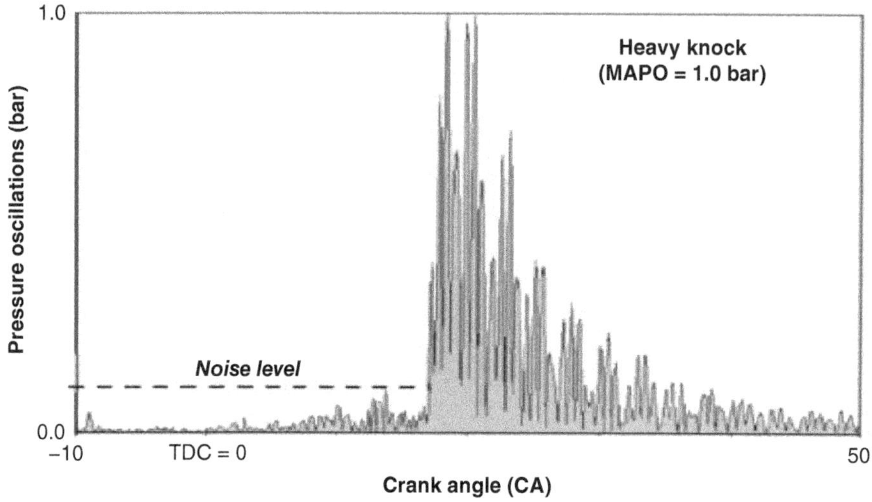

Figure 4.29 Pressure traces with heavy knock [50] / with permission of Elsevier.

valve closing timing (120° CA BTDC) decreased knock intensity. The results of this study agreed with most of the published results of prior studies that optimizing injection pressure within a specific range, lowering equivalence ratio or inlet pressure, and retarding ignition timing influenced combustion temperature and significantly reduced knock intensity. Knock intensity, in this study, was assessed based on changes in cylinder pressure and knock intensity index using various pressure-related parameters such as maximum amplitude of pressure oscillations (MAPO), integral modulus of oscillations (IMPO), and maximum pressure rise rate (MPRR). The knock index gives a quantified value of knock severity. The impact of knock (k_p), calculated from MAPO, is an indicator of the intensity generated by knock [51]. The value of k_p is determined from the following relation:

$$k_p = \frac{1}{N}\sum_1^N \max_{\theta_i,\ \theta_{i+\omega}} |\tilde{p}|$$

where N represents the number of computed cycles, $\tilde{p}$ is the filtered pressure in the cylinder, (θ) represents the crank angle at knock occurrence, and ω is the duration of knock combustion. The larger the k_p, the greater the tendency and intensity of knock. The study carried out a detailed investigation on the effect of many significant parameters on knock.

As part of this study, Figures 4.30a–d reflect the effect of some parameters on the maximum pressure in the cylinder, maximum combustion temperature, and maximum heat release rate (MHRR), etc.

Luo [52] investigated the intensity and frequency of knock in a port-fuel injected (PFI) hydrogen engine. By using a thermodynamic dual-zone model, the researchers observed that the temperature of the air–fuel mixture at the end of combustion was 1000 to 1100 K, which was approximately 200 K higher than for the corresponding gasoline operation. Figure 4.31a shows the inlet pressure and cylinder pressure variation indicating normal and knock conditions at a speed of 3000 rpm. As shown in the figure, backfire occurred in the middle timing of the intake stroke of the 59th cycle at an equivalence ratio of 0.61 and with a speed of 3000 rpm. In the 60th cycle, combustion knock followed backfire [52].

Figure 4.31b shows that backfire occurs in the 144th cycle in the early intake stroke at $\phi = 0.75$ and at a speed of 3500 rpm. Combustion knock occurred at 145 cycle. When the air–mixture is burned in the inlet pipe, temperature increases and hot gases enter the engine cylinder, raising

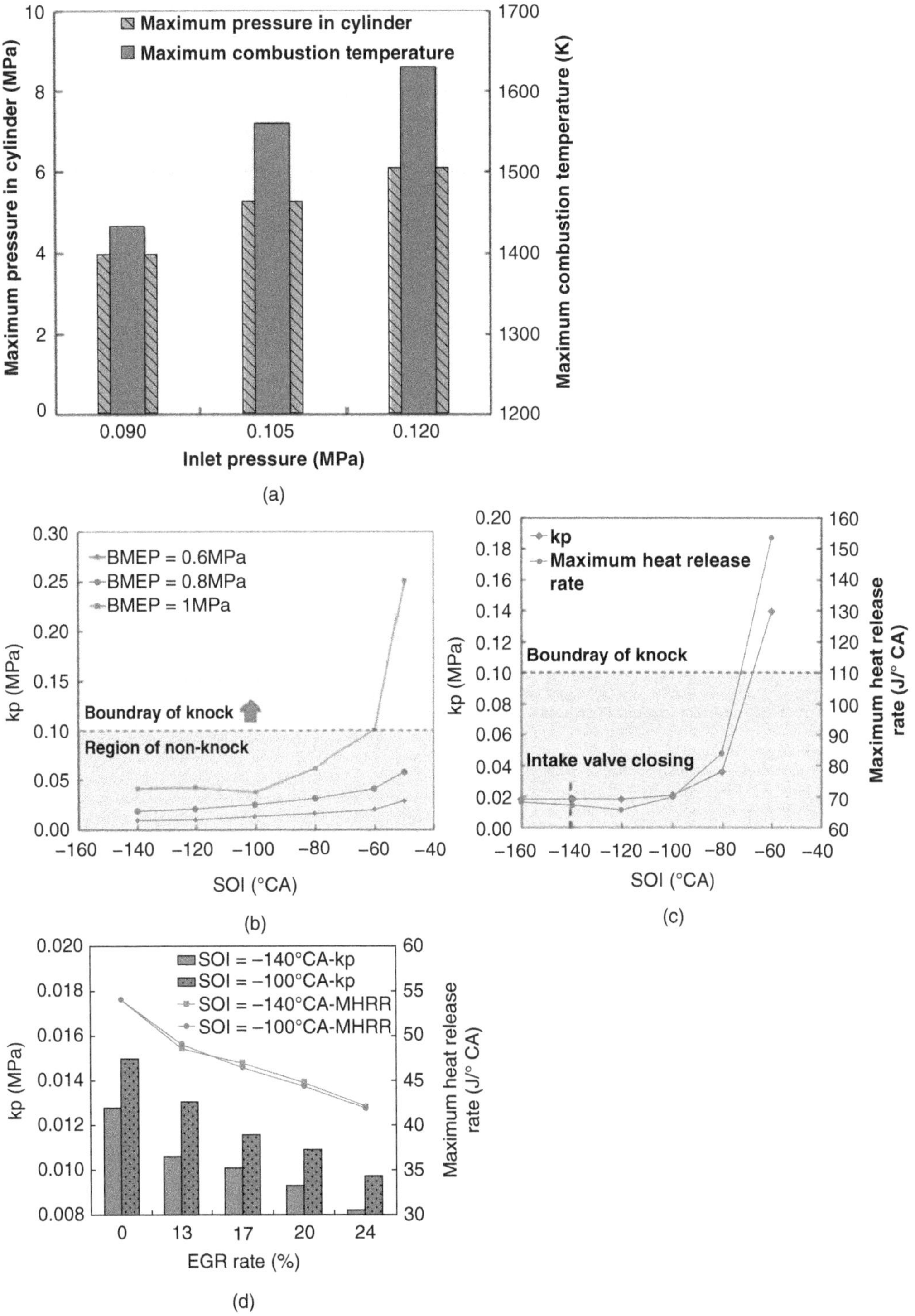

Figure 4.30 (a) Combustion pressure and temperature at 0.6 equivalence ratio [51] / with permission of Elsevier. (b) Variation of k_p with SOI at | different BMEP [51] / with permission of Elsevier. (c) k_p and maximum heat release rate with SOI [51] / with permission of Elsevier. (d) k_p and the maximum heat release rate with EGR at different SOI [51] / with permission of Elsevier.

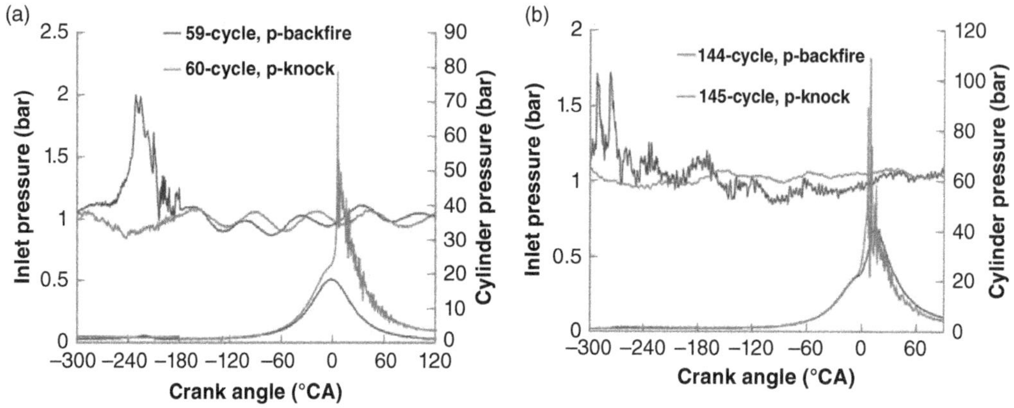

Figure 4.31 Knock and normal pressure (a) at 3000 rpm and (b) at 3500 rpm. Please visit the companion website to view a color version of this image.

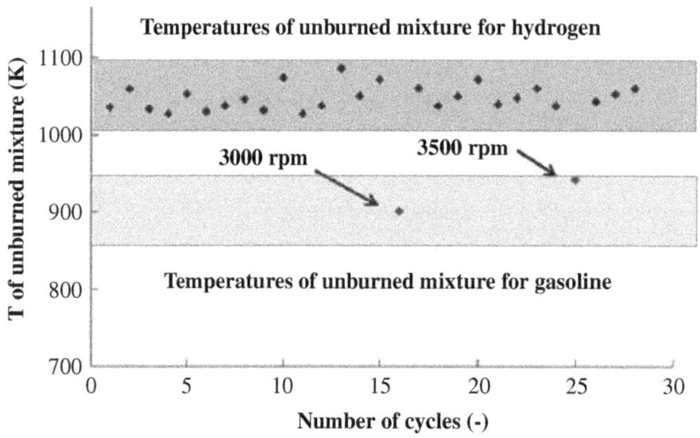

Figure 4.32 Temperature of unburned air–hydrogen mixture as knock is generating [52] / with permission of Elsevier.

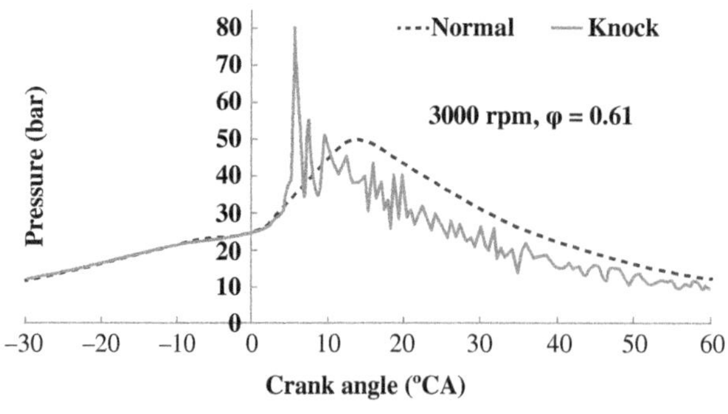

Figure 4.33 Temperature of unburned mixture for the working conditions of occurring knock at 3000 rpm [52] / with permission of Elsevier.

the temperature of the cylinder components. Due to the large amount of heat released during the compression stroke, the temperature rises further at the end of the compression stroke. This leads to a situation in which the possibility of combustion knock caused by hot spots is greater than self-ignition through compression. Figure 4.32 shows the temperature of an unburned air–hydrogen mixture as combustion knock occurs.

Figure 4.33 shows combustion knock taking place early in the compression stroke, indicating that hot spots can lead to knock with even higher probability than self-ignition [52].

Researchers studied knock intensity at higher speed ranges through fast Fourier transform (FFT), which is a powerful tool to analyze signals and understand clearly their frequency content. The knock intensity was evaluated by analyzing the frequency content of the engine's acoustic or vibration signals collected by accelerometers. Filtering and noise reduction were done to obtain more accurate data. The sequence of steps adopted to analyze the combustion knock pressure wave is shown in Figure 4.34.

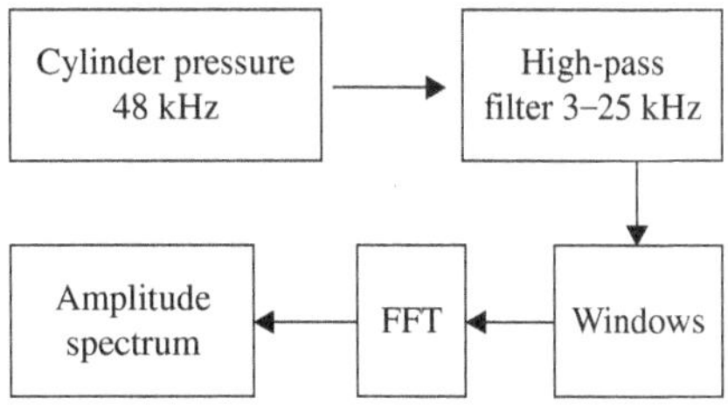

Figure 4.34 Diagram for getting amplitude spectrum [52] / with permission of Elsevier.

The FFT transform window starts at the angle of combustion knock and ends at 40° CA after that. Figure 4.35 shows FTT accelerometer changes under different knock-intensity conditions. Curves generated under conditions of mild knock intensity with KI = 0.35 and severe knock with KI = 20.9 are shown in the graph. A significant conclusion from FFT amplitude showed that while the pressure wave from light knock spreads in a radial direction, the severe knock spreads along the circumferential direction [52].

Knocking combustion continues to be a critical factor in engine design, and there are discrepancies between experiments and simulation models. Usually, since higher temperature occurs close to the flame front, it is most likely to cause autoignition, but many experimental researchers have observed that autoignition occurs away from the flame front in the end gas zone. Analysis of intermediate radicals and species needs to be studied further to understand and identify clearly the type of intermediate radicals and species linked to the cause, intensity, and duration of knock combustion. The analysis in the present study [53] involved cylinder pressure data averaged over 300 consecutive cycles of each experimental series. Figure 4.36 represents a condition of mild knock generated from unstable deflagration combustion in a hydrogen engine operating with a compression ratio of 10.

Figure 4.37 depicts the typical knock phenomenon occurring due to the spontaneous combustion of a hydrogen–air mixture at the end of combustion at a compression ratio of 12. During this phase, combustion was not that severe, but the pressure oscillation due to knock caused by autoignition in hydrogen engines was substantially higher. Figure 4.38 shows the pressure oscillations in a knocking condition due to autoignition. By comparing pressures at the start and end of combustion and analyzing the pressure trace, the mass fraction burned (MFB) profile at a particular crank angle can be calculated. Thus, the MFB profile, calculated from the absolute pressure trace (Figure 4.39), shows the progressive burning of the fuel in the combustion process.

Figure 4.35 FFT accelerometer changing with frequencies at different KIs [52] / with permission of Elsevier.

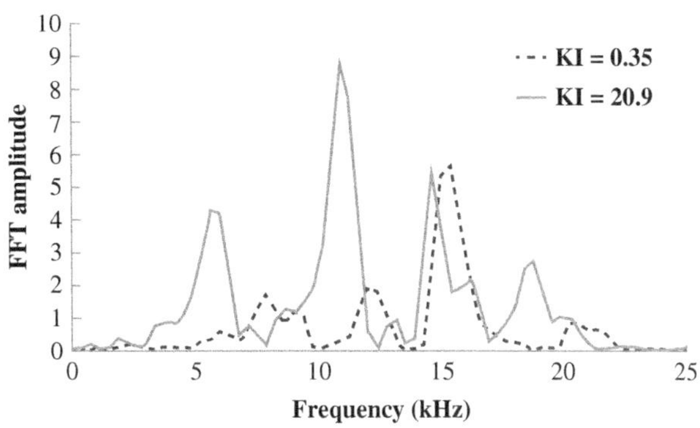

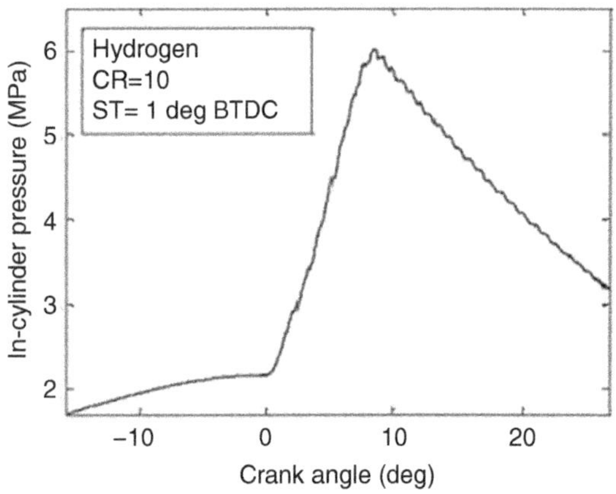

Figure 4.36 Engine operation showing mild knock at CR 10 [53] / Czestochowa University of Technology.

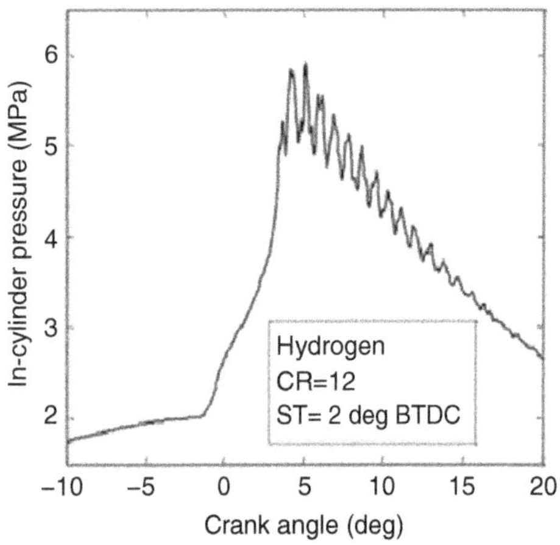

Figure 4.37 Knock, due to spontaneous combustion at CR 12 [53] / Czestochowa University of Technology.

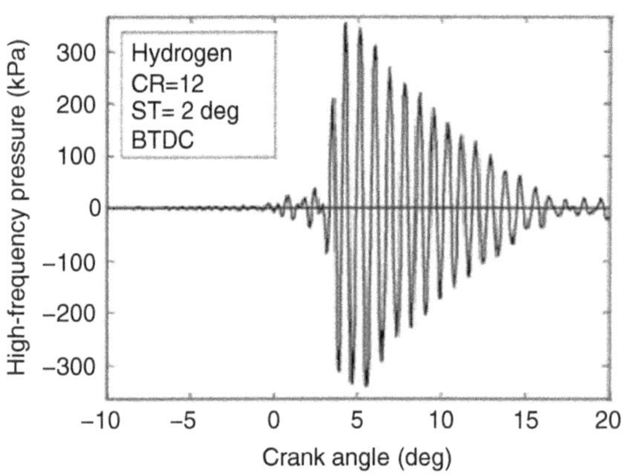

Figure 4.38 Pressure oscillation due to knock [53] / Czestochowa University of Technology.

Roy and co-workers [54] studied the initiation and development of knock in a hydrogen engine by using visualization and a spectrum analysis technique in a rapid compression and expansion machine (RCEM). Bright high-speed direct images in the end gas region clearly indicated the autoignition location. Figure 4.40 shows the schematic diagram of the specially designed RCEM with a high-speed photography and spectroscopy system used there [54].

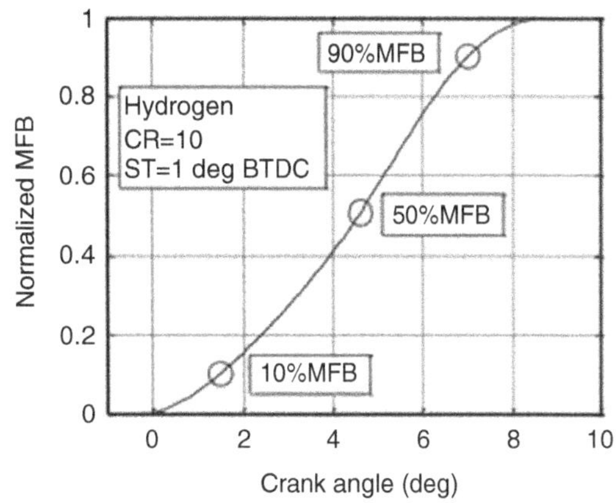

Figure 4.39 MFB profile caused by autoignition at CR 10 [53] / Czestochowa University of Technology.

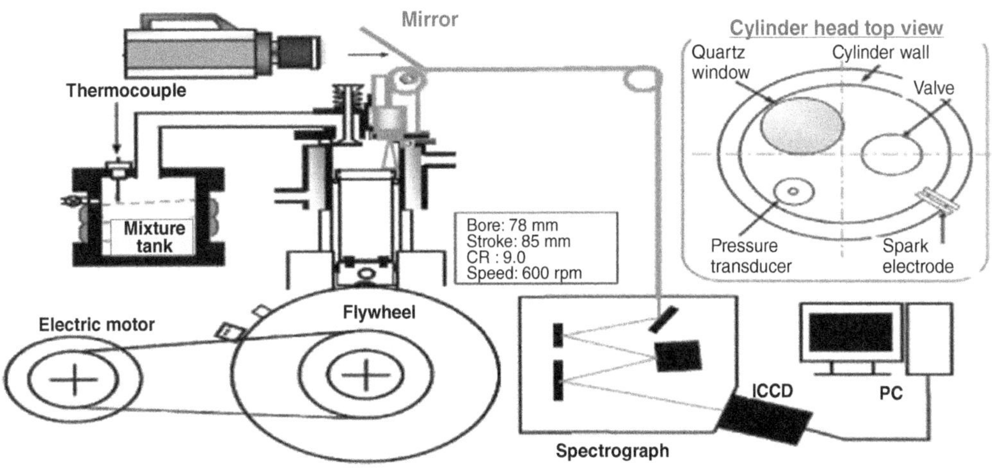

Figure 4.40 Specially designed rapid compression-expansion machine [54] / Forschungszentrum Julich.

To conduct tests on knocks, a homogeneous mixture of hydrogen, oxygen, and argon (H_2—O_2—Ar) was introduced to a mixture tank and engine cylinder as shown in Figure 4.40. Different rates of energy release in the H_2—O_2—Ar mixture resulted in different patterns of pressure variation during knocking combustion. The unburned fuel and mixture composition in the end gas was also affected during knocking. Knock conditions were identified in the pressure–crank angle diagram in Figure 4.41. In this study, the researchers adopted a digital Butterworth band-pass filter having 2.5 kHz cut-off frequency, and they calculated the knock intensity (KI) as the peak of in-cylinder pressure oscillation. Figure 4.42 indicates filtered pressure traces against time. As evident in the figure, the rate of pressure rise indicates a discontinuity at the crank angle where knock occurs. Again, after that band-pass-filtered signals increase just after knock.

In Figure 4.43, the upper images correspond to the normal combustion cycle with hydrogen and air whereas the lower images show the knocking combustion cycle using a hydrogen, oxygen, and argon ($H_2 + O_2 +$ Ar) mixture. High-speed photography shows the blue color flame propagating into the end gas where a condition of pre-knock oxidation develops. The unburned end gas region is shown in a visibly white color. Comparison of knock images for the same ignition timing shows

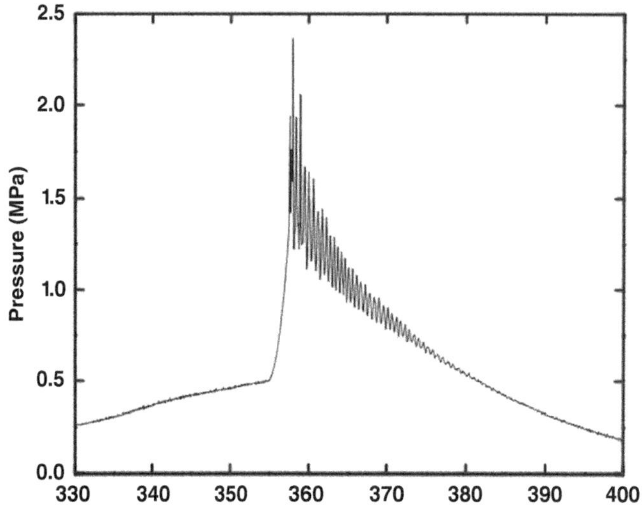

Figure 4.41 Pressure-crank angle with pressure oscillation [54] / Forschungszentrum Julich.

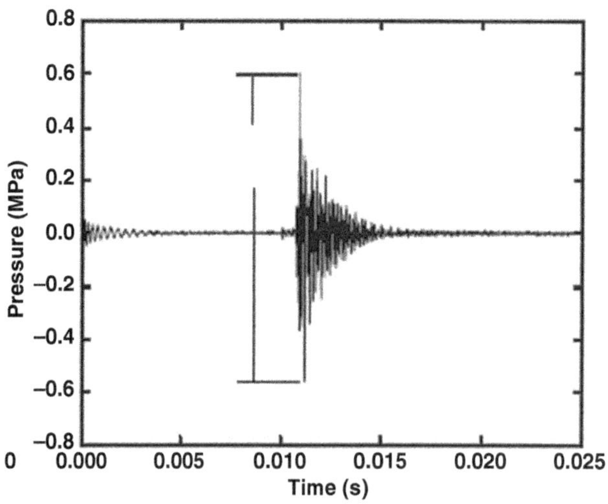

Figure 4.42 Pressure oscillation vs. time indicating intense knock [54] / Forschungszentrum Julich.

cycle-to-cycle variation as in Figure 4.44. High-speed color images generated during the tests show that in spite of fluctuation from cycle to cycle, the location of autoignition remained always in the end gas region close to the cylinder walls.

UV spectra of OH* radical emissions during knocking combustion Figure 4.45 show two distinct emission bands just before autoignition. Before autoignition, compression of the air–fuel mixture in the engine leads to increased pressure and temperature, which causes thermal excitation of the OH* radicals. As the flame front propagates, the end-gas in the combustion chamber is further compressed, leading to increased pressure and temperature. After autoignition, emission intensity of the OH* radical increases. This is because the pressure and temperature continue to rise in the knocking combustion process, which increases the production of these radicals.

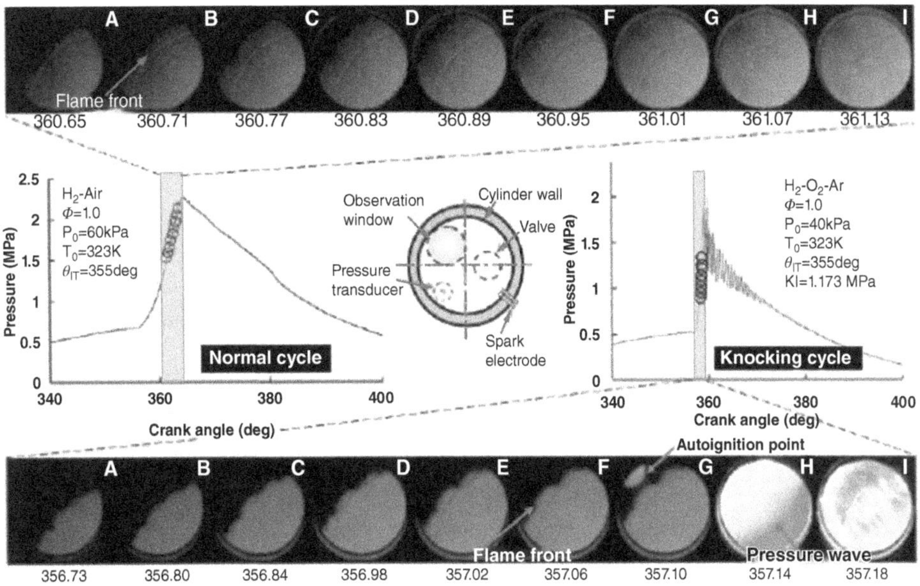

Figure 4.43 Normal and knocking cycles with pressure history [54] / Forschungszentrum Jülich. Please visit the companion website to view a color version of this image.

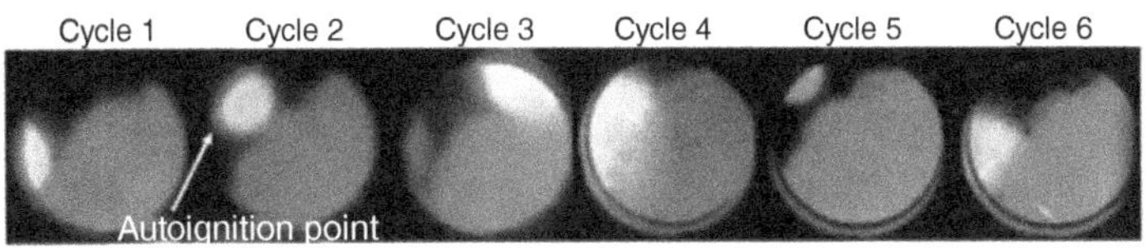

Figure 4.44 Cycle-to-cycle combustion images with autoignited kernel in the end-gas region [54] / Forschungszentrum Jülich. Please visit the companion website to view a color version of this image.

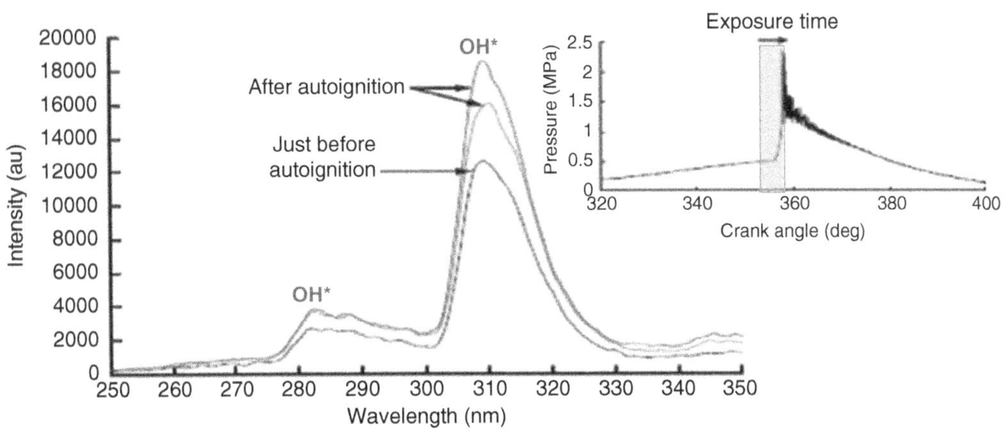

Figure 4.45 Experimental typical OH* spectra at just before and after knock combustion. Adapted from [54]. Please visit the companion website to view a color version of this image.

Jennifer [55] studied knock behavior of a lean-burn, hydrogen-enhanced engine with the objective of suppressing knock by lean hydrogen-enhanced engine operation. Selected tests that were performed to assess octane enhancement effect on both H_2 and CO showed enhanced octane number and reduced knock. One major objective of this study was to explore the possibility of suppressing knock by lean hydrogen-enhanced engine operation. The plasmatron is a fuel reformer used to convert hydrocarbon fuels into a mixture of hydrogen and carbon monoxide. The heart of a plasmatron is a plasma torch that generates extremely high temperatures by ionizing a mixture of gases and creating plasma, a highly energized and electrically conductive state of matter. In a plasmatron, a plasma arc is employed to partially oxidize a mixture of fuel and air.

Tests were carried out with two mixtures such as gasoline with plasmatron reformate and gasoline with hydrogen added. In the hydrogen addition tests, the fraction of hydrogen energy matched corresponded to the plasmatron study. This study shows that a part of the energy from hydrogen to the engine lowers the octane number of the base fuel to reach the audible knock limit. This trend is reflected in Figures 4.46 and 4.47 with two sets of data points raised in two

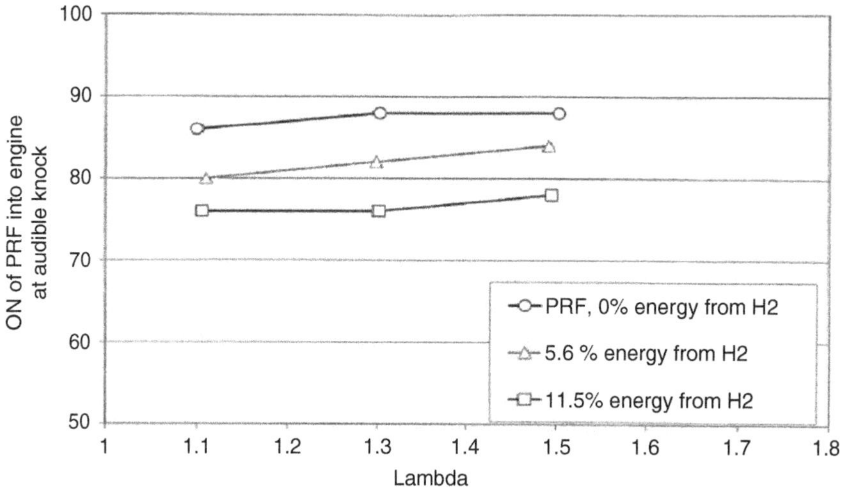

Figure 4.46 Measured octane number of primary reference fuel at audible knock with increasing H_2 addition. All data points are 6.7 bar NIMEP [55] / Massachusetts Institute of Technology.

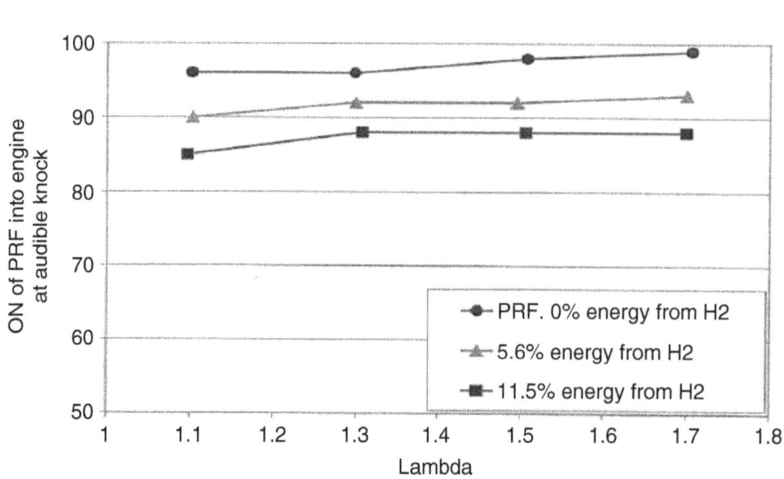

Figure 4.47 Measured octane number of primary reference fuel at audible knock with increasing H_2 addition. All data points are 8.5 bar NIMEP [55] / Massachusetts Institute of Technology.

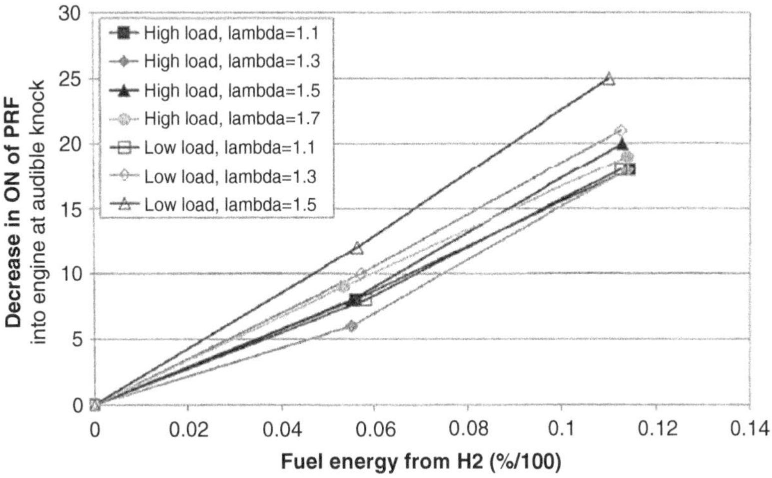

Figure 4.48 Decrease in octane number of the primary reference fuel at audible knock with plasmatron addition for several operating conditions [55].

different conditions of net indicated mean effective pressure (NIMEP) of 6.7 bar and 8.5 bar. The horizontal axis in Figure 4.48 shows the energy fraction from H_2, which is found in 15% and 30% plasmatron reformate fractions. Since hydrogen has a higher octane number, the overall octane number increases when hydrogen is blended with gasoline. It has been estimated that 11% of the fuel energy from hydrogen enhances the overall octane number by 10.

Li et al. [56] in their numerical solution method observed that at a high compression ratio, the effects of the equivalence ratio on the tendency to knock become more pronounced. Engine operation at a higher compression ratio of 17.5 increased the heat release rate (HRR) and knock intensity (KI) with an increase in equivalence ratio. The compression ratio of 11 demonstrated in earlier knock limited spark timing compared to a higher compression ratio of 17.5. Their paper also discusses the significance of signal energy ratio (SER) methods to identify the crank angle at which pressure fluctuations begin in the hydrogen engine and thus determine knock intensity (KI) and knock onset (KO).

4.8 Backfire: Causes and Character

Backfire or backflash refers to a condition of premature autoignition that takes place when for some reason the flame propagates back into the intake manifold. Backfire can occur even in a gasoline engine, but more often it is referred to in hydrogen engines because backfire in hydrogen engines has resulted in very severe consequences. In certain operating conditions, hydrogen engines exhibit abnormal combustion symptoms such as (i) extraordinarily high pressure rise; (ii) preignition in the combustion chamber and its subsequent advancement leading to backfires in the intake manifold; and (iii) occasional backfires with very lean hydrogen–air mixtures or during idling. Usually, the intensity of backfire with a lean mixture ($\lambda = 3$) is low and does not stop the engine. As soon as the intake valve opens, a fresh charge is ignited. This raises the intake and in-cylinder pressure. Sometimes, preignition heats up the combustion chamber and creates a situation leading to backfire. Table 4.1 shows the sources of preignition backfire and steps to avoid it.

Table 4.1 Backfire sources and possible solutions [59] / with permission of Elsevier.

Source of residual in the intake port	Source of ignition	Possible solutions
Former cycle	Exhaust gas at high temperature	Optimum injection timing, short valve overlap conditions
Former cycle	Hot spots at high temperature	Appropriate injection timing
Current cycle	Exhaust gas at high temperature	Advance the intake valve closure, postpone the intake valve opening, delay hydrogen injection
Current cycle	Hot spots at high temperature	Delay hydrogen injection

In the evolution of hydrogen engine technology, backfire was almost an unsolvable problem; many researchers abandoned their efforts as they could not identify its cause nor could they control it. But now, after numerous experiments and theoretical analyses, its causes and control techniques have been broadly identified. The low minimum ignition energy (0.02 mJ) of hydrogen is a property with pros and cons. On one hand, it enables easy ignition due to the low energy spark. At the same time, such a low energy increases the susceptibility of the system to surface ignition. It is important to balance the advantageous feature of easy ignition with the threats of surface ignition, causing preignition, backfire, and rapid pressure rise. Spark plugs and exhaust valves are the two vulnerable components to attain the hottest stages during intake strokes. Thus, it is strongly recommended that cold-rated spark plugs and cooled exhaust valves should be used in a hydrogen engine. There could be situations, when, inadvertently, fresh charge could be ignited by low minimum ignition energy resulting in backfire. Low quenching distance coupled with low energy sparks make it possible for hydrogen–air mixture flame to escape easily. It has been reported that hydrogen–air flame escapes far easily than hydrocarbon–air flame, even from a nearly closed intake valve [57].

4.8.1 Comprehensive Monitoring of Backfire

Overall smooth engine operation of an engine depends upon a number of factors such as fuel quality, engine load and speed, injection timing, equivalence ratio, valve timing, spark timing and ignition system characteristics, coolant temperature, intake air temperature and pressure, turbocharging, etc. Many of these factors are interrelated. Simulation and exhaustive testing of modern hydrogen engine management systems, with the use of sensors and feedback mechanisms, have made it possible to continuously monitor these parameters in real time, ensuring reliable engine operation without backfires.

A general trend in an engine's response to avoid backfire limits cannot be easily established as it depends upon intricacies of individual engine designs and operation. Backfire has been studied by experiment and using CFD analysis [58]. Backfire is usually avoided by lean engine operation, exhaust gas recirculation, intake air-cooled by liquid, etc. Some studies have observed that increasing the depth of crevice volume, optimizing coolant temperature, and managing blow-by gas have proven effective in reducing backfire. A crevice around the spark plug electrode is not a cause of concern, but combustion in the crevices of the piston top land, second land, and ring groove could induce backfire. Thus, increasing the top land crevice volume and controlling the blow-by gas are key strategies. All these methods described above are only short-term successful

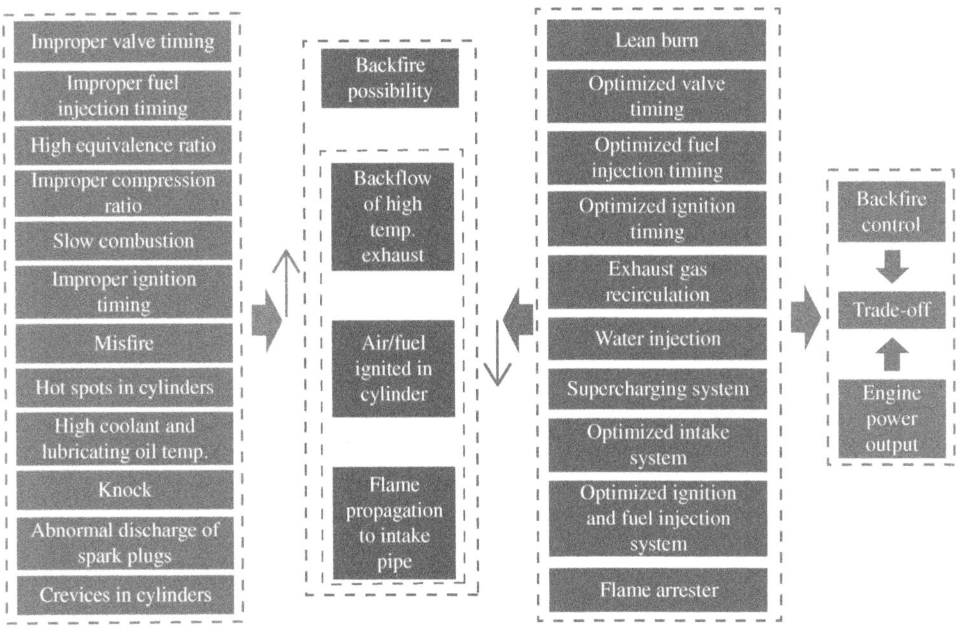

Figure 4.49 Backfire factors and appropriate control measures [59] / with permission of Elsevier.

solutions to some extent and do not qualify as long-term solutions. Appropriate fuel induction technique is the ultimate answer to eliminating this major threat and qualify the hydrogen engine for regular use. Gao and Wang et al. [59] have reported causes of backfire and discussed multiple control strategies, shown in Figure 4.49.

Hot spots and high temperature residual gases promote backfires. Lean engine operation, apart from reducing backfire tendency, brings down NOx concentration. Thus, a combination of lean mixture and low combustion temperature curtails hot spots and lowers backfire possibility. Lee et al. [60] adopted combined strategies of retarding intake valve opening with lean-burn technology to prevent backfire.

A hydrogen engine can achieve stable combustion in an ultra-lean zone where the temperature is below the NOx generation levels. Lee et al. [60] adopted combined strategies of retarding intake valve opening with lean-burn technology to prevent backfire and attain high power and efficiency without NOx emissions. It is evident in Figure 4.50 that equivalence ratios limiting backfire take a decreasing trend as engine speed is increased. There is a significant drop across the engine speed range of 1600–2000 rpm compared to 1000 rpm. This decrease in equivalence ratios indicates less fuel energy delivery to the engine, resulting in a decline in engine power output.

4.8.2 Optimized Valve Timing: Effect on Backfire

Appropriate inlet and exhaust valve timing reduces exhaust gas backflow and residual high-temperature exhaust gas in the cylinder. Retarding the intake valve opening delays the contact between high-temperature exhaust gas and fresh hydrogen–air mixture, allowing further cooling of the remaining exhaust gases. This delay reduces volumetric efficiency and power output.

The BFL equivalence ratio is a crucial parameter because it defines the strength of the mixture (equivalence ratio) immediately prior to the occurrence of backfire with varying intake valve

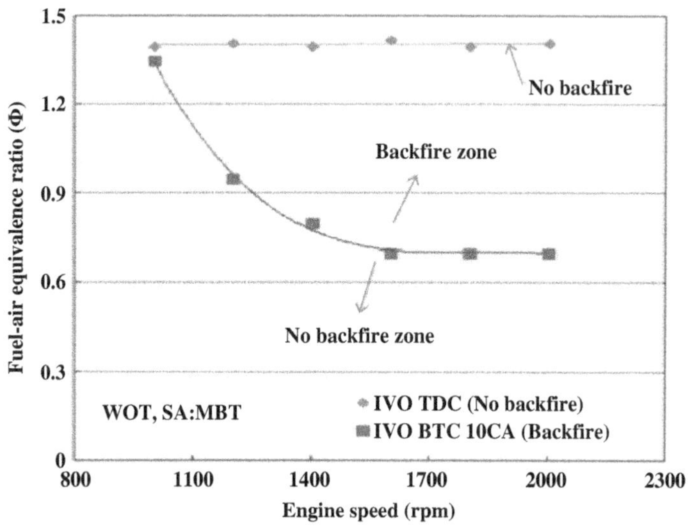

Figure 4.50 Lean-burn technology controlling backfire [60] / with permission of Elsevier.

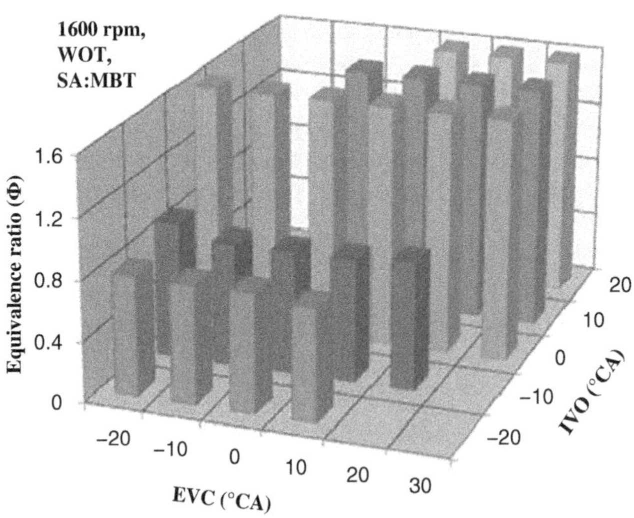

Figure 4.51 Exhaust valve closing timing and intake valve opening timing versus fuel–air equivalence ratio with respect to backfire [60] / with permission of Elsevier.

opening (IVO) timing and exhaust valve closing (EVC) timing. It was observed during the tests that BFL equivalence ratio increased when the IVO was retarded from the BTDC 20° CA. Even after TDC and in the rich zone of the mixture (with equivalence ratio beyond stoichiometric and above with $\Phi = 1.5$) there was no backfire [60]. Figure 4.51 considers backfire and shows exhaust valve closing timing and intake valve opening timing with respect to the equivalence ratio. Such a situation occurs if either the flame from hot spot in combustion chamber and piston crevice volume has cooled down or "burns until the intake valve opens again due to retardation of IVO" [60]. Residual high-temperature gas in the cylinder causes backfire due to the influence of residual gas on heat sources within the combustion chamber, such as chamber walls and crevice volumes.

Figure 4.52 shows the backfire limit equivalence ratio determined under different exhaust pressures. It is relevant to mention here that varying exhaust pressure is equivalent to changing the amount of residual gas because as exhaust pressure decreases, the amount of residual gas also decreases.

Huynh et al. [61] carried out studies with specific emphasis to achieve high performance in a hydrogen engine without backfire symptoms. Figure 4.53 shows a history of pressure curves of in-cylinder pressure. Channel 2 shows in-cylinder pressure curves; normal combustions, instances of backfire occurrence, and cycles of misfiring due to backfires and pressure curve for the inlet pipe as backfire are shown in Channel 3.

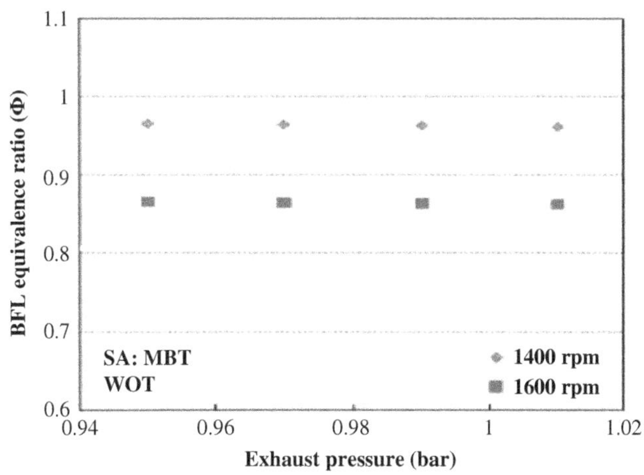

Figure 4.52 BFL equivalence ratio versus exhaust pressure [60] / with permission of Elsevier.

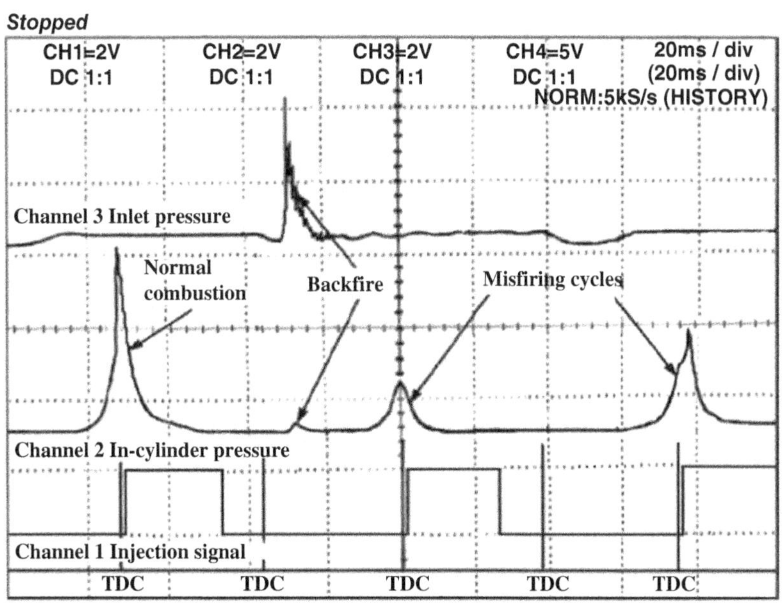

Figure 4.53 History of cylinder pressure showing backfire [61] / The American Society of Mechanical Engineers.

4.8.3 Injection Strategy

Hong et al. [62] studied the effect of injection timing on backfire of a port injection hydrogen engine. As is known, it is possible to control backfire by decreasing the temperature of the combustion chamber. This approach of the investigation is shown in Figure 4.54. The cooling process by the fresh hydrogen–air mixture involves two distinct stages. Initially, during the first stage, cooling occurs concurrently with the overlap between the exhaust valves and intake. Following this, the second cooling stage occurs shortly after the exhaust valve closes when fresh air continues to flow into the cylinder. Intake port injection can be done at this stage when the combustion temperature is kept less than the autoignition temperature of hydrogen (576 °C).

If the injection timing of hydrogen fuel is significantly earlier than the end of this second cooling stage, it may lead to preignition and subsequently escalate into backfire. Therefore, optimal injection timing is essential to ensure excellent hydrogen–air mixing and avoid backfire. As engine speed increases, a larger proportion of the exhaust gas remains in the cylinders. Such a situation needs more cooling effects from fresh air to prevent backfire.

Menaa et al. [63] studied numerical 2D and 3D dynamic simulation using CFD code ANSYS Fluent of hydrogen injection parameters in a single-cylinder H_2–diesel dual fuel engine. The simulation model was also verified with experimental data. The study involved optimization of timed manifold injection (TMI) through maximum valve lift and optimized injection parameters such as injection pressure, injection duration, and injection timing. Figure 4.55 shows the position of the gas injector along with the geometry of inlet and exhaust duct optimizing the hydrogen TMI (timed manifold injection) system.

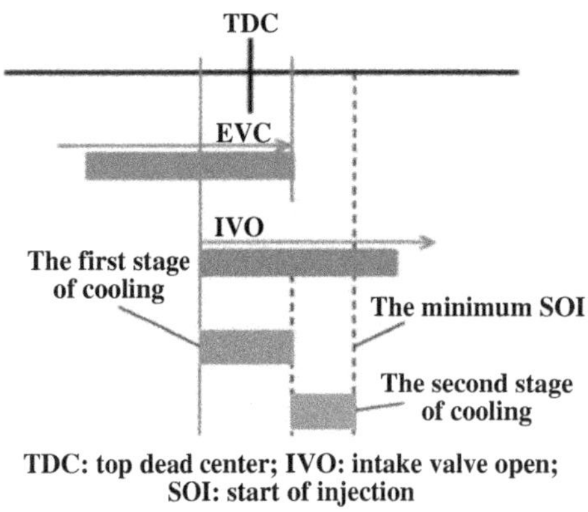

Figure 4.54 Principles of controlling backfire by adjusting hydrogen injection timing [62] / SAE International. Please visit the companion website to view a color version of this image.

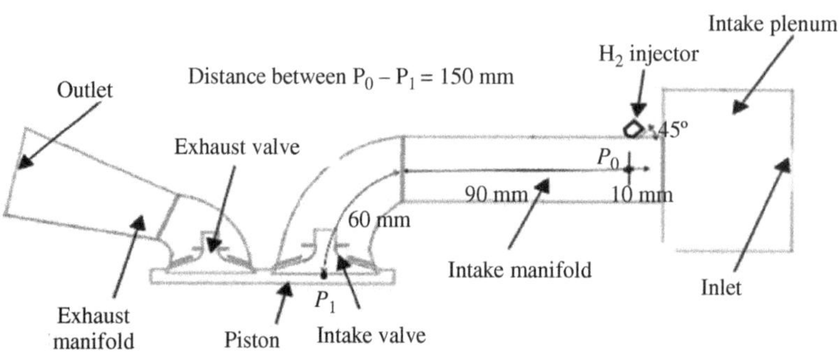

Figure 4.55 Inlet and exhaust duct geometry with gas injector position [63] / with permission of Elsevier.

The tool developed for the analysis was based on experimental data on valve lift, inlet valve closing, and starting of injection, etc. Figure 4.56 shows that increased hydrogen injection pressure leads to an increase in mass flow rate and decreases injection duration. In order to attain this objective, 63 cases were analyzed involving three maximum valve lift values (MVL10.6, MVL8.0, and MVL6.0 mm) in combination with three intake valve closing timings (IVC70, IVC50, and IVC30 CA ABDC) across seven different engine speeds ranging from 1000 to 2500 rpm. The terms MVL 10.6, MVL 8.0, and MVL 6.0 stand for maximum valve lift of 10.60 mm, 8 mm, and 6 mm, respectively. Similarly, IVC 70, IVC 50, and IVC 30 CA ABDC indicate intake valve closing 70°, 50°, and 30° CA after bottom dead center, respectively. This study on the effects of varying valve lift and closing timings was useful for selection of optimal valve lift.

Figures 4.57 and 4.58 show in-cylinder mass and backflow mass for the valve lift law MVL10 IVC70 and MVL10 IVC30 at 90% of load and 42.5% hydrogen energy share using 3D simulation.

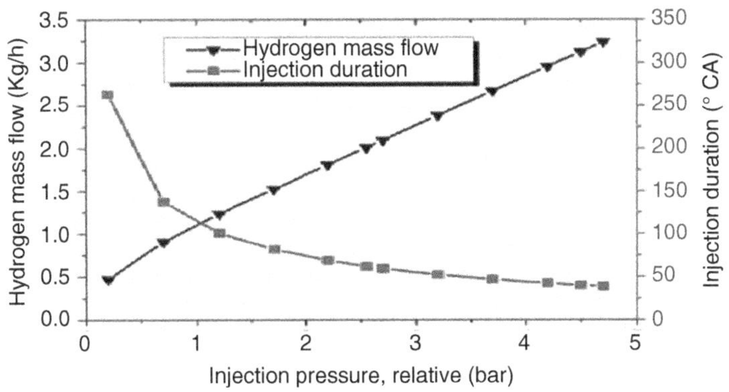

Figure 4.56 Characteristics curve of the hydrogen injector [63] / with permission of Elsevier.

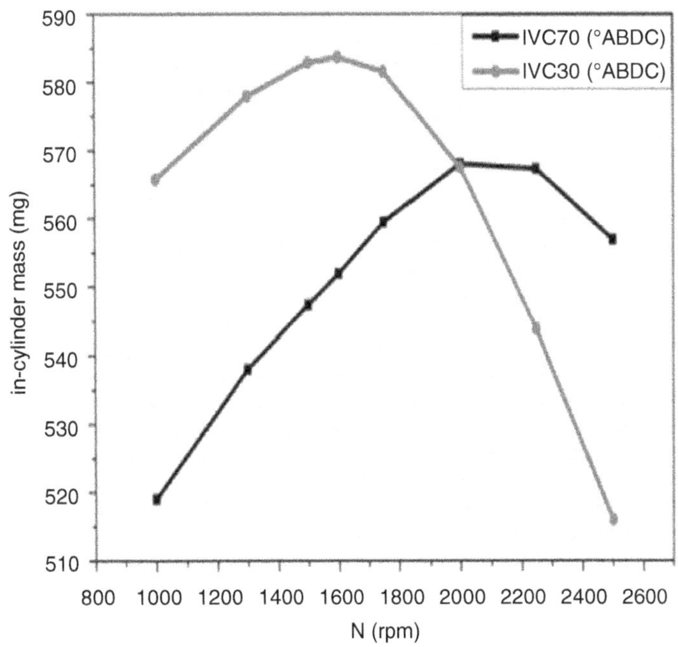

Figure 4.57 In-cylinder mass flow speeds [63] / with permission of Elsevier.

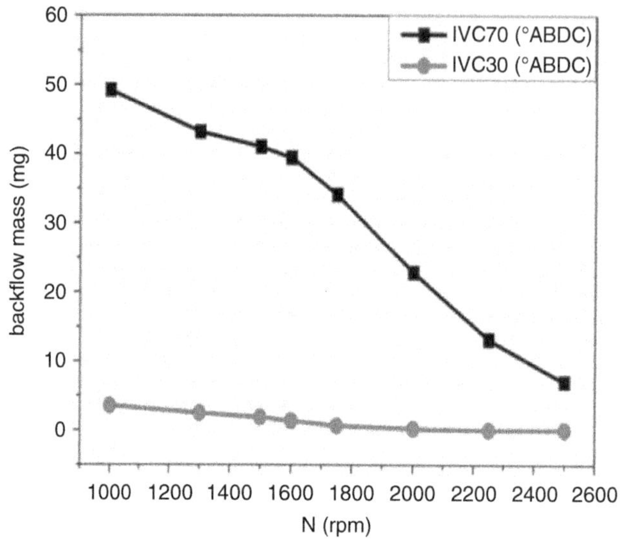

Figure 4.58 Backflow mass at different speeds [63] / with permission of Elsevier.

These graphs of 3D results agree with 2D simulations for engine speeds up to 2000 rpm. It has also been reported in this study [63] in the case of diesel start of injection (DSOI) retard higher than 8° CA, combustion-related parameters such as peak in-cylinder pressure, cylinder temperature, the rate of pressure rise, as well as NOx emissions are comparable or even less than those of a standard diesel engine.

Any situation that causes flame reflux from either the hotspot or the crevice volume into the intake valve during the valve overlap period can induce backfire, more so in the rich zone. Yan et al. [64] developed an appropriate electronic control system with three parts such sensors, electronic control unit (ECU), and actuator.

Figures 4.59 and 4.60 show the box diagrams of hardware for the control system and the PC monitoring system, respectively. In this study a subprogram for ignition control and subprogram for backfire control were developed as guiding strategies for the control system software and control system design.

The monitoring system consists of upper and lower machine modules. The lower machine is hydrogen engine ECU. The serial communication interface (SCI) is designed in ECU to transfer data to the upper machine. The upper machine, which uses PC-based engine monitoring software, executes a series of important functions with ECU such as data transmission, real-time display, online adjustment, and data retention. Postponing ignition timing is effective in reducing the tendency of backfire. However, if the ignition timing is too much postponed, it results in lower thermal efficiency and brings down the power output. So, optimization of ignition timing is crucial, as reported by many researchers. Water injection, which has been successfully tested in hydrogen engine for backfire control decades ago, has many disadvantages such as power loss, cylinder corrosion, and inefficient lubrication. The study made important observations that an optimal and rational integration of both ignition timing and water injection should be made to get the benefits in a hydrogen engine.

Xin and Ji et al. [65] built a four-cylinder, four-stroke gasoline direct injection (GDI) spark ignition (SI) engine to operate on hydrogen with a port fuel injection system. The study aimed at detecting the location of backfire and its intensity. They performed unthrottled operation in an

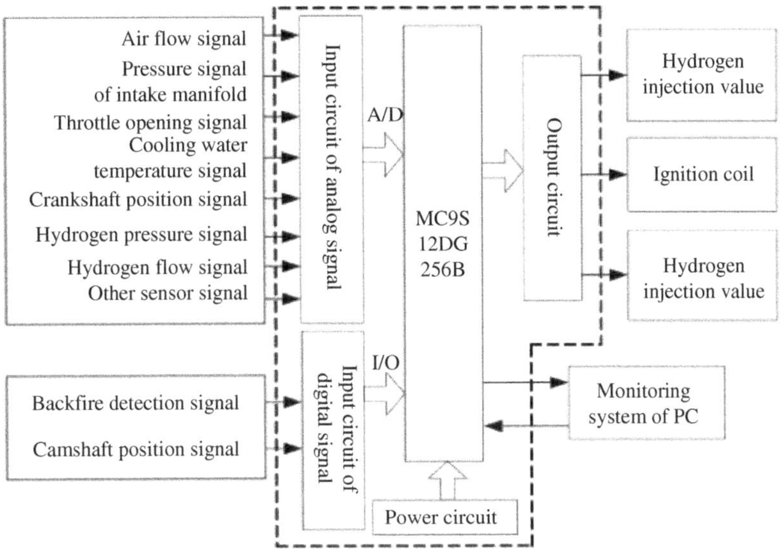

Figure 4.59 Box diagram of hardware for the control system [64] / with permission of Elsevier.

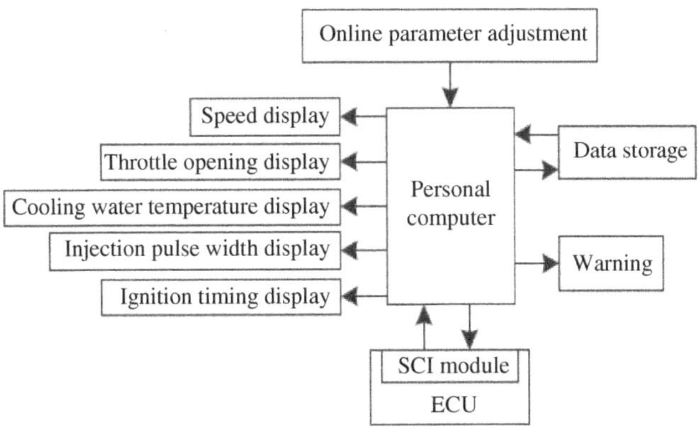

Figure 4.60 Monitor system structure [64] / with permission of Elsevier.

engine under lean burn conditions at a speed of 1000 rpm and at an intake pressure of 100 kPa. Two separate manifold absolute pressure (MAP) sensors were installed on intake manifolds of the first and fourth cylinders respectively for monitoring backfire. A cylinder pressure sensor was installed on the fourth cylinder to evaluate detection accuracy of the MAP sensor. To achieve port fuel injection, four injectors were installed in the intake manifold. An appropriately designed electronic control unit was installed to control ignition timing and the injector's pulse width. The start of hydrogen injection was scheduled at 300° CA before top dead center (BTDC). Table 4.2 shows the "backfiring event judgement logic" [65]. This table shows the location and magnitude of the peak pressures cylinders during each cylinder's backfire [65].

There have been many recent studies highlighting the indispensable role of hydrogen in "decarbonisation pathways for hard-to-electrify transport sectors such as long-haul trucking, international maritime, and aviation [66]" and the use of hydrogen in the transport sector is significant in transitioning to net zero emissions (NZE). Many experimental studies and theoretical analyses are centered on the most prominent combustion anomaly such as backfiring, which is the major setback. Duan et al. [67] used CFD models to identify causes of backfire in a hydrogen engine with port fuel injection at higher loads and observed that the temperature and gas mixture concentration near intake valves were chiefly responsible for causing backfire.

There have been several analytical and experimental studies on valve overlap, variable valve timing residual gas fraction, controlled auto ignition (CAI), homogeneous charged compression ignition (HCCI), etc. for application in various engine configurations [68–70]. The combined influence of valve timing and injection timing is very crucial to the cause of backfire (Figure 4.61). Sometimes, hydrogen-rich zones are formed near the valve seats before the intake valve opens and, on opening the intake valves, ignition takes place due to high-temperature exhaust gas. On the other hand, in the case of very late injection timing hydrogen fuel continues to be delivered after closure of the intake valves. This leads to an accumulation of rich mixture around the seats of the

Table 4.2 Backfiring event judgement logic [65] / with permission of Elsevier.

Sensor	Range of peak pressure				Relative value	Backfiring position
	Intake stroke	Compression stroke	Expansion stroke	Exhaust stroke		
1# MAP					$MAP_{1\#} < MAP_{4\#}$	**Backfiring of first cylinder**
4# MAP						
1# MAP					$MAP_{1\#} \geq MAP_{4\#}$	**Backfiring of second cylinder**
4# MAP						
1# MAP					$MAP_{1\#} \leq MAP_{4\#}$	**Backfiring of third cylinder**
4# MAP						
1# MAP					$MAP_{1\#} > MAP_{4\#}$	**Backfiring of fourth cylinder**
4# MAP						

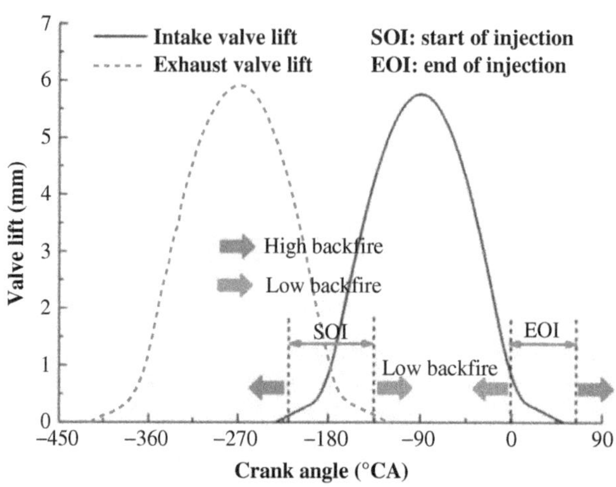

Figure 4.61 Possibilities of backfire caused by improper hydrogen injection timing at TDC 180° CA [59] / with permission of Elsevier. Please visit the companion website to view a color version of this image.

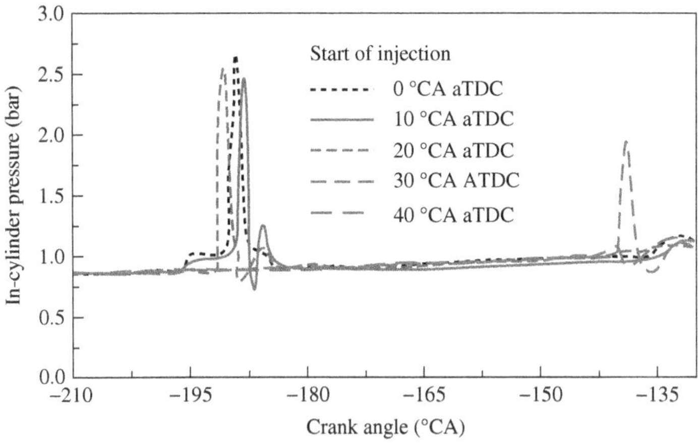

Figure 4.62 Backfire caused by the early fuel injection [71] / with permission of Elsevier. Please visit the companion website to view a color version of this image.

intake valves. This accumulated mixture is ignited at the start of the intake process in the following cycle [59]. Valve timing can only partially help in reducing the backfire, as design of the injector and injection duration along with the engine operating conditions can restrict the flexibility of adjusting the injection timing.

In an effort to investigate the effect of injection timing on backfire, the engine's injection timing was varied from 0° ATDC to 40° ATDC at MBT spark timing and at an equivalence ratio of 0.82 [71]. Delaying hydrogen injection until 40° ATDC eliminated backfire, as seen in Figure 4.62. This timing, at an equivalence ratio ($\phi = 0.82$ there), is known as "backfire limiting injection timing" (BLIT). BLIT optimizes injection strategy to minimize backfire risk by allowing intake air to cool the residual exhaust gases more effectively. This is because delay in injecting fuel increases the cooling duration of residual exhaust gases by intake air.

4.8.4 Equivalence Ratio

Lean hydrogen engine operation has been investigated by many researchers for backfire-free operation with low NOx emission. It has been reported [72] that even a small amount of hydrogen added to gasoline reduces cyclic variability and ensures smooth combustion at low equivalence ratio. Combined hydrogen–gasoline injection with exhaust gas recirculation has been investigated to evaluate improvement due to lean burn characteristics [73]. Studies have explored the lean burn combustion of hydrogen–methanol dual fuel engines [74], as well as the lean burn characteristics of turbocharged, opposed rotary piston hydrogen engines operating at low speeds [75]. Some related discussions have addressed these aspects in Chapters 8 and 9.

Research by Liu et al. [76] on the variations in laminar flame speed with equivalence ratio in hydrogen–air mixtures is shown in Figure 4.63, reproduced from Gao [75]. This graph shows that laminar flame speed varies with equivalence ratio (ϕ), and its peak value lies between 1.5 and 2.0. Such a trend is valid for hydrogen–air flames at different initial pressures. The hydrogen flame propagation speed was approximately 0.2 m/s at an equivalence ratio of about 0.3 and increased to 3.1 m/s at an equivalence ratio of about 1.8. It was observed in this study that just as maximum laminar flame speed occurs at a particular equivalence ratio, similarly the maximum value of laminar flame speed also corresponds to a definite value of initial pressure. For an initial pressure lower than one atmosphere, laminar flame speed increased along with increased initial pressure, but, for an

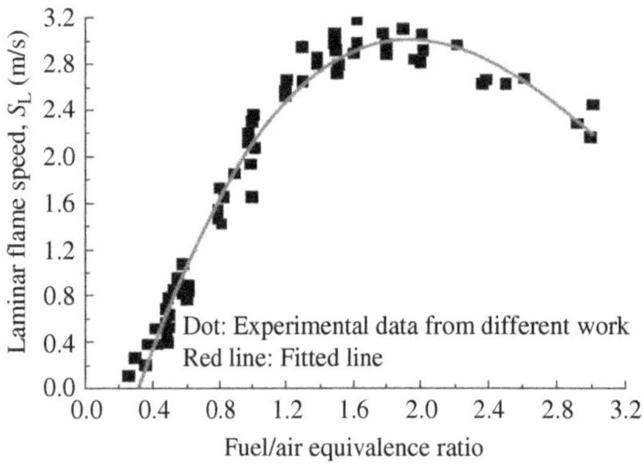

Figure 4.63 Variations of laminar flame speed (*SL*) with equivalence ratios [76] / with permission of Elsevier.

initial pressure of greater than one atmosphere, it showed a decreasing trend with increase of initial pressure. Another significant observation of this study showed that stability of hydrogen–air flame increases with increasing equivalence ratio, whereas it decreases with increase in initial pressure.

4.8.5 Compression Ratio

It would be too broad a statement to say that compression ratio causes backfire. In fact, compression ratio influences several engine-operating conditions that contribute to backfire. A high compression ratio leads to high combustion temperature and pressure inside the engine cylinder, which causes backfire.

There is evidence that engines operating at lower compression ratios with relatively more residual exhaust gases heat fresh hydrogen–air mixtures quickly, causing backfire. As shown in Figure 4.64 [77], the residual gas fraction exhibits a decreasing trend along with increase in compression ratio and decreases with clearance volume. Due to the high clearance volume and lower compression ratio ranges, the residual gas fraction is about 25% of the highest among the three compression ratios.

4.8.6 Slow Combustion of Mixture, Cold Start Condition

Saito et al. observed that "there is an extremely slow burning cycle that continuously burns even at the exhaust stroke" [78] and that the stage of mild combustion, due to the slow burning, continues even after the exhaust stroke. Due to slow burn, a fresh hydrogen–air mixture is ignited at the start of the intake stroke. Conditions of very slow burning that caused backfire were detected in this study by measuring pressure at the intake and exhaust ports and using a high-speed CCD camera to visualize the process. Wei et al. [79] designed a constant volume combustion bomb (CVCB) equipped with a perforated plate to study the effect on combustion of a hydrogen–air mixture diluted with argon, nitrogen, and carbon oxide in different proportions. Researchers used high-speed schlieren photography to capture significant features of combustion such as flame propagation, compression wave formation, and pressure oscillation. Figure 4.65 shows flame propagation for a 5 bar H_2–air mixture and a 5 bar H_2–air mixture with an additional 2 bar of inert gas. The laminar flame passes through the perforated plate, initiating a jet flame that subsequently develops into a turbulent flame.

Furuhama et al. [80] performed several exhaustive studies on backfire during idling and cold start engine conditions. The reason for frequent backfires was similar to that discussed earlier.

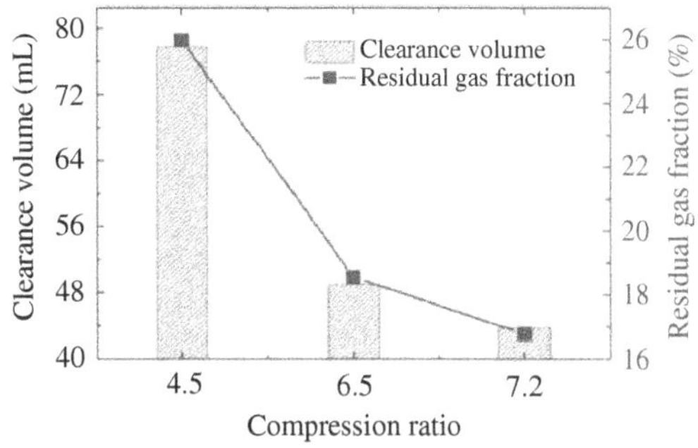

Figure 4.64 Variance of clearance volume and residual gas frictions with compression ratio [77] / with permission of Elsevier.

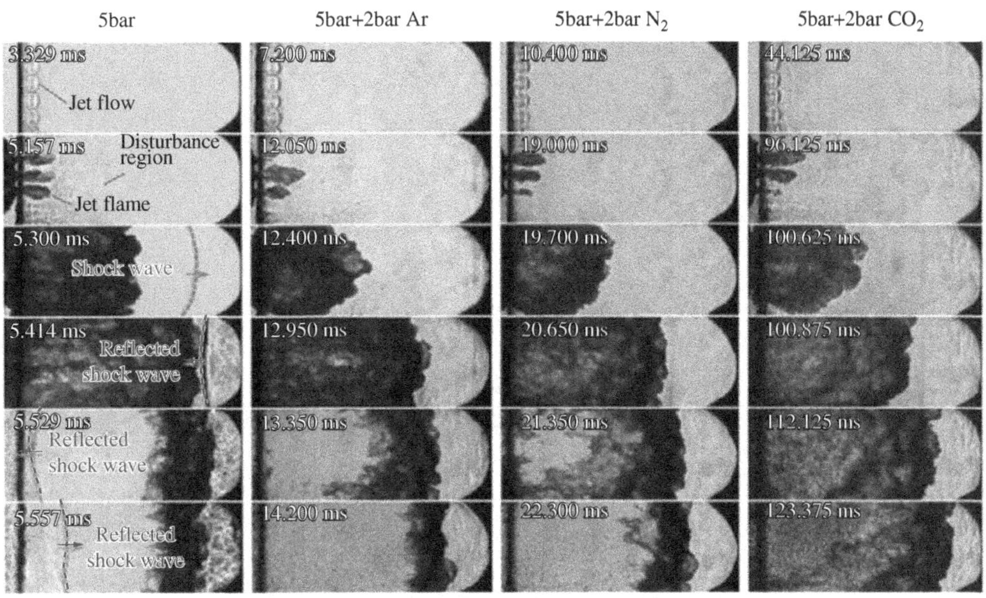

Figure 4.65 High-speed schlieren images of flame propagation under different initial pressures of 5 bar H_2–air mixture and 5 bar H_2–air mixture with additional 2 bar inert gas [79] / with permission of Elsevier. Please visit the companion website to view a color version of this image.

The higher amount of residual exhaust gas during idle conditions slowed down the combustion rate. This led to a situation in which some amounts of unburned fuel remained inside the cylinder after the combustion process. Under such circumstances, combustion was also extended to the start of the intake process. As shown in Figure 4.66, when the hydrogen engine was running with an air–fuel ratio (λ) less than 1.5 and with a speed of less than 1700 rpm, the engine stopped abruptly without backfire symptoms [80]. In such a case, since the combustion was completed too early, the engine's flywheel was unable to overcome the compression work and stopped. A different picture emerges from Figure 4.67, where preignition occurred at a higher engine speed of 2100 rpm and advanced cycle by cycle sequentially toward the valve opening time, leading to backfire.

Backfire due to very lean mixtures is rare and occurs only when the firing energy is inadequate to stop the engine and thus causes rough running. As discussed earlier, some residual gases remaining in the intake or exhaust system from the previous cycle are exposed to sparks or other hot spots and lead to backfire, which usually takes place in the intake or exhaust manifold rather than in the combustion chamber. The possibility of backfire during cold start conditions arises because the

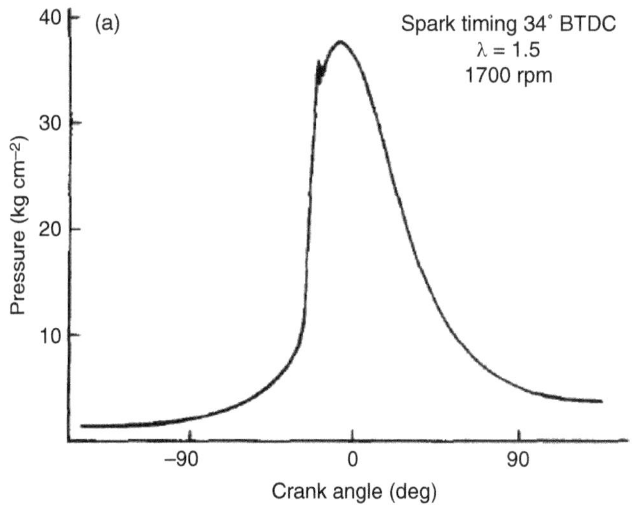

Figure 4.66 Extremely high-pressure rise without preignition [81] / with permission of Elsevier.

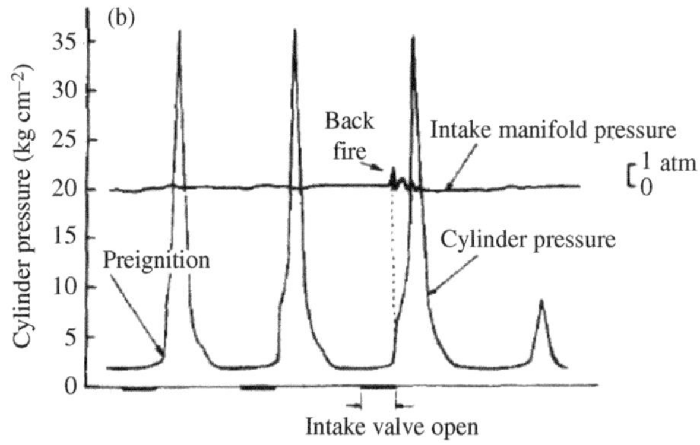

Figure 4.67 Process leading to backfire with preignition [80] / with permission of Elsevier.

low temperature of the cylinder walls leads to quenching effects and prevents hydrogen gas from complete combustion. The unburned hydrogen gas undergoes slow combustion and post-oxidation at the end of the combustion process. This slow burning and subsequent oxidation releases a significant amount of heat, increasing the likelihood of a backfire.

Cycle-by-cycle variation (CCV) in engines poses challenges for engine control. Ma et al. [81] investigated the impact of hydrogen addition on cycle-by-cycle variation (CCV) in a six-cylinder throttle body injection natural gas (NG) engine. Two fuels, compressed natural gas (CNG) and HCNG (with 20% hydrogen by volume), were used for comparative analysis. Hydrogen addition resulted in simultaneous reduction in the duration of flame development as well as flame propagation. CCVs in flame propagation duration led to improved combustion stability. With hydrogen addition, combustion phases were more stable, and the ignition timing could be effectively optimized. NOx emissions obviously increased by hydrogen addition at a given lambda and ignition timing. However, due to the fast burn rate of hydrogen, ignition timing was retarded for MBT operation of the engine and thus there was no increase in the NOx emission level. Figure 4.68 shows the standard deviation between ignition and 50% MFB. Standard deviation of 0–50% burn duration is often considered for combustion phasing fluctuation, as a CA of 50% is conventionally

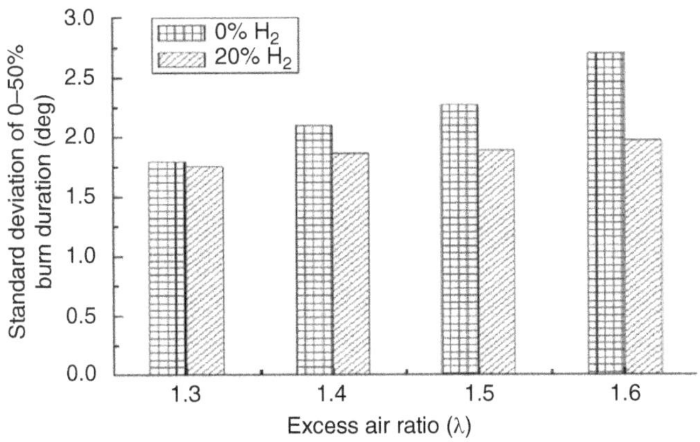

Figure 4.68 Standard deviation of combustion duration at MBT at various excess air ratios [82] / MDPI / CC BY 4.0.

a standard parameter in combustion phasing analysis. As is evident in the figure, the standard deviation increases from approximately 1.75° CA to 2.7° CA when λ increases from 1.3 to 1.6.

Saaidia et al. [82] analyzed CNG–hydrogen blends through numerical simulation using SWFS code for CFD. The optimal ratio for smooth engine operation was identified as 40% of hydrogen in the blend. Severe backfire problems, ranging from operational disturbances to complete engine shutdown [82], were encountered beyond 50%. Niu et al. [83] performed tests to evaluate the effect of hydrogen proportion on the performance of a dual fuel SI engine using hydrogen direct injection. The engine was operated in lean-burn mode at a constant speed of 1500 rpm with varying hydrogen fractions with varying excess air ratios. At a spark advance of 19° CA and an excess air ratio of 1.2, increased hydrogen fraction led to higher in-cylinder pressure and heat release rate. An addition of 7.2% hydrogen increased the in-cylinder pressure by 42%. At retarded MBT, both flame development duration and combustion duration decreased with increase of hydrogen addition under lean-burn conditions, even though this effect was not significant under stoichiometric conditions.

4.8.7 Exergy Analysis

Proper ignition timing's significance cannot be underestimated in its relevance to backfire in hydrogen engines. Extreme care is needed to ensure optimal ignition timing under high engine speed, low engine load, lean-burn, high EGR, and cold start conditions to avoid problems of different character. Lee and Kim [84] performed an exergy analysis on the basis of second law of thermodynamics considering variation in excess air ratio.

Tests were conducted on a commercial automotive spark ignition engine modified for hydrogen operation at 1500 rpm with various spark timings and excess air ratios under a wide-open throttle. The study also included a comparative evaluation of experimental results with results of zero-dimensional two-zone modeling both the first law (energy) and second law (exergy) of thermodynamics. Parameters such as increase in exhaust exergy loss and decreased heat transfer exergy loss by retarded ignition timing at the same excess air ration were calculated. The analysis showed that the trend of results with exergy analysis agreed with the energy analysis results. Moreover, irreversibility showed an increasing trend with retarded spark timing because of prolonged combustion duration leading to reduction in combustion temperature. Results showed that as the air excess ratio increased, efficiency improved, with values of 40.64% and 43.02% at $\lambda = 1.4$ and 2.0, respectively. Beyond $\lambda = 1.6$, increased air ratio and retarded ignition prolonged combustion and brought down the in-cylinder temperature. According to the second law of thermodynamics, this resulted in decreased heat transfer loss and increased exhaust loss. Delaying ignition also increased combustion irreversibility, reducing efficiency. Conversely, advancing spark timing increased thermo-mechanical (TM) exergy, leading to exergy loss in heat transfer.

4.8.8 Misfire

Misfire and backfire are two prominent undesirable features of combustion. Misfire refers to a condition in which ignition of fuel–air mixture does not take place at the intended time. On the other hand, backfire occurs when combustion of mixture takes place outside the combustion chamber, mostly in the intake or exhaust system. Sometimes misfire develops conditions for backfire to occur in the following cycle because of accumulated unburned fuel in the cylinders. Verhelst et al. [85], in their study on a bi-fuel hydrogen/gasoline engine, established a relationship between misfire and backfire. The researchers observed that spark plug accommodated with the pressure sensor was not an appropriate one for hydrogen operation (with higher heat ranges), and they believed that use of an inappropriate spark plug was responsible for causing misfires and subsequently led to a backfire in the following cycle. Figure 4.69 shows misfire and Figure 4.70 shows the occurrence of backfire. Backfire cycles are shown in bold. Figure 4.71 shows a combination of both diagrams [85]. During misfire, the air–fuel mixture in the cylinder is not ignited properly, so the character of combustion is erratic. This might lead to a situation in which the air and hydrogen mixture gets accumulated in the cylinder. In the absence of proper combustion, the heat generated from the combustion process is absent and the temperature of a misfired cycle can be very low, perhaps about 5000 °C. The possibility of a hot spot forming at such low temperature is ruled out. Because of the misfire and accumulation of unburned fuel, the equivalence ratio of the following cycle increases. With no thermal energy source to trigger ignition, misfire alone cannot cause abnormal combustion.

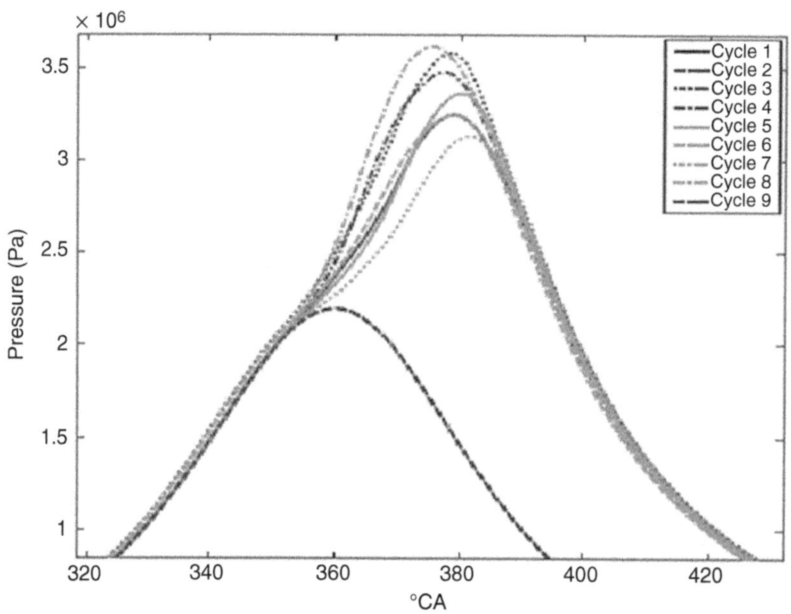

Figure 4.69 Cyclic variation in in-cylinder pressure and the effect of misfires due to a failing spark plug for hydrogen at wide-open throttle (WOT) [85] / with permission of Elsevier.

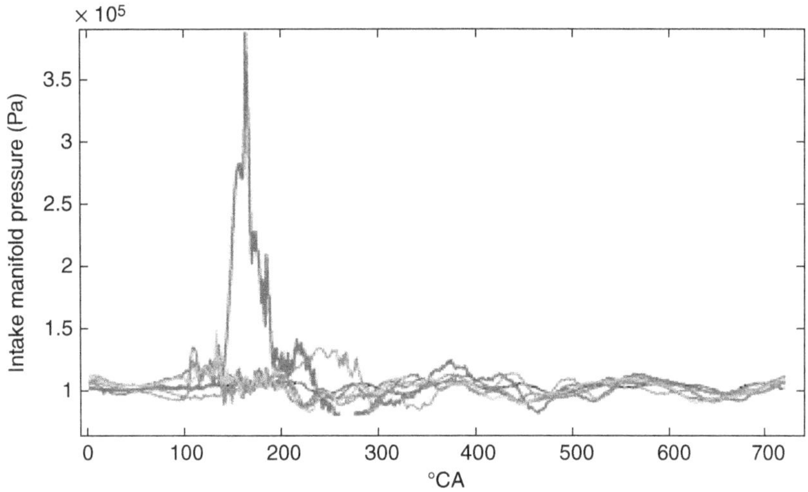

Figure 4.70 Backfire due to misfires: hydrogen at WOT. Intake manifold pressure measurements of seven consecutive cycles [85] / with permission of Elsevier. Please visit the companion website to view a color version of this image.

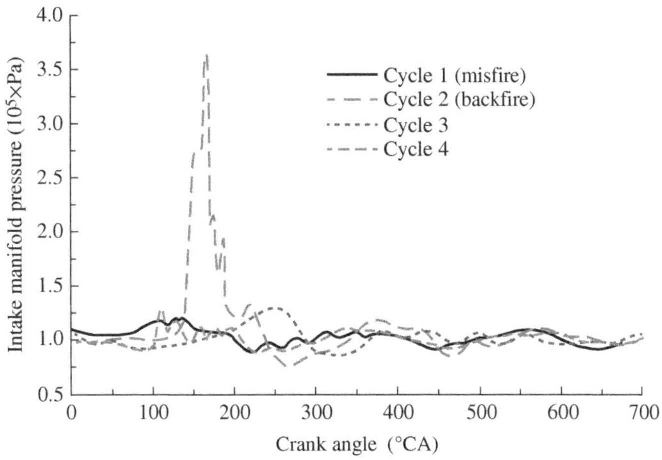

Figure 4.71 Backfire caused by misfire under WOT conditions; TDC: 180° CA [85] / with permission of Elsevier.

MacCarley's research [86] suggested that excessive temperatures in the combustion chamber were mainly responsible for induction ignition leading to backfire in a hydrogen engine. The most probable locations studied were large thermal mass surfaces like cylinder walls, valves, pistons, or spark plugs. Backfire could also have resulted from low thermal mass sites with cyclically varying temperatures, such as surface deposits, suspended pyrolysis products, or hot residual exhaust gases. Thus, methods to control backfire depend upon maintaining acceptable low average temperatures on large thermal mass surfaces and ensure convective cooling for small thermal mass sites before fuel delivery to the cylinder. Steps adopted included improved cooling of cylinder heads, pistons, and valves, and use of very cold spark plugs to achieve lower average temperatures. Design and development of a high-speed injection valve for electronic fuel injection proved successful in eliminating backfire [87, 88]. Researchers at Cornell University have continued analytical study and research in the laboratory on hydrogen engines to identify practical system development solutions to control backfire [89–92]. Homan adopted LIRIAM (late injection rapid ignition and mixing) [88] and did an exhaustive study of hydrogen engines and eliminated backfire. The corresponding equivalence ratios to achieve minimum NOx emission under these conditions were also identified. Physical location of the spark plug was crucial to the hydrogen jet for smooth combustion.

The combined effect of delayed ignition timing and water injection at the intake was discussed earlier, but assessment of optimum water injection is also critical to avoid structural challenges due to cylinder corrosion and deteriorating lubrication performance [64]. Liu et al. [93] used CFD simulation to study the effect of injection timing on mixture formation in manifold injection hydrogen engines under varying conditions of equivalence ratio and engine speed. The analysis shows that in a manifold injection hydrogen engine the likelihood of backfire is intricately tied to the concentration of hydrogen residual at the intake ports. Backfire decreases as the concentration of hydrogen residual in the intake ports becomes leaner. Injection timing and injection duration were found to be crucial in the occurrence of backfire.

Research pursuits for years on hydrogen engines at the Indian Institute of Technology, Delhi put a lot of emphasis on the long-term development strategy of appropriate fuel induction mechanisms to avoid the backfire problem. It was observed during several tests that timed fuel injection, whether mechanically or electronically controlled, allows for delayed fuel delivery, facilitating pre-cooling of small thermal mass sites and residual exhaust gases. So an appropriately designed timed injection system was successful in providing an eventual solution to backfire. Three different injection system designs were developed based on hydraulically operated, mechanically actuated, and electronically controlled solenoid-actuated systems to eliminate backfire [94–97].

4.8.9 High Coolant and Lubricating Oil Temperature

The thermal status of coolant and lubricating oil plays vital roles in maintaining the temperature of the cylinder walls and engine block. This information provides relevant guidelines to estimate the in-cylinder heat losses and to assess the situation leading to the formation of hot spots during combustion. During cold start conditions, when the temperature of both the coolant and lubricant are relatively low, hot spots are less likely to form around the exhaust valve seat or in the spark plugs resulting in hot spot-induced backfire. A high port temperature heated up the mixture and caused backfire in cylinder 4. Figure 4.72 [98] shows that the intake air temperature of cylinder 4 increases rapidly because of backfire followed by cylinder 2. Other cylinders, with relatively far less temperature, were less affected by backfire. Backfire, which occurred close to the mounting of cylinder 4, resulted in melting plastic material and generated a hole through which the hydrogen–air mixture leaked. This study also examined the effect of speed and cylinder-to-cylinder variation due to backfire in a multi-cylinder hydrogen engine.

Figure 4.73 shows the coolant flow in engine block and cylinder head where coolant flows through cylinder 1 to cylinder 4 in a definite sequence [98]. Water for cooling the combustion chamber is supplied from the inlet at the lower side of cylinder 1. It flows from engine block and cylinder as shown in the figure and exits from the head of cylinder 4. Any amount of cooling water, which somehow or other could not pass through this route, cooled by the radiator went to the inlet at the lower portion of cylinder 1. As cool water came in contact with the heat of combustion, it led to a situation in which "temperature of intake port of cylinder 4 near the exit portion becomes higher than that of the other cylinders" [98].

This study highlights some critical observations for multi-cylinder studies with hydrogen. It is important to install a sensor at the appropriate location in multi-point injection to detect backfire.

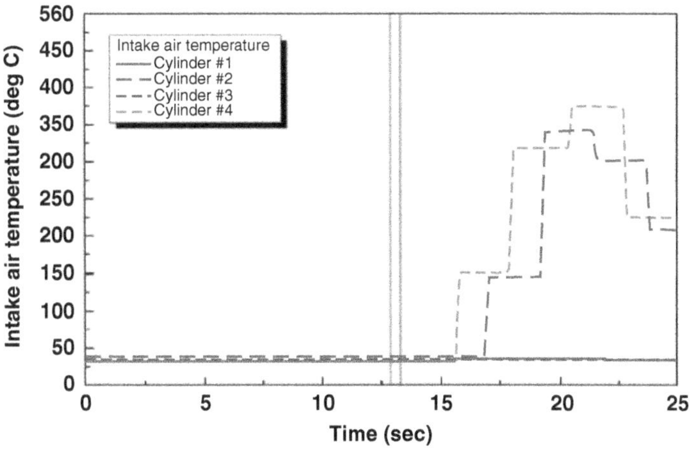

Figure 4.72 Intake air temperature changes in time for each cylinder with the gradual increase in hydrogen fuel input at an engine speed of 2000 rpm [98] / with permission of Elsevier.

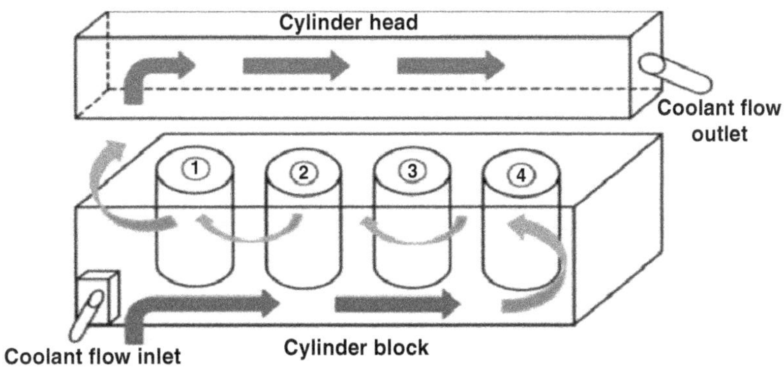

Figure 4.73 Coolant flow directions in the cylinder block and cylinder head [98] / with permission of Elsevier.

Optimum lubrication is crucial. A thin wall, formed between the surfaces, ensures conditions such that direct metal-to-metal contact is avoided, and effect of friction and chances of subsequent wear are substantially reduced. In this context, lubricant plays a critical role. If the oil temperature is very low, it becomes thick and cannot flow to lubricate engine components. On the other hand, thin lubricating oil has low viscosity and cannot effectively lubricate. Inadequate lubrication leads to direct metal-to-metal contact that often results in overheating and high friction, resulting in damage to the cylinder wall.

4.8.10 Abnormal Discharge of Spark Plugs

Kondo et al. studied backfires encountered due to "abnormal discharge in the intake stroke" [99]. Hydrogen flames have lower ion concentrations because of hydrogen's low ignition energy. So even a small amount of residual exhaust energy from the previous cycle, in a sensitive area, can ignite fresh hydrogen–air mixture.

Modification of the ignition system was not successful in preventing residual energy and overcome backfire at an excess air ratio around λ of 1.6. This study also agrees with some other researcher results that a cold spark plug with a -platinum tip very much reduces the chances of the tip of the spark plug igniting the air–fuel charge. The basic design approach of hot-rated spark plugs is to maintain a certain amount of heat to avoid situations of high carbon deposits. Hydrogen, which contains no carbon, does not need hot spark plugs.

Platinum-tip spark plugs should be avoided for hydrogen engines because platinum being a catalyst can cause hydrogen to oxidize with air. Sometimes, combustion in crevices extends into the intake process, causing burning gas to contact the fresh hydrogen–air mixture. The mixture gets ignited, causing a flame propagating into the intake manifold. Such an undesirable combustion phenomenon usually occurs under cold start, idle, or low load engine conditions. It has been adequately emphasized that an effective injection system has relevant injection characteristics such as injection pressure, injector actuation mechanism, injection timing, injection duration, and jet characteristics, which are extremely crucial strategies to avoid backfire in hydrogen engines [61, 67, 100–103]. Some research shows that for engines running at lower speeds, the injection should be timed early and the mixture be rich. Liu et al. [93] designed a hydrogen injector to evaluate the injector characteristics and to assess the amount of hydrogen injected at a pressure of 2 bar over a period of injection duration of 24 ms. Figure 4.74 shows the injected hydrogen mass following a linear increasing trend with longer injection duration, depending on the crank angle position and engine speed. One of the major objectives of this study was to avoid backfire conditions with a narrow range of hydrogen injection duration limited to 24 ms. The parameters would be different in the case of a high-speed engine aiming at higher amounts of hydrogen delivery into the engine cylinder per cycle to have higher power output.

Liu's simulation results, as depicted in Figure 4.75, show that in the event of delaying the end of hydrogen injection, the hydrogen fraction (residual mixture at the intake port) drops down initially and then increases rapidly. It was also observed that if the end of injection was scheduled earlier than 500°, the hydrogen fraction in the intake port remains almost the same. But if the injection ends after 520°, delaying the timing will leave more hydrogen residual in the intake port after the intake valve closes. Low engine speed and rich mixture cause early injection end.

4.8.11 Injection System: Timing, Pressure, and Duration

The contribution by Duan and team [67] to control backfire and ensure improvement in power output in a port-injected hydrogen engine was briefly highlighted earlier. Pressure fluctuations are reliable indicators of backfire. The pressure–crank angle diagram generated in normal engine

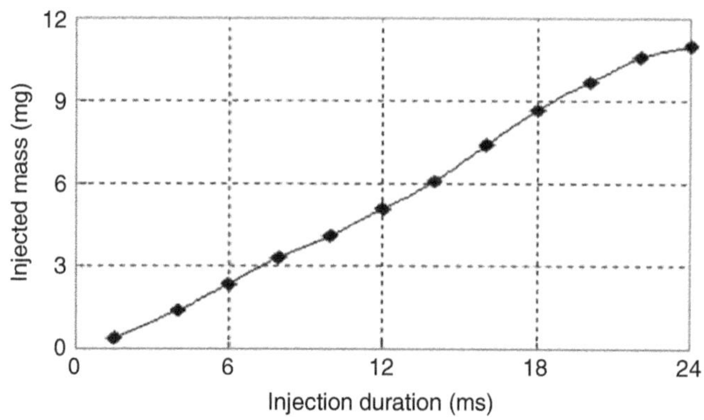

Figure 4.74 Hydrogen injection mass at 2 bar [93] / with permission of Elsevier.

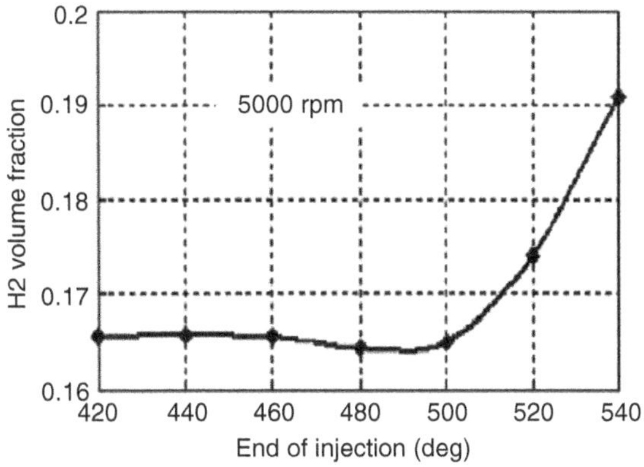

Figure 4.75 Hydrogen residual at intake port at 5000 rpm [93] / with permission of Elsevier.

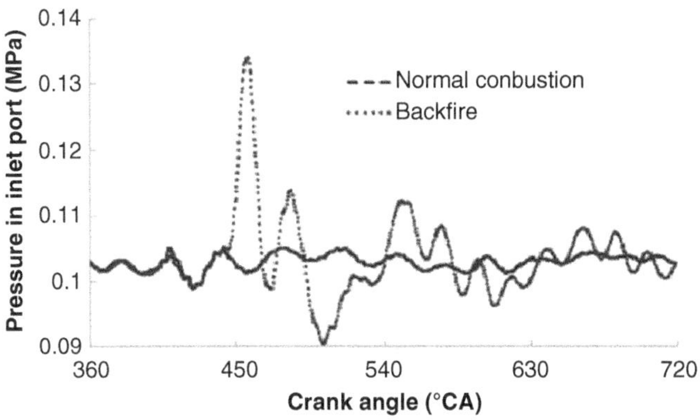

Figure 4.76 Pressure fluctuations in the inlet with and without backfire [67] / with permission of Elsevier.

combustion tests shows slow and mild fluctuation in the pressure in the inlet, with the maximum value limited to 0.11 MPa. In sharp contrast to normal combustion, a cycle exhibiting backfire shows a rapid change in inlet pressure variation (Figure 4.76) and has a higher maximum value close to 0.136 MPa [67].

Two sets of experiments with two different operating conditions were conducted to ascertain backfire causes. In one set of tests, the engine operated at 3000 rpm with an equivalence ratio of 0.65 and continuous hydrogen injection. The injection pressure was 0.2 MPa, scheduled at an injection angle of 141° CA. In the other set of tests, the engine operated at 5000 rpm with the continuous injection angle set at 210° CA. The equivalence ratio (0.65) and injection pressure (0.2 MPa) remained the same for both test sets. Peak pressure and pressure rise rates were measured at intervals of 30° CA, starting from injection at 168° CA until 468° CA. Figure 4.77 demonstrates that the start angle of the hydrogen injection valve is strongly correlated with backfire.

In the case of backfire, an extremely small or large start angle increases the maximum pressure and pressure rise rate inside the inlet. Late hydrogen injection timing raises the maximum pressure and rate of pressure rise, largely leading to backfires of higher intensity. Pressure changes are minimum when the start angle is in the medium range, and it indicates no backfire. Engine speed also affects backfire. Figure 4.77 shows that at 3000 rpm and a fuel–air ratio of 0.65, backfire is avoided when the start angle is between 258° and 378° CA. At 5000 rpm with the same fuel–air

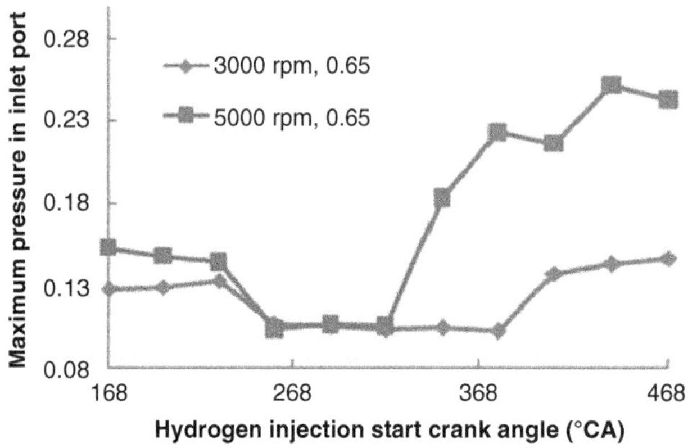

Figure 4.77 Variation of maximum pressure at the inlet port with different hydrogen injection start angles [67] / with permission of Elsevier.

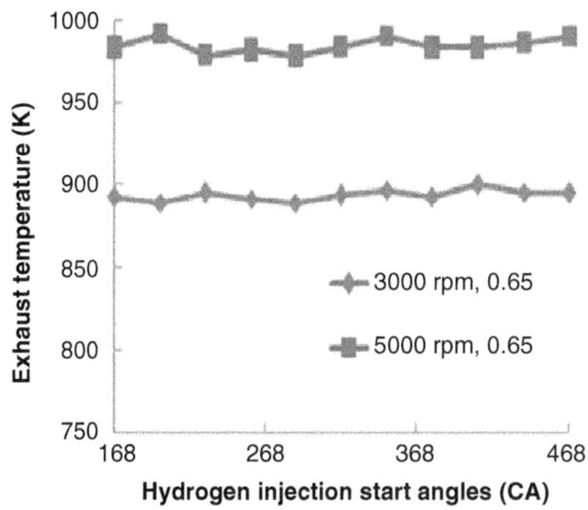

Figure 4.78 Exhaust temperature at different hydrogen injection start angles [67] / with permission of Elsevier.

ratio, the range narrows to 258–318° CA. High rotation speeds require a narrow start angle range to prevent backfires whereas backfire becomes more intense at higher speeds due to increased maximum pressure and pressure rise rate.

Figure 4.78 shows exhaust temperature variations with different hydrogen injection timings under two conditions. At 3000 rpm, exhaust temperature ranged from 889 to 901 K, while at 5000 rpm, it was between 978 and 993 K. Both exceeded the hydrogen autoignition temperature of 858 K. A higher exhaust temperature at 5000 r/min led to more intense backfire, confirming that high-temperature exhaust gas backflow was the cause of backfire.

Using an inlet port fuel injection system, the researchers analyzed backfire under high loads with a computational fluid dynamics model optimizing temperature and mixture concentration in the intake valve. The analysis aimed at optimizing injection timing and hydrogen injection pressure.

Simulated data of local equivalence ratios and local temperatures are shown in Figure 4.79. Figure 4.80 shows both with respect to the start of injection. These data refer to the experimental conditions of speed (3000 and 5000 rpm), equivalence ratio of 0.65, and injection angle between 168° and 468° CA. At a speed of 3000 rpm and during the hydrogen start between 258° and 378° CA,

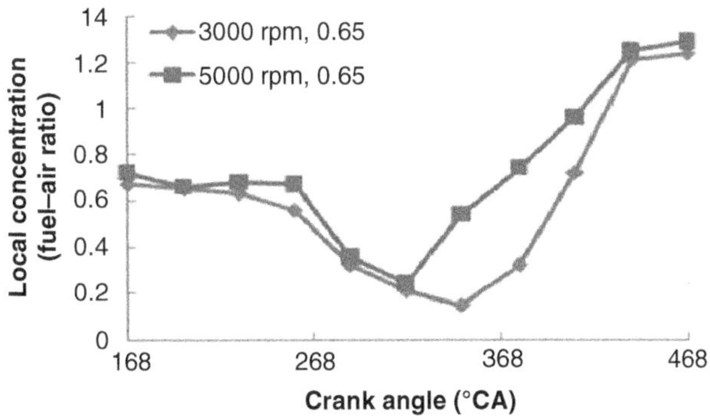

Figure 4.79 Local concentration at different hydrogen injection start angles [67] / with permission of Elsevier.

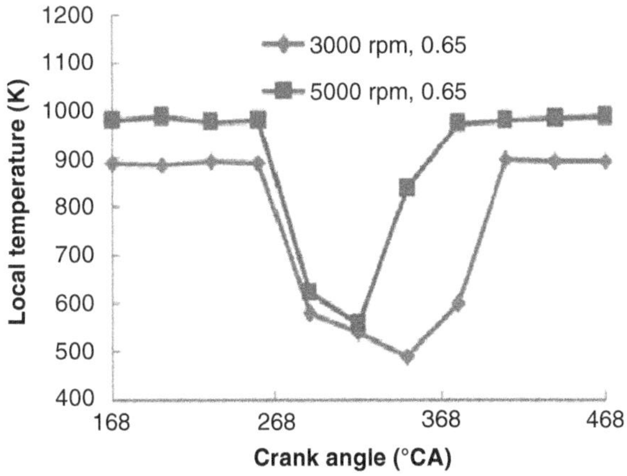

Figure 4.80 Local temperature under different hydrogen injection start angles [67] / with permission of Elsevier.

local concentration was less than 0.5, and local temperature was below 600 K. There was no backfire under this condition. At an injection start angle of 348° CA, both local temperature and local concentration were low, which indicates that 348° CA is the most desirable optimized timing to avoid backfire. On the other hand, at a speed of 5000 rpm, the backfire-free zone was within the range of 258° to 318° CA. The local concentration was less than 0.5 and local temperature was below 680 K. In this case, as in the figure, 318° CA is the optimum angle for injection start where local concentration is low enough (0.24) and temperature is 560 K.

The potential of lean burn to control backfire has been well documented. Lean burn can be accomplished by lowering the combustion temperature and by adjusting the injection timing, which can be accomplished by reducing the injection duration and either by retarding the start of injection or by advancing the end of injection. Of course, these adjustments need to be scrupulously carried out with appropriate valve timing to avoid improper distribution around the valve seats. Depending on the engine configuration, high injection pressure and multiple injectors may be necessary to achieve the desired fuel injection rate

Figure 4.81 shows that at 5000 rpm with a stoichiometric mixture injected at 0.35 MPa, the duration is around 280° CA reduced to 208° CA at 0.5 MPa. Higher injection pressure ensures adequate hydrogen supply to the engine cylinder under high load conditions with shorter injection

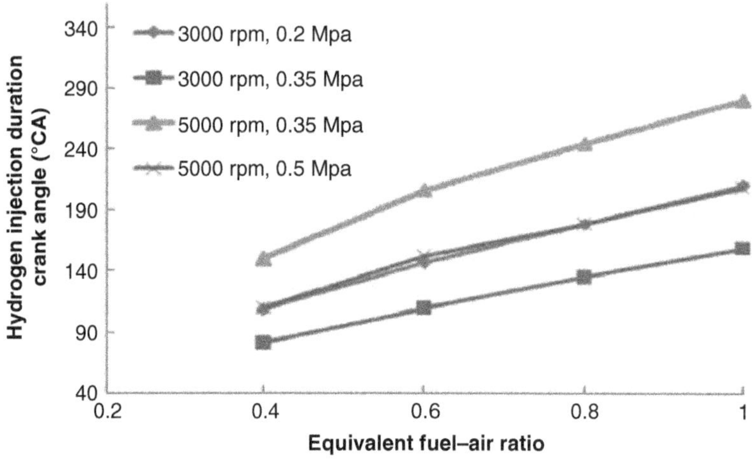

Figure 4.81 Hydrogen injection duration at different injection pressures [67] / with permission of Elsevier.

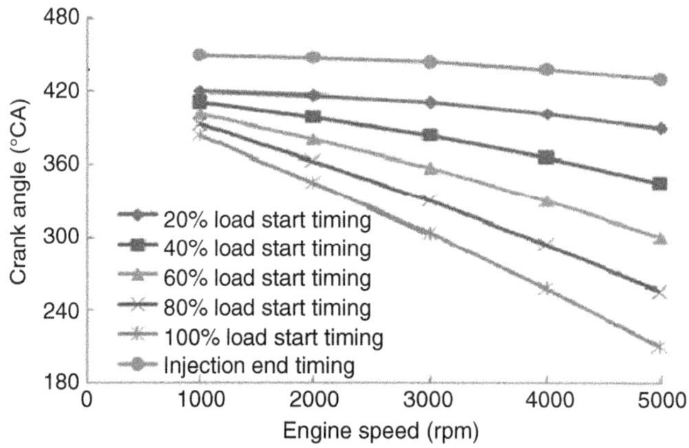

Figure 4.82 Optimized injection duration under different speeds and loads [67] / with permission of Elsevier.

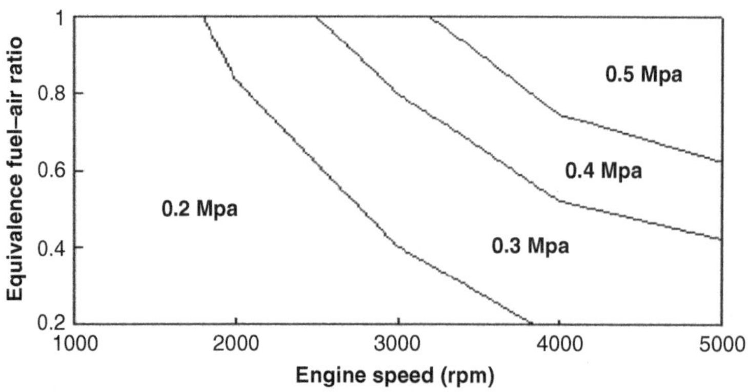

Figure 4.83 Injection pressure optimized under different speeds and loads [67] / with permission of Elsevier.

durations. Optimizing injection timing at high pressures allows the engine to achieve higher power output and operate at higher speeds without backfiring. Appropriate injection timing controls local temperature and equivalence ratio and prevents backfires. Injection duration, defining the start and end of injection, investigated through simulation, as shown in Figure 4.82 [67].

Early injection timing ensures complete entry of an adequate amount of hydrogen into the engine cylinder. The severe thermal environment of the engine cylinder makes it difficult for the injection valve to survive longer and operate effectively. Therefore, depending upon engine operating conditions, optimum injection strategies need to be adopted. As shown in Figure 4.83, engines that operate at low speeds and with low loads should be operated at low injection pressures, while engines operating at high speeds and loads should be operated at high injection pressures.

Concluding Remarks and Perspectives

Undesirable combustion in a hydrogen engine is a complex combination of preignition, knock, and backfire. Sometimes one of these problems gives rise to another. Severe thermal environments have been one of the major causes at specific locations during engine cycle operation. Most of the research activities on hydrogen engines have focused on reducing the temperature level only in the cylinder. Preignition is a random event. It has not been possible to measure accurately the time, place, and intensity of preignition. In the case of hydrogen engines, it often leads to backfire. Extensive research has brought out that parameters such as equivalence ratio, compression ratio, start of injection (SOI), injection timing, and increased injection pressure are crucial to the phenomenon of knock in hydrogen engines. Equivalence ratio has a more pronounced effect on knock than inlet pressure and ignition timing. The heightened pressure facilitates a more even distribution of the hydrogen–air mixture, acting as a preventive measure against knock events. Lowering equivalence ratios or inlet pressures and retarding ignition timing has influenced knock intensity. Exhaust gas recirculation (EGR) has been a useful method to reduce combustion temperature in the cylinder, lowering knock intensity.

Backfire is mistimed fuel firing, which usually occurs inside the cylinder due to fuel ignition before the intake valve closes. This results in firing from the inlet side. The sources of preignition may be certain hot spots in the cylinder, such as crevice volumes, hot spark plugs, hot valves, etc., improper valve timing, and fuel injection timing. Most backfire factors, excluding abnormal spark plug discharges, ultimately hinge on the combustion chamber temperature during gas exchange. Typically, factors leading to backfire initially manifest as preignition, eventually developing backfire events. Lean-burn strategies, adjustments in fuel injection timing and spark timing, and implementation of effectual EGR decrease the combustion temperature and mitigate the formation of hot spots. Additionally, adjusting intake strategies, such as separating intake pipes and connecting to individual intake ports, coupled with effective hydrogen injection prove efficient in controlling backfire. An appropriately designed fuel injection system has been effective to avoid problems with backfire. Multi-objective optimizations and application of intelligent neural network methods provide a comprehensive means of controlling backfire in hydrogen internal combustion engines.

References

1 Strehlow, R.A. (1984). *Combustion Fundamentals*. New York: McGraw Hill.

2 Lewis, B. and Von Elbe, G. (1961). *Combustion Flame and Explosion of Gases*, 2e. New York: Academic Press.

3 Das, L.M. (1996). Hydrogen-oxygen reaction mechanism and its implication to hydrogen engine combustion. *International Journal of Hydrogen Energy* 21 (8): 703–715.
4 Drell, I.L. and Belles, F.E. (1957). Survey of hydrogen combustion properties. *Technical Report* 1383, NACA–National Advisory Committee for Aeronautics.
5 Jagadish, D., Nageswara Rao, A.V., and Kumar, M.S. (2023). Engine combustion and emission analysis using optical methods: An overview. *Materials Today: Proceedings*. https://doi.org/10.1016/j.matpr.2023.05.588.
6 Iijima, T. and Takeno, T. (1986). Effects of temperature and pressure on burning velocity. *Combustion and Flame* 65: 35–43.
7 Dowdy, D.R., Smith, D.B., Taylor, S.C. et al. (1990). Laminar burning velocities and stretch effects in hydrogen/air mixtures. *23rd Symposium (International) on Combustion* 23: 325–332.
8 Qin, X., Kobayashi, H., and Niioka, T. (2000). Laminar burning velocity of hydrogen-air premixed flames at elevated pressure. *Experimental Thermal and Fluid Science* 21: 58–63.
9 Kwon, O.C. and Faeth, G.M. (2001). Flame/stretch interactions of premixed hydrogen-fueled flames: measurements and predictions. *Combustion and Flame* 124: 590–610.
10 Edmondson, H. and Heap, M.P. (1971). The burning velocity of hydrogen–air flames. *Combustion and Flame* 16: 161–165.
11 Andrews, G.E. and Bradley, D. (1973). Determination of burning velocity by double ignition in a closed vessel. *Combustion and Flame* 20: 77–89.
12 Liu, D.D.S. and MacFarlane, R. (1983). Laminar burning velocities of hydrogen–air and hydrogen–air-steam flames. *Combustion and Flame* 49: 59–71.
13 Milton, B.E. (1984). Laminar burning velocities in stoichiometric hydrogen and hydrogen–hydrocarbon gas mixtures. *Combustion and Flame* 58: 13–22.
14 Aung, K.T., Tseng, L.K., Ismail, M.A. et al. (1995). Brief communication, response to comment by S.C. Taylor and D.B. Smith on laminar burning velocities and Markstein numbers of hydrocarbon/air flames. *Combustion and Flame* 102: 526–530.
15 Law, C.K. and Kwon, O.C. (2004). Effects of hydrocarbon substitution on atmospheric hydrogen–air flame propagation. *International Journal of Hydrogen Energy* 29: 867–879.
16 Wierzba, I. and Kilchyk, V. (2001). Flammability limits of hydrogen–carbon monoxide mixtures at moderately elevated temperatures. *International Journal of Hydrogen Energy* 26: 639–643.
17 Dahoe, A.E. (2005). Laminar burning velocities of hydrogen–air mixtures from closed vessel gas explosions. *Journal of Loss Prevention Process Industries* 18: 152–166.
18 Kwon, O.C., Tseng, L-K., and Faeth, G.M. (1992). Laminar burning velocities and transition to unstable flames in H2/O2/N2 and C2H8/O2/N2 mixtures. *Combustion and Flame* 90: 230–246.
19 Chu, H., Berger, L., Grenga, T. et al. (2023). Effects of differential diffusion on hydrogen flame kernel development under engine conditions. *Proceedings of the Combustion Institute* 39: 2129–2138.
20 Ilbas, M., Crayford, A.P., Yılmaz, I. et al. (2006). Laminar-burning velocities of hydrogen–air and hydrogen–methane–air mixtures: An experimental study. *International Journal of Hydrogen Energy* 31: 1768–1779.
21 Liu, X., Ji, C., Gao, B. et al. (2013). A laminar flame speed correlation of hydrogen–methanol blends valid at engine-like conditions. *International Journal of Hydrogen Energy* 38 (35): 15500–15509.
22 Yan, Y., Liu, Z., and Liu, J. (2022). Hydrogen enriched ammonia engines: Assessment of hydrogen concentration in the fuel feed from the laminar flame speed viewpoint. *Proceedings of the ASME 2022 ICE Forward Conference*. Indianapolis, Indiana, USA. October 16–19. V001T02A002, ASME. https://doi.org/10.1115/ICEF2022-88682.

23 Yan, Y., Liu, Z., and Liu, J. (2023). An evaluation of the conversion of gasoline and natural gas spark ignition engines to ammonia/hydrogen operation from the perspective of laminar flame speed. ASME. *Journal Energy Resources Technology* 145 (1): 012302. https://doi.org/10.1115/1.4054754.

24 Wu, M.-S., Kwon, S., Driscoll, J.F. et al. (1990). Turbulent premixed hydrogen/air flames at high Reynolds numbers. *Combustion Science and Technology* 73: 327–350.

25 Kido, H., Nakahara, M., Nakashima, K. et al. (2003). Turbulent burning velocity of lean hydrogen mixtures. SAE Technical Paper No. 2003-01-1773.

26 Chen, J.H. and Im, H.G. (2000). Stretch effects on the burning velocity of turbulent premixed hydrogen/air flames. *28th Symposium (International) on Combustion* 211–218.

27 Im H.G. and Chen, J.H. (2002). Preferential diffusion effects on the burning rate of interacting turbulent premixed hydrogen-air flames. *Combustion and Flame* 131: 246–258.

28 Heywood, J.B. and Vilchis, F.R. (1984). Comparison of flame development in a spark-ignition engine fueled with propane and hydrogen. *Combustion Science and Technology* 38: 313–324.

29 Giménez, B., Melgar, A., Horrillo, A. et al. (2021). A correlation for turbulent combustion speed accounting for instabilities and expansion speed in a hydrogen-natural gas spark ignition engine. *Combustion and Flame* 223: 15–27.

30 Pielecha, I., Merkisz, J., Urbański, P. et al. (2022). A numerical study of the effect of hydrogen fuelled turbulent jet ignition engine. SAE Technical Paper No. 2022-01-1007. https://doi.org/10.4271/2022-01-1007.

31 Al-Baghdadi, M.A.R.S. (2004). Effect of compression ratio, equivalence ratio and engine speed on the performance and emission characteristics of a spark ignition engine using hydrogen as a fuel. *Renewable Energy* 29 (15): 2245–2260.

32 Tang, X., Kabat, D.M., Natkin, R.J. et al. (2002). Ford P2000 hydrogen engine dynamometer development. SAE Technical Paper No. 2002-01-0242.

33 Yip, H.L., Srna, A., Yuen, A.C.Y. et al. (2019). A review of hydrogen direct injection for internal combustion engines: towards carbon-free combustion. *Applied Sciences* 9 (22): 4842.

34 Li, H. and Karim, G.A. (2004) Knock in spark ignition hydrogen engines. *International Journal of Hydrogen Energy* 29 (8): 859–865.

35 White, M., Steeper, R.R., and Lutz, A.E. (2006). The hydrogen-fueled internal combustion engine: a technical review. *International Journal of Hydrogen Energy* 31 (10): 1292–1305.

36 Poursadegh, F., Brear, M., Hayward, B. et al. (2023). Autoignition, knock, detonation and the octane rating of hydrogen. *Fuel* 332: 126201.

37 Malenshek, M. and Olsen, D.B. (2009). Methane number testing of alternative gaseous fuels. *Fuel* 88: 650–656.

38 Kawahara, N. and Tomita, E. (2009). Visualization of auto-ignition and pressure wave during knocking in hydrogen spark-ignition engine. *International Journal of Hydrogen Energy* 34: 3156–3163.

39 Nakada, T., Itoh, T., and Takagi, Y. (1993). Application of CARS to development of high compression ratio spark ignition engine. SAE Technical Paper No. 932644. http://doi.org/10.4271/932644.

40 Li, Y., Gao, W., Li, Y. et al. (2022). Numerical investigation on combustion and knock formation mechanism of hydrogen direct injection engine. *Fuel* 316: 123302.

41 Rockstroh, T., Kolodziej, C., Jespersen, M. et al. (2018). Insights into engine knock: Comparison of knock metrics across ranges of intake temperature and pressure in the CFR engine. *SAE International Journal of Fuels and Lubricants* 11 (4): 545–561.

42 Swarts, A., Yates, A., Viljoen, C. et al. (2004). Standard knock intensity revisited: Atypical burn rate characteristics identified in the CFR octane rating engine. SAE Technical Paper No. 2004-01-1850. https://doi.org/10.4271/2004-01-1850.
43 Manzoor, M.U., Yosri, M., Talei, M. et al. (2023). Normal and knocking combustion of hydrogen: a numerical study. *Fuel* 344: 128093.
44 Tang, X., Stockhausen, W.F., Kabat, D.M. et al. (2002). Ford P2000 hydrogen engine dynamometer development. SAE Technical Paper No. 2002-01-0242.
45 Jaura, A., Ortmann, W., Stuntz, R. et al. (2004). Ford's H2RV: An industry first HEV propelled with a HD2-fueled engine–a fuel efficient and clean solution for sustainable mobility. SAE Technical Paper No. 2004-01-0058.
46 Natkin, R.J., Tang, X., Whipple, K.M. et al. (2002). Ford hydrogen engine laboratory testing facility. SAE Technical Paper No. 2002-01-0241.
47 Natkin, R.J., Denlinger, A.R., Younkins, M.A. et al. (2012). Ford 6.8L hydrogen IC engine for the E-450 shuttle van. SAE Technical Paper No. 2007-01-4096.
48 Ciatti, S.A., Wallner, T., and Ng, H. et al. (2006). Study of combustion anomalies of H2-ICE with external mixture formation, ICES2006-1398. *Proceedings of ICES2006 ASME Internal Combustion Engine Division 2006 Spring Technical Conference May 7–9*, Aachen, Germany.
49 Diéguez, P.M., Urroz, J.C., Sáinz, D. et al. (2018). Characterization of combustion anomalies in a hydrogen-fueled 1.4 L commercial spark-ignition engine by means of in-cylinder pressure, block engine vibration, and acoustic measurements. *Energy Conversion and Management* 172: 67–80.
50 Brecq, G., Bellettre, J., and Tazerout, M. (2003). A new indicator for knock detection in gas SI engines. *International Journal of Thermal Sciences* 42: 523–532.
51 Lai, F.-Y., Sun, B.-G., Wang, X. et al. (2023). Research on the inducing factors and characteristics of knock combustion in a DI hydrogen internal combustion engine in the process of improving performance and thermal efficiency. *International Journal of Hydrogen Energy* 48: 7488–7498.
52 Luo, Q.-H. and Sun, B.-G. (2016) Inducing factors and frequency of combustion knock in hydrogen internal combustion engines. *International Journal of Hydrogen Energy* 41: 16296–16305.
53 Szwaja, S. (2011). Knock and combustion rate interaction in a hydrogen-fueled combustion engine. *Journal of KONES Powertrain and Transport* 18 (3): 431–438.
54 Roy, M.K, Kawahara, N., and Tomita, E. (2010). Visualization of knocking combustion in a hydrogen spark-ignition engine. *18th World Hydrogen Energy Conference 2010– Proceedings WHEC 2010*, Essen, Germany.
55 Topinka, J.A. (2003). Knock behavior of a lean-burn, hydrogen-enhanced engine concept. MS thesis. Massachusetts Institute of Technology.
56 Li, Y., Gao, W., Zhang, P. et al. (2021). Influence of the equivalence ratio on the knock and performance of a hydrogen direct injection internal combustion engine under different compression ratios. *International Journal of Hydrogen Energy* 46: 11982–11993.
57 Das, L.M. (1994). *Combustion and Emission Characteristics of Hydrogen Engines*. Florida, USA: Hydrogen Energy Progress X.
58 Dhyani, V. and Subramanian, K.A. (2019). Fundamental characterization of backfire in a hydrogen fuelled spark ignition engine using cfd and experiments. *International Journal of Hydrogen Energy* 44 (60): 32254–32270.
59 Gao, J., Wang, X., Song, P. et al. (2022). Review of the backfire occurrences and control strategies for port hydrogen injection internal combustion engines. *Fuel* 307: 121553.
60 Lee, J., Lee, K., Lee, J. et al. (2014). High power performance with zero NOx emission in a hydrogen-fueled spark ignition engine by valve timing and lean boosting. *Fuel* 128: 381–338.

61 Huynh, T.C., Kang, J.K., Noh, K.C. et al. (2008). Controlling backfire for a hydrogen-fueled engine using external mixture injection. ASME. *Journal of Engineering for Gas Turbines and Power* 130 (6): 062804.

62 Hong, M., Zhang, J., Li, X. et al. (2008). Effect of injection timing on backfire of port injection hydrogen engine. SAE Technical Paper No. 2008-01-1788.

63 Menaa, A., Lounici, M.S., Amrouche, F. et al. (2019). CFD analysis of hydrogen injection pressure and valve profile law effects on backfire and pre-ignition phenomena in hydrogen diesel dual fuel engine. *International Journal of Hydrogen Energy* 44 (18): 9408–942.

64 Yang, Z., Wang, L., Zhang, Q. et al. (2012). Research on optimum method to eliminate backfire of hydrogen internal combustion engines based on combining postponing ignition timing with water injection of intake manifold. *International Journal of Hydrogen Energy* 37: 12868–12887.

65 Xin, G., Ji, C., Wang, S. et al. (2022). Monitoring of hydrogen-fueled engine backfires using dual manifold absolute pressure sensors. *International Journal of Hydrogen Energy* 47: 13134–13142.

66 Ampah, J.D., Jin, C., Afrane, S. et al. (2024). Race towards net zero emissions (NZE) by 2050: Reviewing a decade of research on hydrogen-fuelled internal combustion engines (ICE). *Journal of Green Chemistry*. https://doi.org/10.1039/D4GC00864B 26: 9025–9047.

67 Duan, J., Liu, F., and Sun, B. (2014). Backfire control and power enhancement of a hydrogen internal combustion engine. *International Journal of Hydrogen Energy* 39 (9): 4581–4589.

68 Leroy, T., Alix, G., Chauvin, J. et al. (2009). Modeling fresh air charge and residual gas fraction on a dual independent variable valve timing si engine. *SAE International Journal of Engines* 1 (1): 627–635.

69 Choi, S., Ki, M., and Min, K. (2010). Development of an on-line model to predict the in-cylinder residual gas fraction by using the measured intake/exhaust and cylinder pressures. *International Journal of Automotive Technology* 11 (6): 773–781.

70 Cavina, N., Ponti, F., Siviero, C. et al. (2004). Residual gas fraction estimation for model-based variable valve timing and spark advance control. *Internal Combustion Engine Division Fall Technical Conference*. 37467: 457–465.

71 Dhyani, V. and Subramanian, K.A. (2018). Experimental investigation on effects of knocking on backfire and its control in a hydrogen fueled spark ignition engine. *International Journal of Hydrogen Energy* 43 (14): 7169–7178.

72 Akif Ceviz, M., Sen, A.K., Küleri, A.K. et al. (2012). Engine performance, exhaust emissions, and cyclic variations in a lean-burn SI engine fueled by gasoline–hydrogen blends. *Applied Thermal Engineering* 36: 314–324.

73 Yu, X., Guo, Z., He, L., Dong, W. et al. (2019). Experimental study on lean-burn characteristics of an SI engine with hydrogen/gasoline combined injection and EGR. *International Journal of Hydrogen Energy* 44 (26): 13988–13998.

74 Gong, C., Li, Z., Yi, L. et al. (2020). Comparative analysis of various combustion phase control methods in a lean-burn H2/methanol fuel dual-injection engine. *Fuel* 262: 116592.

75 Gao, J., Tian, G., Ma, C. et al. (2021). Lean-burn characteristics of a turbocharged opposed rotary piston engine fuelled with hydrogen at low engine speed conditions. *International Journal of Hydrogen Energy* 46 (1): 1219–1233.

76 Liu, Q., Chen, X., Shen, Y. et al. (2019). Parameter extraction from spherically expanding flames propagated in hydrogen/air mixtures. *Journal of Hydrogen Energy* 44 (2): 1227–1238.

77 Salvi, B.L. and Subramanian, KA. (2016). Experimental investigation on effects of compression ratio and exhaust gas recirculation on backfire, performance and emission characteristics fuelled spark ignition engine. *Journal of Hydrogen Energy* 41 (13): 5842–5855.

78 Saito, S. and Adachi, M. (2001). Study on the process of backfiring. SAE Technical Paper No. 2001-01-1881.

79 Wei, H., Xu, Z., Zhou, L. et al. (2018). Effect of hydrogen-air mixture diluted with argon/nitrogen/carbon dioxide on combustion processes in confined space. *Journal of Hydrogen Energy* 43 (31): 14798–14805.

80 Furuhama, S., Yamane, K., and Yamaguchi, I. (1977). Combustion improvement in a hydrogen fueled engine. *Journal of Hydrogen Energy* 2 (3): 329–340.

81 Ma, F., Wang, Y., Liu, H. et al. (2008). Effects of hydrogen addition on cycle-by-cycle variations in a lean burn natural gas spark-ignition engine. *Journal of Hydrogen Energy* 33 (2): 823–831.

82 Saaidia, R., Jemni, M.A., and Abid, M. (2018) Simulation and empirical studies of the commercial SI engine performance and its emission levels when running on a CNG and hydrogen blend. *Energies* 11: 29.

83 Niu, R., Yu, X., Du, Y. et al. (2016). Effect of hydrogen proportion on lean burn performance of a dual fuel SI engine using hydrogen direct-injection. *Fuel* 186: 792–799.

84 Lee, S., Kim, Y., Lee, J. et al. (2023). Energy and exergy analyses of hydrogen-fueled spark ignition engine with various air excess ratios and ignition timings. *Fuel* 349: 128588.

85 Verhelst, S., Demuynck, J., Sierens, R. et al. (2010). Impact of variable valve timing on power, emissions and backfire of a bi-fuel hydrogen/gasoline engine. *Journal of Hydrogen Energy* 35 (9): 4399–4408.

86 MacCarley, C. (1981). A study of thermally induced backfiring in hydrogen fueled engines, and methods for backfire control. *Proceedings of the Intersociety Energy Conversion Engineering Conference*, Atlanta, Georgia. OSTI ID: 5359502. osti.gov.

87 MacCarley, C.A. and VanVorst, W.D. (1980). Electronic fuel injection techniques for hydrogen powered I.C. engines. *International Journal of Hydrogen Energy* 5: 179–203.

88 MacCarley, C.A. (1981). Development of a high speed injection valve for electronic hydrogen fuel injection. *Proceedings of the 3rd World Hydrogen Energy Conference, Tokyo, Japan*, (eds. T.N. Veziroglu, K. Fueki, and T. Ohta). New York: Pergamon Press.

89 Homan, H.S. (1978). An experimental study of reciprocating internal combustion engines operated on hydrogen. PhD thesis. Cornell University.

90 De Boer, P.C.T., McLean, W.J., and Homan, H.S. (1976). Performance and emissions of hydrogen fueled internal combustion engines. *International Journal of Hydrogen Energy* 1: 153–172.

91 Fagelson, J.J. (1975). Hydrogen as a reciprocating engine fuel: An analytical study. Master's thesis. Cornell University.

92 Homan, H.S., De Boer, P.C.T., and McLean, W.J. (1978). The effect of fuel injection on nox emissions and undesirable combustl0n for hydrogen-fueled piston engines. SAE Paper 780945.

93 Liu, X.-H., Liu, F.-S., Zhou, L. et al. (2008). Backfire prediction in a manifold injection hydrogen internal combustion engine. *International Journal of Hydrogen Energy* 33 (14): 3847–3855.

94 Das, L.M. (1987). Studies on timed manifold Injection in hydrogen operated spark ignition engine: Performance, combustion and exhaust emission characteristics. PhD thesis. Indian Institute of Technology, Delhi, India.

95 Das, L.M. (2007). The potential of hydrogen as an alternative fuel for the transport and power sector in India. *International Journal of Environmental Studies* 64 (6): 749–759.

96 Das, L.M. (2016). Chapter 7: Hydrogen-fueled internal combustion engines. In: *Compendium of Hydrogen Energy*, Volume 3: *Hydrogen Energy Conversion*, pp. 177–217. Elsevier.

97 Das, L.M. (1990). Fuel induction techniques for a hydrogen operated engine. *International Journal of Hydrogen Energy* 15 (Il): 833–842.

98 Park, C., Kim, Y., Choi, Y. et al. (2019). The effect of engine speed and cylinder-to-cylinder variations on backfire in a hydrogen-fueled internal combustion engine. *International Journal of Hydrogen Energy* 44 (39): 22223–22230.

99 Kondo, T., Iio, S., and Hiruma, M. (1997). A study on the mechanism of backfire in external mixture formation hydrogen engines-about backfire occurred by cause of the spark plug. SAE Technical Paper No. 971704. https://doi.org/10.4271/971704.

100 Cheng, X., Baigang, S., and Zhen, H. (2017). Investigation on jet characteristics of hydrogen injection and injection strategy for backfire control in a port fuel injection hydrogen engine. *Energy Procedia* 105: 1588–1599.

101 Mathur, H.B. and Das, L.M. (1991). Performance characteristics of a hydrogen fuelled S I engine using timed manifold injection. *International Journal of Hydrogen Energy* 16 (2): 115–127.

102 Welch, A., Mumford, D., Munshi, S. et al. (2008). Challenges in developing hydrogen direct injection technology for internal combustion engines. SAE Technical Paper No. 2008-01-2379.

103 Verhelst, S. and Wallner, T. (2009). Hydrogen-fueled internal combustion engines. *Progress in Energy and Combustion Science* 35 (6): 490–527.

5

Modeling and Simulation Studies on the Hydrogen Engine

A society grows great when old men plant trees whose shade they know they shall never sit in.
Greek proverb

5.1 Introduction

Combustion modeling is a complex task, involving interdisciplinary analysis in several areas such as thermodynamics, fluid dynamics, reaction kinetics, and numerical methods.An approach to modeling depends upon research goals to predict performance, emission, combustion, or any other relevant factors. Methodologies commonly used for modeling include chemical kinetic models, computational fluid dynamics (CFD), combustion models including flow aspects such as turbulence, finite element analysis (FEA), emission modeling, and multidimensional modeling. In modeling studies, uncertainty and sensitivity analyses need to be carried out. This is necessary to quantify the impact of variation on model inputs and to identify the factors critical to the complexity of combustion behavior. Moreover, it is always desirable that modeling results should be validated against experimental data to ensure the model's predictive accuracy.

Abnormal combustion is the most crucial factor in the development of hydrogen engines. Causes, consequences, and control techniques of preignition, knocking, and backfire have been elaborately discussed in Chapter 4. Some significant theoretical studies, simulation models, and fault diagnosis on hydrogen combustion in engines are discussed in the present chapter. In their study, Dang and Wang [1] used a comprehensive approach to hydrogen engine fault diagnosis, integrating the strengths of a self-organizing maps (SOM) neural network and a multi-agent system (MAS). Initially, the combustion characteristics of hydrogen in an engine were studied by analyzing pressure data obtained from tests on the engine. This was achieved through the application of SOM neural network. Subsequently, a MAS study was taken up to scrutinize the topological changes within the hydrogen engine, focusing on abnormal combustion symptoms. The results of this analysis were in agreement with 44 sets of pressure data samples generated from engine tests. The SOM-MAS fusion model is useful diagnosing preignition faults in hydrogen engines. In another study on abnormal combustion, Dang and Wang [2] used an ant colony optimization-back propagation (ACO-BP) algorithm with a nonlinear mapping model

Hydrogen Engines: Design, Performance Evaluation, Combustion Analysis, and Exhaust Emissions, First Edition.
Lalit Mohan Das.

correlating hydrogen engine speed and load to the optimal ignition advance angle. Utilizing an experimental dataset, a MAP (mean average precision) diagram was prepared for the optimized model through simulation. The results of the ACO-BP algorithm exhibited high calculation accuracy and was even better than BP neural networks and genetic algorithm-back propagation (GA-BP). The results of this analysis are quite promising for further study on backfire. In their study to optimize engine conditions for near-zero NOx emission, Cheng et al. developed a highly accurate simulation model through the meticulous calibration of hydrogen "critical injection" [3] and the implementation of high-level local grid refinement. An improved piston ring design using low viscosity, low-ash lubrication oil was implemented to control oil consumption and particle emissions. The investigation had focused on the effects of injection timing by employing modeling techniques to analyze the mixing process, combustion, and NOx emissions. NOx emissions remained near-zero levels, and brake thermal efficiency improved from 40.4 to 41.7% with a change of SOI from 200° CA BTDC to 120° CA BTDC at $\lambda = 2.67$. Early injection accelerates turbulence and increases higher NOx emissions in the cylinder, in zones of high turbulent kinetic energy. Even though delayed injection weakens turbulence, still it generates localized NOx concentration in the cylinder. With excessively delayed injection, mixtures with $\lambda < 1.5$ are accumulated and often lead to a sudden NOx surge. Change in start of injection (SOI) showed irregular power and fuel economy distribution. Optimal performance and near-zero NOx emission with maximum power and maximum efficiency were achieved when the control variable was adjusted to end of injection (EOI) and maintained at −80° CA. It resulted in a maximum brake thermal efficiency of 41.9% and a 1.5% increase in efficiency. Thus, optimal injection timing was crucial for zero NOx emission.

Wu et al. [4] investigated modeling a hydrogen engine for optimizing direct injection strategies, with the turbulent Schmidt number as a crucial factor influencing predictions of hydrogen–air mixing. On an analysis based on alignment with experimental planar laser-induced fluorescence (PLIF) data, the Schmidt number 0.5 was proposed for hydrogen engines. The correct design of tumble plates in intake ports was identified for a more accurate assessment of in-cylinder flow fields. Simulation studies on an injection system with different injection pressures and timing clearly brought out the crucial role of high-pressure injection in fuel–air mixing. Residence time and injection momentum were very relevant for a homogeneous mixture. Even though there was a broad satisfactory agreement between experimental and simulation results, challenges were encountered due to the effect of hydrogen diffusion layers and local mixture concentration fluctuations. With their focus on emission control, Jayakrishnan et al. [5] studied a single-cylinder hydrogen engine with a peak power output of 4.8 kW at 3600 rpm. A one-dimensional model for the single-cylinder engine successfully predicted trends in brake torque, brake power, brake thermal efficiency, and NOx emissions at different speeds. In the analysis, Woschni's model underpredicted heat losses from a hydrogen engine and subsequently, a modified heat transfer model with a heat transfer coefficient multiplier of 2.2 was introduced, and that led to accurate predictions of measured parameters such as power output and NOx emissions. The study examined the effect of spark timing on torque and NOx formation and established a correlation between torque trends and in-cylinder pressure profiles. Optimal torque occurred when the peak in-cylinder pressure was between 15 and 25° CA ATDC. NOx concentration in the exhaust increased with spark advance at all engine speeds. Furthermore, the research confirmed that increasing the equivalence ratio led to higher torque output and resulted in higher NOx emissions from the exhaust.

5.2 Chemical Kinetics Models

Chemical kinetics models simulate the chemical reactions involved in a hydrogen combustion process for which the detailed reaction mechanism is used. In this analysis, it is necessary that the reaction mechanism captures accurately the typical combustion features such as wide flammability limit and diffusivity. It is possible to get detailed insights into species concentration, reaction rate, and ignition timing. If a new dedicated hydrogen engine has to be designed, it is often necessary to develop computational fluid dynamics (CFD) models to simulate hydrogen combustion. This would be necessary to predict the optimal engine design and operating conditions, which significantly influence the engine's performance and combustion characteristics.

Accurate hydrogen chemical kinetic submodels provide very relevant information to CFD models to evaluate reaction rates in engine conditions. Usually, temperature and pressure conditions in an IC engine are relatively much higher than those in usual experimental combustion studies. The fuel–air mixture will encounter stringent temperature and pressure environments. Marinov et al. [6] analyzed hydrogen combustion under conditions similar to internal combustion engines by adopting a chemical kinetic model. Parametric variation of the laminar flame speed under typical high pressure and temperature conditions was analyzed at different equivalence ratios—a common feature in spark ignition engines. Predicted NO concentration was observed to be closely linked to the peak temperature attained in the cycle. NO production mostly was by the reaction N2 + O = NO + N. It was also evident in this analysis that a small amount of hydrogen added to natural gas leads to a reduction in NOx emissions from a free-piston engine. Figure 5.1 shows the effect of the equivalence ratio on laminar flame speed and oxides of nitrogen. The data shown in this figure correspond to an unburned gas temperature of 900 K and 40 atm. Data have been computed for ranges of lean equivalence ratio (0.2–0.7) and maximum flame speed was about 837 c/s. It is well-known that NOx formation depends upon the residence time of burned gases and temperature level, which is high near the top dead center (TDC). As is evident in the graph, there is a sharp increase in the NOx emission level beyond an equivalence ratio of 0.5. In such a condition, increased oxygen content and higher combustion temperatures promote formation of thermal NOx through the Zeldovich mechanism.

It is also observed that the burned gas temperature was 1930, 2150, and 2330 K, for equivalence ratios of 0.4, 0.5, and 0.6, respectively. In this study, the researchers also analyzed the sensitivity of

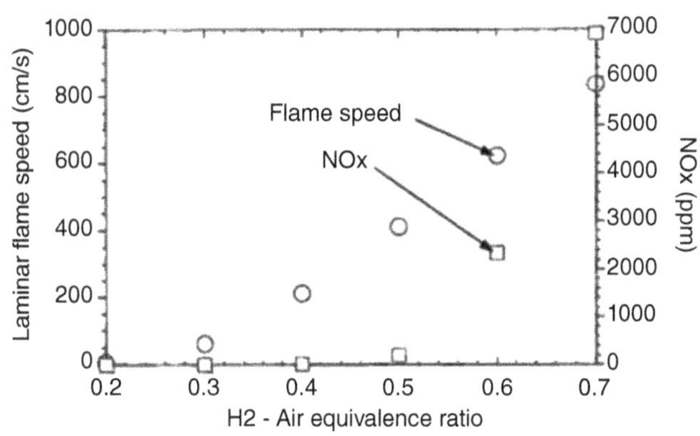

Figure 5.1 Laminar flame speed and NOx emissions versus equivalence ratio [6] / American Chemical Society.

the laminar flame speed and NOx production on the unburned gas temperature under conditions when pressure and equivalence ratio are maintained constant at 40 atm and 0.4, respectively. The results of this analysis shown in Figure 5.2 distinctly indicate that both flame speed and oxides of nitrogen exhibit an increasingly exponential trend in the unburned gas temperature range between 700 and 1200 K.

The correlation between the laminar flame speed and pressure in fuel-lean conditions has been described using several empirical models. The equation used for laminar flame speed in fuel-lean conditions in this study is given by the relation

$$S_u = P^{-0.2}$$

where S_u is the laminar flame speed.

Figure 5.3 shows the variation in laminar flame speed with pressure. The curve has been generated at a temperature of 900 K and at an equivalence ratio of $\phi = 0.4$. It is evident in the graph that laminar flame speed shows a decreasing trend as the pressure increases from 1 to 70 atm. The correlation between pressure and flame speed is due to the combined effect of chemical kinetics and the transport process, but it is believed that chemical kinetics greatly influences the difference in pressure exponent between the ultralow equivalence ratio flame and the stoichiometric flame.

Kéromnès et al. [7] carried out experiments on the oxidation of syngas (synthetic gas), which is primarily a mixture of hydrogen (H_2) and carbon monoxide (CO), and subsequently simulated

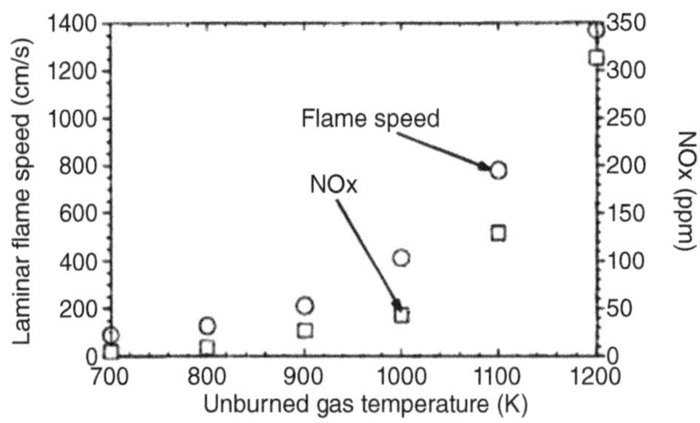

Figure 5.2 Laminar flame speed and nitrogen oxides variation with unburned gas temperature at 40 atm and ϕ = 0.4 [6] / American Chemical Society.

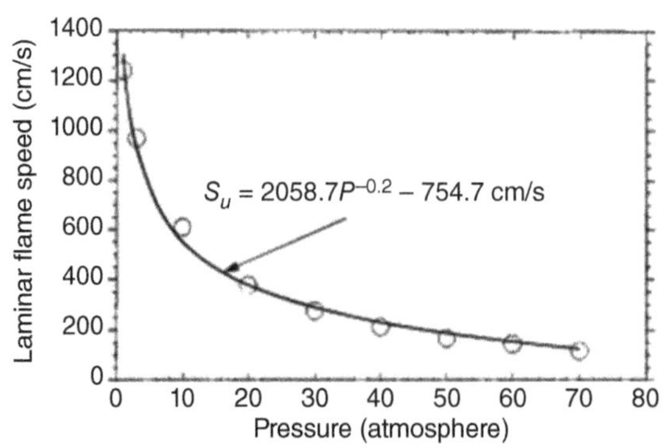

Figure 5.3 Shows the effect of variation in pressure on laminar flame speeds (at 900 K and ϕ = 0.4) [6] / American Chemical Society.

with a kinetic model. In this study, ignition delay times for $H_2/CO/O_2/N_2/Ar$ mixtures were measured with two rapid compression machines and shock tubes. Measurements were carried out in pressure ranges from 1 to 7 bar, with temperatures from 914 to 2220 K and at an equivalence ratio over a very wide range between 0.1 and 4. It was observed that ignition delay showed a decreasing trend with increase in temperature, pressure, and equivalence ratio. In the syngas mixture with CO concentration lesser than 50%, hydrogen chemistry dominantly governs reactivity of the mixture. In a mixture with higher CO concentration, the inhibiting effect of CO in slowing down the overall reactivity was glaringly observed. In syngas, where both H_2 and CO are present, CO consumes some of the available oxygen but does not release as much heat as hydrogen releases upon combustion. This can affect overall combustion efficiency. The reaction sequence

$$H_2 + HO_2 \leftrightarrows H + H_2O_2 \text{ followed by } H_2O_2(+M) \leftrightarrows OH + OH(+M)$$

had a greater influence on "hydrogen ignition under high pressure and intermediate temperature conditions" [7]. In chemical kinetics, the rate constant is a fundamental parameter that quantifies the rate at which a chemical reaction occurs. It is an essential component of the rate equation, which describes how the concentration of reactants changes over time.

Liu and Karim [8] carried out an analytical investigation on a HCCI hydrogen engine to assess how changes in the swirl velocity of the intake mixture have an influence on the combustion process. The researchers developed a turbulent transient 3D predictive computational model taking into account the detailed chemical kinetics for oxidation of hydrogen. It was evident in this analysis that circumferential velocity of the mixture within the engine cylinder showed an increasing trend with an increase in the initial intake swirl ratio. Radial and axial velocities were influenced by intensity of the combustion and the initial intake swirl ratio. It was observed that the delay period for autoignition was increased due to an increase in the initial flow swirl ratio or speed. This also resulted in extending the combustion period. However, the NOx concentration was reduced. Analysis also identified optimum values of the initial swirl ratio and engine speed for an engine under specific conditions of temperature, pressure, equivalence ratio, and compression ratio. The analysis aimed at high thermal efficiency and low NOx emissions while drastically reducing knocking symptoms. A rational coordination of adjustment of intake temperature, equivalence ratio, swirl ratio, engine speed, and initial intake pressure was useful in controlling the time to autoignition and combustion duration of a hydrogen HCCI engine at any compression ratio. Thus, it is clear that predictive CFD modeling with detailed chemical kinetics is an effective tool to assess the influence of velocity, temperature, pressure, and several other properties on hydrogen-operated HCCI configurations. Such modeling could provide guidelines for areas of improvement. Relationships among several parameters and effects are shown in Figures 5.4 through 5.7.

Intake swirl refers to rotational motion imparted to the air–fuel mixture as it enters the combustion chamber of an engine. Intake swirl ratio is defined as mixture uniform rotational velocity relative to that of the rotational speed of the engine. It is evident in Figure 5.6 that there is a decreasing trend in NOx emission as the initial swirl ration increases. Formation of NOx is temperature dependent, so the lower level of NOx emission is primarily due to the reduction in maximum temperature. As shown in the graph, indicated thermal efficiency decreases with the increase in the initial swirl ratio, and this is because of the increased heat loss and later ignition. Figure 5.7 shows the pressure oscillations during knock. Severe oscillations of the cylinder pressure rising with high frequency are evident in the graph around an equivalence ratio of 0.8.

Wang and colleagues [9] studied the 19-reaction H_2 oxidation chemical kinetic model by optimizing it and quantifying uncertainties. They started by estimating uncertainties in reaction rate

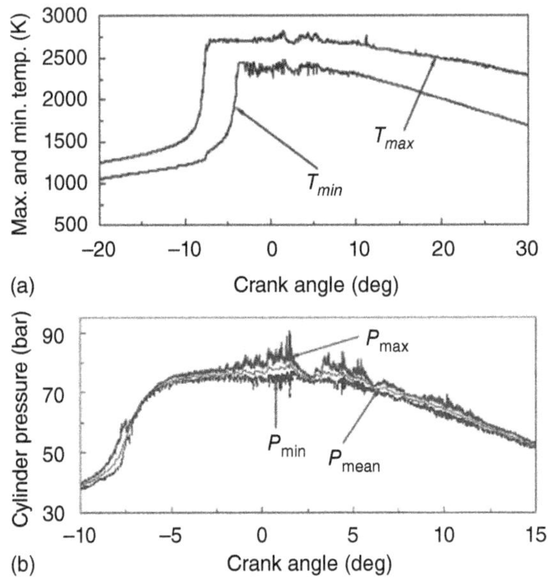

Figure 5.4 Variation of cylinder temperature (a) and pressure (b) with crank angle at knock conditions [8] / with permission of Elsevier.

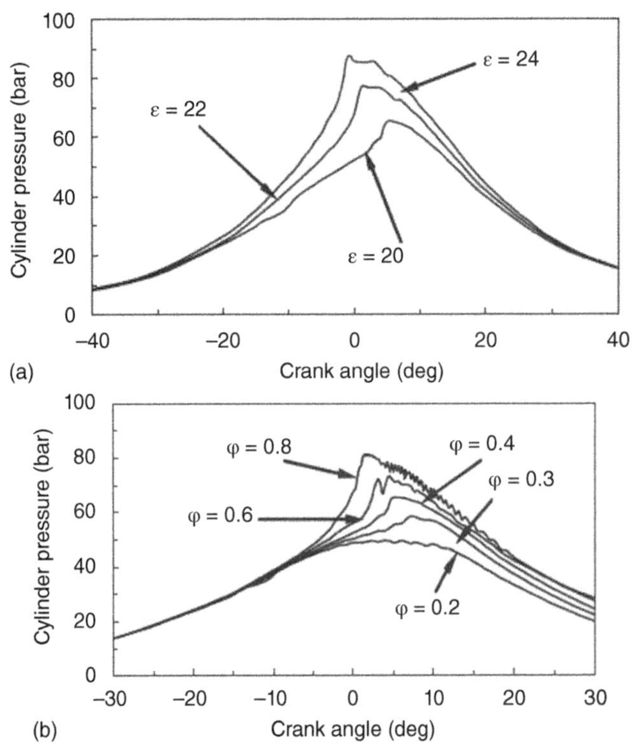

Figure 5.5 Effect of compression ratio (a) and equivalence ratio (b) on combustion (ε stands for compression ratio and φ for equivalence ratio).

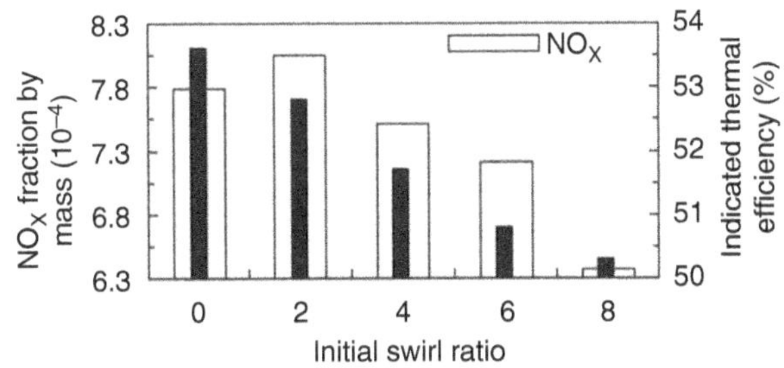

Figure 5.6 Effect of initial swirl ratio of NOx emission and efficiency [8] / with permission of Elsevier.

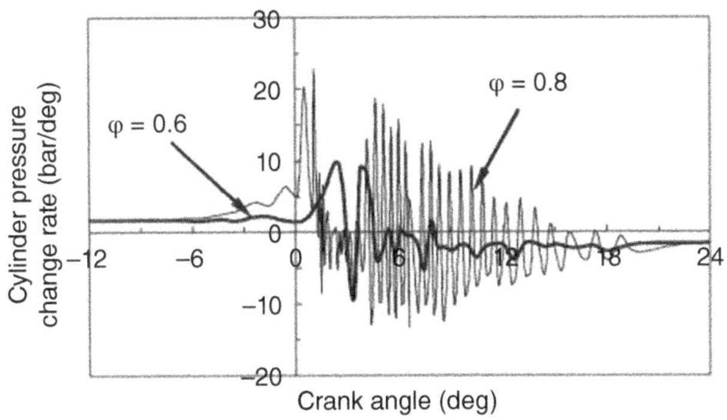

Figure 5.7 Cylinder pressure rise change rate during knock conditions [8] / with permission of Elsevier.

constant (RRC) parameters using direct measurements and reviews. A deterministic approach was initially used, followed by a probabilistic treatment to narrow uncertainty intervals and expand the temperature range validity for RRCs with high uncertainty levels. Two quantities, discrepancy measures and uncertainty contributions, were introduced to assess these uncertainties. Monte Carlo simulations and polynomial regression were then used to create a response surface for optimization. Subsequently, the focus was narrowed to 10 key reaction channels for further refinement. Probability density functions were computed based on discrepancy measures for four channels to reduce large uncertainty intervals. Data for training the model were meticulously gathered from various experiments such as shock tubes, rapid compression machines, jet stirred reactors, plug flow reactors, and premixed laminar flames etc. Inconsistent experimental targets were identified and removed from consideration. Final model optimization involved minimizing the discrepancy measure. The optimized model correctly predicted ignition delay times, laminar flame speeds, and concentration profiles with minimal discrepancy. The application of statistical methods proved effective in treating complex reaction systems, facilitating the development of predictive reaction models with quantified valid ranges. The optimized kinetic model exhibits promising predictive performance for H_2 combustion data within and beyond the tested range.

Yakovenko and Kiverin [10] have made an exhaustive study on numerical modeling of hydrogen combustion. The discussion does not make specific reference to engines, but it deals with several technical approaches to hydrogen combustion. As numerical modeling is a very powerful tool for analyzing combustion, technical discussions can be useful for a comprehensive analysis of hydrogen combustion in engines. Two different approaches were adopted to analyze gaseous combustion. One was based on a full system of Navier–Stokes equations taking into consideration the dynamics of the compressible reactive medium. This analysis considered the set of Navier–Stokes equations of the multicomponent reactive hydrogen–air mixture. Relevant significant parameters such as compressibility, viscosity, heat transfer, multicomponent diffusion, real equation of state, and detailed chemical kinetics were considered in the Navier–Stokes model, and equations were solved by using the CABARET (compact accurately boundary-adjusting high-resolution technique) numerical scheme and conventional CPM (coarse particles method). The second approach utilized low Mach number approximation and was implemented on the basis of a fire dynamics simulator (FDS) numerical scheme. The FDS computational package is very useful for modeling large-scale compartment fires.

The paper [10] also discussed the benefits and the shortcomings of the low Mach number approach. The combustion analysis for laminar flame velocity test, spherical bomb test, and multidimensional modeling of combustion development inside the rectangular closed vessel were carried out to identify the limitations of low Mach number approximation. It was observed from this study that in open systems with weak interaction between compression waves and flame fronts, the low Mach number approximation closely matches compressible models. Closed vessel scenarios, heavily influenced by compression wave propagation, showed discrepancies in the low Mach number model, particularly in underestimating pressure rise dynamics. Interaction between compression waves and flame fronts accelerates or decelerates locally, intensifying combustion and hastening pressure build-up. Multidimensional simulations highlight that combustion intensity is not the primary factor impacting compression waves. Even mild compression waves can induce instability in mixtures prone to instability like lean hydrogen–air mixtures, altering combustion dynamics significantly. The paper also highlights the application potential of various computational techniques, which could be valid for specific aspects of the analysis of hydrogen combustion in engines. The numerical reactive gas-dynamics (NRG) package contains models that could be used for heat transfer, multicomponent diffusion, and chemical kinetic schemes.The NRG package contains the traditional low-order coarse particles method (CPM). CPM is a robust versatile computational approach that can solve a wide range of problems related to gas dynamic flows. However, while its high numerical dissipation makes it extremely suitable for studying accurately the general flow characteristics, its ability to reproduce small-scale flow features like acoustic and vortical perturbations is restricted [10].

CABARET is the most suitable numerical scheme for modeling complex flows in reactive gaseous mixtures. Compact computational stencil is one of its most distinctive features in conducting simulations to optimize computational efficiency and performance. The analysis also shows broadly the capabilities of low Mach numbers in FDS approach compared to the compressible model with CABARET and CPM numerical schemes. Particularly in open systems, the low Mach number approach gives results that closely agree with those of the compressible model. This is so because, in open systems, there is no intense interaction between compression waves and the flame front. A very significant conclusion from this analysis shows that in lean hydrogen–air mixtures, even low-intense compression waves can “trigger instability development.” This can lead to multidimensional evolution of the flame front, which in turn can result in a “substantial change in the dynamics of combustion process” [10].

5.3 Thermodynamic Models

The thermodynamic modeling done by J. J. Fagelson, W. J. McLean, and P. C. T. de Boer [11] was one of the early attempts at studying modeling of hydrogen combustion in engines using a two-zone quasi-dimensional model. The researchers used a semi-empirical combustion model

$$U_t = A\,Re^B\,U_l$$

where A and B are constants; Re is the Reynolds number based on piston diameter, mean piston speed, and burned gas properties; and U_t and U_l are the turbulent and laminar burning velocities, respectively.

Their analysis had assumed spherical flame propagation, and the effect of heat transfer was neglected. The NOx emissions were determined by integrating the chemical rate equations resulting from the extended Zeldovich mechanism over the boundary conditions determined by the cycle analysis. Laminar burning velocity was calculated from an overall second-order reaction with an estimated activation energy. The model developed by the researchers was validated against measurements with varying equivalence ratios and ignition timing. Hydrogen has a very rapid burning rate, which results in high temperature and pressure—this leads to fast formation and destruction rates of NO. In this analysis, the peak value of NO concentration was predicted for a lean mixture of an equivalence ratio of 0.8. Rapid decomposition of NO during the expansion stroke brought down the level of NOx emission in the rich mixture during the expansion stroke. In this analysis, exhaust gas recirculation (EGR) was found to be more effective in controlling NOx levels in the lean zone compared to stoichiometric mixture operation. Moreover, EGR was effective in moderating the burning rate, of course with loss in power and very marginal loss in efficiency. Supercharging is one of the conventional routes to compensate for power output.

Prabhu-Kumar et al. [12] developed an analytical model to study the performance and emission characteristics of a supercharged spark-ignited hydrogen engine. In this study, the model predictions were verified by the experimental data generated by one of the authors in tests done at Aachen Technical University. For the analytical model, they used semi-empirical turbulent flame speed suggested by Fagelson [11].

As evident in Figure 5.8, in these tests, the computed data for prediction of the NO level was comparable with the test data generated in the engine. Figure 5.9 shows the pressure–crank angle diagrams at a lean operating condition with an equivalence ratio of 0.4 under various supercharging pressures. As shown in the figure, the peak pressure for naturally aspirated engines is substantially lower compared to supercharging conditions. Maximum cylinder pressures are 90, 110, and 130 atm for supercharged conditions of the engine at a supercharging pressure of 0.8, 1.2, and 1.6 atm, respectively. Thus, the rate of pressure rise gives an indication of burning velocity.

In an effort to improve power output and reduce NOx emission level, Maher et al. [13] considered the compression, combustion, and expansion processes along with exhaust emissions of a supercharged hydrogen engine. For the analysis, the combustion chamber was divided into burned and unburned zones separated by the flame front. Uniform pressure was assumed all through the cylinder charge. Thermodynamic equations of state and energy, mass, and volume balance were used to predict cylinder pressure and temperature data for the burned and unburned zones. The turbulent flame front speed (ST) was modeled by a semi-empirical formula as discussed earlier and suggested by Fagelson.

Figures 5.10 and 5.11 show the comparative simulation results for measured power output data generated by operating the hydrogen engine and the unsupercharged gasoline engine under the same operating conditions of a compression ratio of 7.5 and an engine speed of 1500 rpm. Moreover, the inlet charge of the hydrogen engine was so adjusted that it produced the same amount of power as developed by gasoline operation for operating ranges of equivalence ratios. From this analytical study it is inferred that supercharging is a more effective method to increase the output of the hydrogen engine compared to increasing its compression ratio. If hydrogen engine operation is limited to an equivalence ratio of 0.6, it can produce the same power output as an unsupercharged

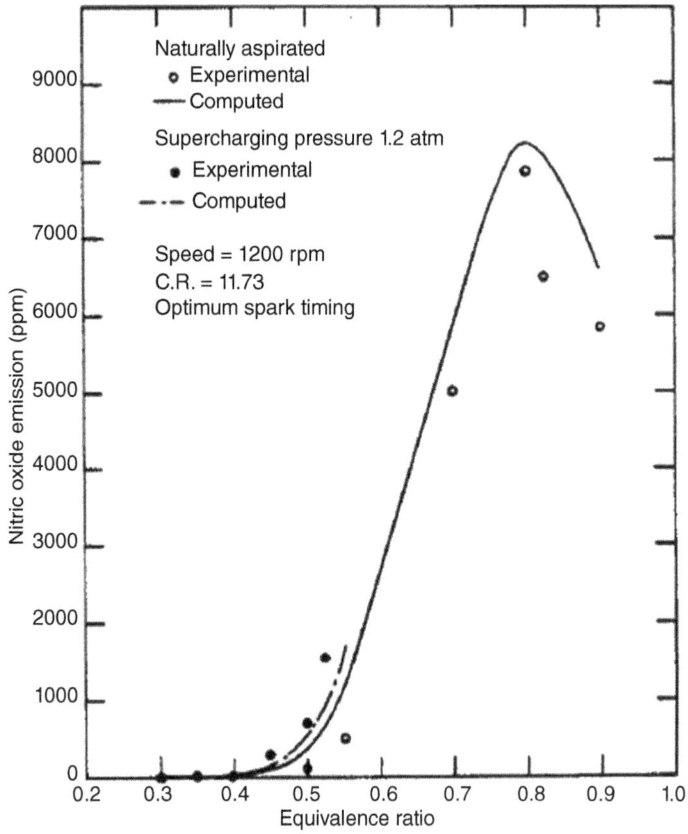

Figure 5.8 Comparison of measured and computed nitric oxide emission in a naturally aspirated and supercharged hydrogen [12] / with permission of Elsevier.

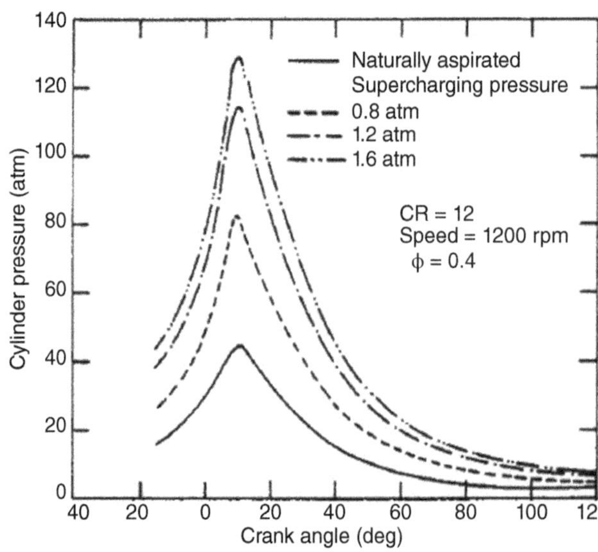

Figure 5.9 Computed cylinder pressure as a function of crank angle degree for naturally aspirated and supercharged hydrogen engines [12] / with permission of Elsevier.

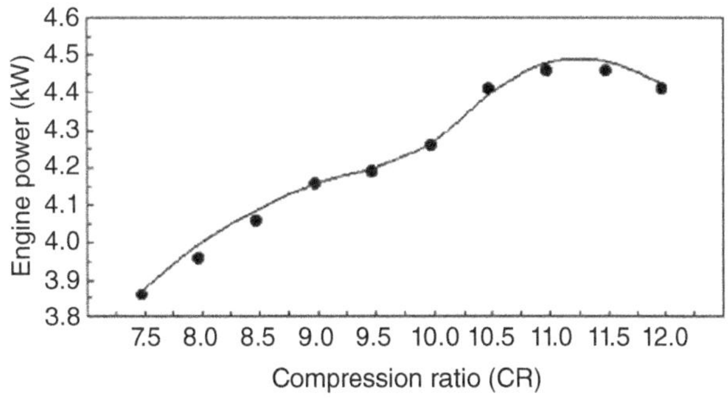

Figure 5.10 Measurements (dots) and predictions (line) of the effect of the compression ratio on engine power for a hydrogen fueled, unsupercharged engine [13] / with permission of Elsevier.

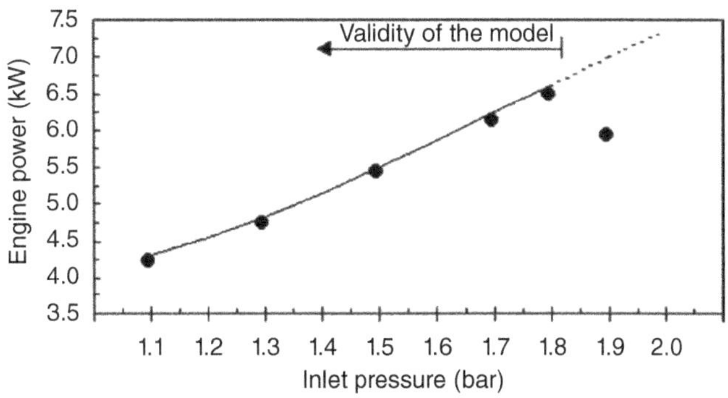

Figure 5.11 Measurements (dots) and predictions (dashed line) of the effect of the inlet pressure on engine power for a hydrogen-fueled engine [13] / with permission of Elsevier.

gasoline engine. At the same time, NOx emissions would be well within the limits and not beyond the level of unsupercharged gasoline operation. In a subsequent paper, Maher et al. [14] used the same model for further studies and analysis on combustion and NOx emission. Simulated data were compared with the test values varying compression ratio, equivalence ratio, and engine speed. Preignition limits were also determined for hydrogen engines. Lean mixture engine operation at relatively high speed was possible without any undesirable combustion phenomena such as backfire and knocking. Figure 5.12 shows the relation between compression ratio and equivalence ratio within the limit of the safe operational zone without preignition problems. An increased compression ratio, as expected, improved efficiency and power output with smooth combustion. For equivalence ratios less than 0.8, NOx emission exhibited an increasing trend with the increase in compression ratio. For equivalence ratios less than 0.8, NOx emissions exhibited an increasing trend with the rise in compression ratio. However, for equivalence ratios greater than 0.8, the NOx level decreased as the compression ratio increased, due to the abundance of oxygen, even despite the high combustion temperature.

NOx formation is more pronounced at higher temperatures because the reaction between nitrogen and oxygen from air to form NO is endothermic. Lean mixtures have lower combustion temperatures because there is an excess of oxygen compared to fuel, resulting in dilution. This limits the availability of reactants. Chemical processes resulting in net NO emission are closely linked to mixture stoichiometry at different stages. A lean mixture tends to produce lower level of NOx emissions in a hydrogen engine. In the case of a lean mixture, the quenching of formation

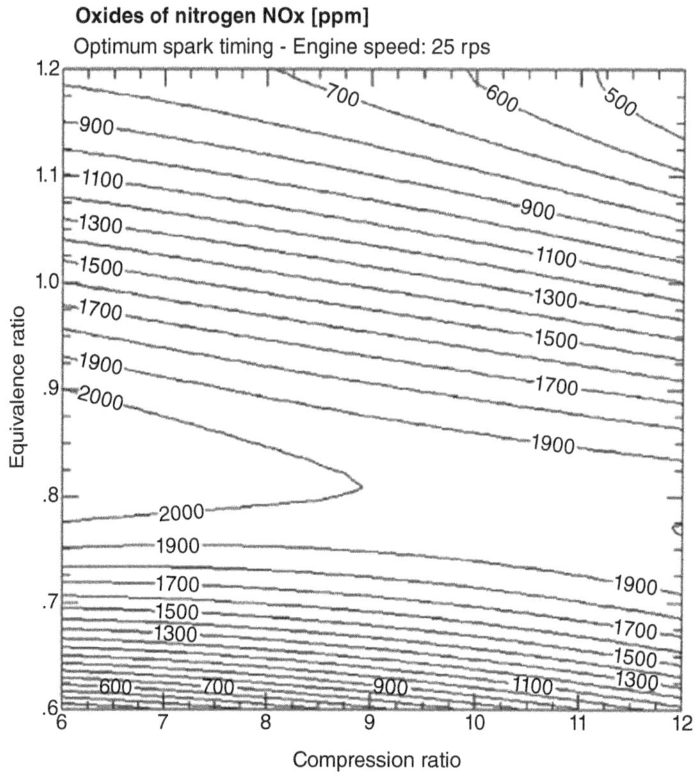

Figure 5.12 Effect of compression ratio and equivalence ratio on the NOx emission, at optimum spark timing for best torque and 25 rps (1500 rpm) engine speed [14] / with permission of Elsevier.

reactions dominates during the expansion stroke for equivalence ratios less than 0.8 and thus NO emissions are controlled. "Quenching" in this context means that the high excess of air cools down the combustion process, restricting formation of oxides of nitrogen, especially nitric oxide (NO). On the other hand, for rich mixtures, NO emissions are controlled by the quenching of decomposition reactions during the expansion process.

Keck [15] carried out tests with hydrogen as well as propane in an optically accessible engine. A turbulent entrainment model was used to compare the analytically predicted trends with the experimentally generated data. Optical observations from the relevant schlieren and laser scattering techniques showed that there always occurs a smooth spherical laminar flame in the initial phase of combustion. Subsequently, the flame front gets distorted with wrinkles by the effect of turbulence as the flame continues to grow. Comparison of schlieren photographs of combustion between hydrogen/air and propane/air mixtures showed the combustion process between the two gases to be radically different. The hydrogen flame was seen to expand constantly at a relatively higher speed right from the onset of ignition and without practically any induction time. The initial expansion speed of hydrogen was approximately 10 times that of propane. This sharp contrast to the expansion speed with propane was evident as there was relatively much lesser increase in speed with propane. As compared to the propane flame, the hydrogen flame shape was more spherical with larger wrinkles. Earlier observations in the laboratory had shown that the turbulent burning zone in hydrogen was substantially thinner as compared to that in propane. All these experimental observations were in good qualitative agreement with the corresponding predicted values.

Researchers at Czech Technical University used zero and multidimensional models for advanced simulation of hydrogen-fueled engines [16, 17]. A zero-dimensional model, referred to as a "0D" model, is a simplified mathematical representation of the combustion process that assumes uniform condition without considering spatial variation. Wiebe's law was used to evaluate the heat release rate, and the extended Zeldovich mechanism was used to calculate the NOx emission level. In this analysis, the CFD model was based on an advanced multizone Eulerian model to simulate and analyze the complex combustion process. In such a multizone analysis, the combustion chamber is divided into several discrete zones where properties can vary. In an Eulerian framework, it tracks the properties of air and fuel within each zone as it moves and undergoes chemical reaction. This study was carried out to analyze the effect of compression ratio, equivalence ratio, spark timing, etc. on an engine's performance and NOx emission characteristics and compare them with corresponding test data. Results of modeled velocity and temperature fields as well as methods of controlling knock and reducing NOx emission levels were discussed.

Kim and Kim [18] used neural network algorithms to examine the technical feasibility of transition from compressed natural gas (CNG) to hydrogen in mass transportation applications. Performance maps, generated from tests on a single-cylinder engine, were subsequently used for the development of a vehicle simulation model. The regression analysis was carried out to identify the operating parameters responsible for NOx emissions. Three-dimensional numerical simulation analysis was adopted to evaluate the combustion characteristics. The combustion submodel was validated with experimental data using in-cylinder pressure traces. The significant results from this study showed that buses fueled by hydrogen could be more efficient than the existing CNG buses. Vehicle driving simulation study suggests that use of a NOx catalyst is necessary to reduce the NOx emissions from the multi-cylinder bus engine, particularly during high load operations. Use of the H_2-SCR catalyst was found to be the most suitable from an engine efficiency point of view.

Ma et al. [19] optimized the significant parameters such as compression ratio, equivalence ratio, and ignition timing, etc. in a comprehensive computer simulation model developed to predict the hydrogen engine's performance. The "filling and emptying" simulation model shown in Figure 5.13 shows the major parts of the engine system. The influence of several significant operating parameters on the dynamic process of the engine was also taken into account. The researchers calculated several important properties and analyzed the simulated engine performance with the calculated properties applied to the mathematical model.

Several aspects of engine configuration and performance were analyzed. The effect of variation in cylinder pressure, combustion regulation with various cylinder diameters, effect of spark advance, and effect of equivalence ratio and compression ratio were analyzed. Simulation results showed that increasing the spark advance angle raises the peak cylinder pressure as shown in Figure 5.14.

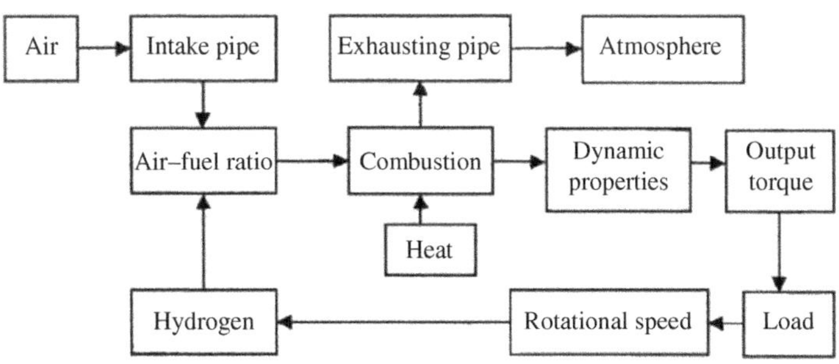

Figure 5.13 Filling and emptying model for hydrogen engine's performance [19] / with permission of Elsevier.

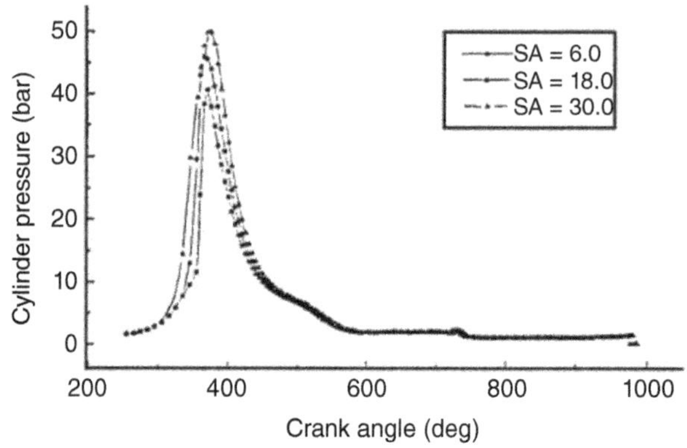

Figure 5.14 Cylinder pressure vs. crank angle with various spark advances [19] / with permission of Elsevier.

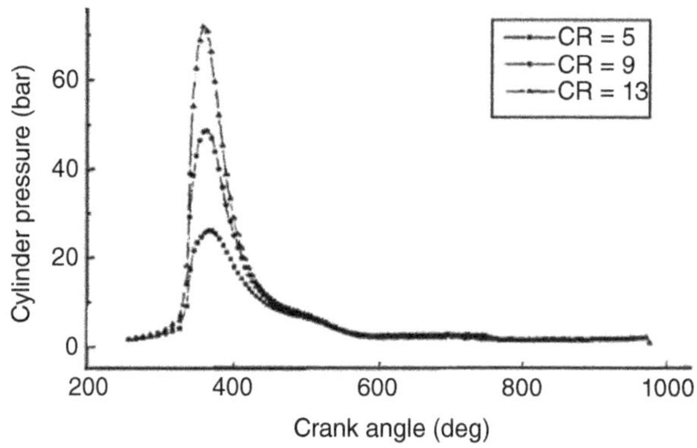

Figure 5.15 Cylinder pressure vs. crank angle with various compression ratios [19] / with permission of Elsevier.

The compression ratio affects the peak cylinder pressure. As seen in Figure 5.15, increase of the compression ratio results in a sharp increase in the predicted peak cylinder pressure. Some of the significant conclusions from this modeling study very closely agree with the well-established published results on hydrogen engines with regard to the influence of lean mixture and compression ratio on improved efficiency.

D'Errico and colleagues [20,21] carried out both experimental and modeling studies in a BMW spark ignition engine operated on hydrogen. A quasi-dimensional combustion model for a hydrogen engine with the capability of predicting the burning rate of combustion of a homogeneous mixture of hydrogen and air was considered for the analysis. The authors adopted full-cycle simulations using 1D thermo-fluid dynamic code using hydrogen's typical physico-chemical properties for calculations. A fractal approach with intricate and irregular pattern was used for evaluating turbulent flame speed. On the other hand, a set of data for laminar burning velocities was generated by applying kinetic simulation code that simulates chemical reactions and behavior of the nonturbulent laminar flames. In this modeling study, 1D simulation was done taking into account unsteady flow in both the intake and exhaust system. This quasi-dimensional combustion model also took into account the transport of chemical species considering the low temperature of cryogenic port injection. The working fluid was assumed to be a mixture of ideal gases including para-hydrogen. Since it was a gas mixture, specific heat was determined not only by temperature, but it also took into account mole fraction of different gases in the mixture. Moreover, hydrogen was in a para state in cryogenic conditions and in this cryogenic state, and the effect of pressure on para-hydrogen concentration was almost negligible. Only temperature has a dominant influence on para-hydrogen concentration. The BMW test rig was appropriately instrumented. One pressure transducer was installed in the intake system to help channel and control flow into the engine cylinder and another was located in the exhaust system. A thermo-quartz was in the combustion chamber to record peak cylinder pressure. A hydrogen mass flow sensor was installed to obtain accurate and real-time data on the mass flow rate of hydrogen. Several thermocouples were installed at different locations to get the correct temperature at those locations. An exhaust gas analyzer was used to measure the concentration level of different gases emitted from the engine. All these instruments were used to generate relevant test data important to validate the simulation model for the hydrogen engine using cryogenic port injection. A comparative evaluation of computed results with experimental data on maximum pressure, turbulent intensity, and pressure pulses is shown in Figures 5.16 through 5.19. Experimental data and analytically predicted results were generally in good agreement for most operating conditions.

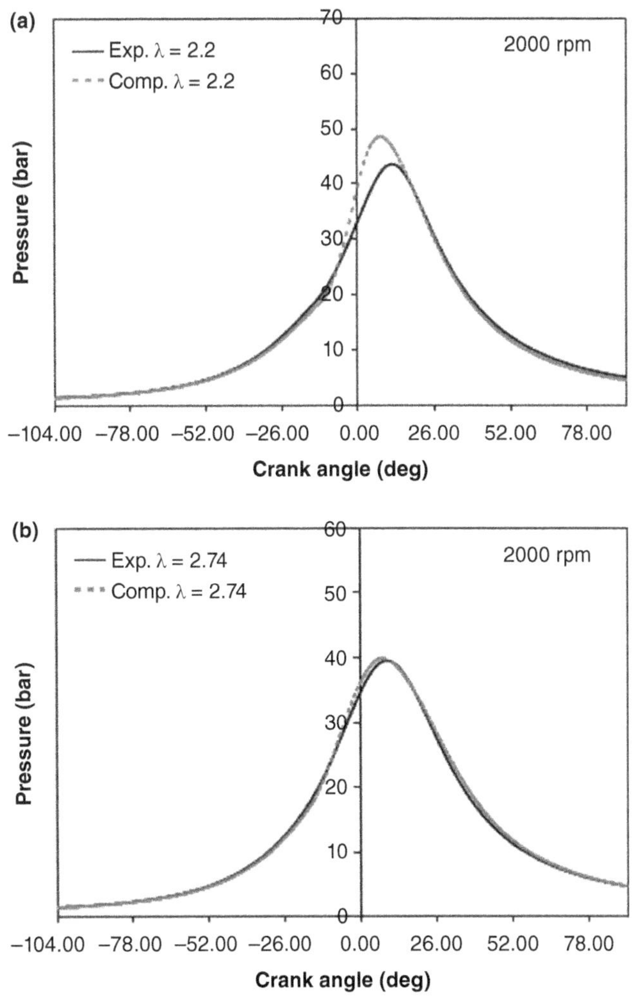

Figure 5.16 Experimental and computed cylinder pressures at 2000 rpm, for (a) stoichiometric and (b) lean mixture conditions with $\lambda = 2.74$ [21] / with permission of Elsevier.

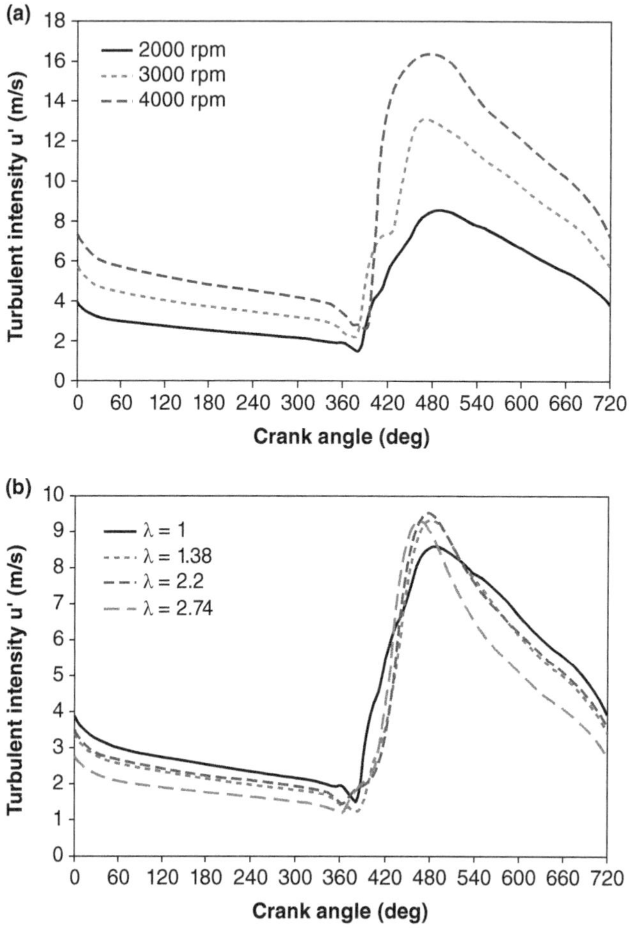

Figure 5.17 Computed turbulent intensity as a function of the engine speed (a) at stoichiometric conditions with $\lambda = 1$ and (b) as a function of different relative air–fuel ratios at 2000 rpm [21] / with permission of Elsevier.

Verhelst and Sierens [22] developed a simulation program for the power cycle of hydrogen-fueled engines using a quasi-dimensional model. In an effort to study combustion more accurately and evaluate the impact of finite turbulent flame thickness, which affects the end of combustion, the researchers adopted a combustion model consisting of two differential equations. One was for "entrainment" mass burning rate and the other for "fully burnt mass burning rate." They compared experimental data with simulation results obtained from a number of turbulent burning velocity models and selected an appropriate relationship correlating laminar burning velocity of hydrogen/air mixtures with pressure development. Thus, correlation correctly developed by the author was calibrated to a single reference condition. Subsequently, simulations were carried out by varying the relevant parameters such as equivalence ratio, ignition timing, and compression ratio, and the results were compared with the reference condition. Variation of these parameters was useful in getting an idea of the changes in pressure, temperature, and mixture composition. This information was useful in carrying out a comparative evaluation between experimental data with simulated results over a wide range of operating conditions.

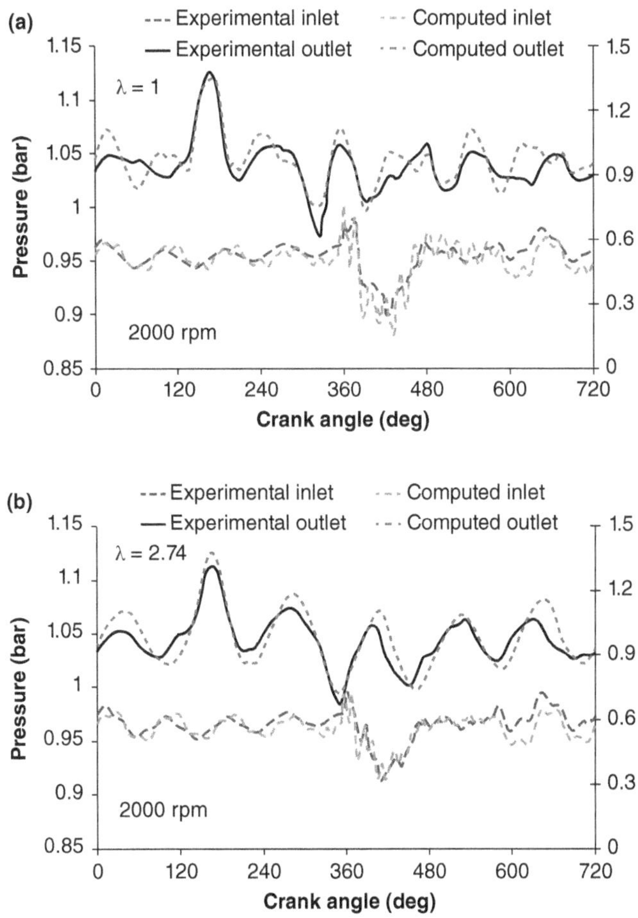

Figure 5.18 Experimental and computed pressure pulses in the intake and exhaust duct systems at 2000 rpm for (a) stoichiometric (λ = 1) and (b) very lean (λ = 2.74) operating conditions [21] / with permission of Elsevier.

Safari et al. [23] developed a thermo-kinetic model using turbulent burning velocity for flame propagation with the objective to also assess methods of controlled NOx emission. Lean burn strategy and the use of both hot and cooled exhaust gas recirculation techniques were theoretically analyzed using relevant simulation code. NOx emissions showed an increasing trend at high loads and high equivalence ratio near stoichiometric conditions. For a compression ratio of 9.5, a peak NOx emission level of about 6800 ppm occurred at an equivalence ratio of $\phi = 0.83$. The analysis for NOx reduction further showed that lean burn strategy is more effective at low load conditions whereas exhaust gas recirculation is preferred at high load conditions. Another significant result showed that cooled EGR produced more power than hot EGR under the same NOx emission conditions. Combustion analysis was carried out by dividing the charge into three distinct zones: the unburned zone, the flame zone, and the burned zone, and detailed kinetics was used for both the burned and unburned zones. The adiabatic flame was assumed to be in thermodynamic equilibrium. Results of this simulation with varying equivalence ratio, ignition timing, and compression ratio were in good agreement with experimental data. Further study and analysis of models for end-gas chemical reactivity will be useful for predicting knock [23].

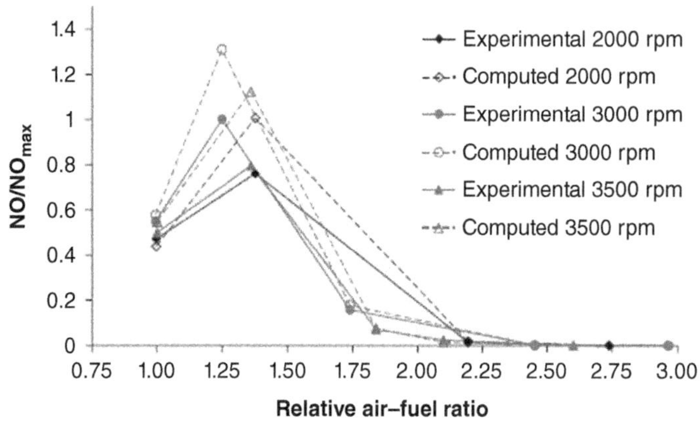

Figure 5.19 Experimental and computed NO cylinder emissions in nondimensional form with respect to the measured maximum value, for a wide range of relative air–fuel ratios and three engine speeds of 2000, 3000, and 3500 rpm [21] / with permission of Elsevier.

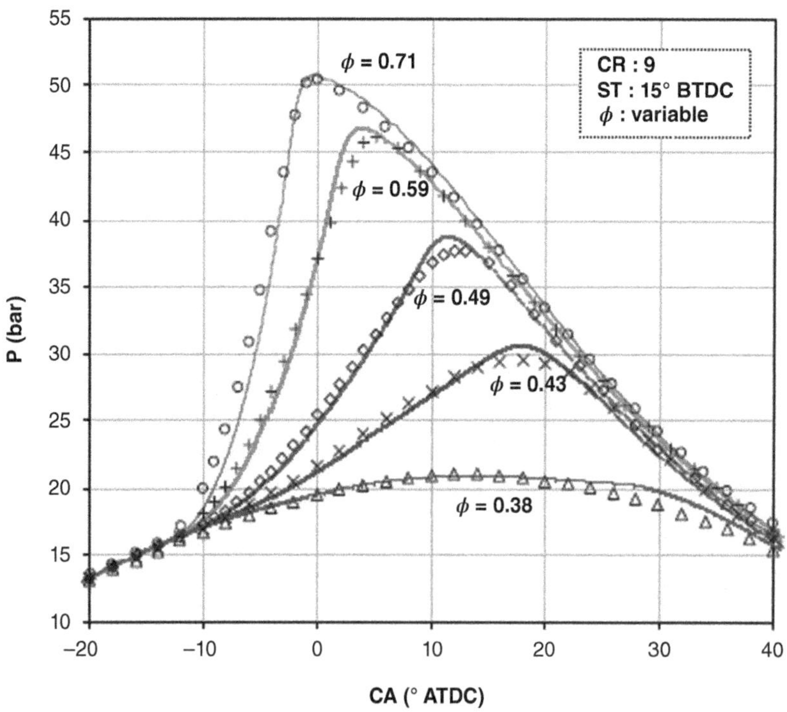

Figure 5.20 Experimental and simulated pressure variation with equivalence ratio; the symbols are experimental data, the solid lines are simulation results [23] / with permission of Elsevier.

Figure 5.20 shows the variation of peak cylinder pressure with changes in equivalence ratio at an ignition timing of 15° BTDC and at a compression ratio of 9. As evident in the graph, the test data are in close agreement with the simulated values. Figure 5.21 shows that the maximum value of NOx is about 6800 ppm, which occurs at an equivalence ratio of ~0.8. There is a drastic reduction in NOx levels at a lower equivalence ratio, with NOx concentration of less than 100 ppm at an

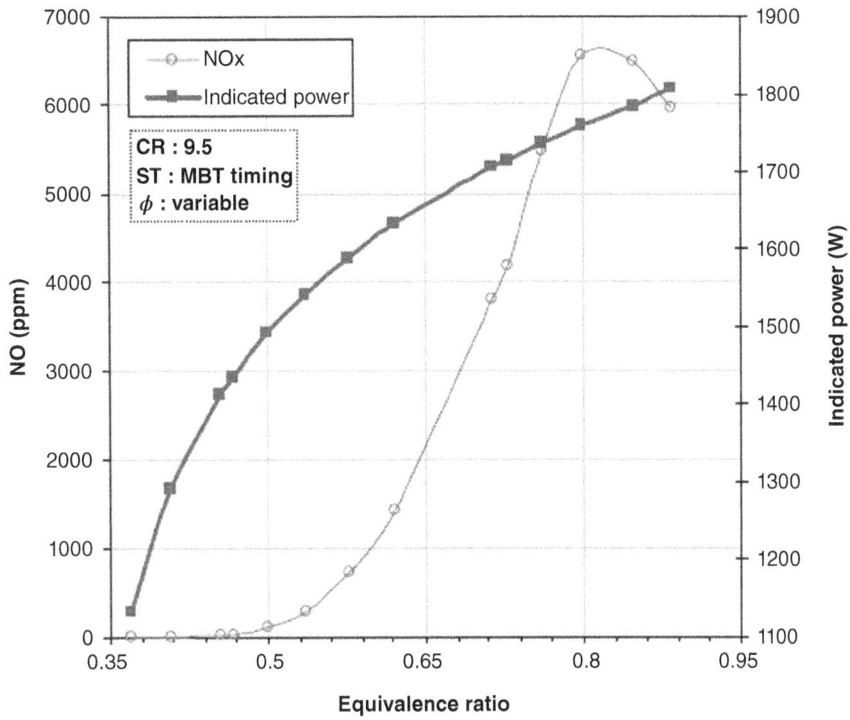

Figure 5.21 NO concentration and indicated power against equivalence ratio [23] / with permission of Elsevier.

equivalence ratio $\phi = 0.5$ and about 10 ppm for $\phi = 0.4$. The curve nearly coincides with the X-axis at a still lower equivalence ratio. However, as shown in Figure 5.21, there is a reduction of power of about 20% if the equivalence ratio is decreased from 0.88 to 0.5. It is well known that the level of NOx formation is highly dependent on combustion temperature, excess oxygen, and residence time. This trend is shown in Figure 5.22. Closer to the stoichiometric condition, the NO formed becomes dissociated later in the cycle, mainly due to lack of oxygen and at high-enough temperatures suitable for the dissociation.

Figure 5.23 shows the concentration of nitric oxide emitted by adopting different dilution strategies such as lean burn operation or adopting hot EGR and cooled EGR. It is clear that the NOx emission level with EGR (both hot EGR and cooled EGR) is substantially lower as compared to lean burn strategy. The difference is larger in the range of equivalence ratio beyond about 0.7. Lean burn strategy is not the desirable mode of engine operation for low NOx emission, and adopting EGR happens to be an effective mode to run the engine at stoichiometric condition to strike a trade-off between NOx and power output. In their analysis [23], with a view to study the effect of the presence of water vapor, EGR flow temperature was set at 110 °C, so the exhaust gas also had water vapor at stoichiometric condition. On the other hand, due to cooling the EGR flow and evacuating the condensate, the cooled EGR flow did not contain water vapor. Another significant observation is that Figure 5.24 depicts the concentration of NOx to be lower in the case of hot EGR as compared to that of cooled EGR, even for the same percentage of EGR. Hot exhaust gases often contain water vapor as water is one of the by-products in this case. When water vapor is introduced into the combustion chamber, it absorbs heat and reduces the combustion temperature. Thus, the mean and maximum combustion temperature with hot EGR are relatively less as compared to those with cooled EGR. Lower peak temperature with hot EGR thus reduces NOx formation.

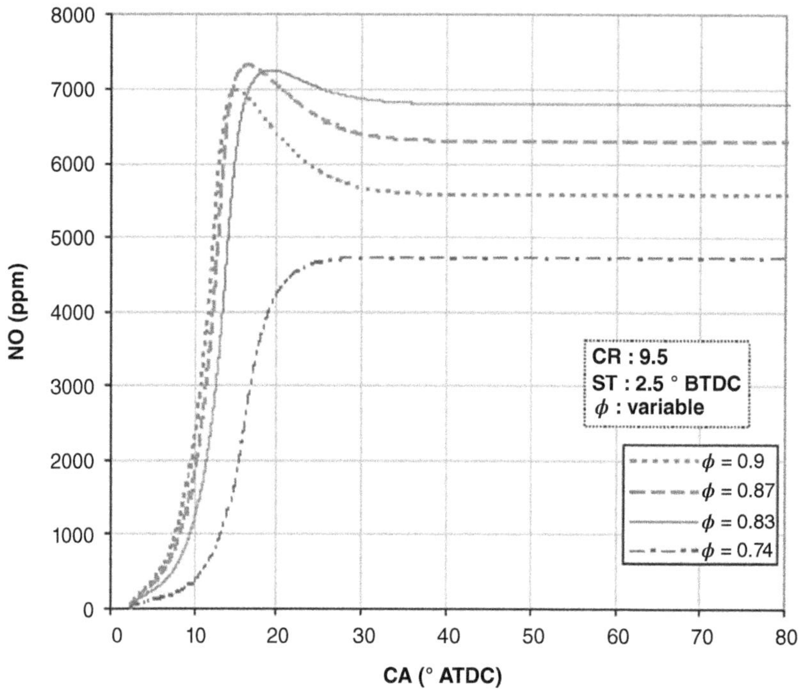

Figure 5.22 NO concentration versus crank angle for changes in equivalence ratio [23] / with permission of Elsevier.

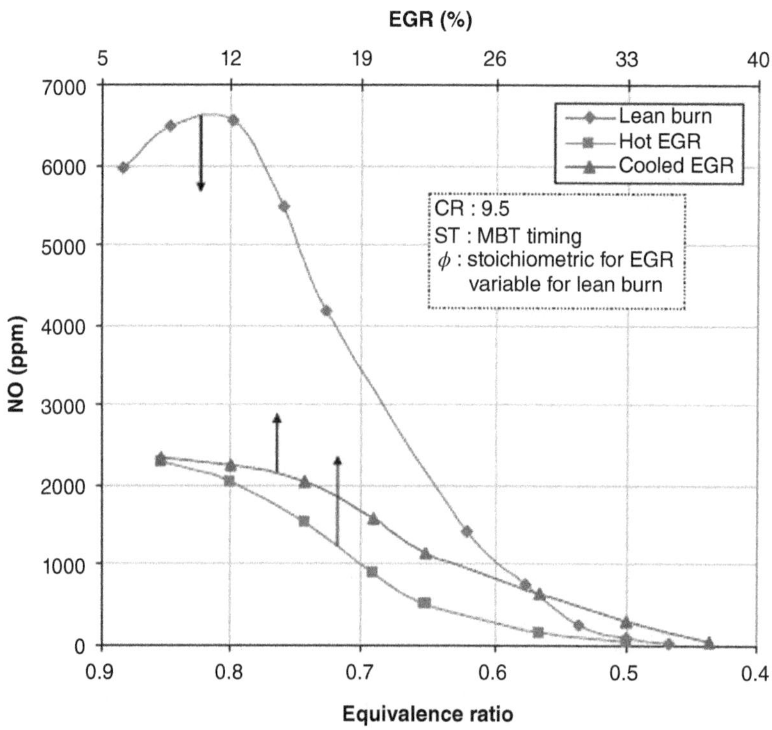

Figure 5.23 Nitric oxide emission for different dilution strategies [23] / with permission of Elsevier.

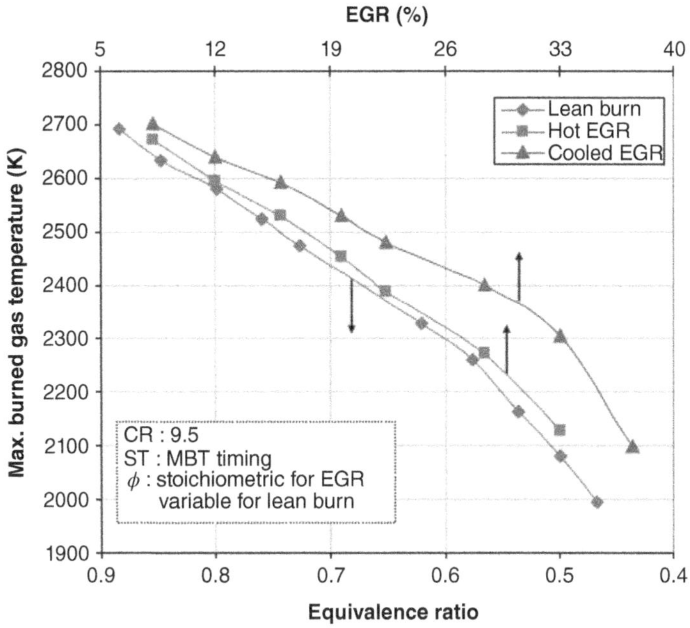

Figure 5.24 The maximum burned gas temperature for different dilution strategies [23] / with permission of Elsevier.

In Figure 5.24, both hot EGR and cooled EGR have a slightly higher peak combustion temperature than that of a lean burn strategy. In lean burn combustion strategy, the excess air (as compared to stoichiometry) lowers the peak temperature during combustion because there is more air available to absorb and dissipate the heat. Some studies have reported that chemically termolecular reactions involving H + O_2 + R could significantly influence combustion pathways that can affect global reactivity and burning velocities of flames. Konnov [24] made a detailed analysis of hydrogen combustion kinetics by introducing four new reactions and updates to three rate constants in the mechanism. These changes led to reduction in calculated burning velocities compared to experimental data and previous model predictions, mainly due to the influence of thermolecular reactions. Subsequently, a revised mechanism, incorporating updated kinetics and transport properties, improved the performance. Measured data for burning velocity for hydrogen flames was observed to be consistent with the values obtained by this analysis.

5.4 Computational Fluid Dynamics (CFD) Models

CFD plays a crucial role in any engine simulation including hydrogen engines to optimize engine design conditions for improved performance, efficiency, and emissions control. Johnson [25] of Los Alamos National Laboratory (LANL) developed a simulation model for designing a hydrogen engine, based on KIVA-3V code of LANL. The analysis was done on a modified eddy-turnover model to validate the intake flow modeling and the turbulence combustion modeling. The adjustments on the KIVA model for hydrogen were very few as compared to what is usually done in the case of modeling hydrocarbon-fueled engines. The model was rigorously tested by applying it to various ignition timings because changes in the ignition timing strongly affects the character of combustion, including turbulence, pressure, and temperature. Results of the modeling were in good agreement with experimental data generated on efficiency and NOx emission of the engine over a wide range of ignition timing. Adgulkar et al. [26] used AVL FIRE as the CFD code and carried out three-dimensional simulation studies for premixed hydrogen combustion. The analysis considered a simplified model of 3D geometry of an engine with a single valve. Thus, for simulation, only the inlet valve was considered. The researchers also tested and validated their analysis of hydrogen combustion in a spark ignition engine with a number of other models such as the eddy breakup model, turbulent flame speed closure combustion model, coherent flame model, and probability density function model. The well-known features of hydrogen combustion being greatly influenced by hydrogen's wider flammability range, high flame speed, and high autoignition temperature were brought out in this analysis. Performance and combustion characteristics of the engine were analyzed with varying operating conditions of speed (from 1200 to 3000 rpm), equivalence ratio (from lean mixture with $\phi = 0.5$ to a little above the stoichiometric with $\phi = 1.2$) and compression ratio of 8, 9, and 10. The effects of these varying operating parameters were evaluated in terms of pressure distribution, temperature distribution, species mass fraction, reaction progress variable, and rate of heat release for a complete cycle. The analysis of the present work agrees well with the reported results of power loss to the tune of almost 12–15% less than that of gasoline operation.

Backfire, identified as one of the major abnormal combustion phenomena in a hydrogen engine, is intrinsically avoided by direct cylinder injection. Wimmer et al. [27] carried out an exhaustive thermodynamic analysis considering losses due to injection during the compression stroke that are quite often ignored with conventional fuels. Results were also analyzed from test bed experiments. It was found that wall heat losses had a strong influence on overall efficiency. Local surface temperatures were measured. Considering all these aspects, this work identified the conditions leading to higher thermal efficiency with the least NOx emissions. The team of researchers from Graz University of Technology and a BMW research group [28] worked on the 3D CFD to investigate the mixture formation and study the complex in-cylinder processes related to combustion. The investigators applied the performance of a standard simulation model for mixture formation. A "user-defined combustion model" with a commercial CFD code was applied to simulation analysis. The results of the 3D CFD analysis were subsequently validated with measurements obtained on an optical engine [28].

5.4.1 Turbulent Flame Speed Closure (TFC) Models

Many engine combustion studies have focused on predicting turbulent flame speed—the speed at which a turbulent flame front propagates through a fuel–air mixture. Turbulent flame speed closure (TFC) models have been used by many researchers for simulation of combustion processes, especially in cases involving premixed and partially premixed flames. In turbulent combustion simulations, it is important to understand how turbulence affects the combustion process. The major steps for an accurate analysis through TFC models involve modeling the turbulent transport of chemical species and heat; modeling the interaction between turbulence and combustion; determination of turbulent flame speed and its impact on the overall flow field; and accounting for turbulence effects on flame stability. The k-ε and k-ω are two common turbulence models in CFD analysis. The (k-ε) turbulence model represents turbulence by tracking two key variables: k (kappa) and ε (epsilon). Kappa is the variable that represents turbulence intensity in the flow. It quantifies how much energy is associated with turbulent eddies at different scales. Epsilon represents the turbulence dissipation rate, which quantifies the rate at which turbulent kinetic energy is converted into thermal energy due to dissipation. The k-omega (k-ω) model addresses some of the limitations of k-ε model, particularly in flows with adverse pressure gradient and near-wall region.

The Reynolds stress model (RSM) is a more advanced and comprehensive turbulence model compared to the k-ε and k-ω models. The major focus of RSM is to predict the full Reynolds stress tensor, which describes turbulent fluctuations in velocity in all three spatial directions. The extended coherent flame model (ECFM) extends the conventional combustion model to account for coherent structures within turbulent flame. The shear stress transport (SST) model is essentially a combination of both k-ε and k-ω models with an ability to analyze wider levels of turbulence. While TFC models are mostly used for simulating turbulent combustion, focusing on the interaction between turbulence and chemical reactions, the SST model is a versatile turbulence model used for a wide range of turbulent simulations on boundary layer behavior and several other complex features.

In his doctoral dissertation, Udo Gerke [29] described numerical analysis and experimental studies on mixture formation and combustion in a direct injection hydrogen engine. While modeling turbulent combustion, laminar flame speed of hydrogen was considered a crucial factor. As an initial step related to combustion, the laminar burning velocity of hydrogen equivalent to engine-relevant conditions was generated by flame speed measurements in a single-cylinder compression machine. The experimental data were generated by tests on optical engines and multi-cylinder engines. The simulation analysis involved application of a 3D-CFD model to a hydrogen engine. Several turbulent combustion models such as models based on turbulent flame speed closure (TFC), on the basis of a flame surface density approach and extended coherent flame model (ECFM), were considered for their capability to predict combustion modes of both premixed and nonpremixed flames simultaneously. In this analysis, the models were integrated with the solver ANSYS CFX software, which has core capability to simulate fluid flow in a 3D environment considering factors such as turbulence, heat transfer, and chemical reactions. The interrelation between injection timing and mixture homogenization degree was analyzed from direct injection mixture formation.This work also generated results from premixed, partially premixed, and nonpremixed hydrogen combustion in homogeneous, stratified, and multi-injection engine configurations. Simulated results of fuel concentration were found to be in agreement with the measured data by planar laser induced fluorescence (PLIF) measurements for various injection timings. It has further been studied that the use of models built on renormalization group (RNG) analysis to compute combustion on the basis of mixture results does not adequately account for turbulent intensity in stratified conditions. As far as the application potential of the ECFM combustion model is concerned, it does project a satisfactory picture for PFI using a prefactor or a coefficient to ensure that the mathematical model matches the observed data.However, for direct injection, the prefactor needs to be readjusted for each operating point to match the test data. A large number of models such as the eddy viscosity turbulence models standard (k-ε) model, RNG (k-ε) model, shear stress transport model, Reynolds stress turbulence model, Zimont model, Bradley model, etc. have been discussed [29] with applicability of these models in different turbulent and combustion conditions.

Anticaglia et al. [30] adopted a numerical investigation with a multi-fidelity computational fluid dynamics approach to develop an existing gasoline engine into hydrogen operation. Researchers used both 1D and 3D simulation to evolve comprehensive strategies for redesigning existing gasoline engines to ensure lean burn hydrogen engines. In view of typical hydrogen combustion characteristics, emphasis was laid on valve timing, intake port design, compression ratio, detailed injection strategy, and turbocharging to achieve improved performance characteristics (power and efficiency) without any undesirable combustion phenomena. The researchers carried out a detailed analysis of each of the key performance indicators (KPI) as shown in Figure 5.25 to evolve the numerical methodology for optimum design conditions (Figure 5.26) to obtain comparable power density with current turbocharged spark ignition engines.

The study [30] centers around a high-performance turbocharged gasoline direct injection (GDI) engine. Commercial software GT-Power (v2016) was used for 1D-CFD simulation. Subsequently steady state 3D-CFD analysis with Star-CCMþ (v15.06) software was used to construct and validate the model. Commercial software CONVERGE CFD (v3.0.22) was used to analyze the combustion process of hydrogen mixtures. The results of this investigation were in close agreement with some of the well-documented performance and emission results on hydrogen engines such as improved thermal efficiency at higher compression ratio, reduced NOx in lean operation, and strong relationship between peak cylinder pressure and NOx concentration. An appropriate injection strategy plays a critical role to ensure optimum performance with least emission and smooth combustion. Injectors with high mass flow and late injection timing can reduce the compression work.

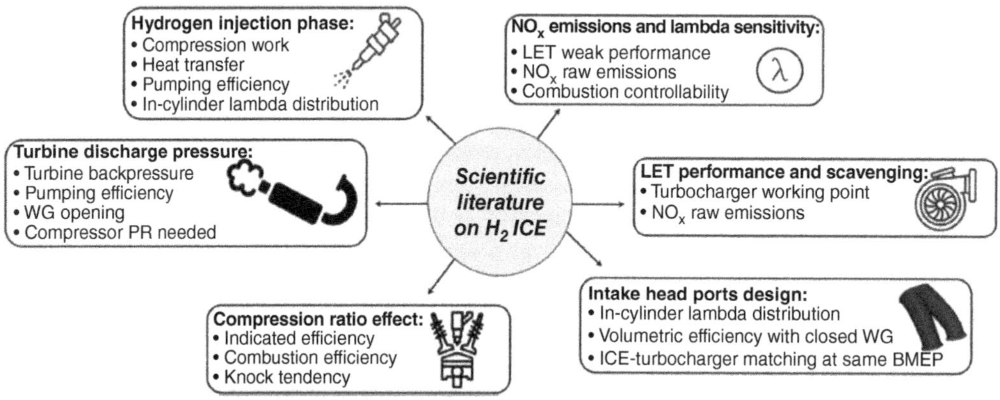

Figure 5.25 Guidelines from the literature for analytical study [30] / with permission of Elsevier.

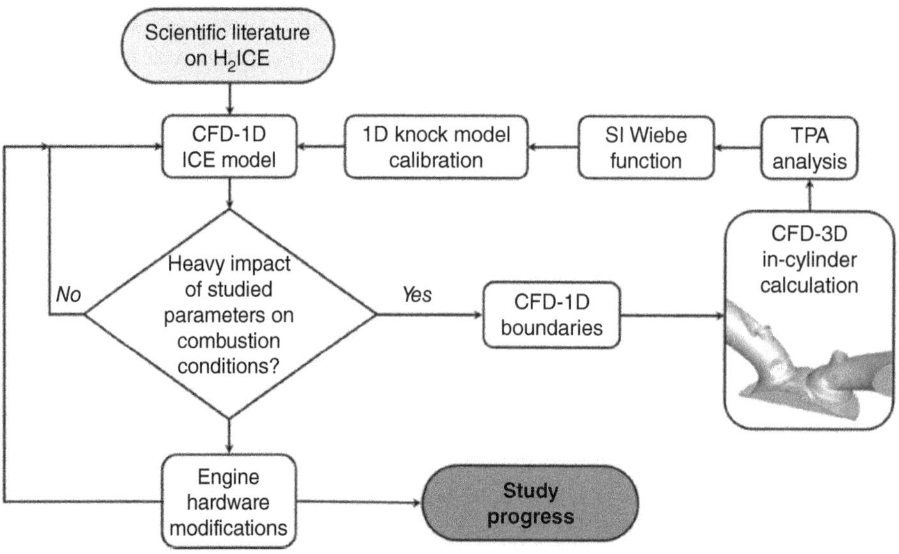

Figure 5.26 Methodology used for the numerical investigation [30] / with permission of Elsevier.

Babayev et al. [31] developed a CFD model with the appropriate numerical methods to study hydrogen combustion in a direct injection (DI) compression ignition (CI) nonpremixed configuration in an engine. The boundary conditions for baseline simulations were set up. The models were validated against both optical hydrogen jet measurements in a constant volume optical chamber and data generated from diesel engine tests. A well-pronounced effect of stratification at the fuel–air interface of the hydrogen jet was observed, which indicates a slim chance of premixing. Despite its wider flammability range and high diffusivity, most of the reactions of hydrogen jets were concentrated in a narrow zone near stoichiometry. In this direct injection compression ignition (DICI) system for a hydrogen engine, the free jet mixing phase was dominant and the heat release rate was significantly high with hydrogen. This shows the dual behavior of hydrogen combustion in a nonpremixed hydrogen combustion process.

Scarcelli et al. [32] investigated mixture formation in a direct injection hydrogen-fueled passenger car SI engine with a centrally located injector using a CFD code. They used a single-hole and a 13-hole nozzle operating at injection pressures of approximately 100 and 25 bar. A quantitative planar laser-induced fluorescence technique was adopted to obtain phase-locked images of the fuel mole-fraction, whereas high-speed schlieren photographs captured single-cycle visualization of the early jet penetration. Numerical data by the analysis from the commercial code Fluent (v6.3.35) were compared with the test data generated from the measurements in an optically accessible engine. Studies were carried out on the specific features of the computational grids in regions where jets were "under-expanded" and this occurred in the "near-nozzle" region where the jets could not achieve the intended characteristics. Simulation results of injection from single hole nozzles in respect of jet penetration and jet revolution were in agreement with the test data obtained from the optical engine. The picture was different, however, in the case of 13-hole nozzles, where jet-to-jet interaction caused one jet to be attracted toward the curved surface of the other jet, leading to interactions like merging or bending of the jets. As a result, all 13 jets merged into a single jet immediately downstream of the underexpanded region. When two jets approach each other, they usually influence each other's flow patterns and trajectories. Such behavior is referred to as the Coanda effect. In the Coanda effect, one jet is attracted toward the curved surface of the other jet, leading to interactions like merging or bending of the jets. It has been relatively difficult to build numerical simulation considering the Coanda effect, particularly for fields near the injector nozzle.

Oxides of nitrogen are the single pollutant of concern in hydrogen engines. Kherdekar and Bhatia [33] used the mass and energy equation and developed a model to predict the NO emission level in a hydrogen-operated spark ignition engine. The transient spark energy delivery model is built on experimental data. Heat transfer was modeled comprehensively, considering forced convection, conduction, and natural convection with extended surfaces. Convective heat transfer from cylinder to coolant was identified as the main resistance. The simulated results on the basis of a kinetic-based approach showing the effect of variables such as engine speed, equivalence ratio, compression ratio, and spark advance on the NO emissions were found to be consistent with the experimental data reported in the literature.

Concentration of NO emission levels depends on the speed of the hydrogen engine. The simulated data for an equivalence ratio of 0.5 shown in Figure 5.27 indicate the variation of NO emissions and peak in-cylinder temperature with respect to speed. The maximum in-cylinder temperature and levels of NO emissions in the figure represent the simulated data for an equivalence ratio.

A closer comparative look at both Figures 5.27 and 5.28 shows that the variation pattern of NO between the two figures is very different. Even though the trend of variation in maximum temperature is similar, the maximum temperature shown in Figure 5.28 is higher than that shown

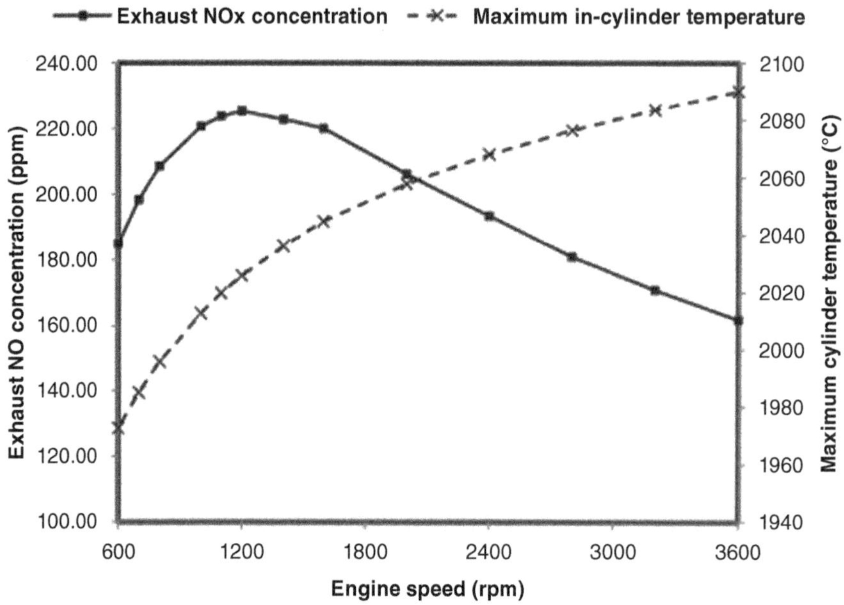

Figure 5.27 Variation of exhaust NO concentration and maximum in-cylinder temperature with engine speed for $\phi = 0.5$ [33] / with permission of Elsevier.

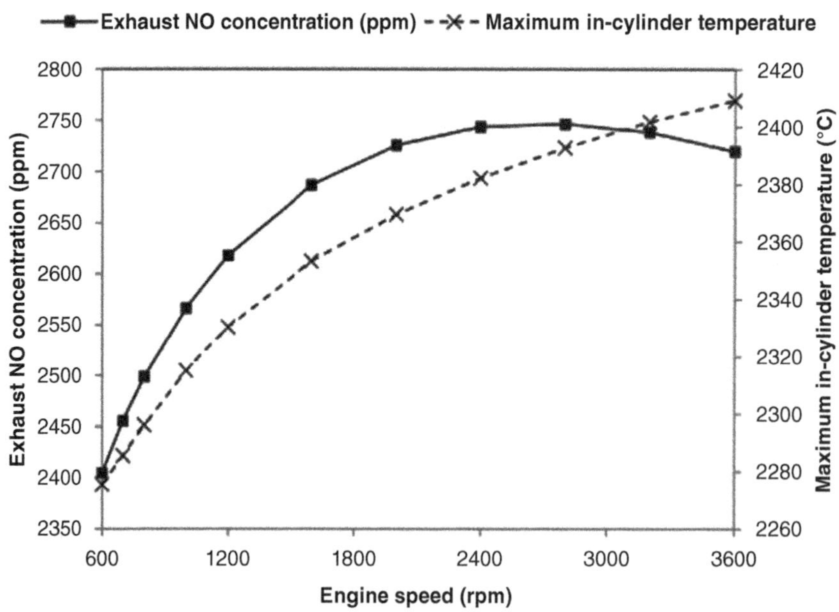

Figure 5.28 Variation of exhaust NO concentration and maximum in-cylinder temperature with engine speed for $\phi = 0.65$ [33] / with permission of Elsevier.

in Figure 5.29. This difference shows that at a constant speed engine operation with higher equivalence ratio results in higher temperature level, which subsequently leads to higher NO emission. In Figure 5.27, the NO emission level exhibits an increasing trend with the increase in engine speed at the normal idling speed in the range of 600–1000 rpm. This increasing trend of

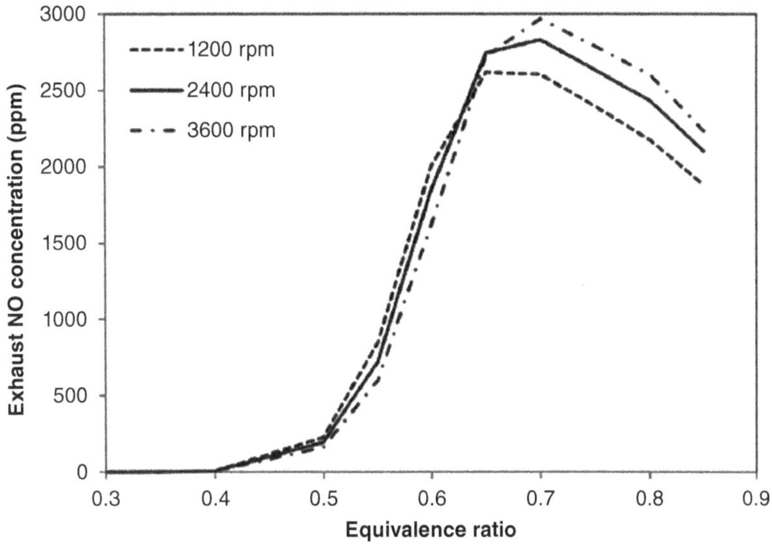

Figure 5.29 Variation of exhaust NO concentration with equivalence ratio at different engine speeds [33] / with permission of Elsevier.

NO emissions with respect to engine speed is exhibited until about 1200 rpm and then takes a downward trend. NO formation has very specific relationship with temperature. Beyond a certain range of speed, the effect of a period of high temperature available for NO formation becomes less pronounced. Thus, the time available for NO formation also decreases. Under such circumstances, despite higher temperatures at increased speeds, NO concentration decreases beyond 1200 rpm due to insufficient time for NO formation. This trend was clearly reflected in the simulation data in Figure 5.27.

Figure 5.29 shows the variation of exhaust NO concentration with equivalence ratio at different engine speeds. Depending upon the experimental conditions, several researchers have reported a maximum value of oxides of nitrogen a little earlier to the stoichiometric value (between 0.7 and 0.9).

Figure 5.30a shows the exhaust NO concentration at different spark advance angles. This is due to the increase in temperature attained with the increase in the angle of spark advance. The compression ratio affects the exhaust NO concentration, with 9–9.5 suggested for optimal power and emissions. Formation mechanism of oxides of nitrogen is highly temperature dependent. Figure 5.30b shows that exhaust NO concentration takes a slow gradual trend within the compression range of 7–9, increasing significantly when the compression ratio increases. This trend is similar for all three operating speeds used for analysis. The authors in their analysis [33] have also observed the same trend: the peak cylinder temperature increases slowly and gradually when the compression ratio increases from 7 to 9. However, the in-cylinder temperature rapidly increases thereafter beyond a compression ratio of 9. This similarity in analysis further justifies the relationship of temperature with the NO emission level.

Researchers have simulated results indicating the effect of the spark advance on the cylinder temperature. From Figure 5.31 it is evident that the peak cylinder temperature increases with increase in the spark advance.

Since NO formation is closely linked to temperature, it is also possible to analyze and correlate concentration of NO emission with spark advance. Figures 5.32a and 5.32b depict in-cylinder temperature and NO concentration during the power stroke for equivalence ratios of 0.65 and

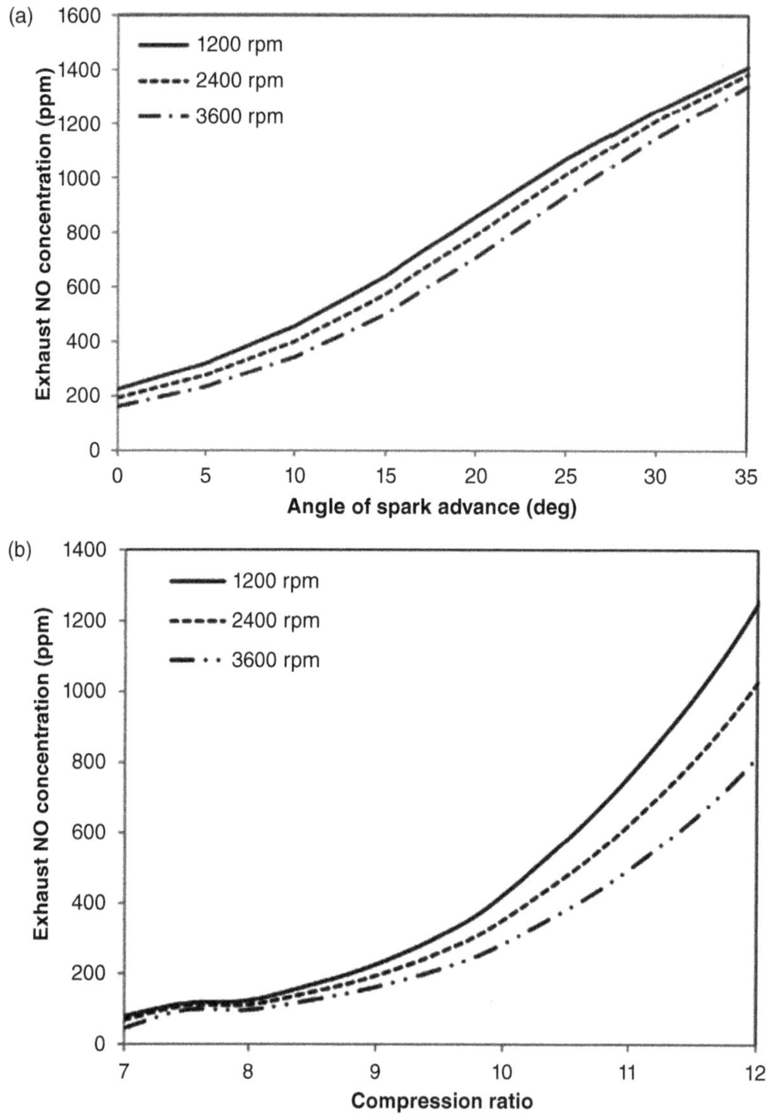

Figure 5.30 Variation of exhaust NO concentration with (a) spark advance and (b) compression ratio [33] / with permission of Elsevier.

0.75, respectively, at 2400 rpm. Increasing the equivalence ratio to 0.75 raises peak in-cylinder temperatures a little above 2500 °C resulting in higher peak NO concentrations, compared to 0.65. However, the in-cylinder NO concentration at the end of the power stroke is lower for 0.75 than for 0.65 because of NO decomposition, which is favored at high temperatures resulting from high equivalence ratios. This illustrates that NOx emissions depend not only on peak in-cylinder temperatures but also on transient temperature variations. Trends in in-cylinder NO concentration variation with crank angle agree with findings by Safari et al. [23]. In that analysis, a global kinetics model for hydrogen combustion and NO formation is used. In-cylinder heat transfer, coolant heat transfer, and spark energy delivery are considered without relying heavily on empirical correlations. The analysis concludes that engine speed and equivalence ratio strongly influence exhaust NO concentration, with higher speeds leading to maximum emissions. The equivalence ratio impacts in-cylinder temperature and NO formation.

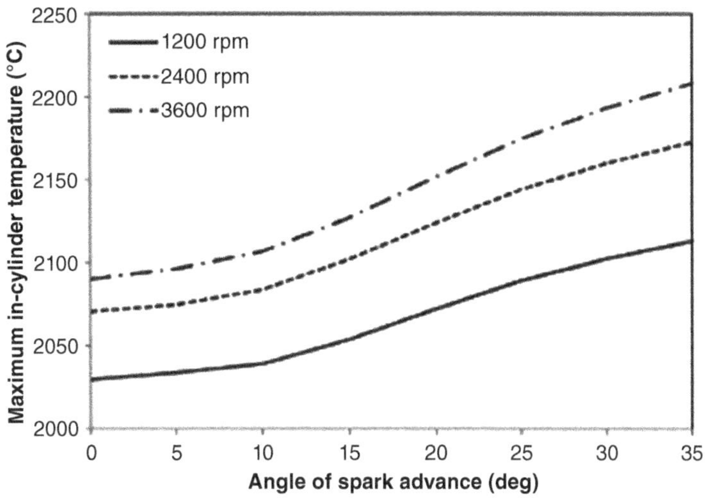

Figure 5.31 Maximum in-cylinder temperature with spark advance [33] / with permission of Elsevier.

Ballerini et al. [34] carried out simulation studies on the 1D thermo-fluid dynamic simulation code with emphasis on simulation of unsteady flows in the entire intake and exhaust system along with the hydrogen injection process. This analysis assumed a mixture of ideal gases as the working fluid, and the specific heats were considered the basis of gas temperature and composition. Figure 5.33 shows the schematic of the two-zone combustion mode. In their analysis of the combustion process, the researchers used a tabulated kinetics approach rather than a correlation-based approach for computing the laminar flame speed for the homogeneous air–hydrogen mixture. Models developed on a tabulated kinetic approach are preferred as they provide a structured and readily accessible database to analyze reaction mechanisms, which allows for a more precise calculation of the laminar flame speed under varying conditions. Excellent agreement of cylinder pressure variation at different mixture compositions between the tabulated kinetic approach and experimental data are shown in Figure 5.34a – 5.34d.

Figure 5.35 shows how closely experimental and simulated data match for pressure trend in lean and stoichiometric mixtures. The paper also discusses the results obtained from simulation on a heavy-duty hydrogen engine. A comparative assessment with experimental data indicates the improvement of the model on the one-dimensional code to simulate a hydrogen-fueled engine. This improved version can be adopted for simulation of more complex configurations such as multi-cylinder or turbocharged engines and could be useful to simulate NOx control. Rawat and colleagues [35] performed simulations by injecting NO into a single-cylinder H_2 engine and showed that injecting NO can reduce its further formation. The reduction was shown to be due to equilibrium limitations and not to the reduction of temperature. The researchers found that high values of equivalence ratio, compression ratio, and load can result in a high degree of NOx reduction. This work was done using a global kinetic model for H_2 combustion and NO formation. Agarwal and co-workers [36] further analyzed the impact of injecting NO using two different microkinetic models and found that the fuel reburning reactions did not play a significant role in the reduction of NO. In another work [37], a detailed reaction pathway analysis was conducted, and the major reactions contributing to the formation of NO in the first phase and those contributing to the reduction of NO in the second phase were found. It was inferred that the mechanism of NOx reduction was not influenced by the addition of NO to the engine cylinder. Figure 5.36 shows the predicted in-cylinder NO concentration profile upon injecting NO.

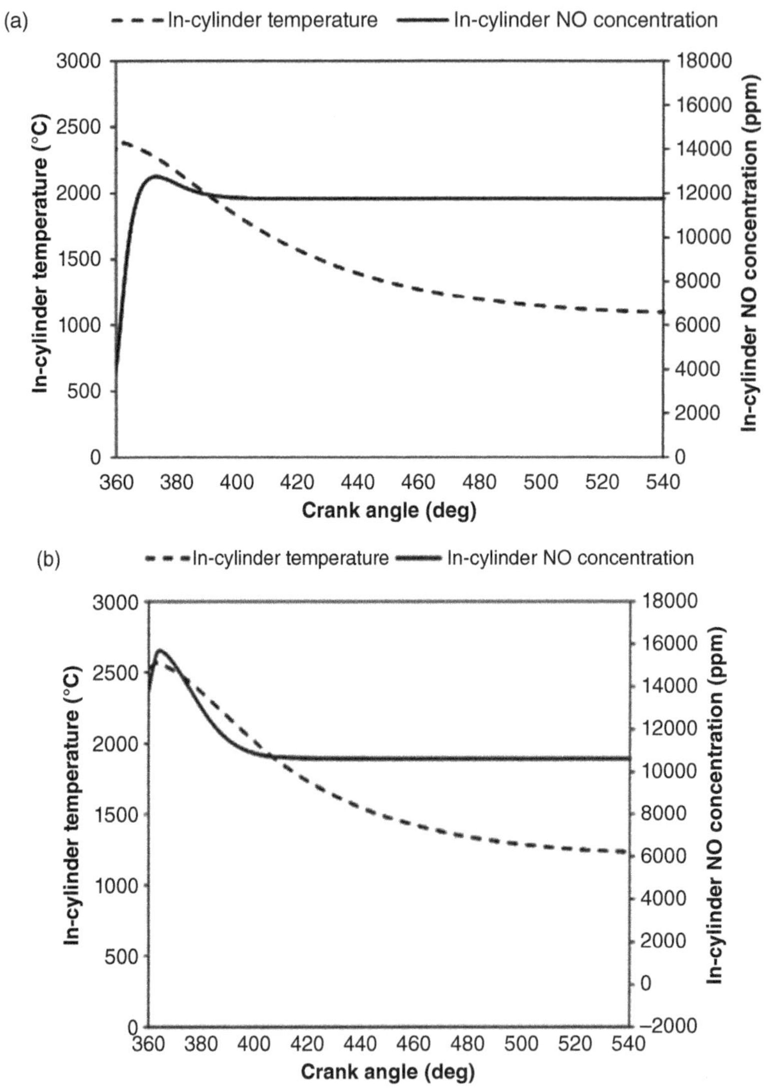

Figure 5.32 In-cylinder temperature and NO concentration during the power stroke at 2400 rpm at equivalence ratios of (a) 0.65 and (b) 0.75, respectively [33] / with permission of Elsevier.

Smirnov and Nikitin [38] did an exhaustive study in on combustion modeling in a hydrogen engine. They developed a predictive model for simulating chemically reacting turbulent flows of both premixed and nonpremixed flows. The mathematical model and numerical tool proved effective in simulating unsteady ignition and combustion processes in various type of engines. The code was developed to simulate chemical reactions in hydrogen combustion to accurately predict the ignition delay times under varying pressures. The code was effective even when the behavior of ignition delay times deviated and did not follow the expected pattern. Initially, a systematic investigation was carried out to verify and validate chemical kinetic models related to hydrogen oxidation by examining the ignition delay times of hydrogen–oxygen mixtures under varying pressure, temperature, and equivalence ratio. The solver developed in the computer program was used for the simulation of premixed combustion (where fuel and oxidizer are mixed before ignition)

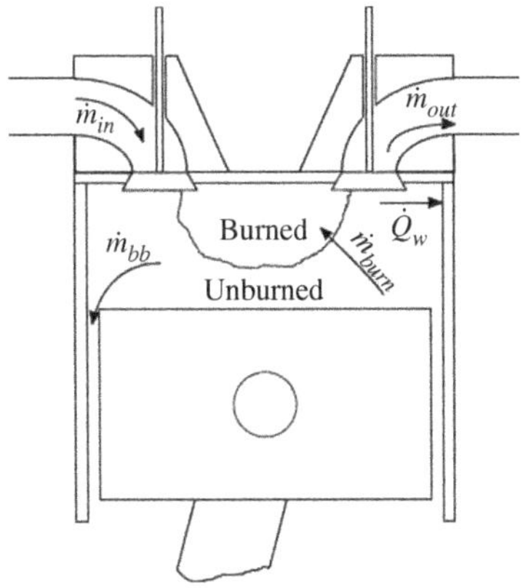

Figure 5.33 Schematic representation of the two-zone combustion mode [34] / IOP Publishing.

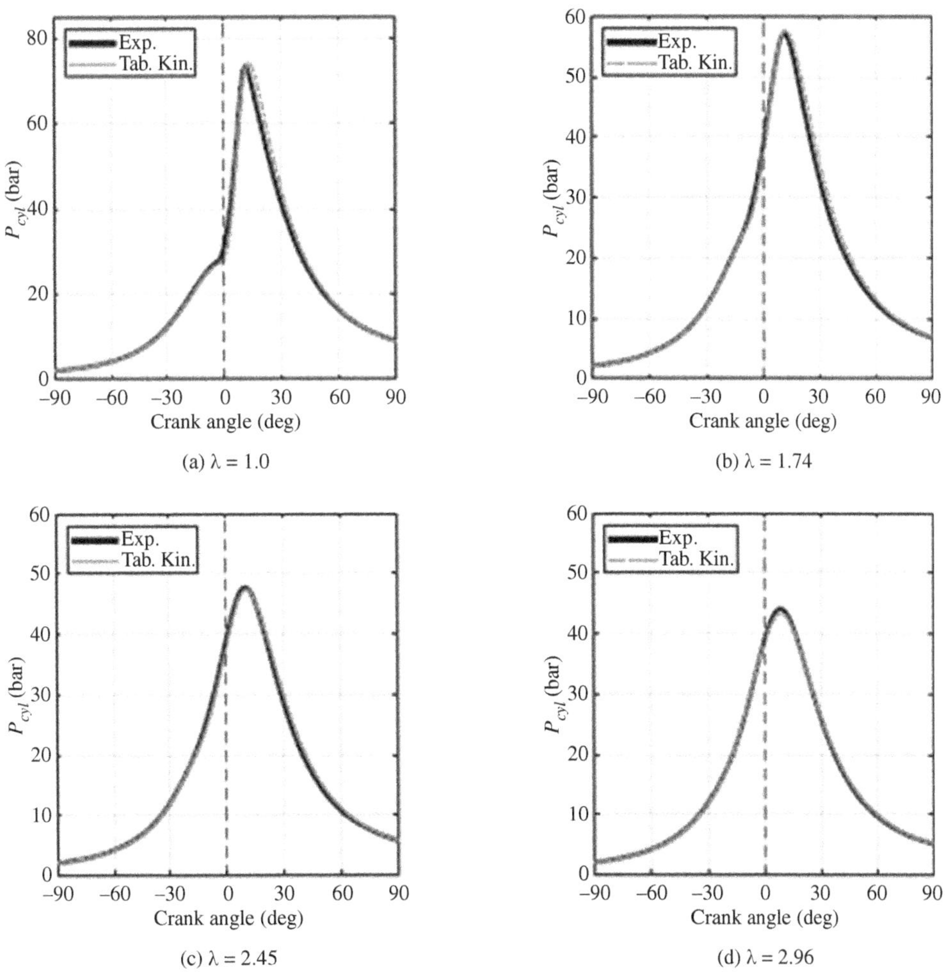

Figure 5.34 Tabulated kinetic and experimental data at four different λ [34] / IOP Publishing.

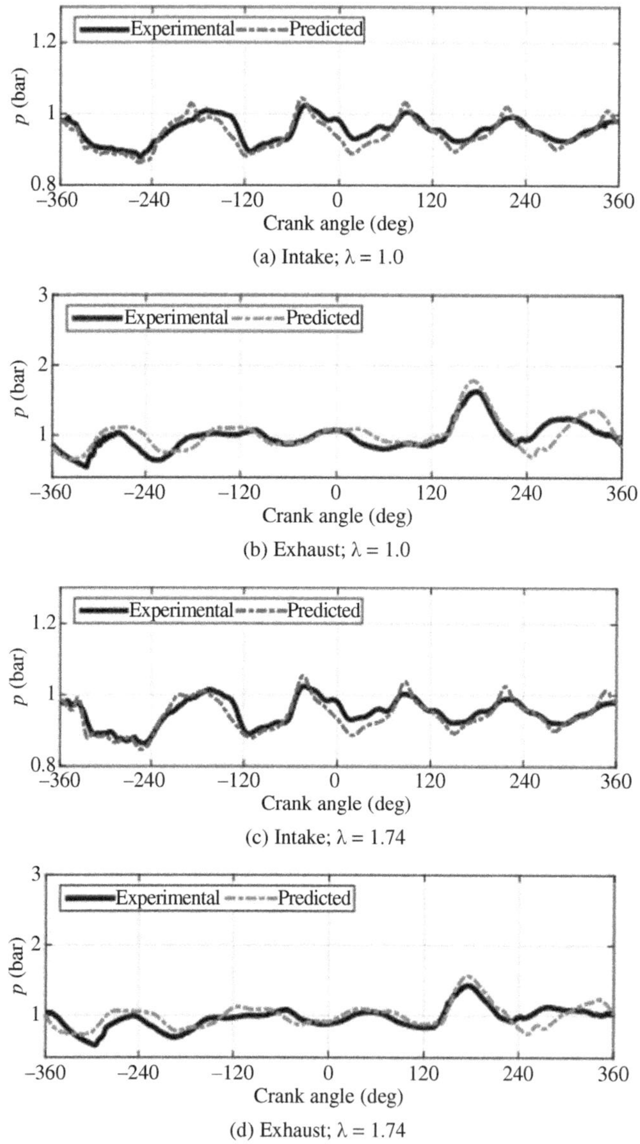

Figure 5.35 Pressure variation in stoichiometric (λ = 1) and lean (λ = 1.7) mixtures [34] / IOP Publishing.

and nonpremixed combustion (where fuel and oxidizer are not mixed beforehand), and also the detonation-related phenomena such as onset, degeneration, and deflagration to detonation transition (DDT). The solver had the ability to model the critical and complex processes in combustion scenario, including slow diffusion, fast kinetic combustion, and detonation. It could also simulate the transition from deflagration (a slower combustion) to detonation (a faster, explosive combustion).

Figure 5.37 shows the variation of ignition delay with equivalence ratio. It is clear from the graph that ignition delay time increased both on the lean and rich sides of the stoichiometric condition. This is the general trend, though there are exceptions. Computer simulation has been used to study how a detonation starts when energy is suddenly released in a small area. Simulation study

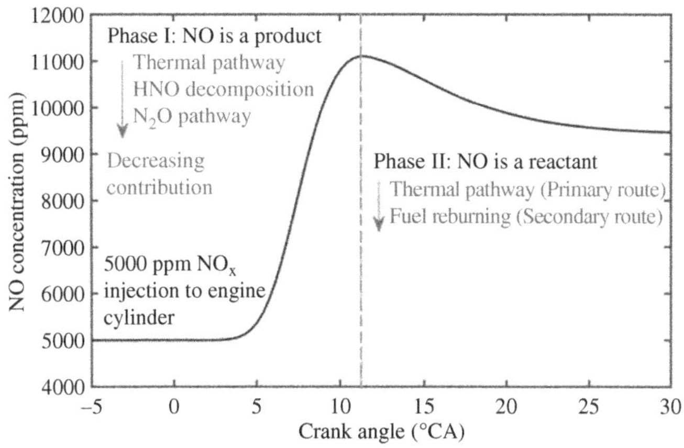

Figure 5.36 Predicted in-cylinder NO concentration profile upon injecting NO [37] / American Chemical Society.

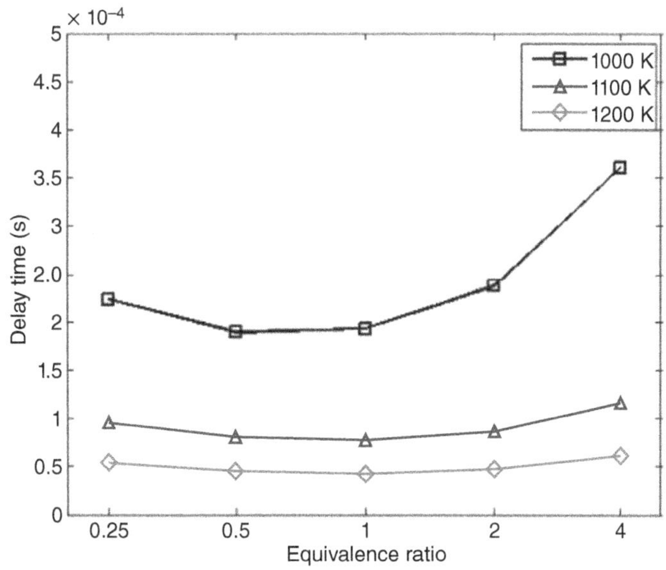

Figure 5.37 Ignition delays for a hydrogen–air mixture as a function of its composition for different initial temperatures [38] / with permission of Elsevier.

of the combustion process in a combustion chamber showed that the combustion process did not remain in a steady state. Initially, when reactants (reagents) were introduced into the combustion chamber, the combustion process behaved more like a diffusion flame. Diffusion flames typically occur when fuel and air mix and burn relatively close to the point of injection. As the combustion process progressed further into the combustion chamber, reactants (fuel and oxygen) were not the primary constraint, and zones of reaction got wider and spread over the entire combustion chamber. In studies related to detonation, it was observed that the detonation wave lost its intensity and became weaker as it moved from a narrow gap into the wider cylindrical chamber. However, subsequently it regained its intensity as it moved further into the wider chamber. A significant feature of this study is the simulation of complex dynamics in transition of deflagration to detonation. The deflagration to detonation transition (DDT) and further transmission of detonation wave into the large combustion chamber give an overall idea of how a pulse detonation engine operates.

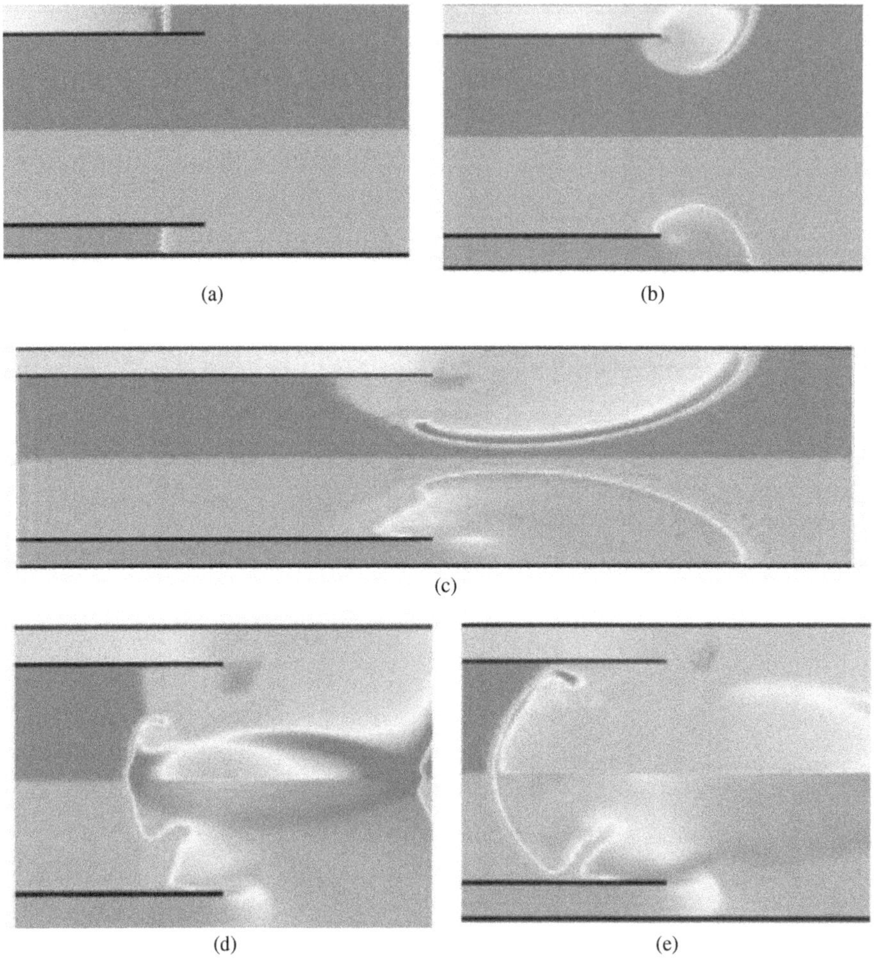

Figure 5.38 Transmitting detonation wave: narrow gap to cylindrical chamber. (H_2:O_2:N_2 = 2:1:6). Times are as follows: (a) 0.2448 ms; (b) 0.2629 ms; (c) 0.2707 ms; (d) 0.2855 ms; (e) 0.2978 ms [38] / with permission of Elsevier. Please visit the companion website to view a color version of this image.

Figure 5.38 shows the successive transmission of a detonation wave from a narrow gap of 20 mm width into a large cylindrical chamber of 200 mm diameter in a diluted hydrogen–oxygen mixture. The authors assumed a model with axial symmetry, and accordingly the upper part of each figure shows pressure fields in a cross section and the lower part illustrating temperature fields [38].

Maurya and Akhil [39] performed simulation of a HCCI combustion engine with 15 different hydrogen mechanisms using a stochastic reactor model to evaluate the combustion parameters. In HCCI engines, chemical kinetics of oxidation plays a crucial role. A comparative assessment was made among the different mechanisms adopted to simulate HCCI combustion by varying significant input parameters such as intake pressure (1.0–3.0 bar), inlet temperature (400–440 K), engine speed (1000–3000 rpm) and relative air–fuel ratio ($\lambda = 2$ to 8) for all mechanisms. Accuracy of the predicted values of crucial combustion characteristics such as start of combustion, peak cylinder pressure, and maximum air release rate was compared with experimental data. Closeness between them was identified for corresponding mechanism. Falfari et al. [40] discussed the significance of 3D CFD simulation in studying the ignition and combustion of a hydrogen-fueled IC engine with its well-identified symptoms of abnormal combustion. Their simulation results are consistent with established experimental results by several researchers showing that direct cylinder injection into the combustion chamber was an effective technique to reduce intensity and frequency of backfire and preignition in the intake. The tendency of detonation and knocking could be eliminated by using a diluted hydrogen–air mixture. Hot areas close to the spark plug often caused preignition, and the use of cold spark plugs proved helpful in preventing this.

Several researchers have observed that apart from injection timing and injection duration, an optimum physical layout and shape of the injector with number of holes, diameter, etc., are relevant to model the process of injection accurately. However, development of an optimum fuel injection system for a hydrogen-specific engine has been a tough task mainly due to the high velocity of the hydrogen jet. Even if test data available in the literature are adequate, they are not consistent in providing a conclusive picture to adopt the design parameters. Such an inadequacy is due to the complexity caused by flame instability at pressures in engines. This is important for predicting specific properties such as laminar flame velocity and the autoignition time based on input parameters. As the ratio of hydrogen and air changes in the mixture, the flame velocity does not consistently increase or decrease; rather, it varies in a more complex manner. Moreover, flame velocity increases as the unburned gas temperature rises but decreases at high pressures. These complex behaviors make it challenging to develop the necessary correlations for robust and reliable CFD models. To develop the CFD simulation, changes in pressure and temperature occurring within the engine's cylinder are to be analyzed as combustion takes place. It is important to track how the flame develops and how this flame produced by burning hydrogen propagates within the combustion chamber. The laminar burn rate of the air–hydrogen mixture inside the cylinder also forms an important aspect of these simulations as it shows how fast the flame spreads through the mixture under controlled, nonturbulent conditions.

Thus, laminar flame speed (LFS), ignition delay time (IDT), an optimum gas injection process, and an appropriate ignition system are extremely important to correctly predict the initiation and development of the hydrogen flame in the combustion chamber. Such an analysis would prevent autoignition of the hydrogen-based mixture due to hot spots on and around the spark plug. Another vital aspect of hydrogen engine modeling that has not been adequately studied is oil–fuel dilution to prevent the formation of oil particulates. It is well known that hydrogen flame has a very low quenching distance (0.64 mm), which often affects combustion in cylinder–piston spaces. As a result, the evaporation of the lubricant increases and forms oil particulates. It is desirable to have a submodel added to the simulation analysis to obtain a more comprehensive calculation.

5.5 Heat Transfer Submodel

A heat transfer submodel is used by many researchers to calculate heat transfer between cylinder gases and combustion chamber walls. Absence of significant amounts of soot and particles in an SI engine reduces the radiation effect. SI engine combustion typically occurs over a relatively short time duration compared to diesel engines. Convection, which involves the movement of hot gases, is more effective at transferring heat quickly during this short period. Due to convection, the instantaneous heat transfer coefficient can be calculated from the equation

$$\frac{dQ}{dT} = hA(T - T_{wall})$$

where

h is the convection heat transfer coefficient,

A is the total wall surface area,

T is the bulk gas temperature and,

T_{wall} is the wall temperature.

The heat transfer coefficient and wall temperature have been averaged over the heat transfer surface. Woschni's equation is widely used to model convective heat transfer from the gas in the combustion chamber to the cylinder walls. There an equation contains two convective terms considering piston motion and convection due to combustion [41]. It is a universally accepted equation to broadly evaluate the heat transfer coefficient.

Woschni's equation is as follows:

$$h = C_1 \left(\frac{p}{p_{ref}}\right)^{0.8} \left(\frac{T}{T_{ref}}\right)^{0.55} \left(\frac{u}{u_{ref}}\right)$$

where

h is the instantaneous heat transfer coefficient (W/m^2K),

p is the pressure in the combustion chamber (Pa),

p_{ref} is the reference pressure (normally taken as atmospheric pressure,

T is the gas temperature in the combustion temperature,

T_{ref} is a reference temperature (normally taken 1000 K),

u is the velocity of gas relative to cylinder wall (m/s), and

u_{ref} is a reference velocity (usually taken as 10 m/s).

C_1 is an empirical constant that depends upon the specific engine and operating conditions. The value of C_1 is determined through experimentation and calibration for a particular engine and operating condition and hence it varies depending upon the engine configuration and combustion strategy.

At this stage of engine research and development, Woschni's equation is a simplified model. It may not be accurate for all engines and especially for hydrogen engine operating conditions. However, this empirical equation can be used as a starting point and refined based on experimental data specific to hydrogen engines. Shudo and Suzuki [42] developed a new heat transfer equation in which they included rate of gas velocity. The new equation showed better applicability for hydrogen combustion than Woschni's equation. Shudo and Nabetani [43], in their investigation of a homogeneous charge hydrogen SI engine using injection, studied the effect of the degree of constant volume and cooling loss on thermal efficiency. They analyzed the cylinder pressure diagram and exhaust composition to evaluate cooling loss to the burning gas. The combustion phase was identified from the pressure diagram, and heat release was estimated from thermodynamic analysis (gas laws and working fluid properties). Approximate cooling loss was determined from the difference between heat released and work done. The exhaust gas composition was analyzed using an exhaust gas analyzer, and its specific heat capacities were calculated. Using these data, the cooling loss was determined from the difference in change in enthalpy of the exhaust gases and the work done during combustion.

The Wiebe function, a zero-dimensional engine model, is used to model the fraction of fuel burned during the combustion process. When fitted to the rate of heat release curve (obtained from the cylinder pressure diagram), it helps to distinguish between different phases of combustion. It was possible to obtain the degree of constant volume burning and constant volume cooling. The hydrogen flame is closer to the surface due to its small quenching distance. When the flame is closer to the surface, the thermal boundary layer is thinner, and this causes more efficient heat transfer from the flame. Thus the shorter quenching distance of a hydrogen flame led to increased heat transfer. Because of hydrogen's high flame speed, the convection effect was more pronounced near stoichiometric combustion, and this feature was advantageous for rapid combustion. Moreover, hydrogen's higher thermal conductivity had some influence on overall heat transfer and the thermal behavior in the combustion process.

Khalaf et al. [44] used one-dimensional gas dynamics for the analysis of the flow and heat transfer process within the in-cylinder of a port-injected automotive hydrogen engine. Speed, equivalence ratio, and injection timing were the major parameters considered for simulation. Speed was varied in the range of 1000–6000 rpm, injection timing from 60°ATDC to 160° ATDC during the intake stroke and the equivalence ratio from ultralean mode ($\phi = 0.2$) to above stoichiometric condition ($\phi = 1.2$). The simulated data were verified with the corresponding experimental values. It is evident from Figure 5.39 that the rate of heat transfer takes up a continuous linear increasing trend with speed for all values of injection timing. This is because of the dominance of forced convection along with increase in speed. Figure 5.40 shows an increase in overall heat transfer rate with increasing engine speed over the entire range of equivalence ratio, from very lean to above stoichiometric. This trend also reflects the influence of forced heat transfer. In this study the predictions of overall heat transfer characteristics, based on one-dimensional gas dynamics, were quite reasonable for the equivalence ratio and engine speed. However, a correct picture on the effect of injection parameters on overall heat transfer could not be obtained in this study as the one-dimensional gas dynamics model adopted for the combustion approach within the Wiebe function did not consider the injection issue.

Shahpouri et al. [45] developed the 0D model to verify the simulation results with the corresponding experimental data for a total hydrogen-operated SI engine as well as for a dual fuel hydrogen engine using hydrogen and diesel. Since the 0D model requires a laminar flame

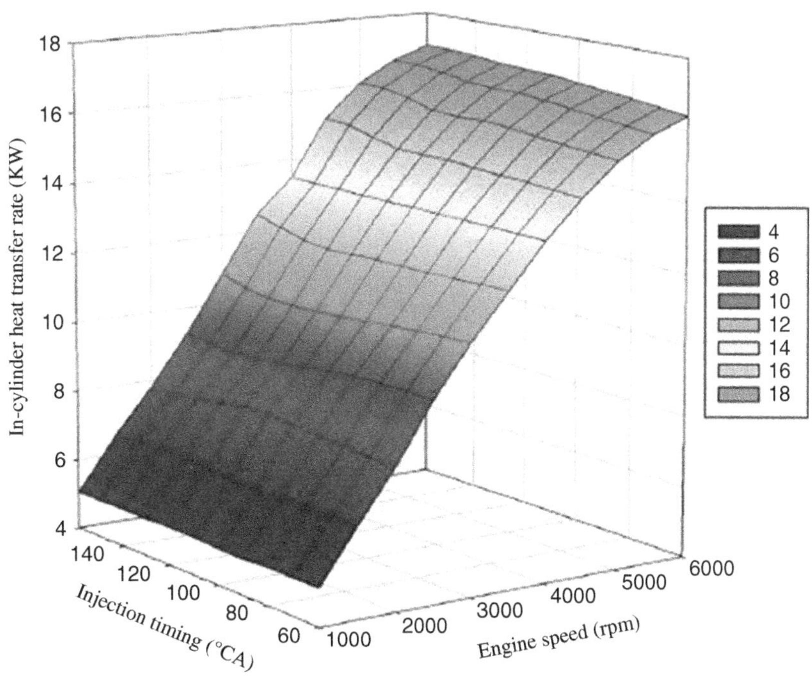

Figure 5.39 Variation of the overall heat transfer rate with injection timing and engine speed [44] / Academic Journals. Please visit the companion website to view a color version of this image.

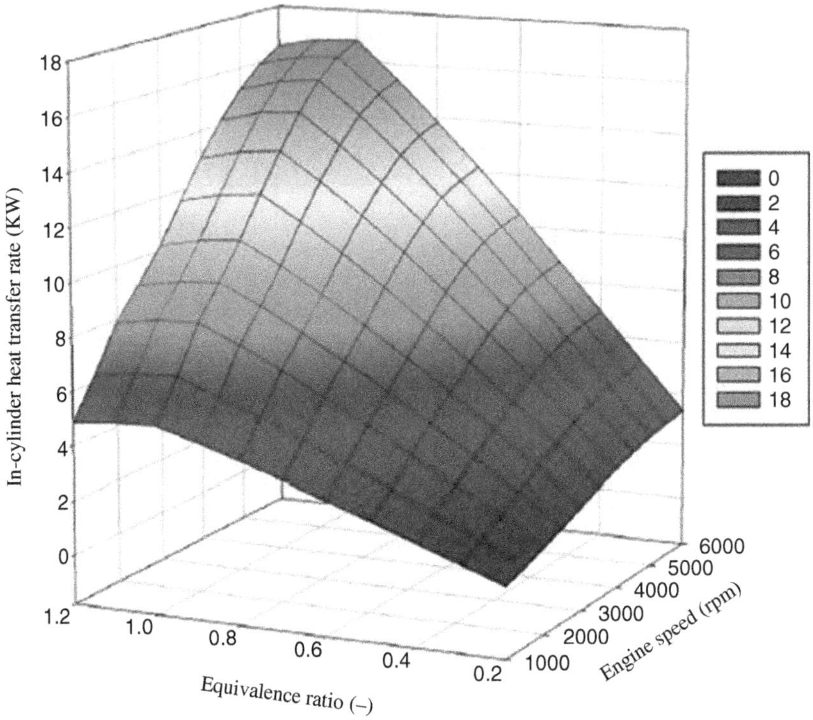

Figure 5.40 Variation of the overall heat transfer rate with engine speed and equivalence ratio [44] / Academic Journals. Please visit the companion website to view a color version of this image.

speed, the neural network based machine learning ANN model was used in this analysis for both engines. These 0D engine models, developed for laminar flame speed, proved quite accurate in both configurations: the total hydrogen operated SI engine and the dual fuel mode diesel engine. Maximum temperature and pressure indicated increasing trend with increasing hydrogen energy fraction in the dual fuel mode. In such operating conditions, the level of NOx emissions was increased. In the analysis of the dual fuel model, the contents in the cylinder were divided into three distinct thermodynamic zones: the main unburned zone, the spray zone containing the injected fuel and entrained gas, and the spray burned zone containing the combustion products. The researchers considered three calibration parameters using experimental data for analysis of the spark ignition engine model. These were turbulent flame speed multiplier, Taylor length scale multiplier, and flame kernel growth multiplier. The Taylor length scale was used to characterize the size of turbulent structures or eddies within the combustion chamber. The dynamics of flame kernel formation and growth provided information to improve ignition and injection strategies. The extended Zeldovich mechanism was used for NOx emissions. In developing the dual fuel model, four additional calibration parameters such as entrainment rate multipliers, ignition delay multipliers, premixed combustion rate multipliers, and diffusion combustion rate multipliers were considered, which are very relevant to the ignition process. Schubert et al. [46] used a quasi-dimensional model considering significant practical features of the system such as turbulent flow in the combustion chamber and the high- and low-pressure parts of the cycle. Results of this heat transfer analysis were compared with different direct injection diesel engines.

Michl et al. [47] developed a simulation model that is capable of describing wall heat transfer within the entire range of calibration. Tests were carried out on a hydrogen engine equipped with

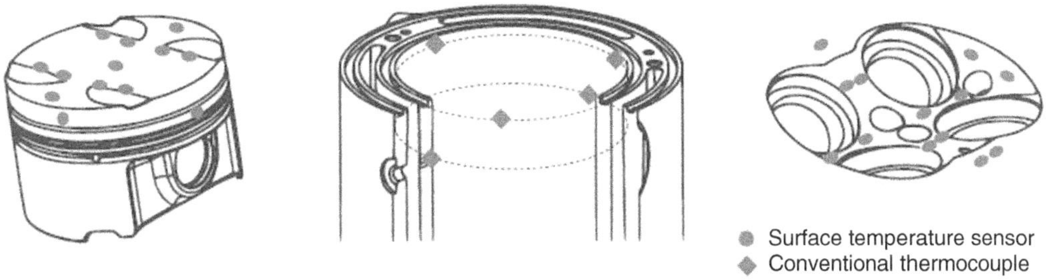

Figure 5.41 Distribution of sensors in the combustion chamber: piston (left), liner (middle), and cylinder head (right) [47] / with permission of Elsevier.

temperature sensors. Temperatures and heat fluxes from various points on every surface of the chamber were measured with a high level of accuracy. A well-established method was followed for converting measured surface temperatures into heat flux. This process involved a sequence of steps. Initially, surface temperatures at varying points within the combustion chamber were measured using sensors (Figure 5.41).

These sensors provided temperature variations with high precision. To convert these measured temperatures into heat fluxes, a Fourier analysis was performed to represent it as an analytical expression to solve the Fourier heat equation. Since thermal diffusivity characterizes how quickly heat diffuses through material, different sensors may have different diffusivity values based on their construction and materials. Therefore, solution of the equation depends on a particular sensor's unique thermal properties. Initially, local heat flux was determined to get an idea as to how much each sensor covers the entire surface area of combustion chamber. By summing up these values, it was possible to obtain the mean heat flux, which already had in it the relative importance of different sensors based on their coverage area. This spatial averaging procedure restricts the validation of the model approach to a single-zone model. A single-zone model, which assumes the entire chamber as one uniform zone, considers much fewer details than a multizone model and so often fails to capture the actual combustion process.

Most of existing heat transfer analyses in engine studies have been built upon empirical heat transfer formulations and therefore do not give accurate results. Many experimental results show that the existing correlations do not match, nor do they provide satisfactory results in various modes of real engine operating conditions. In their analysis, Nefischer et al. [48] suggested some additional features to the heat transfer model proposed by Schubert et al. [46]. Their paper [48] also discussed an empirical hypothesis on turbulence generated during the combustion process. An elaborate description of the turbulent kinetic energy was done from which the characteristic velocity was evaluated. Direct hydrogen injection affects turbulent kinetic energy. An analysis on the influence of injection on turbulence and on the in-cylinder flow field was used to predict the wall heat transfer for a hydrogen engine. The turbulence model developed to predict the heat release rate by the combustion process was validated by measurement of the wall heat flux and measured value of in-cylinder pressure.

Verma and Das [49] studied spark advance modeling of hydrogen-fueled SI engines based on combustion descriptors. Position of the pressure peak (PPP) and 50% mass fraction burned (MFB) were adopted as combustion descriptors to evaluate the efficiency of combustion. The CFD software ANSYS FLUENT was used for numerical simulation. The simulated results were validated with experimental results. Empirical rules for PPP and 50% MFB, which held good for best spark advance in gasoline engines, were not valid for hydrogen. A closer examination of the experimental data suggested that PPP of 8–9° after piston top dead center (ATDC) and position of 50%

MFB is 0–1° ATDC were the optimum conditions for maximum brake torque (MBT). This range might vary slightly depending upon the engine design, but the relationship broadly remains the same. These empirical rules were also useful for evaluating spark advance timings for the working range.

Shahpouri et al. [50] predicted laminar flame speed (LFS) by an artificial neural network (ANN) machine learning (ML) method. This study deals with the relatively newer method of gray box modeling for hydrogen engines. Efforts for calibration were reduced, and the hydrogen engine was optimized for least NOx and soot emissions with the hardware-in-the-loop (HIL) set-up. In engine testing, HIL set-up refers to a method of testing and validating complex software systems on a specially equipped test bench that receives data inputs from physical devices. Emissions from engines were reduced by optimal model-based combustion control [50]. Calibration is a time-consuming and resource-intensive process. On the other hand, HIL proved to be an effective way of connecting real hardware (the engine and its components) to the simulation for testing and thus evaluating various control strategies and parameters to evolve the optimal settings without the need for calibration or full-scale testing. Both HIL and MCC techniques rely on fast and accurate models of NOx and soot emissions.

With the objective of developing models to predict NOx and soot emission in a hydrogen–diesel engine, different modeling techniques such as artificial neural networks (ANN), support vector machines (SVM), and Gaussian process regression (GPR) are often used. SVM is a machine learning algorithm used for classification and regression tasks. It finds the best hyperplane that separates data into different classes (in classification) or predicts numerical values (in regression). Gaussian process regression is a probabilistic machine learning technique used for regression models. SVM is known for its effectiveness in high-dimensional spaces and ability to handle complex data relationships, whereas GPR is particularly useful when dealing with uncertainty in the data and can provide probabilistic predictions. R^2 is a statistical measure of goodness of fit in regression analysis. It represents the proportion of variance in the dependent variable explained by independent variables in a model. These models are categorized as "black box" or "grey box" based on the amount of internal knowledge of the engine's working they incorporate. A black box model does not rely on knowledge of internal workings of the system it models. It takes input data and produces output predictions without explicitly understanding the underlying mechanisms. On the other hand, a gray box model combines elements of both black box and white box (knowledge-based) modeling. It may incorporate some knowledge of the system's internal structure while still being data driven.

Figure 5.42 shows comparative picture of 79 different cases of NOx and soot data generated experimentally and predicted by modeling analysis [50]. It was observed that the average covariance, which indicates how closely the experimental data matches the predicted values, was 2.1 for NOx emission and 5.7 for soot emission. Lower covariance shows better agreement between experimental and predicted data. Experimental data on NOx and soot emissions were also comparable with corresponding data on gray box models.This may be due to the fact that the gray box model had considered additional information from the physical model with machine learning laminar flame speed (PMLS) including NOx prediction using the Zeldovich mechanism.

In this study [50], for the dual fuel engine using hydrogen and diesel, both black box and gray box emission models were developed. Test data were utilized to develop a machine learning (ML) model whereas the development of the gray box modeling used both experimental data from the test rig and the output from PMLS. Gray box emission models, developed with SVM and GPR methods with various features, achieve high accuracy in predicting NOx and soot emissions, as indicated by the test value of R^2 exceeding 0.99. Models that accurately predicted emissions in a

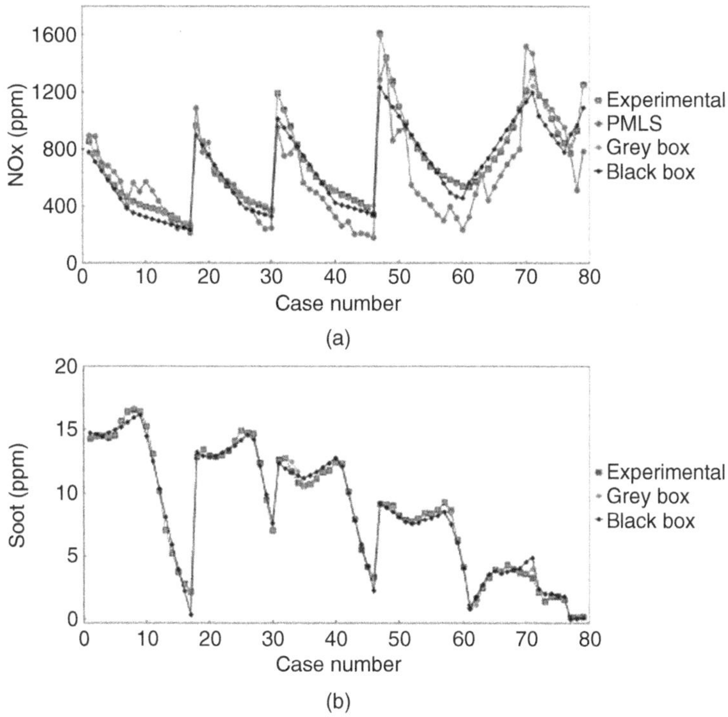

Figure 5.42 Experimental and predicted NOx and soot emissions. PMLS stands for physical model with machine learning laminar flame speed. Average covariance of the experimental measurement for NOx and soot emissions is 2.1 and 5.7, respectively [50] / with permission of Elsevier.

hydrogen–diesel engine were tested in hardware-in-the-loop (HIL) set-ups, where precision is crucial for optimizing engine performance and reducing emissions. Grey box models have additional information and predict more accurate emissions than black box models. On the other hand, black box emission models predicted NOx and soot emissions with a reasonably high level of accuracy as indicated by R^2 test values higher than 0.95. Black box models are much faster than gray box models. Thus, in the absence of necessary computational power, a black box model seems to be preferred for model-based hydrogen combustion analysis. In short, a black box model is a fast and efficient model, but gray box models are a better choice if accuracy is of paramount importance.

Tang et al. [51] did a simulation study on the combustion and NOx emission of a hydrogen engine by adding nitrogen and argon at various rates to the intake gas. This modeling study was on a diesel engine modified into spark ignition mode with intake pipe injection. Injection of hydrogen was scheduled into the intake pipe and mixed with air before entering the cylinder. The simulation was done for the period when the intake valve closed (at −147°) and the exhaust valve opened (132°). The RNG (renormalization group) k-epsilon model is an improvement on the standard k-ε model that includes the principles of renormalization group theory for more accurate prediction of turbulence. The extended Zeldovich mechanism was adopted for NOx emissions. The well-known O'Rourke–Amsden (OA) model was used for simulating interaction of combustion and turbulence near walls. Thus, the laws of boundary conditions were analyzed using OA heat transfer models.

Figure 5.43 shows a comparative picture of the experimental data with the simulated values. The simulation was carried out with the intake valve closed at an engine speed of 1800 rpm, and an equivalence ratio of 0.5. Other conditions included cylinder temperature of 393 K and initial

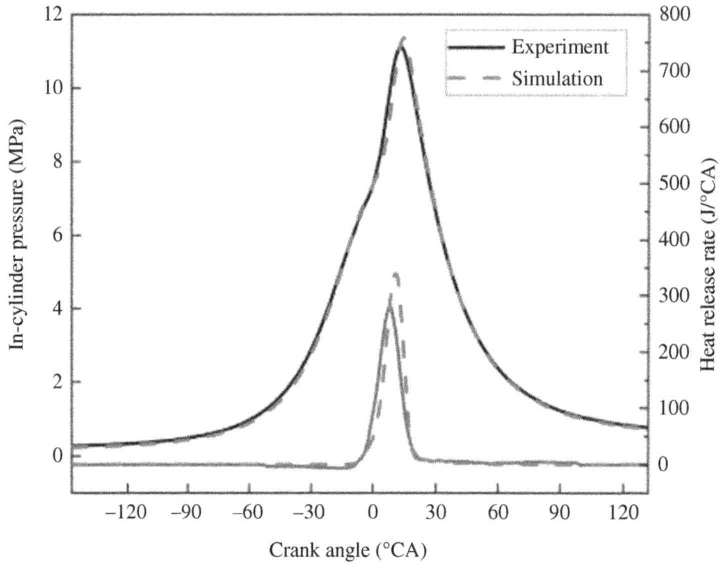

Figure 5.43 Comparison of experimental and simulated data of in-cylinder pressure and heat release rate [51] / with permission of Elsevier.

cylinder pressure of 0.234 MPa. The ignition advance angle was 12° CA BTDC. Despite the fact that nitrogen dilution decreased NOx emission levels, it had the shortcoming of decreasing thermal efficiency. Using argon along with nitrogen as a dilution gas helped improve thermal efficiency. The researchers investigated the effect on emissions and combustion by adding N_2 and Ar in various proportions to the intake gas. The proportions tested included 100% nitrogen, 80% nitrogen with 20% argon, both nitrogen and argon at 50%, nitrogen 20% with argon 80%, and 100% argon. With the condition of dilution rate increasing from 5% to 20% for 100% nitrogen, NOx emission decreased from 255 ppm to 3 ppm. However, the rate of combustion was reduced, resulting in a drastic fall in thermal efficiency from 44.4% to 22%. This study showed that an optimum configuration, striking a balance between performance characteristics having thermal efficiency (47.6–49.3%) and emission (NOx level between 250 and 500 ppm), can be achieved with a dilution rate of 17–27% and an argon proportion of 80–90%.

Figure 5.44 shows the variation in NOx emissions with respect to dilution rate. Dilution rate in context of this work is defined as

$$\mathrm{DR}_{dilution\ gas} = \frac{M_{dilution\ gas}}{M_{dilution\ gas} + M_{air} + M_{H_2}} * 100\%$$

where

$M_{dilution\ gas}$ is the mass of the dilution gas with the corresponding mass fractions,
M_{H2} is the mass of hydrogen, and
M_{air} is the mass of the air dilution rate (DR).

It is evident in Figure 5.45 that indicated thermal efficiency shows an increasing trend in dilution rate from 5% to 25%. The decreasing trend reflected in cases 1, 2, and 3 is due to excessive dilution of the gas inside the cylinder.

Krebs and Biet [52] used a quasi-dimensional, multizone cylinder model for lean hydrogen combustion in an SI engine with both an entrainment model and reaction kinetics. The analysis considered various piston bowl geometries for flame front propagation. Reaction kinetics were

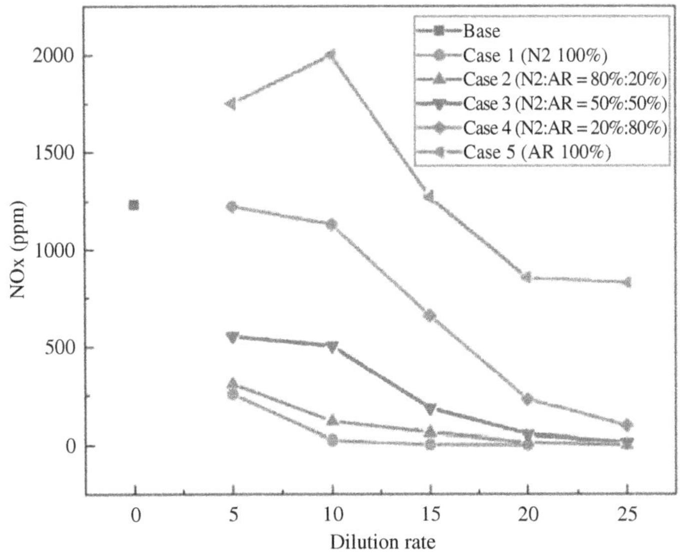

Figure 5.44 Impact of the dilution gases on NOx emission [51] / with permission of Elsevier.

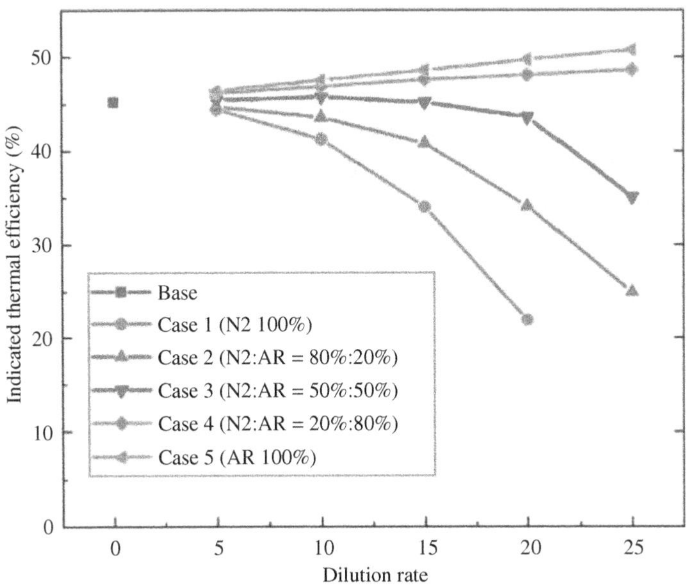

Figure 5.45 Effect of the dilution gases on the indicated thermal efficiency [51] / with permission of Elsevier.

computed online based on composition of unburned gas. Laminar flame speed was calculated online, using a one-dimensional free flame problem with the direct use of gas composition of the unburned zone. Simulation results demonstrated insignificant NOx emission levels when the air-to-fuel equivalence ratio exceeded 2.2 without exhaust gas recirculation. As far as modeling and simulation studies on hydrogen engines are concerned, undesirable combustion phenomena have not been widely studied.

Sometimes simulated values do not agree with experimental data. Direct injection avoids backfire. Lean operation improves efficiency and reduces NOx emission. Chen and his co-researchers [53] developed a three-dimensional transient calculation model coupled chemical reaction mechanism and studied direct injection with lean burn in a hydrogen engine. Reliability and accuracy of the results of the analysis were compared with experimental data. The study identified several significant features of injection strategies. Hydrogen injection angles varied with different injection positions for maximum thermal efficiency and also for nitric oxide to meet the Europe V emission standard.

Several research activities on modeling and simulation are being pursued on hydrogen along with conventional and alternative liquid and gaseous fuels in both SI and CI engines. A major part of the discussion in this chapter deals with engines fueled with only hydrogen. A few of the recent research pursuits on modeling hydrogen fuel with other fuels and in different engine configurations are briefly discussed here [54–61]. Wang et al. [54] experimented with a modified CNG engine where nitrogen was injected into the intake manifold. Increasing the nitrogen dilution ratio brings out the NOx emission level substantially in the range of 17–81%, but increasing the dilution ratio had a serious impact on combustion parameters. Peak cylinder pressure and maximum heat release rate indicated a decreasing trend with increased combustion delay. Results of this work on CNG can be useful for obtaining some general guidelines for operating a hydrogen engine with good performance and low emissions.

Through the use of numerical modeling, combustion in a CI engine was studied with a late compression/early power stroke direct gaseous hydrogen injection ignited by a diesel pilot at up to 99% hydrogen energy share [55]. A 3D CFD numerical study investigates a diesel pilot-ignited, late high-pressure direct hydrogen injection engine comparing it to methane. NOx emissions were higher with hydrogen combustion than methane, but carbon-based emissions were lower. Hydrogen showed faster ignition and higher heat release rates due to its wider flames and quicker penetration. High load performance was better with hydrogen, but low load performance deteriorated due to unstable combustion beyond 95% gaseous energy share. Tao Hai et al. [56] designed a novel multi-task transfer learning (TL) approach (MLS-SVR) and studied emissions from a hydrogen-enriched natural gas engine (HENGE) engine. Multi-output least squares support vector regression (MLS-SVR), whose accuracy could be monitored using some graphical and numerical methods, was found to be an effective tool for predicting simultaneously CH_4, CO, and NO emissions in a HENGE system.

Hydrogen Wankel rotary engines (WREs) are receiving the attention of researchers because of their excellent power and emissions characteristics. Wang et al. [57] performed 1D modeling and simulation for parametric analysis on the performance and exhaust emission characteristics of a hydrogen Wankel rotary engine (WRE). Simulation in the WRE also analyzed the effect of various leakage patterns on volumetric efficiency. Effects of intake port closing (IPC) timing and exhaust port opening (EPO) timing were investigated on the performance and emission characteristics of an engine. Analysis based on the Spearman rank correlation (SRC) coefficient was used to calculate statistical relevance from the model input parameters [57].

Researchers have used virtual engine calibration models that use thermodynamic simulation to predict in-cylinder pressure for ethanol, hydrogen, and natural gas as fuel [58]. That study formulated an optimization problem based on known engine data and used combustion and heat transfer model parameters as decision variables. Combustion and heat transfer optimization were analyzed with two different models: a zero dimensional Wiebe model and a quasi-dimensional fractal geometry model. It was possible to have an accurate pressure curve prediction with a coefficient of determination more than 0.99. This analysis was found to predict the peak pressure and the crank angle corresponding to 50% of mass fraction burned with a maximum deviation of 2.5% and 1.5° CA, respectively [58].

A thermodynamic model developed to reduce marine fleet emissions using three promising alternative fuels—methanol, ammonia, and hydrogen—has been investigated [59]. Performance in turbocharged and intercooled homogeneous charge compression ignition engines (HCCI) at different ambient temperatures and equivalence ratios showed ammonia to have the highest engine performance, followed by hydrogen and methanol. Moreover, the majority of exergy destruction rates (ranging from 65.26% for ammonia to 84.02% for hydrogen) within the engine occurred in the HCCI engine configuration.

Varuvel and co-researchers [60] investigated two predictive techniques, the artificial neural network (ANN) and response surface methodology (RSM), for optimizing a hydrogen-supported biodiesel engine on PCCI operations [60]. Experimental results were used in this study to develop the ANN and RSM model. Various fuel blends were analyzed, and responses such as rate of pressure rise, brake thermal efficiency (BTE), and emissions (NOx, HC, CO, smoke) were predicted. The ANOVA method identified influential factors and interactions. RSM optimized input parameters, yielding observed optimum values of 2.381 kW load and 12 MJ/kWh brake specific energy consumption (BSEC). Predicted responses under these conditions include 4.166 bar/degree CA pressure rise rate, 29.067% BTE, 6.902 g/kWh NOx, 0.638 g/kWh HC, and 6.052 g/kWh CO. RSM demonstrated accuracy superior to that of ANN based on correlation coefficient (R_2) and mean square error (MSE) values, concluding its effectiveness in response prediction. Overall analysis of this study recommends response surface methodology as the appropriate artificial intelligence tool to optimize the engine.

Hydrogen enrichment in biogas was tested as a possible technique to improve the limitations associated with a biodiesel–biogas dual fuel engine [61]. A simpler simulation methodology was adopted to optimize hydrogen injection parameters for improved performance and safe operation of the engine. In addition, exergy analysis and irreversibility quantification of various dual fuel engine processes were studied.

A single-cylinder diesel engine was modified to operate in dual fuel mode using *Jatropha curcas* biodiesel as the pilot fuel and biogas as the main fuel. An electronic control unit was developed to facilitate 5–20% hydrogen enrichment in biogas using the timed manifold injection (TMI) technique. A three-dimensional computational fluid dynamics-based simulation methodology was used for optimal selection of TMI parameters to ensure efficient and safe operation of the engine. Subsequently, the optimized injection conditions were used in tests to evaluate performance and emission characteristics of the engine at low and high engine loads. Engine performance was analyzed based on energy and exergy analyses, whereas hydrocarbon (HC), carbon monoxide (CO), oxides of nitrogen (NOx), and smoke emissions were analyzed for emission characterization. It was observed from the simulation studies that an increase in injection angle directed the flow toward the combustion chamber and increased turbulent kinetic energy (TKE), leading to better mixing. Late injection timing was an effective approach to avoid backfire. However, a timing that is too late may result in poor mixture formation. Therefore, an optimum injection timing should be selected with TMI of hydrogen. In this work, a combination of injection angle at 60° and injection pressure at 150 kPa offered good mixture formation. Experimental results showed that at 20% hydrogen enrichment, exergy efficiencies of the dual fuel engine increased from 8.4% to 10.1% at low load and from 23.3% to 25.5% at high load. However, maximum reductions in HC and CO emissions of 35.6% and 50.0%, respectively, were calculated at low load. Figures 5.46 and 5.47 show the exergy balance under low-load and high-load conditions, respectively. In these figures,

Xq = exergy transfer rate with wall heat losses
Xw = rate of exergy transfer with engine work output
Xex = rate of exergy transfer rate with exhaust gases
Xd = rate of exergy destruction

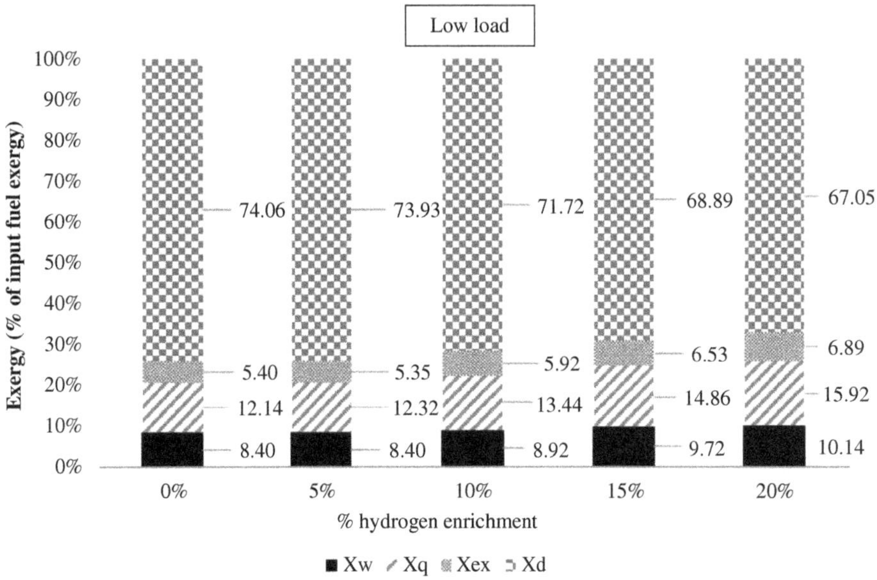

Figure 5.46 Exergy balance for biodiesel–biogas dual fuel engine with various hydrogen enrichment percentages at low-load conditions [61] / The American Society of Mechanical Engineers.

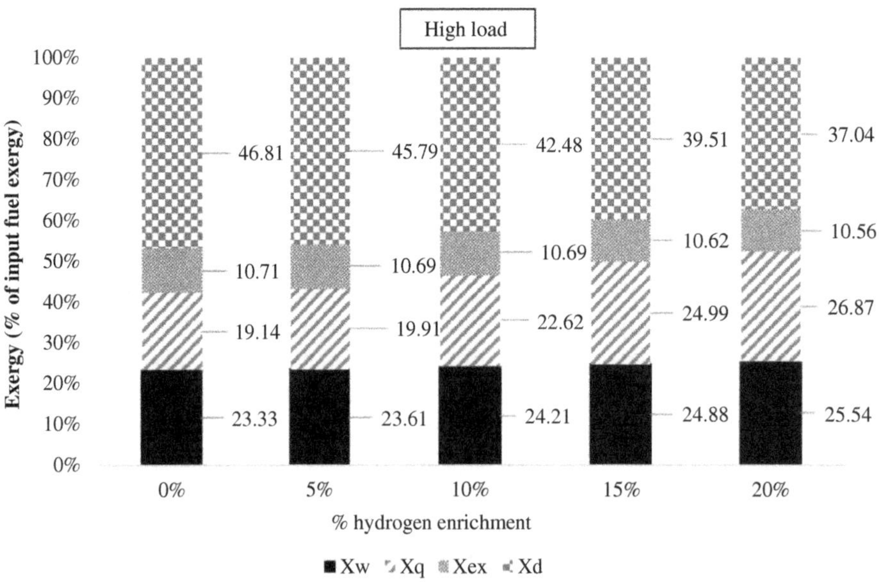

Figure 5.47 Exergy balance for biodiesel–biogas DF engine with various hydrogen enrichment percentages at high-load conditions [61] / The American Society of Mechanical Engineers.

Aliramezani et al. [62] studied thermodynamic modeling of a partially stratified charge engine with hydrogen–methane blends in ultra-lean mode [62]. This study focused on flame propagation to predict characteristics of spark ignition engines running on hydrogen–methane blends. The study involved using a micro direct injection of pure fuel or a fuel–air mixture near the spark plug to create a partially stratified charge that can improve engine performance. This approach was validated using experimental data with natural gas under lean conditions. The model was extended to predict engine performance over a range of hydrogen molar concentrations (0, 15, 30, and 45%). It was observed that partially charged stratification increase indicated mean effective pressure and reduced specific fuel consumption. Moreover, higher hydrogen content further improved the stratification effect on engine performance. The model offers the flexibility for analyzing the effect of mixture distribution and fuel constituents on engine performance.

Concluding Remarks and Perspectives

At the present stage of hydrogen engine development, modeling studies are essential to evaluate the optimized conditions for widespread real-world implementation. By simulating various scenarios, it is possible to decide on the design parameters related to good performance, low emission, stable combustion, and long-term durability of the prototype. The temperamental combustion characteristics of hydrogen engines still pose some challenges for safe and efficient use. State-of-the-art models for reaction mechanisms are developed on the complex chemistry of hydrogen combustion with detailed kinetics of hydrogen oxidation. These models are useful for predicting ignition delay times, flame propagation, and emissions, etc.

Modeling of predictive tools is essential to assess performance, emissions, and combustion under varying engine loads and speeds. Among these, modeling combustion in hydrogen engines has been the most critical area in understanding and optimizing the complex combustion process. Preventing preignition in hydrogen engines is crucial and needs careful design of the combustion chamber, spark plug location, and ignition-timing control. Problems of backfire are well known in damaging the intake manifold, throttle body, engine valves, and several other sensitive components in the intake system. Knocking generates extremely high pressure and temperature spikes within the combustion chamber, which causes immense damage to piston and cylinder apart from disrupting the normal combustion process. One of the major objectives of modeling and simulation studies is to prescribe the engine designs, ignition strategies, and fuel injection that will minimize risk of these undesirable combustion phenomena.

CFD simulation is the backbone of hydrogen engine combustion modeling. Advanced CFD models enable researchers to simulate the flow of hydrogen and air within an engine's combustion chamber, predict temperature and pressure distributions, and track combustion reactions in detail. These simulations help in optimizing the design of combustion chambers, fuel injection strategies, and emission control systems. Turbulence plays a crucial role in the combustion process. Advanced turbulence models, such as large eddy simulation (LES) and Reynolds-averaged Navier–Stokes (RANS), capture turbulent flow characteristics in the combustion chamber. These models predict with reasonable accuracy the formation and propagation of hydrogen flames. Modeling has already been done on formation and emissions of NOx in the hydrogen engine in high-temperature combustion.Computational models have been developed to simulate the injection strategy, spray formation, and subsequent mixing to ensure the uniform distribution of the fuel–air mixture within the combustion chamber.

References

1 Dang, J. and Wang, L. (2022). Research on diagnosis of pre-ignition of hydrogen engine based on SOM-MAS. *International Journal of Hydrogen Energy* 47 (82): 35112–35121.
2 Dang, J. and Wang, L. (2021). Optimization control of hydrogen engine ignition system based on ACO-BP. *International Journal of Hydrogen Energy* 46 (78): 38903–38912.

3 Zhao, F.C., Sun, B.G., Yuan, S. et al. (2024). Experimental and modeling investigations to improve the performance of the near-zero NOx emissions direct-injection hydrogen engine by injection optimization. *International Journal of Hydrogen Energy* 49: 713–724.
4 Wu, B., Torelli, R., and Pei, Y. (2023). Numerical modelling of hydrogen mixing in a direct-injection engine fueled with gaseous hydrogen. *Fuel* 341: 127725.
5 Krishnanunni, J., Bhatia, D., and Das, L.M. (2017). Experimental and modelling investigations on the performance and emission characteristics of a single cylinder hydrogen engine. *International Journal of Hydrogen Energy* 42: 29574–29584.
6 Marinov, N.M., Curran, H.J., Pitz, W.J. et al. (1998). Chemical kinetic modeling of hydrogen under conditions found in internal combustion engines. *Energy & Fuels* 12 (1): 78–82.
7 Kéromnès, A., Metcalfe, W.K., Heufer, K.A. et al. (2013). An experimental and detailed chemical kinetic modeling study of hydrogen and syngas mixture oxidation at elevated pressures. *Combustion and Flame* 160 (6): 995–1011.
8 Liu, C. and Karim, G.A. (2008). A simulation of the combustion of hydrogen in HCCI engines using a 3D model with detailed chemical kinetics. *International Journal of Hydrogen Energy* 33 (14): 3863–3875.
9 Wang, H., Slavinskaya, N., and Haidn, O. (2022). A comprehensive kinetic modeling study of hydrogen combustion with uncertainty quantification. *Fuel* 319: 123705.
10 Yakovenko, I. and Kiverin, A. (2023). Numerical modeling of hydrogen combustion: approaches and benchmarks. *Fire* 6 (6): 239. https://doi.org/10.3390/fire6060239.
11 Fagelson, J.J., McLean, W.J., and De Boer, P.C.T. (1978). Performance and NOx emissions of spark-ignited combustion engines using alternative fuels — quasi one-dimensional modeling. I. Hydrogen fueled engines. *Combustion Science and Technology* 18 (1–2): 47–57.
12 Prabhu-Kumar, G.P., Nagalingam, B., and Gopalakrishnan, K.V. (1985). Theoretical studies of a spark-ignited supercharged hydrogen engine. *International Journal of Hydrogen Energy* 10 (6): 389–397.
13 Al-Baghdadi, M.A.S. and Al-Janabi, H.A.S. (2003). A prediction study of a spark ignition supercharged hydrogen engine. *Energy Conversion and Management* 44 (20): 3143–3150.
14 Al-Baghdadi, M.A.S. (2004). Effect of compression ratio, equivalence ratio and engine speed on the performance and emission characteristics of a spark ignition engine using hydrogen as a fuel. *Renewable Energy* 29 (15): 2245–2260.
15 Keck, J.C. (1982). Turbulent flame structure and speed in spark-ignition engines. *19th Symposium (International) on Combustion* 19: 1451–1466.
16 Takats, M., Macek, J., Polasek, M. et al. (1998). Hydrogen fueled reciprocating engine as an automotive prime mover? Fisita World Automotive Congress, Paris, France. Paper No. F98T/P693.
17 Polasek, M., Macek, J., Takats, M. et al. (2002). Application of advanced simulation methods and their combination with experiments to modeling of hydrogen fueled engine emission potentials. SAE Technical Paper No. 2002-01-0373.
18 Kim, S. and Kim, J. (2023). Assessing fuel economy and NOx emissions of a hydrogen engine bus using neural network algorithms for urban mass transit systems. *Energy* 275: 127517.
19 Ma, J., Su, Y., Zhou, Y. et al. (2003). Simulation and prediction on the performance of a vehicle's hydrogen engine. *International Journal of Hydrogen Energy* 28 (1): 77–83.
20 D'Errico, G., Onorati, A., Ellgas, S. et al. (2006). Thermo-fluid dynamic simulation of a S.I. single-cylinder H2 engine and comparison with experimental data. ASME Spring Engine Technology Conference, Aachen, Germany. Paper Number ICES2003-1311. https://doi.org/10.1115/ICES2006-1311

21 Errico, G.D., Onorati, A., and Ellgas, S. (2008). 1D thermo-fluid dynamic modelling of an SI single-cylinder H2 engine with cryogenic port injection. *International Journal of Hydrogen Energy* 33 (20): 5829–5841.

22 Verhelst, S. and Sierens, R. (2007). A quasi-dimensional model for the power cycle of a hydrogen-fuelled ICE. *International Journal of Hydrogen Energy* 32 (15): 3545–3554.

23 Safari, H., Jazayeri, S.A., and Ebrahimi, R. (2009). Potentials of NOX emission reduction methods in SI hydrogen engines: simulation study. *International Journal of Hydrogen Energy* 34 (2): 1015–1025.

24 Konnov, AA. (2019). Yet another kinetic mechanism for hydrogen combustion. *Combustion and Flame* 203: 14–22.

25 Johnson, N. (1997). Hydrogen as a zero-emission, high-efficiency fuel: uniqueness, experiments and simulation. 3rd International Conference ICE97, Internal combustion engines: experiments and modelling, Naples, Italy.

26 Adgulkar, D.D., Deshpande, N.V., Thombre, S.B. et al. (2008). 3D CFD simulations of hydrogen fuelled spark ignition engine. ASME Spring Engine Technology Conference, Chicago, Illinois. Paper No. ICES2008-1649.

27 Wimmer, T., Wallner, J., Ringler, F. et al. (2005). H2-direct injection a highly promising combustion concept. SAE Technical Paper No. 2005-01-0108.

28 Messner, D., Wimmer, A., Gerke, U. et al. (2006). Application and validation of the 3D CFD method for a hydrogen fueled IC engine with internal mixture formation. SAE Technical Paper No. 2006-01-0448.

29 Gerke, U. (2007). Numerical analysis of mixture formation and combustion in a hydrogen direct-injection internal combustion engine. Dissertation submitted to the Swiss Federal Institute of Technology, Zurich, for the degree of Doctor of Science (Diss. ETH No. 17477).

30 Anticaglia, A., Balduzzi, F., Ferrara, G. et al. (2023). Feasibility analysis of a direct injection H2 internal combustion engine: numerical assessment and proof-of-concept. *International Journal of Hydrogen Energy* 48 (83): 32553–32571.

31 Babayev, R., Andersson, A., Dalmau, A.S. et al. (2021). Computational characterization of hydrogen direct injection and non-premixed combustion in a compression-ignition engine. *International Journal of Hydrogen Energy* 46 (35): 18678–18696.

32 Scarcelli, R., Wallner, T., Matthias, N. et al. (2011). Mixture formation in direct injection hydrogen engines: CFD and optical analysis of single- and multi-hole nozzles. *SAE International Journal of Engines* 4 (2): 2361–2375. https://doi.org/10.4271/2011-24-0096.

33 Kherdekar, P V. and Bhatia, D. (2017). Simulation of a spark ignited hydrogen engine for minimization of NOx emissions. *International Journal of Hydrogen Energy* 42 (7): 4579–4596.

34 Ballerini, A., Cerri, T., Marinoni, A.M. et al. (2022). Advances in 1D thermo-fluid dynamic simulation of SI hydrogen-fueled engine, ATI Annual Congress (ATI 2022). *Journal of Physics: Conference Series* 2385: 012055. IOP Publishing. https://doi.org/10.1088/1742 6596/2385/1/012055.

35 Rawat, K., Srivastava, V., and Bhatia, D. (2019). Kinetic analysis of the role of selective NOx recirculation in reducing NOx emissions from a hydrogen engine. *Chemical Engineering Journal* 377: 120143. https://doi.org/10.1016/j.cej.2018.10.080.

36 Agarwal, S., Srivastava, V., Kherdekar, P.V. et al. (2021). Reduction of nitrogen oxides by injecting nitric oxide into a hydrogen engine: A micro-kinetic analysis. *SAE International Journal of Engines* 14 (4): 467–492. https://www.jstor.org/stable/27099222.

37 Agarwal, S., Kherdekar, P.V., and Bhatia, D. (2022). Reaction pathway analysis for investigation of the NOx reduction mechanism during selective NOx recirculation in a hydrogen engine. *Energy & Fuels* 36 (10): 5337–5351. https://doi.org/10.1021/acs.energyfuels.2c00411

38 Smirnov, N. and Nikitin, V.F. (2014). Modeling and simulation of hydrogen combustion in engines. *International Journal of Hydrogen Energy* 39 (2): 1122–1136.

39 Maurya, R.K. and Akhil, N. (2017). Comparative study of the simulation ability of various recent hydrogen combustion mechanisms in HCCI engines using stochastic reactor model. *International Journal of Hydrogen Energy* 42: 11911–11925.

40 Falfari, S., Cazzoli, G., Mariani, V. et al. (2023). Hydrogen application as a fuel in internal combustion engines. *Energies* 16: 2545. https://doi.org/10.3390/en16062545.

41 Woschni, G. (1967). A universally applicable equation for the instantaneous heat transfer coefficient in the internal combustion engine. SAE Technical Paper No. 670931.

42 Shudo, T. and Suzuki, H. (2002). New heat transfer equation applicable to hydrogen fuelled engines. ASME Fall Technical Conference, New Orleans, Louisiana. Paper No. ICEF2002-515.

43 Shudo, T. and Nabetani, S. (2001). Analysis of degree of constant volume and cooling loss in a hydrogen fuelled SI engine. SAE Technical Paper No. 2001-01-3561.

44 Rahman, M., Hamada, K., Kadirgama, K., et al. (2012). Cycle analysis of in-cylinder heat transfer characteristics for hydrogen fueled engine. *Scientific Research and Essays* 7 (8): 891–902. http://www.academicjournals.org/SRE. https://doi.org/10.5897/SRE10.1168. ISSN 1992-2248. ©2012 Academic Journals.

45 Shahpouri, S., Norouzi, A., Hayduk, C. et al. (2022). Modeling of a single-fuel hydrogen spark ignition and a dual-fuel diesel-hydrogen engines. *Proceedings of Combustion Institute – Canadian Section*, Spring Technical Meeting, The University of Ottawa, May 16–19.

46 Schubert, C., Wimmer, A., and Chmela, F. (2005). Advanced heat transfer model for CI engines. SAE Technical Paper No. 2005-01-0695.

47 Michl, J., Neumann, J., Rottengruber, H. et al. (2016). Derivation and validation of a heat transfer model in a hydrogen combustion engine. *Applied Thermal Engineering* 98: 502–512.

48 Nefischer, A., Hallmannsegger, M., Wimmer, A. et al. (2009). Application of a flow field based heat transfer model to hydrogen internal combustion engines. SAE Technical Paper No. 2009-01-1423.

49 Verma, S. and Das, L.M. (2018). Spark advance modeling of hydrogen-fueled spark ignition engines using combustion descriptors. *Journal of Engineering for Gas Turbines and Power*. 140 (8): 081501. Paper No: GTP-16 1007. https://doi.org/10.1115/1.4038798.

50 Shahpouri, S., Gordon, D., Hayduk, C. et al. (2023). Hybrid emission and combustion modeling of hydrogen fueled engines. *International Journal of Hydrogen Energy* 48 (62): 24037–24053.

51 Tang, Y., Xing, K., Huang, H. et al. (2024). Simulation study of N2/Ar dilution gas to improve thermal efficiency and reduce NOx emissions in hydrogen engines. *International Journal of Hydrogen Energy* 49: 636–651.

52 Krebs, S. and Biet, C. (2021). Predictive model of a premixed, lean hydrogen combustion for internal combustion engines. *Transportation Engineering* 5: 100086.

53 Chen, W., Lu, C., Zuo, Q. et al. (2024). Combustion characteristics analysis and performance evaluation of a hydrogen engine under direct injection plus lean burn mode *Journal of Cleaner Production* 470: 143323.

54 Wang, Z., Zuo, H., Liu, Z. et al. (2014). Impact of N2 dilution on combustion and emissions in a spark ignition CNG engine. *Energy Conversion and Management* 85: 354–360.

55 Ramsay, C.J. and Koratage, R.D.K. (2024). Numerical modelling of a heavy-duty diesel-hydrogen dual-fuel engine with late high pressure hydrogen direct injection and diesel pilot. *International Journal of Hydrogen Energy* 49 Part B: 674–696.

56 Hai, T., Kadir, D.H., and Ghanbari, A. (2023). Modeling the emission characteristics of the hydrogen-enriched natural gas engines by multi-output least-squares support vector regression: comprehensive statistical and operating analyses. *Energy* 276: 127515.

57 Wang, H., Ji, C., Shi, C. et al. (2022). Modeling and parametric study of the performance-emissions trade-off of a hydrogen Wankel rotary engine. *Fuel* 318: 123662.

58 Ayad, S.M.M.E., Vago, C.R.L., Bachelor, C.R.P. et al. (2021). Cylinder pressure based calibration model for engines using ethanol, hydrogen and natural gas as alternative fuels. *Energy Reports* 7: 7940–7954.

59 Djermouni, M. and Ouadha, A. (2023). Thermodynamic analysis of methanol, ammonia, and hydrogen as alternative fuels in HCCI engines. *International Journal of Thermofluids* 19: 100372.

60 Varuvel, E.G., Seetharaman, S., Bai, F.J.J.S. et al. (2023). Development of artificial neural network and response surface methodology model to optimize the engine parameters of rubber seed oil–hydrogen on PCCI operation. *Energy* 283: 129110.

61 Verma, S., Kumar, K., Das, L.M. et al. (2021). Effect of hydrogen enrichment strategy on performance and emission features of biodiesel-biogas dual fuel engine using simulation and experimental analyses. *Journal of Energy Resources Technology* 143 (9): 092301.

62 Aliramezani, M., Chitsaz, I., and Mozafari, A.A. (2013). Thermodynamic modelling of partially stratified charge engine characteristics for hydrogen-methane blends at ultra-lean conditions. *International Journal of Hydrogen Energy* 38 (25): 10640–10647.

6

Laser Diagnostics, Optical, and Other Sensing Techniques

I do not know what I may appear to the world, but to myself I seem to have been only like a boy playing on the sea-shore, and diverting myself in now and then finding a smoother pebble or a prettier shell than ordinary, whilst the great ocean of truth lay all undiscovered before me.

Isaac Newton

6.1 Introduction

It has been widely discussed in earlier chapters that hydrogen has many desirable properties for its effective use in engines. To utilize these properties for optimal benefits, it is necessary to identify and quantify the impact of some of the typical combustion characteristics. Many engine studies on conventional petroleum fuels have utilized several advanced techniques, but such studies on hydrogen engines are few and far between. However, besides traditional instruments, laser diagnostics and some optical techniques have recently emerged that provide a more precise perspective on combustion phenomena in hydrogen engines. These methods offer nonintrusive, high-resolution approaches to analyze hydrogen combustion processes and generate data for optimizing engine performance and reducing emissions. Laser-based techniques, known for their regulated illumination, high spatial clarity, and temporal resolution, play a key role in these studies. Non-laser optical and other sensing and imaging techniques also contribute to comprehensive understanding of complex combustion dynamics. The choice of technique depends on the specific parameters to be investigated and the characteristics of the combustion system.

6.2 Key Laser Diagnostics and Optical Methods

There are several laser-based techniques and optical methods for studying and analyzing combustion-related activities in engines. Planar laser-induced fluorescence (PLIF) enables one to visualize the distribution of specific species within the combustion chamber. In hydrogen combustion studies, PLIF is employed to monitor spatial distribution of important species like OH radicals or H_2 molecules. Laser-induced breakdown spectroscopy (LIBS) is a rapid and versatile technique suitable for both qualitative and quantitative elemental analysis. The spectroscopic

Hydrogen Engines: Design, Performance Evaluation, Combustion Analysis, and Exhaust Emissions, First Edition.
Lalit Mohan Das.

technique involves focusing a laser onto a small region of a sample, causing the material to break down into plasma. The emission spectrum of the resulting plasma is analyzed to identify the elemental composition of fuel. LIBS is useful for analyzing the composition of combustion products and studying the formation of pollutants. Schlieren imaging is an optical method used to visualize density gradients in transparent media. It can be employed to observe shock waves, flame fronts, and other flow phenomena and understand the dynamic behavior of the combustion process. Raman spectroscopy, a powerful technique for identifying and quantifying molecular species, is useful in analyzing the composition of reacting gases. It provides information on temperature, pressure, and species concentrations during combustion. Rayleigh scattering is used to measure gas density and temperature. It helps to visualize density variations within the combustion chamber and understand the thermodynamics of the combustion process. Spark-induced breakdown spectroscopy (SIBS) involves use of a spark discharge to generate a plasma in combustion gases. The spark induces a breakdown of the sample into a plasma state, and the emitted light from the plasma is analyzed to determine elemental composition of the sample. It is particularly useful for real-time and in-situ monitoring of elements in the combustion chamber. Both SIBS and LIBS share the basic principle of inducing a breakdown of a sample into a plasma for spectroscopic analysis. The difference between them lies in the method used to adopt the breakdown. SIBS uses a spark discharge, while LIBS typically uses a high-intensity laser pulse. Particle image velocimetry (PIV) is used to measure fluid velocity and study flow patterns. This is useful in identifying turbulence within the combustion chamber. UV-visible spectroscopy is a nonintrusive technique used to study the absorption and emission of light by molecules in the combustion process. It helps in identifying and quantifying various chemical species present during combustion. High-speed cameras are used in capturing rapid and dynamic events during combustion and obtaining information on flame propagation, ignition, and other transient phenomena. Endoscopic imaging involves inserting optical probes into the combustion chamber, providing a direct view of the combustion process. This technique is extremely useful in collecting in-situ data from within the engine. Infrared (IR) imaging is nonintrusive. Infrared cameras capture the thermal radiation emitted during combustion and help in studying temperature distributions, flame shapes, and combustion stability. Acoustic sensors detect pressure variations and combustion-related acoustic signals, which helps to investigate combustion stability and identify combustion anomalies. In some studies electrochemical sensors, based on ion selective electrodes, are used to measure the concentration of ions in the combustion chamber. It helps in getting information about flame ionization and combustion reaction. The flame emission spectroscopy technique involves analyzing the characteristic emission lines from the combustion products using optical spectrometers. It does not require laser sources and it is used to identify and quantify elements present in flames. Microscopic imaging techniques, such as high-speed microscopy, help one to observe combustion at a microscopic level and have a better understanding of ignition and flame propagation. X-ray imaging techniques are applied to visualize the internal structure of combustion systems and study the interaction between fuel and air, and combustion chamber geometry. Magnetic resonance imaging (MRI), widely used in combustion studies, has the capability to provide detailed spatial and temporal information about the distribution of certain chemical species. Traditional white light imaging uses broadband light sources, such as incandescent or halogen lamps, to illuminate a scene. This technique is commonly used for macroscopic observations, and it does not involve the use of lasers. Colorimetry involves capturing and analyzing the flame color. Changes in flame color can indicate variations in temperature and chemical composition. Colorimetry is used in visualizing and characterizing the combustion process.

6.3 Mixture Formation in a Hydrogen Engine

A review of the literature shows that experimentation has been done on several fuel induction techniques like port fuel injection, timed manifold injection, and direct injection. Direct injection spark ignition (DISI) is the most desirable option because it intrinsically precludes backfire by design. Adequate relevant literature is not available on DISI engines, particularly in the later phase of the compression stroke with stratified charge. For a fuel such as hydrogen, with wide range of flammability and low minimum ignition energy, it is important to have a clear picture of fuel distribution and flame propagation under varying operating conditions, involving injection timing, ignition timing, and the fuel–air equivalence ratio. Performance, emission characteristics, and combustion characteristics are very closely linked to mixture distribution. There have been few studies using flow visualization and simulation techniques to gain insight into the fuel–air mixing process in hydrogen engines [1–4]. Flow visualization provides significant information to analyze movement and distribution of fuel and air. Optical methods, with nanosecond response time, possess the unique ability to capture data at individual points in the combustion chamber without altering the flow field and thus can measure the fuel-to-air ratio in an engine.

Engine combustion processes are not identical and vary from one cycle to the next. Such variations occur due to several factors—fuel concentration being one of the most significant. In fact, concentration of fuel around the spark plug and movement of the air–fuel mixture within the combustion chamber mostly regulate the combustion process. A number of techniques such as Raman scattering, infrared (IR) absorption, planar laser-induced fluorescence (PLIF), laser-induced breakdown spectroscopy (LIBS), etc. have been developed to measure the local equivalence ratio of an SI engine, giving information on localized fuel concentration. LIF measurement is a reliable technique that provides a strong signal on two-dimensional fuel concentration data at a particular time and makes it easier to understand the transient behavior of the time-sensitive combustion phenomena.

6.4 Laser-Induced Fluorescence (LIF), Planar Laser-Induced Fluorescence (PLIF), and Particle Image Velocimetry (PIV)

All three techniques (LIF, PLIF, and PIV) have some commonalities, particularly when applied to engine combustion studies. All three methods are noninvasive in character and do not interfere with the combustion process. Kaiser et al. [1] used acetone planar laser-induced fluorescence (PLIF) to measure the hydrogen concentration field. The PIV technique was also widely used in a variety of engine studies including assessment of flow properties [2–4] and also to evaluate the behavior of in-cylinder flow, which includes intake-induced swirl [5, 6], flow-spray interaction [7], and pattern of flow field near the spark plug [8].

6.4.1 Planar Laser-Induced Fluorescence (PLIF)

In a hydrogen engine, CFD simulations have been rigorously validated against experimental data generated from PIV and PLIF techniques [9, 10] from an optically accessible engine. These validated CFD results provide technical guidance to optimize the design of injector nozzle geometry. Kaiser and White [1] also used PIV to investigate in-cylinder mixture formation. While acetone PLIF was used to measure the hydrogen concentration field, PIV measured the velocity vector

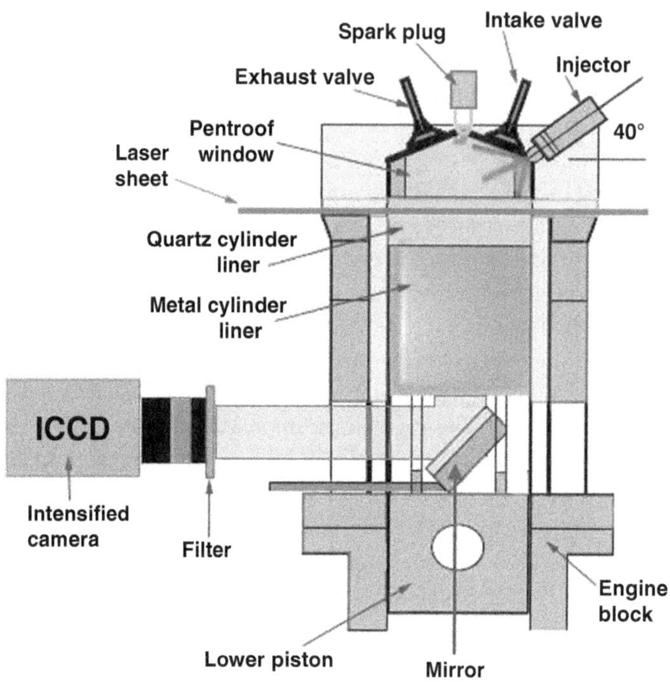

Figure 6.1 Optical engine experimental set-up [1].

field. The researchers conducted tests with three-injection timing after intake valve closure. These post-IVC injections, depending on their timing, are referred as early, intermediate, and late injection. Figure 6.1 shows a schematic layout of the optical engine test set-up. By using an optical piston with a 45° UV-visible mirror within the extended piston region or by using a partial quartz liner, it was possible to gain optical access to the cylinder.

Figure 6.2 shows the field view as seen from the piston window. The outer dashed line in the figure corresponds to the diameter of the cylinder bore and the inner solid line indicates the piston window. The injector is at the top of the image and the spark plug is slightly away from the center. Tests were conducted in both pre-IVC (prior to intake valve closure) and post-IVC (after intake valve closure) conditions. The baseline homogeneous mixture was formulated with a pre-IVC configuration for PLIF calibration experiments and to compare with post-ivc injection. Calibration tests were done to convert the fluorescence signals produced by acetone to the actual equivalence ratio in the combustion mixture. Since the fluorescence signal from acetone changes with temperature, correction was made to ensure accurate measurement of equivalence ratios by adjusting the temperature fluctuations. In case of post-IVC injection, if the start of injection was synchronized with intake valve closure (IVC), results obtained through PLIF were similar to those for pre-IVC injection. A retarded start of injection (SOI) from IVC caused increased mixture inhomogeneity leading to a spatial concentration of a small volume of high hydrogen levels close to the injector. When fuel is injected late in the cycle, the rich fuel turbulent region is distinctly separated from the calm lean fuel zone.

Figure 6.3 shows the equivalence ratio field for three post-IVC injection timings: early, intermediate, and late injection. The left columns indicate a single instantaneous image, whereas the mean over 100 images is shown in the center column, and corresponding root mean square (RMS) is shown in the right column. A distinct difference in RMS color intensity is evident.

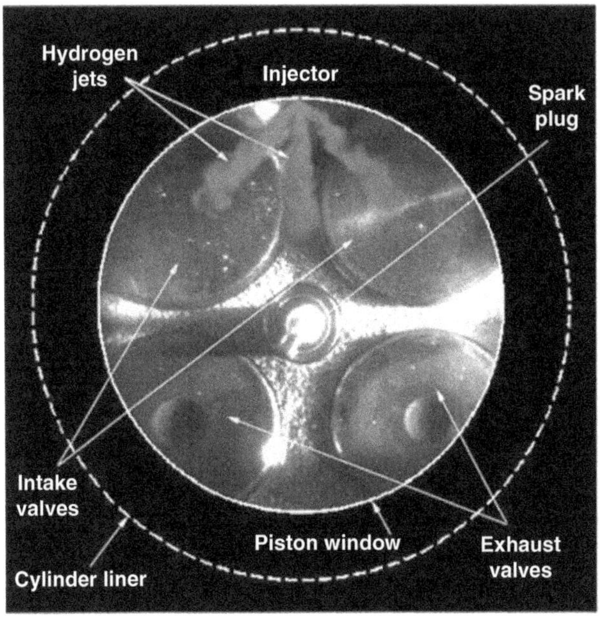

Figure 6.2 Field view as seen from the piston window [1] / SAE International.

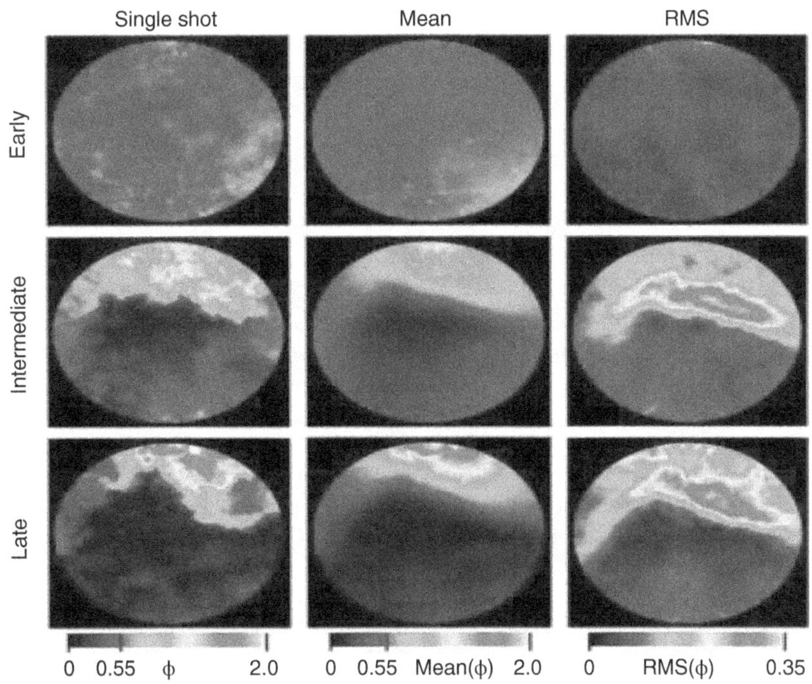

Figure 6.3 Equivalence ratios by PLIF for different post-IVC injection timings [1] / SAE International. Please visit the companion website to view a color version of this image.

Scarcelli et al. [9] studied 3D numerical simulation of in-cylinder processes in an engine fueled by hydrogen direct injection (DI). The computational results, obtained using the commercial code Fluent, were compared with experimental data generated from laser-based measurements from an optically accessible engine. The simulation accounted for intake-port geometry and a supersonic hydrogen jet injection into an engine with cylinder geometry typical of a passenger car spark-ignited engine into which gaseous hydrogen was injected through a high-pressure single-hole nozzle. Flow fields in the vertical, central symmetry plane were compared numerically and experimentally for various crank angles during the compression stroke, with and without fuel injection. The simulation predicted fairly well the flow patterns inside the cylinder without fuel injection. However, simulation during the gas exchange phase consistently showed some distinct disagreement between the simulated calculations and the experimental results all through the compression stroke. In the analysis, the tumble ratio was observed to be underestimated at the beginning of the compression ratio, but it gradually overpredicted as compression progressed. In terms of fuel distribution, the simulation correctly predicted jet penetration, interaction with cylinder walls, and location of fuel-rich cloud during compression. However, mixing near the jet head was also underestimated. Such conflicting trends show some inadequacies in the turbulence model in capturing how energy moves from larger to smaller scales.

In their study, Scarcelli et al. [10] also validated the CFD code Fluent, version v 6.3.35, for simulating and analyzing mixture preparation in the combustion of the engine with cylinder geometry typical of passenger car spark-ignited engines, with a centrally located injector. They used single-hole and 13-hole nozzles for fuel injection at pressures of approximately 100 bar and 25 bar, respectively. Quantitative planar laser-induced fluorescence (PLIF) captured images of fuel mole-fraction, providing detailed information about the fuel distribution pattern over time in an optically accessible engine. The high-speed schlieren technique was used to visualize movement and penetration of the fuel jet during the initial stages. Results of simulation were broadly in good agreement with optical experiments. Results were quite accurate in the numerical analysis of the fuel jet penetration into the combustion chamber and its evolution over time.

Simulation analysis of injection from the 13-hole nozzle showed that all the jets merged into a single effective jet shortly after leaving the underexpanded region. Such a situation, when hydrogen jets interact with the cylinder walls and existing flow field and also among themselves, is essentially as a result of the Coanda effect [11]. This phenomenon makes numerical simulations more challenging because it needs a higher level of detailed simulation and grid. Thus, design of the injector nozzle geometry is very crucial as it creates conditions for efficient fuel distribution and desired stratification at ignition time. The number and diameter of holes in the injector nozzle are responsible for evolution of each jet and its broad profile. Besides the nozzle, effectiveness of the overall injection system depends upon a number of factors such as injection pressure, injection duration, needle lift, etc.

6.4.2 Laser-Induced Fluorescence (LIF)

LIF and tracer–LIF are techniques that use fluorescence to detect and analyze specific molecules within the samples. In the case of tracer-LIF, additional substances with enhanced fluorescence properties (tracers) are introduced to improve detection sensitivity. This method is applied to the study of hydrogen engines, where specific tracers like triethylamine (TEA) and trimethylamine (TMA) are used for investigation. The working principle of laser-induced fluorescence (LIF) and tracer-LIF depends on selected excitation of the tracer substance. A specific selected tracer substance is targeted for excitation by exposing it to laser light of a specific wavelength to match the absorption characteristics of the tracer substance. As the tracer substance absorbs laser light, its electrons are excited and energized to a higher energy state. Following this temporary excited state, the electrons in the tracer substance return to their lower energy states, releasing the absorbed energy in light emission. This emitted light has a different wavelength from absorbed laser light. Light emitted is then detected by an intensified charge-coupled device (ICCD) camera with the capacity to capture and amplify low-light-level signals. Laser illumination slices are then directed into the chamber where the tracer substance is located. The laser illumination slices are controlled and focused laser beams are selectively illuminated by the laser to capture detailed images, so it is possible to ensure precise and targeted excitation of tracer through these slices. Compression ignition (CI) engines can adopt higher ranges of compression ratio and can contribute to improved efficiency, but compression ignition of hydrogen is difficult to achieve because of the high autoignition temperature of hydrogen (about 576 °C).

Project "H2BVplus" (working title: "Highly Efficient and Clean Combustion System for H2–IC Engines for Automotive Applications") was funded by the Austrian Federal Ministry for Transport, Innovation and Technology. The work was carried out with close cooperation between the Institute for Internal Combustion Engines and Thermodynamics at Graz University of Technology in Graz, Austria; BMW Group Research and Technology in Munich, Germany; Hoerbiger Valve Tec GmbH in Vienna, Austria; and HyCentA Research GmbH in Graz, Austria. The project generated an exhaustive study with lots of important information on system design, simulation, performance, emission, and combustion. However, for the present, only details related to laser induced fluorescence (LIF) are discussed here. Heindl et al. [12] produced significant results using 3D CFD simulation (the FLUENT software package in version 6.3.26). Experiments were conducted with an optical research engine, equipped with appropriate instruments and techniques to deliver "information from small measuring volumes or 2D images from so called illumination slices" [12]. The engine, because of its glass components, had some restrictions on load and speed. Due to light refraction, such engines usually have flat piston designs. In their studies on H_2-fueled SI engines, with direct injection at injection pressures above 100 bar, smooth engine operation could not be achieved due to the Coanda effect, which caused overstratification during late injection leading to knocking phenomena limiting both ignition timing and the compression ratio. However, to facilitate direct injection (DI) operation of a CI engine, the engine is designed with two possible injector mounting positions: a central position and a lateral position. The engine's compression ratio can be varied by using piston bowls of different dimensions. Moreover, the cylinder head is equipped with bores to accommodate both a glow plug and a spark plug to ensure reliable ignition of the hydrogen/air mixture in the event compression ignition does not occur. The layout of the combustion chamber H_2 CI engine is shown in Figure 6.4.

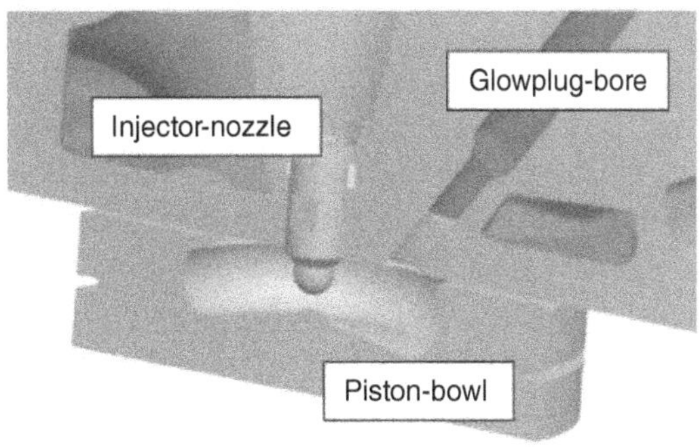

Figure 6.4 Layout of the combustion chamber for a hydrogen CI engine.

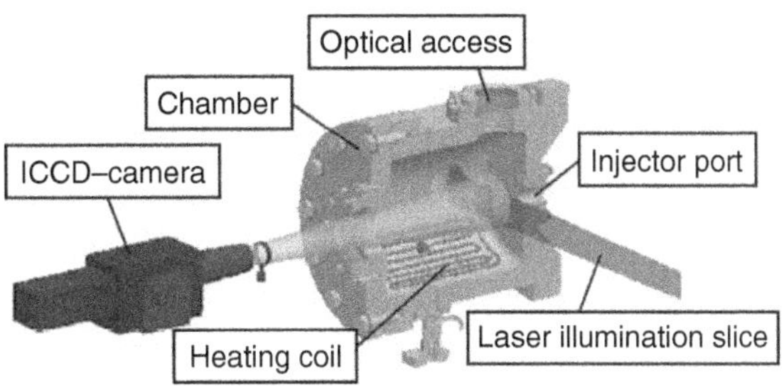

Figure 6.5 Optical chamber set-up [12].

Figure 6.5 shows the optically accessible high pressure/high-temperature chamber set-up used for studying and analyzing the

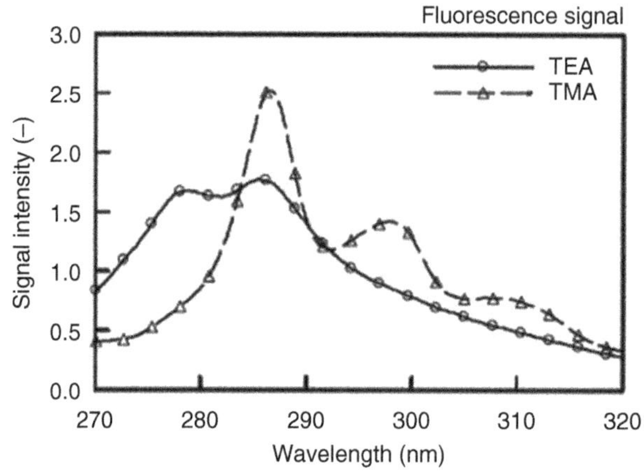

Figure 6.6 Fluorescence emission spectra for TEA and TMA [12] / Springer Nature.

jet pattern generated by the injector nozzle. A heating element was installed to warm up the medium inside the unit. There were four windows, 100 mm diameter each, to facilitate optical accessibility.

Figure 6.6 displays the fluorescence emission spectrum for an excitation wavelength of 248 nm for triethylamine (TEA) and trimethylamine (TMA).

The selection of this wavelength is based on absorption characteristics of TEA and TMA. Even at the same excitation and wave length, the intensity of the LIF signal is influenced by ambient pressure and temperature. A correlation between tracer concentration, signal intensity, and ambient conditions is necessary to accurately interpret the data and obtain a precise quantification of the results.

Blotevogel et al [13] studied PLIF for mixture formation in a hydrogen engine. By using the PLIF technique, detailed information on both process development and measurement of the fuel–air ratio were generated and used to optimize the mixture formation. Subsequently, these optimized PLIF data were verified with Raman scattering measurements. In Raman scattering the fuel–air ratio is determined directly from the concentration measurements of the different gas species. Good agreement of PLIF results with direct Raman measurements indicated that using PLIF it is possible to achieve a quantitative fuel–air ratio measurement during the mixture formation process of hydrogen engines. Kirchwegen and co-researchers [14] used PLIF in a direct injection hydrogen engine to investigate jet patterns during injection, the stratification of the charge at the ignition point, and the flame propagation during combustion. The research engine installed at the Institute for Internal Combustion Engines at Graz University of Technology had optical access through a quartz-glass liner and a window in the piston. Triethylamine as a tracer was mixed with hydrogen at 200 ppm and provided a strong stable LIF signal even at a maximum pressure of 200 bar.

Salazar et al [15–17] carried out research on several aspects of a hydrogen engine on mixture preparation, direct in-cylinder injection, in-cylinder flow characteristics, fuel distribution, etc. One of their studies aimed at evaluating the accuracy and precision of using quantitative PLIF to measure acetone vapor as a fuel tracer in a hydrogen-fueled optical engine [15]. The study involved detailed calibration and correction procedures to generate reliable data. Specific attention was given to accurate measurement and to address problems like temperature effects, linearity of fluorescence with laser energy, drift in acetone-vapor concentration, spatial inhomogeneity in the calibration measurements, noise, etc. The researchers aimed to minimize errors and uncertainties in the measurement process, with particular attention to calibration accuracy in the equivalence ratio and the impact of refractive beam steering on precision. In a series of studies with an optically accessible hydrogen engine with direct cylinder injection and using PLIF of acetone as the tracer, the researchers carried out tests with "six different combinations of nozzle design and injector geometry, each for three injection timings after the intake valve closure" [16]. Both symmetric nozzles (6 holes and 13 holes) and asymmetric nozzles (single hole and 5 holes) were tested with early injection, intermittent injection, and late injection, which showed very different scenarios. Complete merging of all the jets achieved by the 13-hole injector was in close agreement with the reported literature. In their studies with an optically accessible direct cylinder injection hydrogen engine, the researchers investigated the interaction of bulk flow and jet-induced fuel convection using PLIF of gaseous acetone as the fuel tracer [17].This study involved a single-hole injector in which the injector orientation and tumble intensity were varied and the injection timing was also correspondingly altered for each of these two parameters. Images captured for each configuration of mixture formation and injection timing provided relevant information on the final mixture distribution pattern. On changing the direction of the injector in the engine, the overall impact of the bulk flow on fuel convection was relatively mild.

In their paper Tomita and Hamamoto et al. [18] discussed fundamental characteristics of fluorescence and application of PLIF to analyze the concentration distribution of a transient hydrogen jet. To correctly measure PLIF for transient jet concentration fields, diacetyl was used as the fluorescence tracer. The fluorescence signal was found to be directly proportional to the tracer concentration. In this study, each individual jet demonstrated a different configuration and concentration distribution even though the averaged fuel jet spread out symmetrically in all directions. Thus, application of PLIF to the flame jet enabled the determination of fuel concentration fields and observation of flame development. Arnold and co-researchers [19] developed a two-dimensional laser-induced fluorescence measurement technique (2D LIF) of the hydroxyl radical distribution to determine absolute OH concentration fields on a single laser pulse basis. This method, which was a combination of 2D LIF and 1D absorption spectroscopy, and the results of the analysis were consistent with numerical model calculations.

It was discussed earlier that triethylamine (TEA) and trimethylamine (TMA) are usually added as a tracer substance because molecular hydrogen finds it difficult to be excited by laser light. The addition of tracer enhances fluorescence intensity to a higher level without affecting the properties of the fuel. In view of the excellent linearity of the LIF signal to the mixture, this approach has proven to be a time-efficient calibration procedure for optimizing the mixture formation process. Investigation by Kirchwegen et al. [14] on several aspects of direct cylinder injection, including jet pattern during injection, charge stratification at ignition points, and flame front propagation during combustion have already been briefly discussed. In their test set-up, an additional injection unit (in addition to DI injectors) was installed in the intake manifold to achieve a well-homogenized mixture. Fifty images for each crank angle investigated were recorded at DI-operating points and these recordings were averaged. Processing images and quantifying fuel injection by normalizing images and compensating for laser light inhomogeneity, etc. were all done in the right sequence to get an overview of the effects of various factors on hydrogen combustion. As mentioned earlier, in the tests, TEA mixed with hydrogen at 200 ppm used as tracer provided a robust LIF signal. In carrying out tests with high pressure, direct cylinder injection, the possibility of mixing the tracer over the entire pressure range has often been a point of paramount importance. With hydrogen, a very small tracer concentration is required, and it is necessary to maintain very low partial pressure and allow the tracer to remain in a gaseous state under all conditions. Because of such advantageous features, TEA is often chosen as the desirable tracer for hydrogen [13].

Some of the significant results from the investigation by Kirchwegen et al. [14] are shown in Figures 6.7 to 6.9. Figure 6.7 shows the jet patterns of a multi-hole injector for fuel-rail pressures of 50 bar and 100 bar. The two single jets are easily distinguished in the initial phase of injection, but the signal is much weaker at lower pressures.

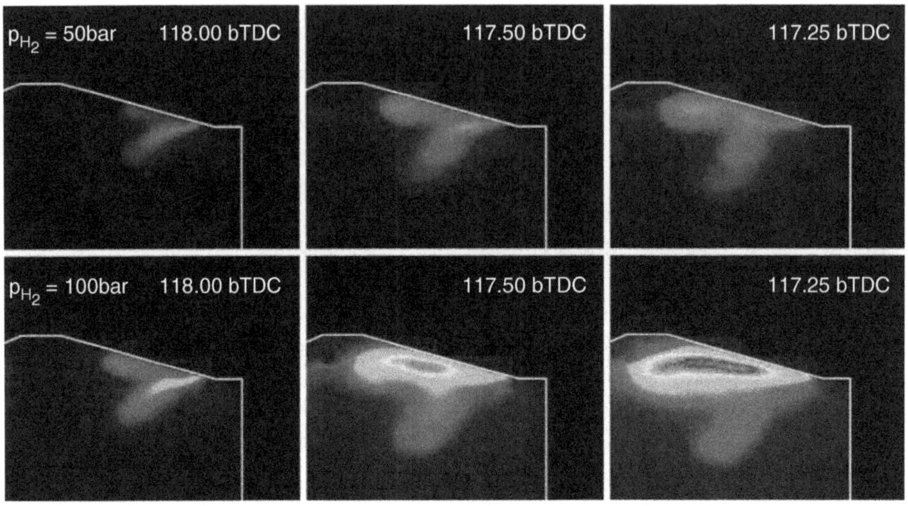

Figure 6.7 Tracer-LIF measurement jet patterns at 50 bar and 100 bar [14] / Springer Nature. Please visit the companion website to view a color version of this image.

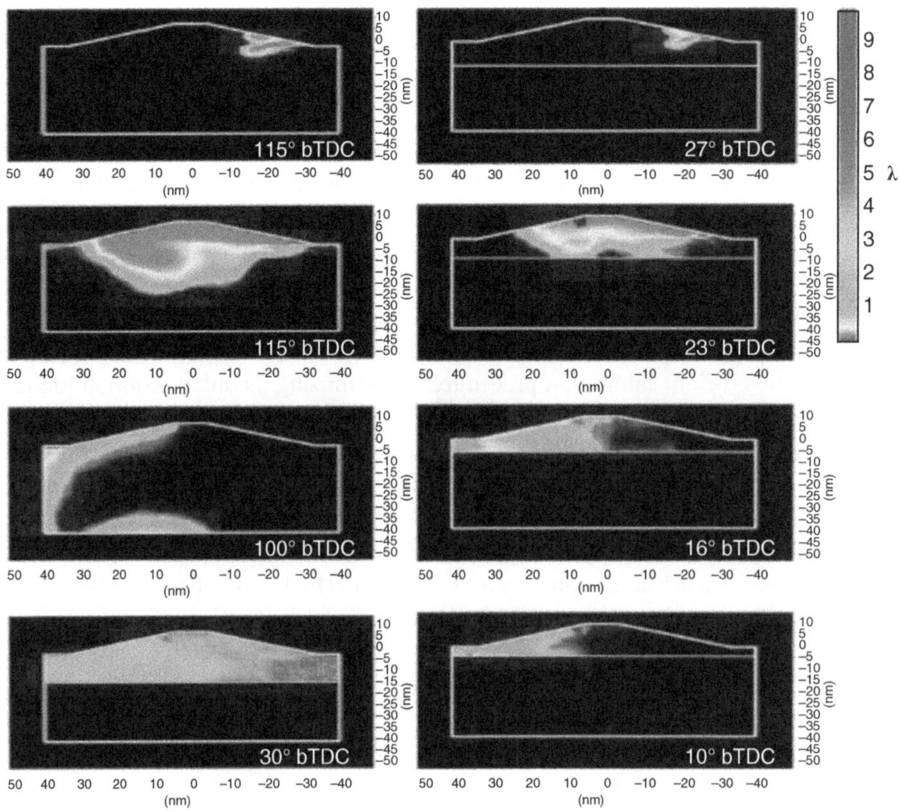

Figure 6.8 Early injection produces a homogeneous charge (left), and late injection (right) leads to stratified charge [14] / Springer Nature. Please visit the companion website to view a color version of this image.

Sometimes it becomes necessary to normalize the direct injection images that might lead to uneven illumination in the laser light sheet because of the superimposition of two laser beams. Figure 6.8 shows the effect of injection timing on the character of the mixture as obtained in quantified tracer-LIF measurements. Early injection on the left shows homogeneous charge; late injection leads to stratified charge as shown on the right.

Homogeneous charge formation at the point of ignition by the tracer-LIF technique helps to visualize burned and unburned zones as the entire combustion chamber is illuminated. Such situations offer the advantage of moving from tracer-LIF to selective LIF as it does not require any changeover or modification in the configuration. Researchers conducted tests to evaluate visualization of the flame front on homogeneous charge with tracer LIF and OH-LIF measurements. In Figure 6.9, the upper image shows the tracer-LIF and the lower image shows the OH-LIF. A closer look at the tracer-LIF in all four cases clearly indicates that the entire combustion chamber is fully illuminated and the flame front is distinctly visible. This indicates effective homogenization of the fuel and the tracer. The image produced is essentially the opposite (negative) of what is seen with OH-LIF. In tracer-LIF, burned zones appear dark. Despite the difference in the extent of illumination, both tracer-LIF and OH-LIF methods provide similar information.

Aleiferis and Rosati [20] conducted an optical investigation with direct injection and port injection on a hydrogen SI engine running with an equivalence ratio between 0.50 and 0.83 at 1000 rpm, and at 0.5 bar intake pressure. To record and analyze the detailed features of the flame front, they applied a laser-induced OH fluorescence on an in-cylinder plane below the spark plug. In-cylinder flame growth was calculated from the crank-angle resolved flame chemiluminescence images of a series of consecutive cycles. Observation made at a larger scale, considering overall appearance of the entire flame, indicated that for an equivalence range from 0.50 to 0.83, the flame growth was almost the same for the port injection and direct injection strategies. In direct cylinder injection (DI), flame expansion speeds peaked at over 35 m/s for ϕ between 0.67 and 0.83, while for ϕ between 0.50 and 0.59, speeds were slightly lower at 25–30 m/s. Hydrogen combustion was slower for port fuel injection with peak expansion speeds of 10–20 m/s across a wider ϕ range of 0.50 to 0.83. Hydrogen's

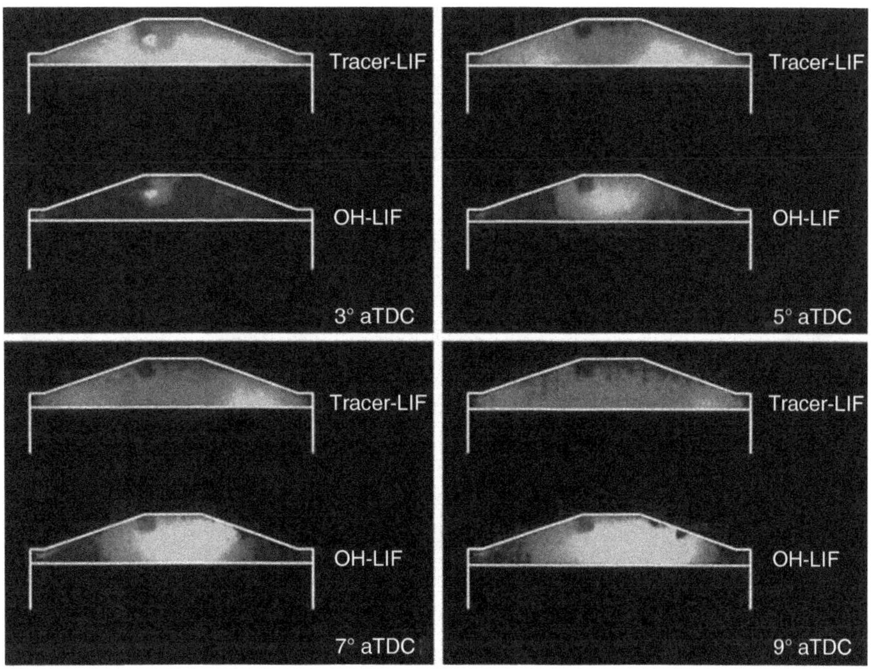

Figure 6.9 Comparison of flame front propagation: tracer-LIF (upper image) and OH-LIF (lower image) [14] / Springer Nature. Please visit the companion website to view a color version of this image.

burning velocity with direct injection was much higher than the corresponding value for the port injection system. Flame stretch refers to deformation of a flame's shape, particularly in response to turbulence occurring during the combustion based on flame size and flame expansion speed. In this study, flame stretch values ranged from 5000 to 10,000 per second for early flames with a radius up to 5 mm. Flame growth in OH-LIF images was mostly symmetric in most of the cycles, while some cycles exhibited distortions and wrinkles at equivalence ratios of around $\phi = 0.50$. It was observed that retarding the start of injection or advancing the spark timing resulted in faster flame growth. Adjustments in timings influenced the effects on OH distribution and flame distortion. Later injection timing increased OH distribution and flame distortion, whereas advancing spark timing led to even more distortion on both large and small scales.

6.5 Particle Image Velocimetry

Kaiser and White [1], in addition to acetone planar laser-induced fluorescence (PLIF), also used particle image velocimetry (PIV).

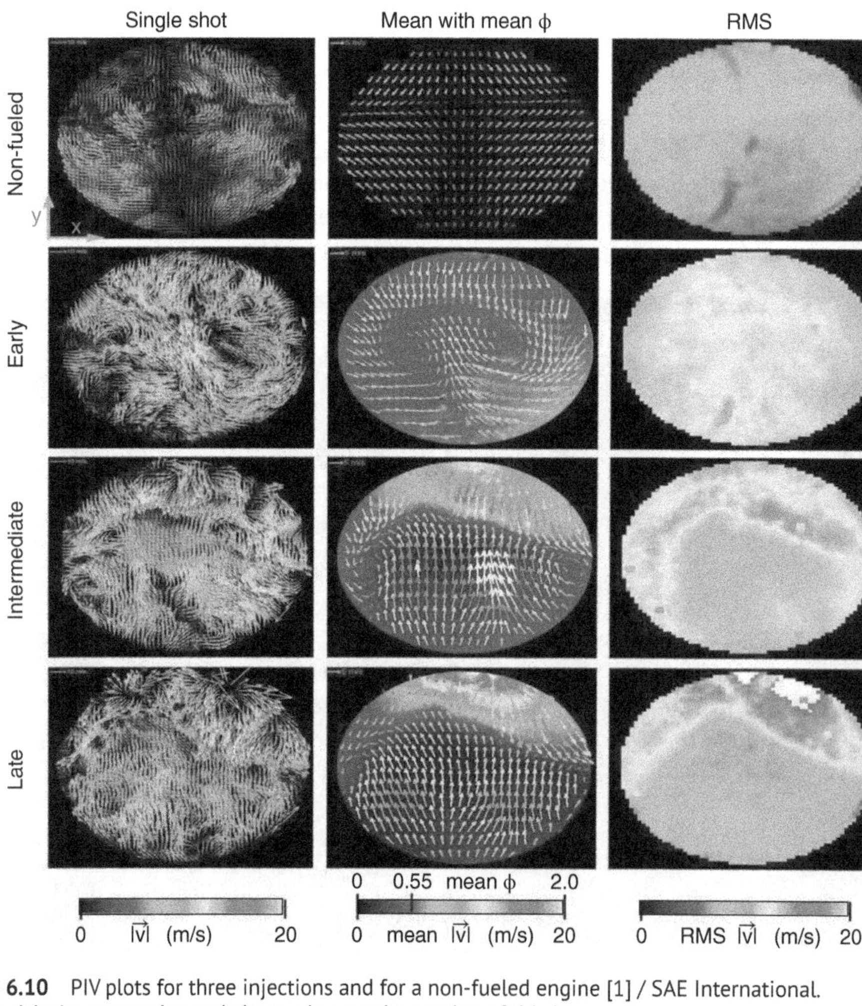

Figure 6.10 PIV plots for three injections and for a non-fueled engine [1] / SAE International. Please visit the companion website to view a color version of this image.

Plots generated for three hydrogen injections for early, intermediate, and late injections and also for non-fueled engine operation are shown in Figure 6.10 in the same format as was done for PLIF in Figure 6.3. The top row shows the velocity data for non-fueled engine operation when there is no hydrogen injection. The left column shows the instantaneous velocity field, the center column represents the mean velocity of 165 fields with the mean equivalence ratio, and the right column shows the RMS value.

Salazar and Kaiser [21] used the PIV technique to study the in-cylinder charge motion during the compression stroke of an optically accessible direct cylinder hydrogen engine. A major objectives of this research was to study how the mean flow field and tumble ratio change with time. The investigators compared scenarios with and without fuel injection. The engine was initially evaluated without injection, both with a standard four-valve pent-roof engine head and with modified intake ports designed to increase tumble. Intake port modification significantly increased tumble flow, raising tumble ratio from 0.22 to 0.70 at bottom dead center (BDC), as shown in Figure 6.11.

In studying mean in-cylinder flow field with fuel injection, the researchers observed that a hydrogen injection from a centrally located single-hole nozzle totally disturbed the intake flow resulting in a sudden reversal of the tumble. Thus, tumble ratio became negative after the injection (Figure 6.12).

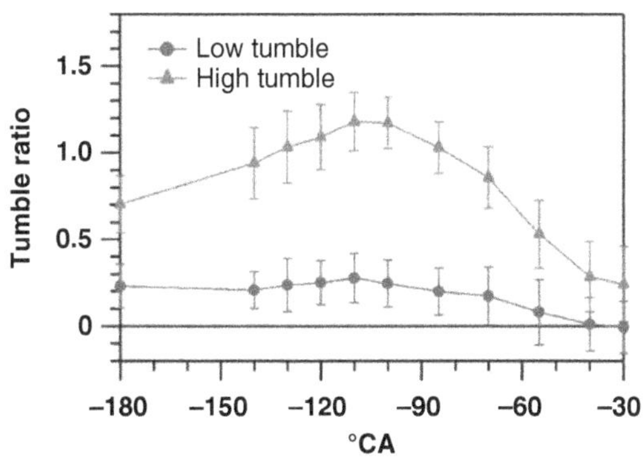

Figure 6.11 Tumble ratio for both intake-flow configurations without fuel injection. The vertical bars represent the limits of cyclic variation in tumble ratio at 90% probability [21].

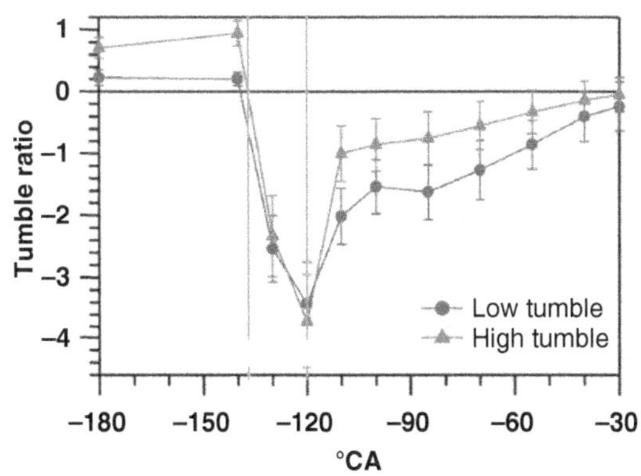

Figure 6.12 Tumble ratio for both intake-flow configurations with fuel injection. The vertical bars represent the limits of variation in tumble ratio at 90% probability [21].

6.6 Laser-Induced Breakdown Spectroscopy (LIBS)

Laser-induced breakdown spectroscopy (LIBS) has some advantages over other laser diagnostic techniques such as Raman spectroscopy, laser-induced fluorescence spectroscopy, etc. Raman and LIF spectroscopy can be effective if the laser excitation wavelength closely matches the absorption wavelength of the target molecules. LIBS does not depend on any single, specific wavelength and can be performed with a range of wavelengths. Instead, LIBS depends on the laser's ability to produce a high enough energy density (or fluence) to create a plasma breakdown, regardless of the exact wavelength used.

Phuoc [22] investigated LIBS to measure ignition by simultaneously igniting and measuring fuel–air ratio of hydrogen–air combustible mixtures. The experimental set-up comprised a high-pressure laser ignition apparatus (HLIA), a laser ignition system, and a LIBS system (Figure 6.13). The HLIA was designed to operate on hydrogen, and H_2-O_2 mixtures at initial pressure of 4 MPa. The laser system had the two single-mode, Q-switched Nd-YAG lasers, and each emitted a 0.6-cm-diameter beam at 1064 nm wavelength with a 5.5 ns pulse duration. Two beam delivery optics directed the lasers into the ignition cell from opposite directions. Each beam delivery system had a coated beam splitter to facilitate reflection of the laser beam and transmit the emission lines with wavelengths from 600 to 900 nm. For spatial analysis, a Multichannel Instaspec IV CCD detection system was installed. Research-grade hydrogen with 99.99% purity and air were precisely delivered using gas flow controllers. This study showed the capability of the LIBS technique in accurate measurement of the fuel–air ratio in a hydrogen mixture.

Shudoa and Oba [23] adopted the LIBS technique to measure hydrogen concentration distribution in a direct injection stratified charge engine. One of the objectives of this study was to understand the relationship between fuel injection conditions and mixture distribution.

The experimental set-up (Figure 6.14) was designed to measure hydrogen jet properties. Before investigating the distribution of a hydrogen jet, the researchers calibrated the correlation between the hydrogen and air mixing ratio and the resulting atomic emissions. Initially, a mixture of hydrogen and nitrogen at a predetermined ratio was used to fill the chamber and was stirred by a fan powered by an electric motor. A Nd-YAG laser was used and the laser beam was focused

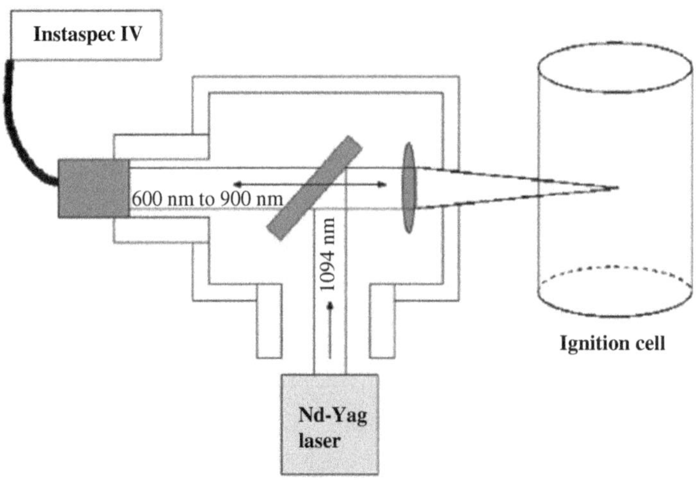

Figure 6.13 Simultaneous LIB ignition and equivalence ratio measurements [22].

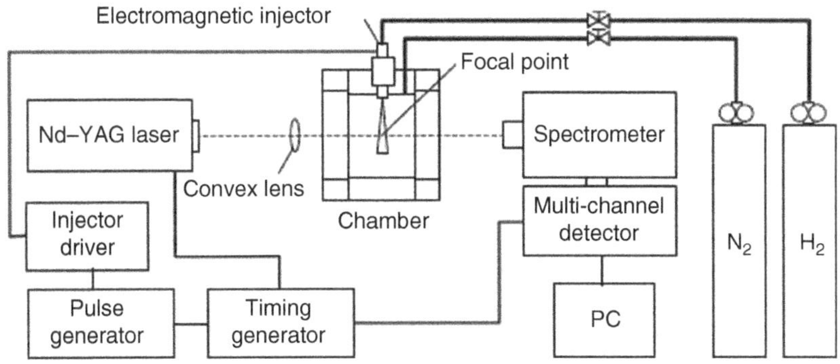

Figure 6.14 Experimental set-up for measuring equivalence ratio distribution using LIBS [23].

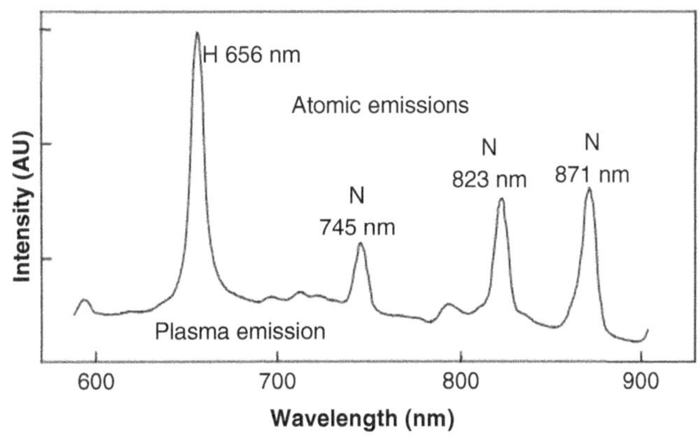

Figure 6.15 Atomic and plasma emissions from a hydrogen–nitrogen mixture using LIBS [23].

within the chamber by a fused silica convex lens. The emitted spectra was analyzed by directing the atomic emissions from a focal point to a spectrometer coupled to a CCD multichannel detector. Thus, all the devices used in the entire experimental set-up such as the laser, spectrometer, and detector, were synchronized with transistor–transistor logic (TTL) signals from a timing generator. Hydrogen was directly injected with an electromagnetic fuel injector with a single nozzle (2 mm hole diameter) into a chamber already filled with nitrogen at a pressure of 0.5 MPa. An electromagnetic fuel injector, connected to an injector driver unit and a pulse generator, was used to control the injection process. Hydrogen was supplied at 5 MPa to the injector, installed on top of the chamber with hydrogen jet direction perpendicular downward from the injector. Similar to the calibration process, a laser beam was directed into the chamber to measure local hydrogen concentration using atomic emissions. To generate reproducible reliable data, average data of ten repeated tests for each specific experimental condition were accepted.

By analyzing the spectrum it is possible to identify the composition and sometimes even the concentration of different elements in the sample. Figure 6.15 displays spectra obtained through LIBS from a hydrogen–nitrogen mixture, over a wavelength range from 600 nm to 900 nm. As evident in the figure, hydrogen has a peak of 656 nm while nitrogen has three peaks of 745 nm, 823 nm, and 871 nm. The intensity of plasma emission, which is weaker than that of atomic emission, spreads over a wider range of wavelength.

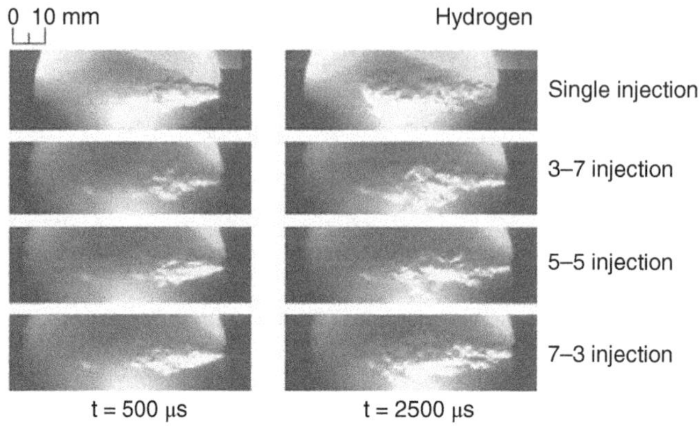

Figure 6.16 Hydrogen with single injection and three sets of split injection [23] / with permission of Elsevier.

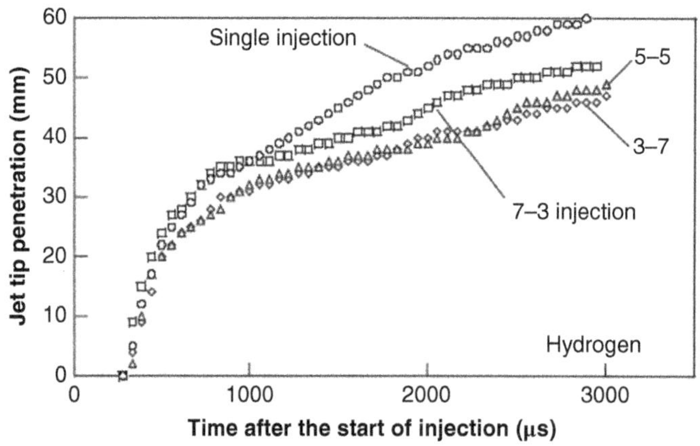

Figure 6.17 Time-series hydrogen jet tip penetration with single injection and three cases of split injections [23].

Figure 6.16 shows schlieren images of hydrogen jets recorded for single injection and three sets of split injection. Figure 6.17 shows time-series tip penetrations of hydrogen jets. It is clear in the figure that tip penetration by split injection is weaker than single injection. A significant observation from this study shows that an optimum setting of split injection of hydrogen can identify the location of high equivalence ratio mixtures near the spark plug and slow diffusion. Such information is useful for reducing cooling loss in a stratified charge hydrogen engine and can lead to improved thermal efficiency.

6.7 Spark-Induced Breakdown Spectroscopy (SIBS)

It has been appropriately mentioned that "SIBS can be regarded as a 'marriage' of two other pulsed plasma techniques for elemental analysis: laser-induced breakdown spectroscopy (LIBS) and traditional spark spectroscopy" [24]. During the process of atomic emission, excited electrons drop to lower energy levels, emitting photons. SIBS uses electric sparks to induce a plasma, which detects the atomic emissions. So like spark spectroscopy, plasma is formed electrically, and just as in LIBS, it is often done in ambient air, with detection timed to avoid bright initial breakdowns. During SIBS, a spectrum of light is generated during the ignition process. This light is emitted when a spark is produced at the electrode of the spark plug. Optical fibers are used to capture the emitted light from this spark. Images captured on a high-speed camera can provide very significant information on flame development, flame propagation, fuel distribution, mixture strength, equivalence ratio, injection timing, ignition timing, etc., which give a distinct picture on combustion in engines.

Roy and co-authors [25] investigated the characteristics of late injection in a stratified DISI engine by spark-induced breakdown spectroscopy (SIBS) to evaluate local fuel–air concentration of a stratified charge engine in the spark gap at time of ignition. Roy et al. carried out investigations on a direct injection hydrogen engine where stratification was generated by the gasoline direct injection (GDI) injector with the jet directed toward the spark plug. A high-speed camera synchronized with the spark was installed in the test set-up to capture the images of jet-guided combustion. A clear picture of the pattern of flame distribution at different intervals, gathered through a series of single shot images, provided information on the time of flame development. Variation in fuel injection timing with respect to ignition timing highly influenced early development of flame and its propagation. Images caught in the high-speed camera showed that combustion started with a whitish blue flame and then propagated through the flammable air–fuel mixture. Tests were carried out at four injection timings relative to ignition timing (Figure 6.18) such as tail ignition (TI), after-tail ignition (ATI), center ignition (CI), and head ignition (HI).

The four modes of ignition with corresponding injection timings for this study are shown in Table 6.1.

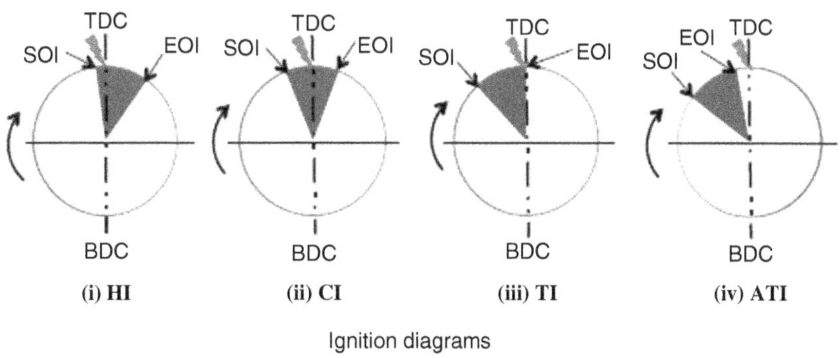

Figure 6.18 Relative timings of fuel injection and spark ignition [25].

Table 6.1 Four modes of ignition.

Tail ignition (TI)	Hydrogen is injected during the compression stroke and ignites at the end of injection.
After-tail ignition (ATI)	Hydrogen is injected during the compression stroke, and end of injection is scheduled at 5° CA before TDC.
Center ignition (CI)	Hydrogen is injected during the later part of the compression stroke, and ignition occurs between the start of injection and end of injection.
Head ignition (HI)	Hydrogen injection occurs during the later part of the compression stroke such that ignition occurs at TDC, close to the start of injection.

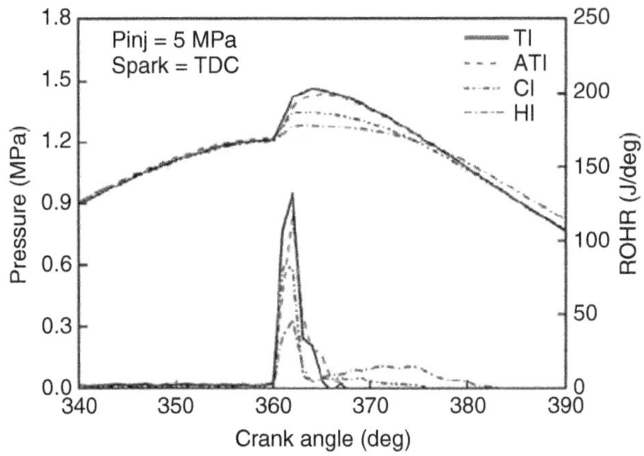

Figure 6.19 Pressure and heat release rate for the four modes of ignition [25].

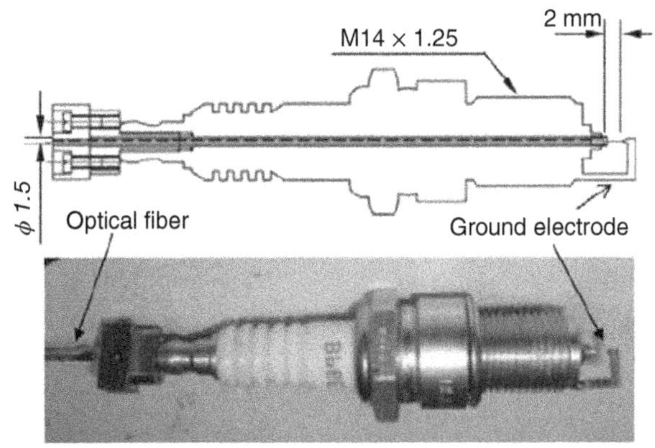

Figure 6.20 Fiber-optic spark plug sensor used in tests [25] / with permission of Elsevier.

Figure 6.19 shows the cylinder pressure and rate of heat release (ROHR) at the four modes of ignition.

Figure 6.20 shows a fiber-optic spark plug with electrode spacing of 2 mm used as a sensor for the SIBS technique in the tests. The local equivalence ratio during stratified charge combustion was measured using SIBS with a charge-coupled device (CCD) spectrometer. The local equivalence ratio, immediately after the spark breakdown, was very rich and there was wide cyclic variation.

In the present analysis, the initial combustion duration was considered as the interval of the crank angle from the ignition start to 5% mass fraction burned. Thus, the measurements of mean local equivalence ratio and initial combustion duration for four ignition modes at global equivalence ratio, $\phi = 0.7$ were also estimated. Figure 6.21 gives a comparative picture of the mean local equivalence ratio and initial combustion duration for the four ignition modes at equivalence ratio at $\phi = 0.7$.

Rahman et al. [26] used a SIBS sensor to measure local fuel concentration of direct injection of hydrogen jet at various locations within a constant-volume vessel. Local fuel concentrations were visualized in a high-speed camera. Very useful information on the effect of spark discharge

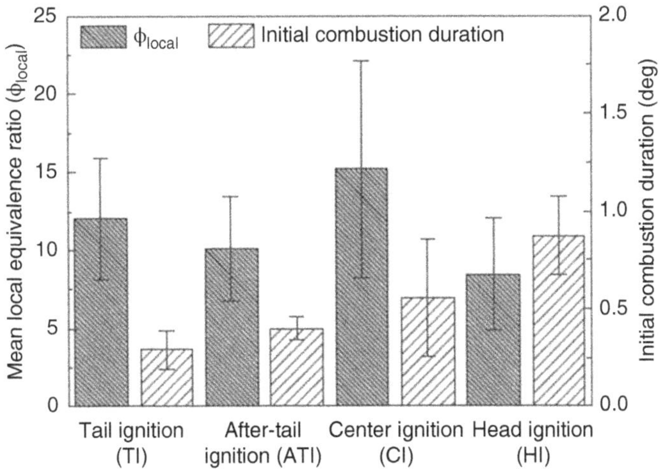

Figure 6.21 Equivalence ratios and initial combustion duration for four ignition modes [25].

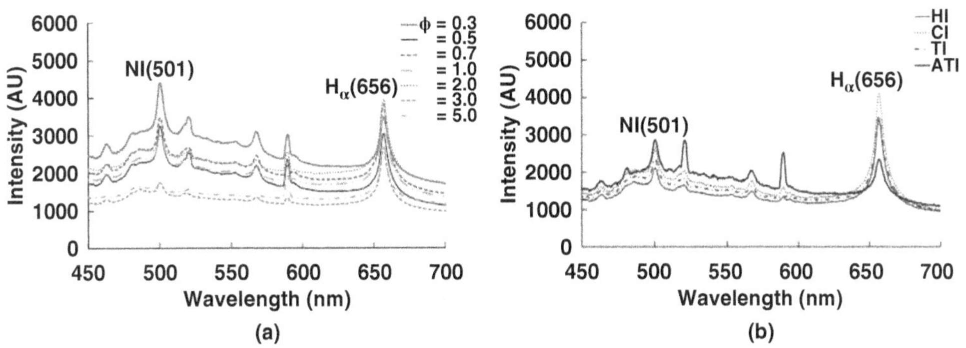

Figure 6.22 (a) Spark emission spectra of the hydrogen/nitrogen mixtures using the SIBS sensor. (b) Emission spectra of a spark from the SIBS sensor at location (0, 16.5) for different spark modes [26].

related to spark timing could be obtained from the direct visualization of the spark discharge. Spectrally resolved atomic emissions from spark-generated plasma determined local equivalence ratios. Figure 6.22a shows emission spectra of hydrogen–nitrogen mixtures measured at different equivalence ratios using a SIBS sensor. Figure 6.22b depicts the emission spectra from the SIBS sensor using the four ignition modes. Two prominent atomic emission peaks were identified as 656 nm corresponding to hydrogen and 501 nm for nitrogen in both figures.

Rahman et al. [27] investigated a quantitative measurement of the concentration of fuel in a direct injection hydrogen engine using SIBS to understand the effects of surrounding atmospheric pressure on ignition timing. For SIBS measurement, a sensor developed from the commercially available M12-type spark plug with minimal changes to electrodes was used. It was possible to collect the emission spectra from the plasma generated by the spark plug through the optical fiber housed in the center electrode of the plug. This fiber-optic spark plug sensor, capable of forming a stable spark in the direct-injection hydrogen engine, provided a distinct picture of local mixture stratification close to the spark gap. A calibration MAP was prepared to evaluate accurately the relationship between air and fuel ratios with both intensity of light emitted during ignition and

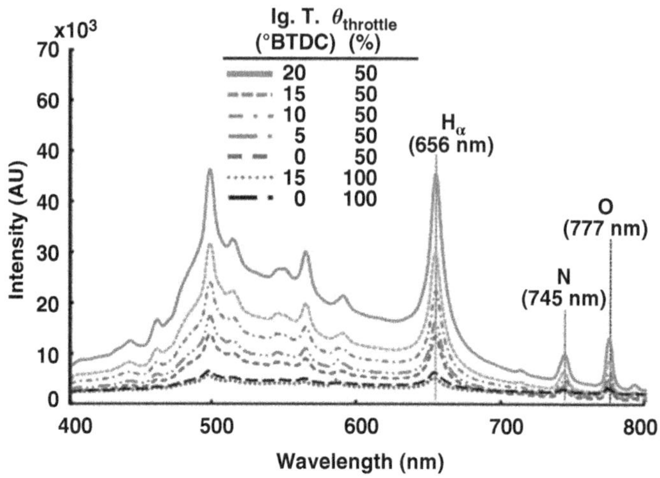

Figure 6.23 Spark emission spectra for hydrogen combustion with λ = 3 and varying ignition timing [27].

pressure at that moment. This approach provided precise quantitative measurements of local fuel concentrations for both port injection and direct-injection strategies. The optical fiber and sapphire window did not degrade or lose their structural integrity in the adverse environment. However, pressure variation had a significant effect on emission intensities, as shown in Figure 6.23.

6.8 Plume Ignition Combustion Concept (PCC)

Oikawa et al. [28] adopted a novel plume ignition combustion concept (PCC) for a direct-injection hydrogen engine. PCC used ignition of the mixture plume either in the middle of the injection period (referred to as mid-injection-ignition PCC) or after the end of fuel injection (referred to as post-injection-ignition PCC).With an objective to optimize PCC, the investigators adopted single- and multi-hole nozzles with a combination of high-speed laser shadowgraph to visualize the propagation of the flame. A DISI engine, under high load condition, significantly reduced NOx emission without any significant drop in thermal efficiency. Hydrogen has a high burning velocity and wide combustible range. Mid-injection ignition (mid-inj-ign) PCC takes advantage of some unique features to improve the combustion performance of hydrogen, especially in a rich mixture. Moreover, the high flow velocity of hydrogen jet generates strong turbulence, which ensures fast burning of hydrogen without misfiring. This study showed that mid-inj-ign PCC with minimizing jet penetration to the chamber wall reduces the NOx emission level even further.

6.9 High-Speed Schlieren Imaging

Salazar and Kaiser [29] adopted the high-speed schlieren imaging technique to study flame propagation in an optically accessible hydrogen engine. They investigated two different intake configurations. One was with low tumble with a tumble ratio of 0.22 and unmodified intake ports. The other configuration involved a modified intake system with a high tumble ratio of 0.70. Tests with fueling upstream the engine had no influence on intake flow. But direct fuel injection from an angled single-hole injector during the compression stroke increased the angular momentum in in-cylinder flow. Flame location and propagation speed from schlieren imaging during combustion, on a single-cycle basis, can be correlated with in-cylinder pressure measurement. In a typical cycle depending upon the flow characteristics, the shape of the flame evolves. In homogeneous fueling with a slender tumble, convection of the flame is relatively less. But the shape of the flame evolves symmetrically in the vertical plane without any major wrinkling in shape. For a homogeneous mixture, increased flame growth along with relatively shorter combustion duration has been reported with increase in tumble ratio from 0.22 to 0.70 [29]. However, for a stratified mixture two distinct regimes were glaringly obvious. The early phase of combustion showed a faster flame growth for high tumble induced by the intake; but in the middle and late phase of combustion with milder tumble, the result was a faster burn rate with significantly reduced combustion. Based on a single cycle, researchers analyzed that early growth of flame has a strong correlation with "crank angle at which 5% of the fuel mass is burned" [29]. It was observed that in direct injection, centroid speeds are relatively higher in the initial phase and then take up a slowly decreasing trend. In sharp contrast to this picture, the centroids, in the case of homogeneous fueling, maintain average constant convective speed.

6.10 Controlled Autoignition (Homogeneous Charge Compression Ignition)

Some typical characteristics of hydrogen like its wide flammability limits, high reactivity, and high octane number have often posed challenges for the hydrogen engine in controlling backfire and maintaining precise autoignition timing. Similarly, hydrogen's high octane number, which offers resistance to knock, makes consistent autoignition difficult. However, HCCI combustion with controlled autoignition offers significant potential for monitoring combustion and achieving near-zero CO_2 and NOx emissions. Aleiferis and Rosati [30] investigated optical studies on controlled autoignition (or homogeneous charge compression ignition) of hydrogen in a direct-injection optical SI engine with a pent roof combustion chamber geometry. To accomplish the objective, preheating of inlet air to a temperature of 200 to 400 °C was done along with residual gas recirculation with negative valve overlap, and hydrogen fueling was set to equivalence ratios between 0.40 and 0.63. Flame chemiluminescence images were captured at specific crank angles for consecutive cycles at 1000 rpm to calculate in-cylinder flame expansion and motion rates. A more accurate and detailed picture of the autoignition pattern was recorded by using planar laser induced fluorescence (LIF) of OH. The effect of mixture preparation on autoignition timing was calculated by single and double (split per cycle) hydrogen injection strategies. Figure 6.24 shows in-cylinder pressure traces for equivalence ratios of $\phi = 0.50$, 0.56, and 0.63 and with intake temperature in the range of 200 to 400 °C. The equivalence ratio ϕ was restricted to 0.63 in the tests to avoid any possible encounter of optical components with stringent environment due to high temperature and pressure. Even though HCCI combustion could be achieved at an equivalence ratio of 0.33, the intake temperature was expected to be beyond 400 °C. Since adequate research has not been done in the area of HCCI with hydrogen in an optical engine configuration, these limits might be expanded in further research.

The first signs of flame chemiluminescence in images (3° CA) and heat release on pressure traces were mildly influenced when intake temperature was raised from 200 to 400 °C. The characteristics of autoignition in hydrogen fuels is widely different from those in hydrocarbons.

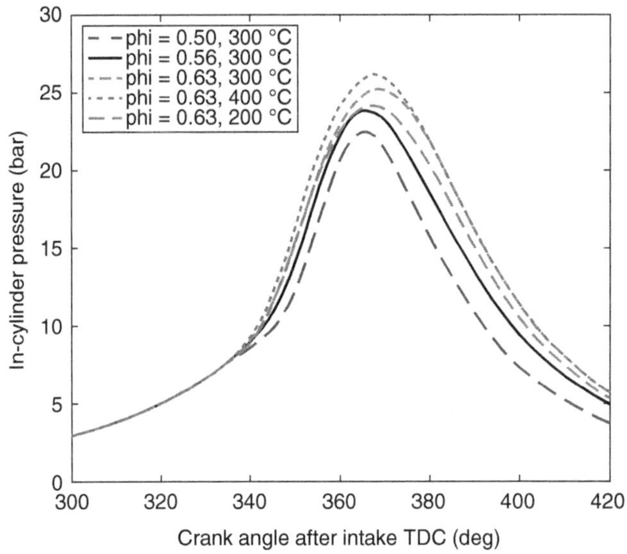

Figure 6.24 In-cylinder pressure traces at different equivalence ratios [30].

Here, autoignition started between intake and exhaust valves near the wall, at an equivalence ratio of 0.5, and subsequently developed into a swirling pattern. At an equivalence ratio of 0.63, autoignition also started on right side of the chamber but not near the walls. It started in a circular shape, and as soon as the flame acquired half the optical bore size it showed a distinct distortion and started expanding toward the left side of the chamber. A cycle-to-cycle basis of analysis shows that the start of autoignition had very restricted spatial variability in the range of $\pm 2°$ CA. The cyclic variation of combustion, following the autoignition, also exhibited a similar pattern of development. Analysis of the data on mixture formation and PIV data indicated that most likely the momentum exchange between hydrogen DI jets and in-cylinder air along with fuel stratification were responsible for the swirling motion of autoignition. Strong fuel stratification was characterized by concentration of major amounts of fuel close to the wall and very much less in the middle. During the autoignition with double-injection strategy for $\phi = 0.50$ and 300 °C, the start of combustion occurred close to the center of the bore at about 10–12° CA earlier compared to single injection and moved symmetrically toward the wall. There were no symptoms of any swirling. A double-injection strategy, when fuel is injected in two bursts, leads to a flame that spreads more slowly compared to injecting it all at once. It was also observed that after hydrogen autoignition started, it developed into a "propagation-like" fashion [30] and not in the characteristic pattern of HCCI combustion having multiple autoignition kernels. Another interesting coincidence was that as the spark-ignited flame front advanced beyond 5° CA after spark timing, autoignition happened ahead of the flame at the same spot where usually autoignition takes place for HCCI.

Rosati and Aleiferis [31] also investigated hydrogen HCCI combustion with port fuel injection (PFI) and direct injection (DI) at 1000 rpm, using optical access to study in-cylinder phenomena through an extended piston and optical crown. To achieve HCCI combustion, air and fuel were mixed and sent into the engine cylinder and the autoignition occurred on its own instead of using a spark. The engine was unthrottled and load was controlled by regulating the mixture quality. Thus, it was possible to avoid big fluid dynamic losses in air intake of a traditional spark-ignited engine. A high-speed camera at 1 or 2° CA resolution was installed to capture combustion images for a series of engine cycles. Techniques such as exhaust gas recirculation or intake air preheating are well-known techniques for varying mixture temperature and obtaining HCCI combustion. However, backfire has been a problem of concern for a fast-burning fuel like hydrogen with wider flammability range and high octane number. In the course of testing, it was necessary to have direct hydrogen injection after intake valve closure to control backfire. Closed-valve direct injection coupled with air-preheating and negative valve overlap to ensure internal EGR could ensure effective HCCI hydrogen combustion even at a low compression ratio.

Concluding Remarks and Perspectives

Broadly speaking, laser diagnostics are part of optical techniques, using lasers as active light sources to gather relevant data. While all laser diagnostics are optical in nature, not all optical techniques use lasers, and they capture light emitted or scattered by combustion. Over the past several years, optical techniques and laser diagnostics have revolutionized engine combustion studies by providing real-time, nonintrusive methods to measure many significant factors related to complex combustion processes. These advanced techniques capture real-time visualization of the detailed combustion event and provide guidelines to adopt optimum combustion strategies. Laser diagnostics possess the distinctive characteristics to provide high-resolution, spatial, and temporal data without interfering in the combustion process itself.

Many of these techniques have matured and some are being developed and are in the experimental stage. Suitable areas of application of some of these techniques are still being studied. Methods such as laser-induced fluorescence (LIF), particle image velocimetry (PIV), and Rayleigh scattering are mature technologies. For example, Raman scattering was widely used in the past for its simplicity. Now, with powerful lasers like Nd-YAG and femtosecond lasers, techniques such as coherent anti-Stokes Raman spectroscopy (CARS) and LIF have become essential. New developments, such as metamaterials, are likely to further improve these diagnostic techniques. Among recently developed techniques, some have experimented with engine combustion research on conventional fuels. Only very few of these have been applied to hydrogen engine combustion. In order to get the utmost benefit from these advanced techniques, a holistic approach is essential. The problem needs to be studied at different levels, starting with generic development to establish fundamental principles, subsequent phenomenological studies to understand the phenomena, and last but not the least, its potential to implement and validate the developed concepts in practical application. For a sustainable fuel like hydrogen, it is vital to study combustion phenomena by delving into chemical kinetics, computational fluid dynamics (CFD), and heat transfer. Advances in instrumentation, driven by laser technology developments, have played a key role. PLIF has been widely used in hydrogen combustion research and is considered a mature and effective technique for understanding combustion characteristics. Visualization of shockwaves, flame front, and flow pattern through schlieren imaging has lots of potential for studying hydrogen combustion in engines. PIV has been successfully applied to provide insights into flow dynamics. Raman high-speed imaging, infrared imaging, and UV-visible spectroscopy are being successfully applied to visualizing fast processes, combustion dynamics, and flame behavior.

At this stage of research and development, it is essential to identify optimum application of specific techniques based on the objective and engine configuration. Application of optical techniques in engine research is indispensable for refining combustion models, guiding the design of advanced combustion systems, and ultimately contributing to the evolution of more environmentally sustainable and fuel-efficient engines.

References

1 Kaiser, S. and White, C M. (2009). PIV and PLIF to evaluate mixture formation in a direct-injection hydrogen-fuelled engine. *SAE International Journal of Engines* 1 (1): 657–668. 2008-01-1034 P 657-668. Paper Number: SAND2008-2318C.

2 Miles, P.C., Hildingsson, L., and Hultqvist, A. (2007). The influence of fuel injection and heat release on bulk flow structures in a direct-injection, swirl-supported diesel engine. *Experiments in Fluids* 43: 273–283.

3 Reeves, M., Garner, C.P., Dent, J.C. et al. (1996). Full-field IC engine flow measurement using particle image velocimetry. *Optical Engineering* 35 (2): 579–587.

4 Bevan K. and Ghandi, J.B. (2004). PIV measurements of in-cylinder flow in a four-stroke utility engine and correlation with steady flow results, SAE Technical Paper No. 2004-32-0005.

5 Buhl, S., Gleiss, F., Köhler, M. et al. (2017). A combined numerical and experimental study of the 3D tumble structure and piston boundary layer development during the intake stroke of a gasoline engine. *Flow, Turbulence and Combustion* 98 (2): 579–600.

6 Zhao, F., Ge, P., Zhuang, H. et al. (2018). Analysis of crank angle-resolved vortex characteristics under high swirl condition in a spark-ignition direct-injection engine. *Journal of Engineering for Gas Turbines and Power* 140 (9): 092807.

7 Stiehl, R., Bode, J., Schorr, J. et al. (2016). Influence of intake geometry variations on in-cylinder flow and flow–spray interactions in a stratified direct-injection spark-ignition engine captured by time-resolved particle image velocimetry. *International Journal of Engine Research* 17 (9): 983–997.

8 Bode, J., Schorr, J., Krüger, C. et al. (2019). Influence of the in-cylinder flow on cycle-to-cycle variations in lean combustion DISI engines measured by high-speed scanning-PIV. *Proceedings of the Combustion Institute* 37 (4): 4929–4936.

9 Scarcelli, R., Wallner, T., Salazar, V.M. et al. (September 2010). CFD and optical investigations of fluid dynamics and mixture formation in a DI-H2ICE. ICEF2010-35084, ASME 2010 Internal Combustion Engines Division Fall Technical Conference (ICEF), San Antonio, Texas, USA.

10 Scarcelli, R., Wallner, T., Matthias, N. et al. (2011). Mixture formation in direct injection hydrogen engines: CFD and optical analysis of single- and multi-hole nozzles. *SAE International Journal of Engines* 4 (2): 2361–2375.

11 Tritton D.J. (1977). *Physical Fluid Dynamics*, Section 22.7, The Coanda Effect. Springer Dordrecht. https://doi.org/10.1007/978-94-009-9992-3.

12 Heindl, R., Gerbig, F., Heller, K. et al. (2009). New and innovative combustion systems for the H_2–ICE: compression ignition and combined processes. *SAE International Journal of Engines* 2 (1): 1231–1250.

13 Blotevogel T., Egermann J., Goldlucke J. et al. (2004). Developing planar laser induced fluorescence for the investigation of the mixture formation process in hydrogen engines. SAE Technical Paper No. 2004-01-1408.

14 Kirchwegen, W., Haslacher, R., Hallmannsegger, M. et al. (2007). Applications of the LIF method for the diagnostics of the combustion process of gas-IC-engines. *Experiments in Fluids* 43: 329–340.

15 Salazar, V., Kaiser, S., and Halter, F. (2009). Optimizing precision and accuracy of quantitative PLIF of acetone as a tracer for hydrogen fuel. SAE Technical Paper No. 2009- 01-1534.

16 Salazar, V. and Kaiser, S. (2009). An optical study of mixture preparation in a hydrogen-fueled engine with direct injection using different nozzle designs. SAE Technical Paper No. 2009-01-2682.

17 Salazar V. and Kaiser S. (2010). Influence of the in-cylinder flow field (tumble) on the fuel distribution in a DI hydrogen engine using a single-hole injector. SAE Technical Paper No. 2010-01-0579.
18 Tomita, E., Hamamoto, Y., Yoshiyama, S. et al. (1998). Measurement of fuel concentration distribution of transient hydrogen jet and its flame using planar laser induced fluorescence method. *JSAE Review* 19 (4): 329–335.
19 Arnold, A., Bombach, R., Käppeli, B. et al. (1997) Quantitative measurements of OH concentration fields by two-dimensional laser-induced fluorescence. *Applied Physics* B 64 (5): 579–583.
20 Aleiferis P.G. and Rosati M.F (2012). Flame chemiluminescence and OH LIF imaging in a hydrogen-fuelled spark-ignition engine. *International Journal of Hydrogen Energy* 37 (2): 1797–1812.
21 Salazar, V. and Kaiser, S. (2011). Interaction of intake-induced flow and injection jet in a direct-injection hydrogen-fueled engine measured by PIV. SAE Technical Paper No. 2011-01-0673.
22 Phuoc, T.X. (2006). Laser-induced spark for simultaneous ignition and fuel-to-air ratio measurements. *Optics and Lasers in Engineering* 44 (6): 520–534.
23 Shudoa, T. and Oba, S. (2009). Mixture distribution measurement using laser induced breakdown spectroscopy in hydrogen direct injection stratified charge. *International Journal of Hydrogen Energy* 34 (5): 2488–2493.
24 Hunter, A.J. and Piper, L.G. (2002). Spark-induced breakdown spectroscopy: a description of an electrically generated LIBS-like Process for Elemental analysis of airborne particulates and solid samples. In: *Laser Induced Breakdown Spectroscopy* (eds. W.A. Miziolek, V. Palleschi, and I. Schechter), pp. 585–614. Cambridge, UK: Cambridge University Press.
25 Roy, M.K., Nobuyuki, K., Tomita, E. et al. (2013). Jet-guided combustion characteristics and local fuel concentration measurements in a hydrogen direct-injection spark-ignition engine. *Proceedings of the Combustion Institute* 34 (2): 2977–2984.
26 Rahman, M.T.A., Kawahara, N., Tsuboi, K. et al. (2014). Visualization and concentration measurement of a direct-injection hydrogen jet in a constant-volume vessel using spark-induced breakdown spectroscopy. *International Journal of Hydrogen Energy* 39 (31): 17896–17905.
27 Rahman, K M., Kawahara, N., Matsunaga, D. et al. (2016). Local fuel concentration measurement through spark-induced breakdown spectroscopy in a direct-injection hydrogen spark-ignition engine. *International Journal of Hydrogen Energy* 41 (32): 14283–14292.
28 Oikawa, M., Ogasawara, Y., Kondo, Y. et al. (2012). Optimization of hydrogen jet configuration by single hole nozzle and high speed laser shadowgraphy in high pressure direct injection hydrogen engines. *International Journal of Automotive Engineering* 3 (1): 1–8.
29 Salazar, V. and Kaiser, S. (2011). Influence of the flow field on flame propagation in a hydrogen-fueled internal combustion engine. SAE Technical Paper No. 2011-24-0098.
30 Aleiferis, P. and Rosati, M. (2012). Controlled autoignition of hydrogen in a direct-injection optical engine. *Combustion and Flame* 159 (7): 2500–2015.
31 Rosati, M. and Aleiferis, P. (2009). Hydrogen Si and HCCI combustion in a direct-injection optical engine. *SAE International Journal of Engines* 2 (1): 1710–1736. https://doi.org/10.4271/2009-01-1921.

7

Design Criteria and Safety Features of a Dedicated Hydrogen Engine

I have not failed. I have just found 10,000 ways that won't work.

Thomas Edison

7.1 Introduction

Most of the research and development activities on the use of hydrogen in internal combustion engines involve converting existing conventional engines to hydrogen operation. By way of extensive studies, it has been possible to identify which parts of existing engine hardware need to be changed or redesigned, and what operating parameters are desirable to ensure safe, prolonged hydrogen operation based on hydrogen properties and engine operating conditions. A "hydrogen-specific engine" or "dedicated hydrogen engine," in the context of this chapter, refers to an engine specifically designed for the typical combustive features of hydrogen. An optimum hydrogen-specific engine would maximize efficiency and power output and establish safe environmental sustainability. Engine design discussed here addresses the development of different components while avoiding hot spots leading to backfires. Certain infamous incidents related to explosive character of hydrogen bombs—the Hindenburg disaster, destruction of hydrogen-inflated airships LZ-4 initiated on 5 August 1908 and occurring thereafter intermittently till 1937, the Challenger space shuttle explosion of 1986, and the Fukushima Nuclear tragedy of 2011—have attached an undeserved stigma to hydrogen as an unsafe fuel. On the other hand, NASA, since inception in 1958, used hydrogen as rocket fuel, as did Centaur, Apollo, and several space shuttle vehicles. Recent human space missions to launch projects to the moon and subsequently to Mars were planned with innovative liquid hydrogen storage. Hydrogen, with an approximate annual production of over 75 million metric tons in several industrial sectors, is being safely carried out for long periods under carefully controlled conditions by skilled operating personnel with the knowledge to handle properly hydrogen's distinctive and temperamental combustion properties.

Different combustion problems in hydrogen engines manifest their influence in diverse ways, affecting overall engine performance. Sometimes one of these problems gives rise to another. Severe thermal environments have been one of the major causes of these effects at specific locations during engine cycle operation. Most of the research activities on hydrogen engines have focused on reducing the temperature level only in the cylinder. Preignition is a random event. It has not been possible to measure accurately the time, place, and intensity of preignition. In the case of hydrogen engines, preignition often leads to backfire.

Hydrogen Engines: Design, Performance Evaluation, Combustion Analysis, and Exhaust Emissions, First Edition.
Lalit Mohan Das.

Abnormal combustion in hydrogen engines has been observed to be a complex combination of preignition, end-gas knock, and backfire. Eliminating these problems requires a complete systems approach of engine design, development, and calibration. Some catastrophic accidents have occurred due to fire, explosion, or ignition system failure. But many of these accidents have taken place largely due to human error, where the impact of hydrogen's combustion characteristics like low minimum ignition energy, wide flammability limit, and high flame speed have been inadvertently ignored. Accidents have been mainly due to inadequate ventilation, undetected leaks, low purity, reaction to air pollutants, failure of safety devices, mechanical failure resulting in embrittlement, etc. Of course, now the overall perception has changed substantially. Enough relevant technical information is available on properties related to safety, character and classification of hazards, and possible safety threats at different stages such as production, transmission, storage, and application [1–9]. Leak detection techniques and sensing devices have been developed and located in sensitive locations. An elaborate discussion of these aspects of safety falls beyond scope of this book, and description here is restricted to safety features in internal combustion engines. For hydrogen-powered vehicles based on internal combustion engines, the complex range of safety stages include safe production, transmission, mode of storage, method of fueling, and appropriate infrastructure at the point of utilization. Safety valves, pressure relief devices, high-pressure regulators, and long tubing extending up to several meters are vulnerable to leakage and need to be checked.

7.2 Safety-Related Properties for the Hydrogen Engine

Thermodynamic properties of hydrogen are given in Chapter 2 compared to common engine fuels including methane and gasoline. Diffusivity, flammability range, ignition energy, ignition temperature, explosion energy, flame temperature, etc. are more are relevant in the context of safety. Hydrogen has a flammability range of 4–75% by volume in air, which is the widest among common gaseous fuels. In the event of a leak or release, if hydrogen reaches the ignition source, it is more likely to be within the flammable range. Of course, hydrogen's buoyancy and diffusivity result in rapid dispersion of gas and reduction of the concentration of the flammable zone.

In view of hydrogen's low minimum ignition energy (0.02 mJ, ordinary static electricity discharges, even those generated by walking across a carpeted room, are enough to ignite hydrogen [10]. Sudden, high-flow hydrogen discharge through relief valves or burst discs can lead to ignition. The autoignition temperature (AIT) of a hydrogen/air mixture is around 585 K. In practice, AIT can vary depending on how the leaking hydrogen or fuel/air mixture is heated. Various nonenclosed and nonadiabatic heating situations, such as hot engine exhaust manifolds or shorted wires, can significantly influence AIT and flammability. The nearly invisible flames of hydrogen have been a safety concern in many situations. Sometimes the diffraction patterns caused by heat may provide some visual indication of the presence of flame, but the flames are not as conspicuous as those produced by conventional fires. The hydrogen fire burns very rapidly, but due to its low emissivity, it radiates very little heat, which stays for a relatively short duration. A person can be closer to a hydrogen fire than to a gasoline fire without being burned. Hydrogen can be explosive at oxygen concentrations between 18 and 59%, whereas gasoline can explode at oxygen concentrations between 1 and 3%. In oxygen-rich environments, gasoline poses a greater explosion risk than hydrogen. Swain et al. [11] studied hydrogen gas leakage in buildings from a computer model. The gas cloud shape and size was predicted using FLUENT 3.03. This study examined how building geometry and passive ventilation affect the formation of combustible gas clouds. It also discussed the leakage rates and duration for which hydrogen can be ventilated from a well-designed building without generating a significant amount of combustible fuel–air mixture.

7.3 Technical Features for Hydrogen Engine Design

Hydrogen engine design involves a multidisciplinary approach considering thermodynamics, materials science, control systems, and safety engineering to achieve desired performance and emissions goals. Actual tests provide guidelines to validate and also suggest further refinement in design wherever necessary. The design features of a hydrogen-specific engine should take into account several factors such as combustion chamber design, ignition system, injection system, cooling system, exhaust system, advanced control system, materials, and lubrication.

Major aspects related to desirable features of an optimum hydrogen engine's performance, emission, combustion systems, along with the model-based analysis, have been discussed in earlier chapters. On the basis of that information, some of the salient guidelines for an optimum hydrogen engine design are highlighted here. The engine should have a high-pressure hydrogen injection system to deliver an appropriate amount of gaseous hydrogen at the optimum point in the engine cycle into combustion chambers to ensure precise hydrogen control flow. Hydrogen has a wide flammability range, but it requires a strong and consistent ignition source. Optimal ignition systems, such as spark plugs or laser igniters, should be used. A higher compression ratio can improve the efficiency of hydrogen engines. Hydrogen combustion generates high temperatures, and so an efficient cooling system is crucial to prevent overheating to ensure the engine's durability. Exhaust systems need to be designed to handle high-temperature, high-velocity combustion gases. Since hydrogen embrittlement has been a problem with some materials, appropriate hydrogen-compatible materials are critical for long-term reliability. Lubrication systems are appropriately designed to handle hydrogen gas. Choice of combustion chamber design and optimization strategies should consider the influence of flame propagation, turbulence, stratification, and other fuel-specific parameters on an engine's operating conditions. The combustion chamber should withstand hydrogen's fast burn rate and high flame speed. Features such as swirl, squish, and multipoint injection can enhance mixing and efficiency. A compact and efficient combustion chamber design helps maximize power output and minimize emissions. CFD simulations are a valuable tool for predicting optimized combustion behavior. Significant combustion parameters such as flame propagation rates, temperature gradients, and in-cylinder pressure dynamics can be analyzed by modeling the hydrogen flame development. Performance characteristics can also be improved by CFD simulation as it can predict the laminar burn rate of a hydrogen–air mixture with a high level of accuracy.

Mostly, hydrogen fuel for engine research uses high-pressure tanks that store hydrogen at 200 bar and 350 bar. Now tanks have been developed to store hydrogen at 70 MPa (700 bar) pressure. For the sake of safety, storage systems should have tanks with in-tank isolation solenoid valves. In engine test rigs, high-pressure regulators should be used to reduce gas pressure to inject gas into the engine. Sensors need to be installed at appropriate locations in the engine area to detect accumulation of hydrogen gas, and the sensors' input is given to the ECU where it is processed and used for controlling the air, fuel, ignition timing, injection timing, and also for engine load and throttle control. In their investigation, Sopena et al. [12] upgraded the electronic control unit (ECU) in the engine test rig from the original to a programmable unit that had the capability to fine-tune and customize engine parameters for a lean mixture. Some modification in sensor setup was tailored to facilitate wide operating range and to meet safety testing requirements. The coefficient of heat dissipation of hydrogen being more than that of CNG and gasoline fuel, heat is carried away by coolant for hydrogen operation. Keeping this in mind, researchers replaced the original lambda sensor with a wideband lambda sensor because it provided a broader range of air–fuel ratio measurements, allowing for more coolant flow as required. The original binary sensor was restricted to controlling mixtures close to stoichiometric conditions.

7.4 Some Property-Based Benefits and Challenges

Hydrogen's high octane number and autoignition temperature enable engine operation at higher compression ratios and thus achieve thermal efficiency and power output. Moreover, its high flame temperature raises efficiency, while the high calorific value improves fuel economy. While hydrogen's low ignition energy and wider flammability limit can enhance efficiency, they also increase the risk of uncontrolled ignition leading to preignition and backfire. Therefore, a rational and judicious application of these properties is essential to prevent undesirable combustion phenomena. Sometimes even minute electrical leakage can cause fire. Backfire and preignition can be eliminated by an appropriate fuel induction technique, and a suitable ignition system can prevent autoignition due to hot spots on and around the spark plug. Stoichiometric hydrogen combustion has high thermal loads on the piston top land and ring grooves. Thus, the first ring should preferably have a hard-anodized coating as well as a ring pack. This set-up is fairly effective in reducing blow-by gas and oil consumption.

7.4.1 High Flame Speed

Hydrogen's high flame speed brings the engine operating cycle closer to the thermodynamic cycle. Better flame stability, accelerated combustion, and high power output even in leaner operating conditions are some of the distinct positive features. Hydrogen's low density occupies about 30% of the combustion space and thus reduces engine power output. Moreover, strategic control of high reactivity and flammability are essential criteria for engine design. Engine operation close to or beyond stoichiometric conditions gives rise to abnormal combustion and generates high NOx levels. Engine operation restricted to an equivalence ratio of about 0.6 avoids preignition and also results in lower NOx emissions. A wider range of flammability limits enables lean engine operation with improved efficiency and reduced NOx emission.

7.4.2 Quenching Distance

Hydrogen's very low quenching distance is one of its salient properties due to which hydrogen flames travel closer to cylinder walls before being extinguished. Hydrogen flames burn rapidly due to their high laminar flame speed, and so the combustion process from ignition to completion occurs very quickly. In an internal combustion engine, the piston–cylinder gap is crucial because, in the absence of sufficient residence time, the short-lived hydrogen flame results in incomplete combustion in the piston–cylinder gap. This often leads to lubricant evaporation.

7.4.3 Diffusivity and Throttling

While hydrogen's high diffusivity facilitates homogenization of the mixture, at the same time it creates difficulty in achieving stratification of the mixture. Similarly, an unthrottled engine avoids pumping loss, but it sometimes causes misfire in low load and idling conditions.

7.4.4 Hydrogen Leaks

Hydrogen fuel lines and systems are more susceptible to leakage compared to other gaseous fuels. Gas joints are a vulnerable location for leakage, and seals should be used on vulnerable parts to prevent hydrogen leakage. Common hydrogen leaks fall into two categories: those involving metal-to-metal seals (e.g. flared or compression joints) and others involving nonmetal seals (e.g. gaskets, packing, and pipe-thread compounds). Chances of leak are relatively less in properly designed leak-checked metal-to-metal joints. However, though factors like thermal expansion and contraction, vibration, and externally applied forces can cause leaks, they are seldom catastrophic. Nonmetal seals may suddenly produce more significant leaks as the flowing leaking gas displaces the sealant material. Brazing material should have a melting point over 538 °C to withstand hydrogen operating conditions [13, 14]. For leaks involving liquid hydrogen (LH_2), vaporization of cold vapor hydrogen to the atmosphere provides a warning because moisture condenses and forms fog. Materials with low melting temperatures like metals, plastics, and elastomers should be avoided in hydrogen fuel line systems to prevent failure and hydrogen release during a fire [15]. Heat from exhaust systems should be kept away from heat-sensitive components to avoid damage or failure.

7.4.5 Octane Number and Knock

Elaborate discussions on several undesirable combustion phenomena are made in Chapter 4 that provide relevant technical guidelines to avoid these problems. A brief discussion is made here on an experimental and numerical study of hydrogen's autoignition, knock, detonation, and the octane rating. Poursadegh et al. [16] carried out an experimental and numerical investigation into hydrogen engine combustion and reported some significant results on octane rating and knock of hydrogen. The numerical analysis approach involved a combination of a calibrated, two-zone combustion modeling and kinetic modeling of the end gas for a numerical analysis. Investigation of the standard research octane number (RON) method indicated a RON of hydrogen in the range of 62–64, which was unusually low. This standard RON method was applicable to very rich hydrogen operation. Moreover, the combustion process happened rapidly, with 50% mass fraction burned (MFB50) occurring before top dead center. Since such features are not at all applicable in production engines, the researchers adopted a modified RON method that was representative of practical engines. Adopting modified tests for hydrogen with $\lambda = 1$, 1.5, and 2, the RON (modified) was observed to be 93.7, 117, and >120, respectively. Autoignition with rapid pressure rise at λ values of 1.5 and 2 under the modified RON condition could be effectively controlled by using small amounts of tetraethyl lead (TEL). There was no rapid rate of pressure rise in engine operation with the stoichiometric modified RON condition, and there was practically no impact of TEL on the combustion. Studies on the modified RON stoichiometric condition showed some distinctly different features from analysis by standard knock intensity method. In stoichiometric conditions, detonation occurred with a compression ratio (CR) greater than the standard knock intensity (SKI). High turbulent flame speeds in the range of 15–20 m/s occurred for stoichiometric mixtures with rapid ignition near the end of combustion. In contrast, modified tests at leaner air–fuel ratios ($\lambda = 1.5$ and 2) showed relatively much slower flame speeds, allowing more time for autoignition to take place during combustion. Results of this study indicate that hydrogen possesses significantly higher resistance to knocking compared to standard gasolines. On the basis of their analysis, the researchers suggested that the existing ASTM octane standards are not well suited for assessing hydrogen-rich fuel blends. The assessment method for evaluating methane number, which assumes that hydrogen is prone to knocking, does not give a correct picture. This phenomenon is sometimes mistaken as autoignition. Such misinterpretation and confusion that might occur when RON is measured as per ASTM standard can be avoided by developing specific methods tailored for hydrogen-rich fuels.

7.5 NOx Emission

Thermal dilution techniques like exhaust gas recirculation (EGR), water injection, SCR, ensuring homogeneity and stratification of the mixture, lean operation, etc. as discussed in the literature have been carried out on experimental research engines with attention focused on NOx control. In a particular test, sometimes there are intrinsic limitations. It is quite likely that there are operating or environmental constraints that restrict experiments to narrow ranges. Therefore, these studies give a partial picture only with reference to some specific conditions even though the test results identify the gaps in a particular study. Thus, a comprehensive data analysis is necessary to formulate rational design conditions.

7.6 Load Control Strategy and Engine Components

Improved efficiency, low NOx emission, and smooth combustion are the basic governing parameters for load control strategy. Power penalty has to be taken into account while operating the engine at a constant equivalence ratio under a throttled condition. Qualitative control is advantageous for hydrogen operation. A wide-open throttle gives the advantage of increased efficiency and should be adopted for idling and at very low loads, at very lean mixtures. A rich mixture leads to higher NOx emission. Lean operation of a hydrogen engine leads to improved thermal efficiency and low NOx emission. Therefore, the load control strategy has to be carried out as per the need of the situation. There are several engine components that need to be judiciously selected for hydrogen application. They may not make a radical change in the overall exterior structure of the engine. Nevertheless, their use in the appropriate location makes a big difference. Broad design features need to consider the spark plug, ignition system, hot spots, injection system, compression ratio, electronic throttle, piston rings and crevice volume, valve seats, lubrication, in-cylinder turbulence, etc.

7.6.1 Spark Plug

Selecting the right spark plug with the appropriate heat rating is essential for hydrogen engine operation. Capacitive spark discharge is a technique used to measure the smallest amount of energy required to ignite a mixture of a fuel and an oxidant. In this method the energy stored in the capacitor is released as a spark, and if the spark has enough energy, it can ignite the substance being tested. The size of the spark gap plays a crucial role: a smaller spark gap requires less energy to create an ignition, while a larger spark gap will require more energy. It has been reported [17, 18] that cold spark plugs need to be used in hydrogen engines so that spark plug electrodes avoid reaching autoignition temperature to avoid conditions leading to backfire. Cold spark plugs, with a longer insulator nose and more extended heat transfer path, dissipate heat more efficiently from the spark plug tip to prevent premature combustion. Since there are hardly any spark plug deposits during hydrogen combustion, cold spark plugs effectively dissipate heat from the combustion chamber and maintain optimum temperature. Spark plugs with a platinum electrode should be avoided, as this can be a catalyst to hydrogen oxidation [19, 20]. Platinum, on the other hand, has been used to oxidize unburned hydrogen in the exhaust [21].

7.6.2 Ignition System

An ignition system should designed to minimize the risk of uncontrolled ignition due to any remaining electrical energy. Sometimes in ignition systems there could be residual energy in some components such as a spark plug or ignition cable. Proper grounding is required to dissipate excess electrical energy into the ground to avoid electrical shocks, fires, or sparks. Flow of electrical energy within the ignition system can also be controlled by using a coil-on-plug system to provide improved ignition performance and reduce the likelihood of interference or induction ignition in adjacent ignition cables [22]. Because of the lower ion concentration of a hydrogen flame, ignition of hydrogen mixtures requires an increased secondary ignition voltage, which can be provided by a high-voltage output ignition system [23, 24]. A coil-on-plug (COP) ignition system satisfies this condition. As an alternative, ignition voltage can be decreased by narrowing down the spark plug gap. A gap of 0.25 mm has been effective for the ignition system [25]. However, in some cases, due to water condensation at the spark plug tip, cold-start problems occur, and it becomes necessary to increase the spark plug gap again.

7.7 Emerging Ignition Technologies

Novel techniques of ignition such as laser-induced ignition, laser-induced plasma ignition, microwave-assisted plasma ignition, radio frequency sustained plasma ignition system, nano-second repetitive pulse discharge (NRPD) ignition, positive corona ignition, and pre-chamber spark ignition systems are emerging fast. Many of these improved ignition systems have brought substantial improvement by increasing the ignition energy and creating multiple ignition sites and have achieved stable combustion with lean or ultra-lean mixtures. Nevertheless, these studies have mostly been on engines with conventional fuels and only a very few on engines with alternative fuels. There is hardly any study on hydrogen engines. In laser-induced ignition (LII), a high-energy laser pulse is used to create a spark within the combustion chamber. The laser energy generates plasma, which serves as the ignition source. Laser-induced ignition can deliver high energy at the precise location and leads to consistent ignition. Laser-induced plasma ignition (LIPI), like LII, uses a high-energy laser pulse to generate plasma within the combustion chamber, which creates a stable and sustained ignition source for an extended period. LIPI generates a high-temperature spark of intense ignition source which is relatively short-lived. A radio frequency sustained plasma ignition system uses radio frequency (RF) energy to generate and sustain a plasma kernel within the combustion chamber, which offers stable and controllable ignition. Microwave-assisted ignition uses microwaves to create highly ionized and stable plasma.

Figure 7.1 (a) Standard spark plug. (b) Radio frequency spark plug. (c) Radio frequency spark plug with a single electrode and with five electrodes [26, 27] / with permission of Elsevier.

Joonsik et al. [26] studied the effect of microwave-assisted plasma ignition on laminar flame development in a constant volume combustion chamber and observed that the lean limit was extended to 0.5 compared to 0.6 for conventional spark ignition. High-speed shadowgraph imaging was captured for early flame kernel development near the spark plug. The flame kernel sizes from the shadowgraph images of the spark ignition system and the microwave-assisted plasma ignition system were compared. Microwave ejection, with its faster flame kernel development, resulted in advanced flame rise time and flame development time. As the microwave ejection timing was retarded, enhancement of combustion due to the microwave-assisted plasma ignition showed a decreasing trend. Studies by Mariani and Foucher [27] showed that the plasma continuous ignition system had been particularly useful for engines with high compression ratios or dilute intake gases. The study noted the effectiveness of the radio frequency sustained plasma ignition system (RFSI) with a radio frequency spark plug (RFSP), shown in Figure 7.1. In a standard spark plug, ignition occurs between the central electrode and the ground electrode, driven by high voltage from the secondary circuit of a conventional ignition coil. The radio frequency spark ignition (RFSI) system, instead of relying on a spark, utilizes an alternating, high-voltage electrostatic field to generate excitation of high-energy electronic states and ionization in the combustion chamber. This process significantly increases the number of free radicals, which play a crucial role in initiating exothermic chemical reactions within the air–fuel mixture that activate the combustion process. The RFSI system delivers much more ionizing energy than conventional ignition systems. Compared to a conventional spark ignition system, RFSI shows improved efficiency and ensures much leaner operation with substantial reduction in cycle variation.

Nanosecond repetitive pulse discharge (NRPD) is a technique used to generate radicals by creating plasma, which is a highly ionized and reactive gas. These plasma discharges lead to the production of radicals.

Merotto et al. [28] applied nanosecond repetitive pulse discharge (NRPD) ignition in their studies on methane–air mixture. NRPD uses high voltage of nanosecond duration electrical pulses across a spark plug electrode that generate multiple sparks in quick succession. Radicals generated in NRPD are chemically reactive species that ensure stable ignition especially in lean laminar conditions due to the increased local concentration of radicals provided by multiple discharges. Thus, there are multiple instances of a nonthermal plasma discharge occurring in a specific region of a combustion chamber or system. NRPD ignition is advantageous when the combustion environment is diluted and also when there is low mixing of air and fuel, possibly due to reduced turbulence in the combustion chamber.

Positive corona ignition is like a low-energy spark ignition process in which the electric field generated by a high voltage causes the surrounding air to ionize, creating a reactive plasma zone that initiates combustion in the engine's air–fuel mixture. It does not necessarily lead to sparks or a complete electrical discharge. This method relies on the gradual buildup of charge on the electrode tip. Even though the method is not combustible, it can be used to ignite some explosive gases. Positive corona ignition reduces the possibility of misfires. The advanced corona ignition system (ACIS) is an alternative suitable option for high-energy ignition for lean mixtures. The ACIS system has some distinct features as compared to a conventional ignition system. It is useful mainly for a lean combustion mixture with lower HC emission and improved cyclic variation at conditions of low load and idling speeds. It does not have any influence on other exhaust gases, and the benefits of the ignition system decline at higher loads [29]. Dale and co-workers [30] discussed the method of inducing a spark kernel motion by the use of a corona wind. This was achieved by designing the spark plug using a voltage potential lower than the breakdown voltage of the gap in the spark plug. This corona wind carried the flame kernel away from both electrodes, reducing heat transfer from the flame to the electrodes. Results of this investigation on a four-cylinder automobile engine also showed that ignition of a lean mixture improved cyclic variation and reduced HC emission by about 25% even though there was practically no effect on other exhaust emissions. The system showed improvement in ignition efficiency and ignition of lean mixtures and was found to improve cyclic variations. However, the ignition system showed a visible declining trend with the increase of engine load and cylinder pressure.

Some recent advanced technologies have been found to improve ignition and combustion in conventional and alternative-fueled engines. Detailed discussions with respect to optical techniques and laser ignition applied to hydrogen combustion are discussed in Chapter 6. A brief description is given below on laser-induced ignition that was investigated on hydrogen engines. Böker and Brüggemann [31] carried out investigations on laser-induced ignition in hydrogen engines to assess the improvement on combustion mainly from an emission point of view. When a high-energy laser pulse on a small area within a fuel–air mixture heated the surrounding gases, it created a localized hot spot. This spot reached the autoignition temperature and spontaneously ignited to form a small flame kernel. The rich energy in the kernel caused ignition even in lean mixtures, and the resulting flame continued to propagate in the absence of external ignition sources. Thus, it was possible to localize the laser ignition to ensure an adaptive control strategy for combustion and to achieve optimum ignition. The researchers also observed that it was possible to control the flame propagation by positioning the optimized ignition. An optimum configuration of a hydrogen engine can be developed by interrelating the laser spark ignitions and mixture formation with the injection system [31]. Azarmanesh and Targhi [32], in their comparative analysis using a quasi-range multi-area method, observed that a laser ignition system had a shorter combustion duration than that of spark plug ignition at various equivalent ratios. At an equivalent ratio of 0.7, laser ignition demonstrated an improvement of thermal efficiency by 8.13% over spark ignition. However, there was relatively much less improvement in thermal efficiency at conditions of the stoichiometric equivalence ratio. This study applied to methane–air combustion in different equivalence ratios can give some broad qualitative trend of results with hydrogen operation.

As far as ignition is concerned, there are a number of unusual technologies that can be useful in the design of the hydrogen engine because of its potential of lean operation with improved efficiency and less emission.

7.8 Fuel Delivery System

Modes of fuel induction techniques such as PFI, TMI, and direct injection were elaborately described in Chapters 3 and 4. The results of each study have been analyzed on the specific engine test rig, and every test rig has its own flexibilities and limitations. In both port fuel injection and

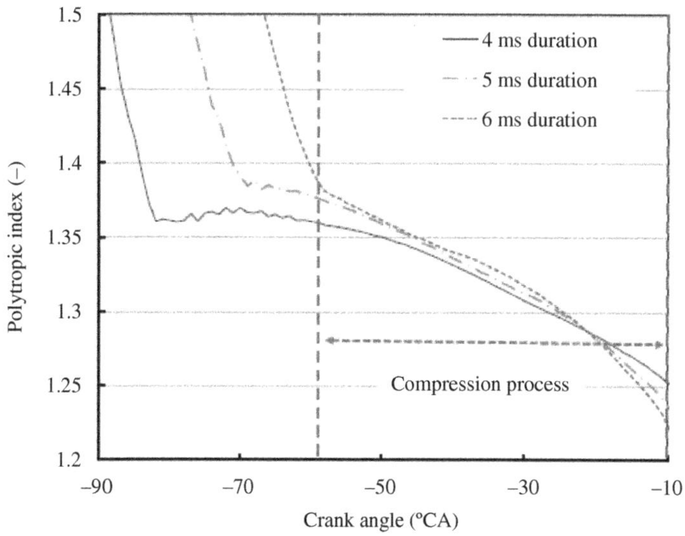

Figure 7.2 The polytropic index changes with the crank angle for various injection durations [33].

manifold injection, regardless of their location and number of injectors, fuel and air are mixed before they enter the combustion chamber. A timed manifold injection (TMI) can be designed so that hydrogen flow commences after opening the intake valve but completed prior to intake valve closure. Thus, the intake manifold does not contain any combustible mixture and the chance of backfire is almost eliminated. PFI also has to time the injection so that air cooling is actuated in the initial phase of the intake stroke, and all the hydrogen needs to be inducted by the end of injection. Direct cylinder injection at the end of the compression stroke means the injector will encounter a severe thermal environment in the combustion chamber, and the possibility of injector leaking is high. Moreover, it is difficult for a homogeneous mixture to be formed at the end of the compression stroke. All these aspects are to be taken into account for an appropriate design of the injection system. Lingzhi et al. [33] investigated the impact of injection duration and injection timing and speed on the polytropic index (PTI) of a direct-injection hydrogen engine. During their tests it was observed that the polytropic index dropped from 2.2 to 1.22 with changes in the crank angle. The polytropic index, as is well known, is an indicator of how pressure and volume change during a thermodynamic process, and it varies during the operation of an engine. The polytropic index is sensitive to the concentration of the hydrogen.

In Figure 7.2, the polytropic index of the direct injection hydrogen engine initially drops down significantly to around 1.37. Following a turning point that corresponds to the end of injection at a specific crank angle, the index stabilizes briefly and then gradually decreases to a minimum of about 1.22. The compression process range spans from a –60° crank angle (CA) to –10° CA, considering the injection (6 ms) and sparking timings. At the onset of compression (–60° CA), the polytropic index for a 6 ms injection duration is slightly higher than others because of the addition of hydrogen. The incompressibility of hydrogen leads to higher compression pressure. However, it subsequently drops to a minimum of 1.23 at a spark angle of –10° CA.

Figure 7.3 presents a comparison of the polytropic index between port fuel injection (PFI) and direct injection (DI). During the compression stroke, spanning from the intake valve closing

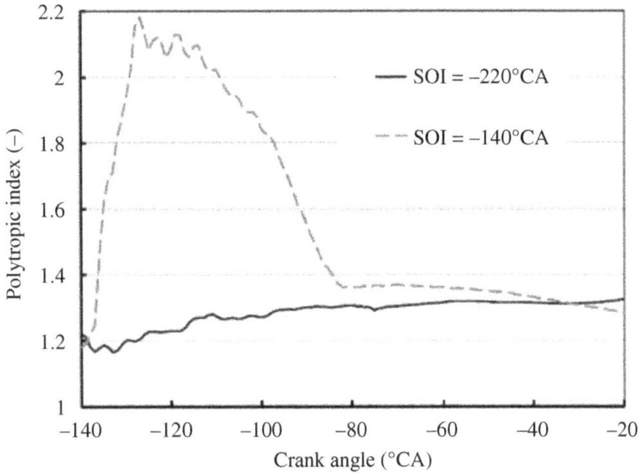

Figure 7.3 Polytropic index comparison between port fuel injection and direct injection in a hydrogen engine [33].

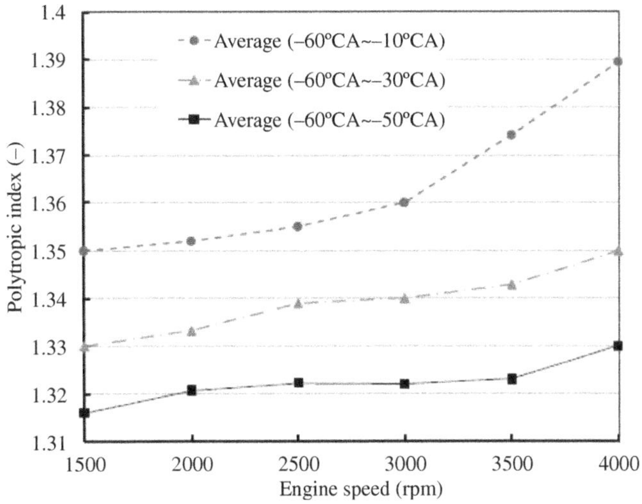

Figure 7.4 Polytropic index changes with the engine speed for different average calculation windows [33].

(–140° CA) to the spark angle (–20° CA), the polytropic index of the PFI engine gradually increased from 1.2 to 1.31. The DI engine exhibited a dramatic initial increase and reached a peak at 2.2, and then, after mild fluctuation, it decreased to –90° CA, which happened to be the end of injection. Subsequently, it steadily dropped to approximately 1.28, showing a distinctly different pattern. Such a trend clearly brings out the crucial role of the SOI and hydrogen injection inducing intense fluctuation in the polytropic index.

Figure 7.4 shows the large difference in average polytropic index changes with speed. In view of the large difference of short-range calculation, the average polytropic index in the study was approximately 1.32 in the –60° CA to –10° CA span.

7.9 Valves

Injectors and valves together play a critical role in the combustion process. Shape, size, and orientation of intake and exhaust valves play a significant role to ensure efficient air flow and gas exchange process. Valve seats are responsible to start and stop the flow, whereas injectors control the flow into the engine. Hydrogen, in view of its low lubricity, cannot provide adequate lubrication for the moving parts, so the valve seat material should be durable and corrosion resistant to maintain proper sealing to prevent excessive wear. Selection of appropriate material against the problem of embrittlement is also important for some materials, mostly at high-pressure conditions.

7.10 Crankcase Ventilation

Even though the basic function of crankcase ventilation is similar in both hydrogen and conventional engines, major design features for hydrogen engines need adequate ventilation mostly from a safety point of view. Crankcase ventilation plays a critical role in ensuring the safety and longevity of the engine. Any unburned hydrogen leaking through the piston rings poses the threat of unscheduled ignition because of hydrogen's low minimum ignition energy. Properly designed positive crankcase ventilation (PCV) is recommended for unthrottled hydrogen operation (high manifold air pressure) and reducing hydrogen concentrations in the crankcase from blow-by gases. A pressure relief valve must be installed on the valve cover as shown in Figure 7.5 [34].

Ignition within the crankcase could lead to anything from a simple startling noise to engine fire. Therefore, effective ventilation is extremely important to prevent the accumulation of hydrogen within the crankcase. Sometimes, due to some reason if there is an accumulation of water vapor in the crankcase that then mixes with the engine oil, this contaminated engine oil can affect the effectiveness of the lubrication system and lead to engine wear. Moreover, the engine generates oil vapors due to high temperature and pressure in the crankcase. Unless properly vented out, these oil vapors can condense and form sludge.

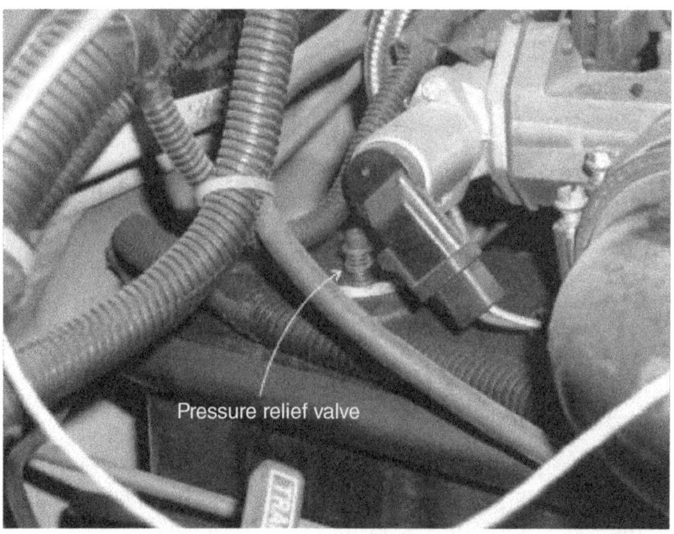

Figure 7.5 A pressure relief valve installed on the engine crankcase [34] / U.S. Department of Transportation.

7.11 Hot Spots

Elimination of backfire due to hot spots is of paramount importance for hydrogen engine design. Tests in the laboratory and a closer view of the literature gives some definite guidelines. Use of cooled exhaust valves, a sodium-filled exhaust valve, or multi-valve engine heads provide additional cooling passages around the valve and avoid hot spots in a combustion chamber. The hollow stem of a sodium-filled exhaust valve is filled with sodium with a relatively low melting point, which allows it to absorb heat from the valve head and transfer it to cooler areas of the engine. A diamond-like coating (DLC) is useful to prevent any scuffing [35] due to inadequate lubrication. Valves and areas with high thermal loads can have additional coolant passages to avoid formation of hot spots. Proper functioning of oil control systems ensures that the engine oil lubricates all critical components, dissipates heat efficiently, and maintains the thermal stability without any hot spot and leads to smooth engine combustion. Avoiding hot spots in a hydrogen engine can also be achieved by delaying fuel induction, which facilitates precooling of air [36]. It has also been reported that hot spots can be avoided by adequate scavenging using variable valve timing, which in turn reduces the residual gas temperatures [37, 38]. Scavenging effectively removes exhaust gases from the combustion chamber during the exhaust stroke. Variable valve timing (VVT) systems, associated with scavenging optimization, allow for precise control of valve timing so that hot spots are avoided by adjusting the timing of the exhaust valve closure and intake valve opening.

7.12 Lubrication System

The lubrication system in a hydrogen engine should be designed based on the unique demands of hydrogen such as high-temperature stability and resistance to oxidation. It should provide excellent lubricity and ensure proper oil circulation, cooling, and filtration. Hydrogen is a clean-burning fuel that produces very little carbonaceous residue, soot, or contaminant particles during combustion and so there is very low wear and tear on engine components, leading to lower lubrication demands in some cases. With reduced lubrication demands, it is crucial to maintain a balance between ensuring adequate lubrication while minimizing energy losses due to friction. In view of its high thermal conductivity, hydrogen effectively transfers heat away from engine components, reducing the risk of overheating and thermal breakdown of lubricants. Hydrogen is a highly reactive gas and sometimes interacts with some materials and lubricants, so lubricants used in hydrogen engines must be carefully chosen to ensure compatibility and prevent chemical reactions that could degrade the lubricant or lead to the formation of harmful by-products. According to Deluchi [39], unlike in gasoline, engine oil is not diluted in hydrogen operation. It is also less likely to form acids that shorten the oil's lifetime. However, this general statement may not be valid over the entire range of engine operation. In fact, at stoichiometric conditions, oxides of nitrogen and water are formed and there is the likelihood of acids being formed. This aspect needs further investigation to have an accurate picture. There are several safety measures that could be adopted while conducting tests inside a laboratory [4], and the results provide relevant guidelines for vehicles operating in the field. It has also been reported by investigators [17, 40] that an ashless lubricant or a low-ash oil grade is recommended to avoid oil carbon formation and to result in less formation of acids. Verhelst and Sierens [41] observed a very high percentage of hydrogen while measuring the composition of the gases in the crankcase. The low density of hydrogen is responsible for such high blow-down volumes in the crankcase gases. Evaluation of the properties of lubricating oil also observed it to have strongly deteriorated. Since there was no major wear in the engine parts as observed from X-ray fluorescence spectrometry, it was the blow-down gases that had caused all the changes in the oil properties [41].

7.13 Piston Rings and Crevice Volumes

A review of the literature shows several efforts that have been made in the past to get rid of backfire, but the role of crevice volume and piston rings in this context have not received adequate attention. Crevice volume refers to the small gaps or pockets within the combustion chamber where unburned or partially burned gases can accumulate. A highly reactive and low-density gas like hydrogen can easily seep into and accumulate in these crevices. There is a possibility that uneven distribution of hydrogen within crevices can create localized hot spots, which can lead to backfire. Careful design of crevice volumes can ensure more controlled combustion and prevent the risk of backfire. In order to prevent backfire, hydrogen flames must be prevented from propagating to the top land. Swain et al. [42] reduced the piston top land clearance to 0.152 mm to quench the flame and avoid backfire. It has been reported that judicious and careful selection of piston rings and crevice volume can ensure backfire-free operation of a hydrogen engine even at stoichiometric mixture [42]. It is always desirable to decrease the piston top land clearance to prevent hydrogen flames from propagating into the land. The smaller quenching distance of a hydrogen flame leads to additional thermal load to the piston top land. Sometimes, high temperature hydrogen flames will come close to the piston top land, and the area is exposed to extreme heat for longer periods. Moreover, sometimes rapid heating and cooling of the piston top land due to the nearby hydrogen flame can cause thermal stress. Application of a special coating to the top piston ring groove area with proper coating material to match the operating conditions of hydrogen has proven effective [43] in regulating thermal load by improving heat dissipation, reducing friction, improving lubrication, and providing thermal insulation to ensure durability of the components. The primary function of piston rings is to provide a tight seal between the piston and the cylinder wall. A good seal prevents the leakage of hydrogen or air from the combustion chamber into the crevice volumes or other areas where backfire could occur. Both crevice volumes and piston rings are integral components in hydrogen engine design to avoid backfire.

7.14 Combustion Chamber

A number of factors including hydrogen's fuel-specific characteristics and engine's optimum operational requirement influence combustion chamber geometry. Several combustion chamber shapes can be used for hydrogen engine such as a torus or toroidal combustion chamber, hemispherical combustion chamber, pent-roof combustion chamber, squish, swirl, etc. There is also variable combustion chamber geometry depending on operating conditions. The Hydrogen 7 introduced by BMW used a modified V12 internal combustion engine that could run on either gasoline or hydrogen. The engine featured a pent-roof combustion chamber design, which is common in many automotive spark-ignition engines. In certain situations, the piston-bowl or toroidal combustion chamber seem more suitable to accommodate and optimize properties such as fast burn rate and high flame speed. The doughnut-like or ring-shaped geometry of a toroidal combustion chamber controls the flow of air and hydrogen during the compression and combustion phases. In some studies, it has been reported that a combustion chamber with a low level of turbulence along with a pancake or disk chamber and axially aligned symmetric intake port have increased engine efficiency [42, 44]. Hydrogen has an exceptionally high laminar flame speed at which a flame front travels through a nonturbulent or quiescent mixture of fuel and oxidizer. Such a feature can lead to efficient and stable combustion. At the same time, the wide flammability range of hydrogen quite often creates flashback and flame blow-off under high turbulence conditions. Moreover, high turbulence often leads to higher combustion temperatures, which causes increased NOx formation. Low turbulence combustion chambers are more suitable for hydrogen engines because of better combustion efficiency by minimizing unburned fuel and improving the mixing of hydrogen and air.

As far as the design of combustion chambers is concerned, pancake or disk chambers with a flat, compact design have a lower surface-to-volume ratio, which reduces heat losses to the surrounding walls of the combustion chamber. The flat and wide geometry of pancake chambers allows better flame propagation and mixing, which facilitates better combustion and also reduces emissions. Axially aligned symmetric intake ports are designed to provide even and consistent distribution of air and fuel to each cylinder in a multi-cylinder engine. This design offers balanced combustion and reduced vibration. Well-designed axially aligned ports can minimize pressure drop, allowing for better airflow into the cylinders and, consequently, better engine performance. Combustion chamber design should consider the specific goals and constraints for specific application. Even though low turbulence is generally suggested, in some specific cases, higher turbulence levels may be desirable to enhance mixing or promote other combustion characteristics. Additionally, advances in computational fluid dynamics (CFD) and engine simulation tools may be used to optimize the specific goals.

Pre-chamber ignition is receiving more attention now because of its ability to reduce the distance of the flame propagation and lead to better ignition stability and more efficient combustion. Spark plugs in the pre-chamber ignite a small amount of the air–fuel mixture, and the flame propagates within this chamber. This initial ignition generates a high-temperature, high-pressure environment in the pre-chamber leading to highly reactive combustion products. These hot gases and particles from the pre-chamber are transferred to the main combustion chamber as rapid turbulent jets through orifices. Thus, the condition of the turbulent jet promotes accelerated combustion and better combustion stability with a wider lean-burn limit [45].

In an effort to evaluate the performance of a pre-chamber spark ignition (PCSI) system, Basha and co-researchers [46] carried out experimental investigation on a constant-volume combustion chamber with optical access. Pre-chamber geometrical parameters and operating conditions have a strong influence on the character of the turbulent flame jet ejected from the pre-chamber. Performance parameters of the PCSI system depend upon these geometrical parameters. Therefore, the researchers carried out an investigation to assess the effect of several relevant parameters such as pre-chamber nozzle hole diameter and volume, equivalence ratio and initial chamber pressure on the combustion, and flame jet characteristics of a hydrogen–air mixture in the PCSI system. In these tests the flame development process was captured using a high-speed camera, and the images were processed in MATLAB to obtain quantitative data. A comparative evaluation between PCSI and a conventional SI system showed that the PCSI system exhibited an early start of combustion, shorter combustion duration, higher peak pressure, and higher peak heat release rate. It was also observed that increasing the initial chamber pressure led to quicker ignition, earlier combustion, shorter combustion process, and higher peak pressure. Improved combustion characteristics were observed for an ultra-lean mixture as compared to a lean mixture. A lean mixture with the PCSI system showed the start of combustion to be 0.7 ms earlier than in a conventional SI system, whereas ultra-lean operation showed it still much earlier by 8.8 ms.

7.15 Throttle, Compression Ratio

One of the inherent advantages of hydrogen engine operation comes from its property of flammability range. Such a wide range permits quality regulation, so the engine power can be controlled by varying the fuel rate without throttling the flow of intake air. An electronic throttle is useful as it can be at wide-open throttle condition wherever possible. Even though a hydrogen engine is capable of unthrottled operation, sometimes, some amount of throttling is required for stable operation. Similar engine behavior on throttling at idle and light load conditions has been reported by several investigators, even though their modes of fueling the hydrogen engine were different [47–50]. A very low value of equivalence ratio 0.26 [49] was necessary when throttling seemed essential for smooth stable engine operation. With increasing ranges from medium to heavy loads, throttling becomes necessary to control NOx emissions. A drive-by-wire with electronic sensors, a control module, and an electric motor to control the throttle position are preferred to a mechanically controlled throttle system. The former uses various sensors to monitor different parameters, including the position of the accelerator pedal, the engine's current speed, and the load on the engine. A control module (often an electronic control unit or ECU) receives input from these sensors and processes the data to determine how much the throttle should be opened or closed. The control module continuously adjusts the throttle position based on the driver's input and the engine's operating conditions.

Hydrogen engines can run smoothly on a much higher compression ratio compared to a conventional gasoline engine due to its high-octane rating and much wider flammability range. A review of the literature shows that tests have been carried out successfully at a compression ratio of about 20 and have achieved a high thermal efficiency with single-digit NOx emission [51] by optimizing the excess air ratio and injection timing. In some research and development projects, hydrogen engines with higher compression ratio have been used in experimental setups. Apart from improving efficiency, an increased compression ratio establishes a stable combustion process in the engine cylinder. By way of increasing the compression ratio, the area-to-volume ratio of the combustion chamber increases. This leads to a lesser amount of residual gases and increases their cooling rate. Such a situation avoids possible backfire due to residual gases.

7.16 Materials

System reliability issues like leakage, abrasion, wear, corrosion, and material compatibility need thorough investigation. Material properties such as strength, hardness, and machinability should be examined before selecting any material for hydrogen use. Selection of materials is crucial. Due to their small size, hydrogen atoms and molecules readily dissolve into and diffuse through materials, causing embrittlement and premature structure failure. It is observed that ferric steels and nickel-based super alloys are highly susceptible, whereas austenitic stainless steels are only moderately affected and low aluminum alloys and copper alloys are relatively unaffected. It should also be remembered that hydrogen is soluble in low-alloy steels and stainless steels and the solubility of hydrogen increases with temperature. Threaded connections and flanges are highly undesirable and should be avoided due to susceptibility to leaks. Flared joints and crushable seal joints are better alternatives. Selection of the welding process and the procedure need to be scrupulously chosen, and welded joints should undergo proper hydrostatic testing and leak checks to ensure a higher level of weld integrity.

The effect of hydrogen in different materials has been studied for several applications [52–57]. Hydrogen reacts with many chemicals. High-temperature hydrogen attack, blistering, and embrittlement have been the major types of hydrogen-induced damage [52]. The problem of embrittlement is the most common among these, and it affects a large number of metallic and nonmetallic materials. In some situations, hydrogen atoms penetrate the material's structure and change its mechanical properties, making it brittle and more prone to cracking or failure. This effect was reported by Johnson [55] who observed some remarkable changes produced in iron and steel by the action of hydrogen and acids. This pioneering research of Johnson sparked an active phase of embrittlement research, and a lot of work has been done since then. Dwivedi et al. [57] studied the problem of hydrogen embrittlement in different materials. High-strength steels, including stainless steels, aluminum alloys, and titanium alloys, are susceptible to embrittlement. Titanium and aluminum alloys, even though they experience some level of embrittlement under specific conditions, are generally not affected to the same extent as certain high-strength steels. Chemical composition, heat treatment, microstructure, etc. often govern the embrittlement sensitivity of steel. Tungsten, molybdenum, copper, nickel, and their alloys demonstrate hydrogen embrittlement at elevated temperatures. Embrittlement has been a serious bottleneck for the use of hydrogen in a number of engineering applications such as pipelines, high-pressure tanks, and the use of steel in the automotive sector and chemical and many other industries.

Depending upon the material and its application area, there have been several methods to avoid embrittlement. Cadmium and nickel plating, as well as cadmium and titanium plating, can provide a protective barrier against diffusion for some applications in steel. The engine's structural components should be strong, lightweight, and resistant to hydrogen embrittlement. High-strength aluminum alloys, stainless steels, or carbon fiber composites are suitable material for hydrogen use. From a safety point of view in engines, it is essential that materials should not contribute to hydrogen leaks or fires. Special attention should be paid to seals, gaskets, and materials used in high-pressure areas. However, a detailed discussion on hydrogen embrittlement is not the objective of this chapter. Some background information on embrittlement has been made here to understand its effect on material used in engines.

Materials such as carbon fiber composites, high-strength aluminum alloys (e.g. 7000 series), and certain stainless steels are commonly used for high-pressure hydrogen storage tanks. Inner liners are often made from polymers like nylon or polyethylene to prevent permeation. Materials for combustion chambers and exhaust components should be able to withstand high temperatures. Heat-resistant alloys like Inconel or specialized ceramics such as silicon carbide (SiC) are suitable for the combustion chamber. Stainless steels with high-temperature resistance, such as the 300 series, should be used for exhaust components. Hydrogen can be corrosive, particularly at high pressures and temperatures. Care should be taken against hydrogen-induced corrosion for valves, piping, tubing, etc. Materials like stainless steel (e.g. 316L), titanium, or high-nickel alloys are suitable for valves, and stainless steel or corrosion-resistant alloys like Hastelloy are suggested for use in piping and tubing. Sealing materials such as gaskets, O-rings, and seals must be compatible with hydrogen and capable of maintaining a tight seal under high pressure. Materials like fluorinated elastomers (e.g. Viton) or specialized hydrogen-compatible seal materials can be used. Wherever there is a heat exchanger, it is suggested that appropriate material needs to be selected with good thermal conductivity and resistance to hydrogen embrittlement. Aluminum alloys and stainless steel are suitable for this purpose. Sometimes, it becomes necessary to facilitate combustion in hydrogen engines by catalyst materials such as platinum (Pt) or palladium (Pd) supported on alumina.

Choice of hydrogen-compatible material to avoid leakage of hydrogen from a safety point of view is of primary concern in a hydrogen engine. A flame traveling backward through the intake system into the hydrogen storage or supply system can cause backfire, explosions, or other hazardous situations. Check valves, built on materials compatible with hydrogen, should be installed in the fuel supply system to allow hydrogen flow in one direction only, preventing backflow. They should resist hydrogen-induced corrosion. Stainless steel and certain high-strength alloys are used for check valves with a tight seal to prevent leakage. Design and installation of mechanical components such as flanges, connections, or joints in transfer lines should be carefully done because these are quite vulnerable to leakage. Cast parts should be avoided as far as possible because these are usually porous and quite often cause diffusion of hydrogen under certain circumstances [58].

Flame arrestors are designed to prevent flames from propagating backward into the hydrogen supply system. These devices are typically installed at critical points, such as the hydrogen tank inlet or vent lines. Material for flame arrestors should be flame resistant to prevent the passage of flames. It should be compatible with hydrogen and should resist hydrogen-induced corrosion. Usually stainless steel mesh, sintered bronze, or specialized flame-resistant materials are suitable for this application. High-quality seals and gaskets made from hydrogen-compatible material must be used at all interfaces and connections to prevent leakage. Similarly, pressure relief devices such as burst disks and relief valves should be chosen from materials that can handle the pressure conditions they encounter.

7.17 Rotary Engine Structure for Hydrogen Operation

Any research and development work on hydrogen has been on the converted gasoline or diesel engines, as these are the commonly used engines. However, hydrogen's low volumetric power density is an inherent constraint of improving the power in a compact structure. This challenge is effectively offset by the unique design features and operational characteristics of the rotary engine. There are only a few studies on the use of hydrogen in rotary engines. This may be due to the fact that reciprocating engines have an established infrastructure and wide range of field applications. Continuous research and development activities are being pursued for improvements and optimizations for reliable and efficient reciprocating engine designs for performance as well as emissions. However, now that the hydrogen infrastructure and demand have started increasing and new technologies and market dynamics are evolving, it is worthwhile discussing the use of hydrogen in the Wankel rotary engine. Meng et al. [59] have made a detailed comparative analysis between performance, emission, and combustion characteristics of the hydrogen-fueled reciprocating engine with those of the hydrogen rotary engine. Some properties of hydrogen could be extremely useful to application in the Wankel rotary engine. Mazda launched several models of rotor engine cars, but the demand was low and production was affected. Some discussions on rotary engines and other augmented hydrogen engine configurations are made in Chapter 9.

7.18 Safety Features for Engine Tests in the Laboratory

After relevant technical discussion that can be useful for designing a hydrogen-specific engine, we now discuss the specific safety steps that should be carried out during tests on a hydrogen engine to generate guideline data for practical system development for field application. Several significant safety steps have already been discussed in Section 7.15 on material compatibility. In developing the hydrogen engine test rig for laboratory tests, mechanical components such as flanges, connections, or joints in transfer lines must be properly designed to eliminate the problems of leakage. Precise operation of the fuel supply system should be ensured with proper adjustments and alignment. It is advisable to avoid cast parts due to their porosity, which could facilitate the diffusion of hydrogen and lead to safety problems. If a large amount of hydrogen is released in a confined area, such as laboratories, it can displace oxygen, leading to asphyxiation risk. As oxygen levels decrease, cognitive functions and coordination get disturbed. Sometimes breathing becomes practically impossible below a certain threshold. All piping and equipment should be cleaned before installation. As an additional safety step, it is better that the fuel supply system be purged with nitrogen to prevent possible contact of hydrogen with air during filling and opening. After the purging, the system should be checked with gas chromatography to locate any traces of oxygen. An emergency fuel shut-off system may be installed just before fuel induction. It is beneficial to spray water in the exhaust to mitigate detonation pressure and lessen the risk of ignition. Sometimes CO_2 is also used along with water sprays.

Hydrogen-specific sensors should be installed at the appropriate location in the lab to monitor gas levels and automatically shut off fuel in case of a leak. These should be calibrated periodically to ensure accuracy and reliability. Sensor placement is crucial because moisture or gas mixtures can affect its sensitivity. Adequate ventilation is essential and in situations where adequate ventilation is not possible, an inert gas like nitrogen, carbon dioxide, or helium can be used to make the environment nonhazardous. Extinguishing powders like ammonium phosphate or potassium chloride can suppress explosions by inertizing hydrogen–air mixtures temporarily [58]. Today's technology enables remote sensing of hydrogen to detect any hydrogen leak. Some hydrogen-fueled test vehicles have been developed with multi-point leak detectors that activate even when the vehicle is inactive or parked [60,61]. One of these systems facilitates automatic opening of side windows, the sunroof, and the trunk lid upon detecting hydrogen leakage [61]. Flame arrestors are systems designed to suppress explosions within hydrogen-containing systems. Designing reliable flame arrestors for hydrogen is challenging due to its small limiting diameter through which explosion can progress. These arrestors function on the principle that the flame gets quenched if sufficient heat can be removed from the gas by the arrestors. The quenching distance of hydrogen is low, and thus flame arrestors designed for hydrocarbon flames will not be suitable for hydrogen flames. Sintered bronze materials are effective for stopping hydrogen flames. Installation of explosion relief systems, such as flame traps or a nonreturn valve in the fuel line to prevent reverse flow to the system, have proven effective in lab tests [58]. In the event of engine misfire, unburned hydrogen must be vented outside through a sealed exhaust system. If the engine abruptly stops for any reason, the hydrogen supply system should be instantly shut off. These design features prevent hydrogen accumulation in the event of sudden engine failure, reducing the risk of accidents.

7.19 Safety Considerations for H_2IC Vehicles

In addition to flammability limit, low minimum ignition energy, autoignition temperature, embrittlement, hydrogen venting, colorless flame, etc., the characteristics of leakage, method of storage, location of the storage device, fueling mode, leak detection device, and refueling the vehicle are significant aspects of vehicular safety. Appropriately designed timed manifold injection or port injection is safe for a hydrogen vehicle, and direct injection is the safest mode of fueling the hydrogen vehicle to preclude the possibility of backfire. During refueling the vehicle with gaseous hydrogen, there should be no air within the hose or vehicle line. Use of an automatic double shut-off mechanism enhances an air-free condition by securely sealing the connection during refueling. This design feature is crucial for the safety of the vehicle as it minimizes the chances of air entering the system [10]. For a hydrogen-operated vehicle based on the IC engine (H_2IC vehicle) there are some additional safety criteria with respect to mode of storage, location of the storage device, refueling, etc. Type III or type IV carbon fiber tanks with aluminum or plastic liners are commonly used for storing hydrogen. These tanks are designed to withstand high-pressure levels, reducing the risk of bursting. Thermally activated pressure relief devices (PRDs) are installed to prevent tank bursts. Regulators are strategically placed to minimize the risk of PRDs being turned off during a crash. Design of a regulator is critical for safety. It is often placed partially inside the tank to minimize the risk of being turned off during a crash. Rapid release of gas and contact with ignition sources can lead to accidents, including autoignition due to static electricity. Installing a solenoid valve inside the tank helps shut off high-pressure gas release. Mechanical energy stored in compressed gas cylinders is significant; during leakage, the impulse from a ruptured tank can generate a horizontal jet, potentially causing acceleration, especially on roads with low friction.

A comparative assessment [62] between hydrogen and gasoline fuel leaks in vehicles was studied by researchers at Miami University [62] as shown in Figure 7.6 (Photos 1–4). Fire from a leak in

Photo 1 - Time: 0 min, 0 sec - Hydrogen powered vehicle on the left. Gasoline powered vehicle on the right.

Photo 3 - Time: 1 min, 0 sec - Hydrogen flow is subsiding, view of gasoline vehicle begins to enlarge

Photo 2 - Time: 0 min, 3 seconds - Ignition of both fuels occurs. Hydrogen flow rate 2100 SCFM. Gasoline flow rate 680 cc/min.

Photo 4 - Time: 1 min, 30 sec - Hydrogen flow almost finished. View of gasoline powered vehicle has been expanded to nearly full screen

Figure 7.6 Effect of leakage and ignition of a hydrogen and a gasoline-powered vehicle [62] / U.S. Department of Energy.

a gasoline engine car and the same kind of leak from a hydrogen engine car were studied. The tests revealed significant differences in their ignition severity. Initially in Photo 2, a long, vertical jet flame was generated after three seconds. After one minute, the hydrogen flame subsided, and the gasoline flame was getting stronger. After one minute and thirty seconds, the hydrogen flame had vanished completely, whereas the gasoline flame was intensifying to a violent form (Photo 4).

BMW has a long record of developing a hydrogen-fueled internal combustion engine since 1979. In an engine developed with liquefied hydrogen storage, the significant aspects were studied with regard to utilization of boil-off, vehicle parking in an enclosed space, and crash behavior of the vehicle. Experimental observations on safety were compared with theoretical analysis. The risk analysis considered some important guidelines related to mechanical strength of the tank under high pressure to prevent leakage. While developing the engine, some hydrogen-specific features were introduced such as selection of appropriate material for valve seat rings, spark plug and ignition system, modified piston, ring assembly, arrangement for installing the cooling circuit, etc. The design conditions considered that the liquid hydrogen tank must not burst. The ignition source was strictly avoided in a certain area, and the possibility of an ignitable mixture being formed inside the vehicle or in any other enclosed space had to be eliminated. BMW also carried out a crash test and found an effective "package affording safety in a crash and an LH_2 fuel supply installation that is resistant to crash" [63]. Some tests were conducted to evaluate the nature and quality of ventilation in a prefabricated garage. However, these aspects of the hydrogen vehicle need to be further investigated to formulate well-defined design features and regulations [63]. Li et al. [64] focused on assessing hydrogen diffusion behavior and explosion risks in mobile hydrogen refueling stations. A 3D full-scale geometry model was established for simulation. Experimental data were generated that validated numerical models. In their analysis, the researchers considered ventilation scenarios within the truck container and ambient wind effects on hydrogen dispersion. Results included hydrogen volume and diffusion distance calculations under severe conditions. Hydrogen concentration in key areas and overpressure from potential explosions were also evaluated. The analysis showed a higher risk with open side door and window, low ambient wind speed, and wind direction from front to rear. Under severe conditions, the average hydrogen volume in the air was 87.9 m^3, with a minimum safe distance of 313.2 m in case of explosion.

Concluding Remarks and Perspectives

Broadly speaking, design criteria must be formulated to maximize efficiency, control emissions, optimize load control strategies to reduce wear, and eliminate leakage at sensitive locations. Since the hydrogen engine development technology has undergone a series of evolutions, the requirement and priorities have been well identified. The problem of backfire, which had been a major stumbling block for decades and the problem for which many researchers had discontinued their research efforts, is no more a point of concern. Well-defined preventive and curative steps have already been identified and successfully applied by researchers working in this area now. One major point that needs to be seriously considered is that present R&D efforts are carried out on converted engine systems that were originally designed to run on liquid petroleum fuels such as gasoline or diesel. Any converted system is bound to exhibit some reliability problems. The research work that is carried out on converted CNG-engines must also adjust for some of the entirely different physico-chemical properties of hydrogen. After carrying out an exhaustive literature survey and observing the test procedures and their results, some efforts were made here to discuss the specific criteria that need to be adopted for the development of an optimum dedicated hydrogen engine. There are a number of crucial subsystems and components that need to be modified for hydrogen operation. Hydrogen's wide flammability limits and high flame velocity allow diverse load control strategies in engine applications. The high autoignition temperature permits increased compression ratios, with current research aiming for brake thermal efficiency exceeding 45% while keeping NOx emissions low. In the injection system, timed injection is necessary, with a focus on high-flow-rate injectors for both port fuel injection (PFI) and direct injection (DI). The intake manifold requires the installation of H_2 injectors for H_2 PFI, coupled with optimization for gas exchange. Proper grounding in the ignition system is essential to prevent uncontrolled ignition, and the use of cold-rated spark plugs is recommended. Direct injection strategies exhibit better volumetric efficiency and avoid problems normally encountered with port fuel injection. However, the challenges of an injection system (actuation mechanism, injection timing, injection pressure) and the ability of the injector to survive without leakage in the hostile thermal environment of the combustion chamber are tough and need to be sorted out. In direct injection (DI) engines, achieving optimal fuel stratification is a critical technical challenge, and piezoelectric injectors satisfy that demand of precise control over the air–fuel mixture during compression. However, there are some problems. Sometimes, small hydrogen atoms diffuse into materials including piezoelectric components and lead to material degradation. Dielectric epoxy coatings, which are used to protect electronic components, sometimes peel off due to the thermal environment or other stresses. Tackling the problem of hydrogen diffusion into piezoelectric components, preventing delamination of the dielectric epoxy coating, taking care of surface blistering, solving the actuator cracking due to tensile loads, and minimizing sliding friction and wear of injector parts are some of the technical challenges likely to be encountered at certain conditions.

Cylinder head design is important to accommodate hydrogen injectors with optimized cooling to avoid hot spots. Use of cooled exhaust valves and appropriate materials is necessary to avoid hydrogen embrittlement. Design consideration needs to aim at reduced oil consumption and minimum blow-by and hydrogen leakage. Modifications in the piston and piston rings may be required in specific engines. Low-turbulence combustion chambers are necessary for the high flame speed of hydrogen. Appropriate compression ratio, turbocharging for lower boost pressure demand, and optimized after-treatment systems are additional design considerations. The crankcase ventilation system should limit hydrogen leakage into the crankcase. Ashless oil should be used to prevent deposits, and an improved blow-by separator should be used. Proper design of engine components with suitable material for long-term durability is important. There are other challenges due to hydrogen slip into the crankcase and high oil consumption to be appropriately designed. Besides all these, robustness of valves and spark plugs in the hydrogen-rich combustion environment is of practical significance.

References

1 Molkov, V. (2012). *Fundamentals of Hydrogen Safety Engineering, Part I.* bookboon.com and Ventus Publishing. ISBN: 978-87-403-0226-4. http://hdl.handle.net/123456789/475.

2 Molkov, V. (2012). *Fundamentals of Hydrogen Safety Engineering, Part II.* bookboon.com. *Fundamentals of Hydrogen Safety Engineering II* (weebly.com). ISBN: 978-87-403-0279-0.

3 Hord, J. (1978). Is hydrogen a safe fuel? *International Journal of Hydrogen Energy* 3 (2): 157–176.

4 Das, L.M. (2024). *Hydrogen Energy: Production, Safety, Storage and Applications*. Chapter 4, 173–219. Chichester: Wiley.

5 ISO/TR 15916 (2004). Basic Considerations for the Safety of Hydrogen Systems. ISO Technical Committee 197, Hydrogen Technologies.

6 Coward, H.F. and Jones, G.W. (1952). Limits of Flammability of Gases and Vapors. Bulletin 503, Bureau of Mines, Washington, DC. https://digital.library.unt.edu/ark:/67531/metadc12662/ (accessed 8 December 2024).

7 National Aeronautics and Space Administration. NSS 1740.16 (1997). Safety Standard for Hydrogen and Hydrogen Systems: Guidelines for Hydrogen System Design, Materials Selection, Operations, Storage, and Transportation, 20546. Washington, DC: Office of Safety and Mission Assurance.

8 Markert, F., Nielsen, S.K., Paulsen, J.L. et al. (2007). Safety aspects of future infrastructure scenarios with hydrogen refuelling stations. *International Journal of Hydrogen Energy* 32: 2227–2234.

9 Rigas, F. and Sklavounos, S. (2008). Hydrogen safety. In: *Hydrogen Fuel: Production, Transport and Storage* (ed. R. Gupta), 537–538. Boca Raton, FL: CRC Press/Taylor & Francis.

10 Hansel, J.G., Mattern, G.W., and Miller, R.N. (1993). Safety considerations in the design of hydrogen powered vehicle. *International Journal of Hydrogen Energy* (18) 9: 783–790.

11 Swain, M.R. and Swain, M.N. (1996). Passive ventilation systems for the safe use of hydrogen. *International Journal of Hydrogen Energy* (21) 10: 823–835.

12 Sopena, C., Diéguez, P.M., Sáinz, D. et al. (2010). Conversion of a commercial spark ignition engine to run on hydrogen and gasoline, performance comparison using hydrogen and gasoline. *International Journal of Hydrogen Energy* 35 (3): 1420–1429.

13 NFPA. (1989). NFPA 50A, Standard of gaseous hydrogen systems at consumer sites. https://entirelysafe.com/complete-list-of-nfpa-standards-and-codes/

14 NFPA. (1989). NFPA 50B, Liquid hydrogen at consumer sites. https://entirelysafe.com/complete-list-of-nfpa-standards-and-codes/

15 Compressed Gas Association. (1991). Standard for hydrogen piping at consumer locations. https://h2tools.org/fuel-cell-codes-and-standards/cga-g54-standard-hydrogen-piping-systems-consumer-sites

16 Poursadegh, F., Brear, M., and Hayward, B. (2023). Autoignition, knock, detonation and the octane rating of hydrogen. *Fuel* 332 (Part 2): 126201.

17 Das, L.M. (2002). Near-term introduction of hydrogen engines for automotive and agricultural application. *International Journal of Hydrogen Energy* 27 (5): 479–487.

18 Kondo, T., Iio, S. and Hiruma, M. (1997). A study on the mechanism of backfire in external mixture formation hydrogen engines—about backfire occurred by the cause of the spark plug. International Spring Fuels and Lubricants Meeting and Exposition, Dearborn, MI. SAE Technical Paper No. 971704. https://doi.org/10-4271/971.

19 Das, L.M. (1996). Hydrogen–oxygen reaction mechanism and its implication to hydrogen engine combustion. *International Journal of Hydrogen Energy* 21: 703–715.

20 Stockhausen, W.F., Natkin, R.J., Kabat, D.M. et al. (2002). Ford P2000 hydrogen engine design and vehicle development program. SAE International Paper No. 2002-01-0240.

21 Olavson, L.G., Baker, N.R., Lynch, F.E. et al. (1984). Hydrogen fuel for underground mining machinery. SAE International Paper No. 840233.

22 MacCarley, C.A. (1981). A study of factors influencing thermally induced backfiring in hydrogen fueled engines, and methods for backfire control. 16th IECEC Conference, Atlanta.

23 Gerbig, F., Strobl, W., Eichlseder, H. et al. (2004). Potentials of the hydrogen combustion engine with innovative hydrogen-specific combustion process. Fisita World Automotive Congress Paper No. F2004V113, Barcelona, Spain.

24 Verhelst, S. and Sierens, R. (2001). Hydrogen engine–specific properties. *International Journal of Hydrogen Energy* 26: 987–90.

25 Lucas, G.G. and Morris, L.E. (1980). The backfire problem of the hydrogen engine. Symposium organized by the University's Internal Combustion Engine Group. London: King's College.

26 Hwang, J., Bae, C., Park. J. et al. (2016). Microwave-assisted plasma ignition in a constant volume combustion chamber. *Combustion and Flame* 167: 86–96.

27 Mariani, A. and Foucher, F. (2014). Radio frequency spark plug: an ignition system for modern internal combustion engines. *Applied Energy* 122: 151–161.

28 Merotto, L., Balmelli, M., Vera-Tudela, W. et al. (2022). Comparison of ignition and early flame propagation in methane/air mixtures using nanosecond repetitively pulsed discharge and inductive ignition in a pre-chamber setup under engine relevant conditions. *Combustion and Flame* 237: 111851.

29 Alvarez, C.E.C., Couto, G.E., Roso, V.R. et al. (2018). A review of prechamber ignition systems as lean combustion technology for SI engines. *Applied Thermal Engineering* 128: 107–120.

30 Dale, D., Checkela, M.D., and Smyba, P.R. (1997). Application of high energy ignition systems to engines. *Progress in Energy and Combustion Science* 23: 379–398.

31 Böker, D. and Brüggemann, D. (2011). Advancing lean combustion of hydrogen–air mixtures by laser-induced spark ignition. *International Journal of Hydrogen Energy* 36 (22): 14759–14767.

32 Azarmanesh, S. and Targhi, M.Z. (2021). Comparison of laser ignition and spark plug by thermodynamic simulation of multi-zone combustion for lean methane-air mixtures in the internal combustion engine. *Energy* 216: 119309.

33 Bao L.-Z., Sun, B.-G., Luo, Q.-H. et al. (2020). Experimental study of the polytropic index of the compression stroke for a direct injection hydrogen engine. *International Journal of Hydrogen Energy* 45 (52): 28196–28203.

34 Hydrogen Use in Internal Combustion Engines. (December 2001). Module 3: Hydrogen use in internal combustion engines. College of the Desert, Palm Desert, California. https://www1.eere.energy.gov/hydrogenandfuelcells/tech_validation/pdfs/fcm03r0.pdf.

35 Natkin, R.J., Tang, X., Boyer, B. et al. (2010). Hydrogen IC engine boosting performance and NOx study. SAE International Technical Paper 1. https://doi.org/10.4271/2003-01-0631.

36 Das, L.M. (1990). Fuel induction technique for a hydrogen-operated engine. *International Journal of Hydrogen Energy* 15 (11): 833–842.

37 Tang, X., Stockhausen, W.F., Kabat, D.M. et al. (2002). Ford P2000 hydrogen engine dynamometer development. SAE International Paper No. 2002-01-0242.

38 Berckmüller, M., Rottengrubber, H.S., Eder, A. et al. (2003). Potentials of a charged SI-hydrogen engine. SAE International Paper No. 2003-01-3210.

39 Deluchi, M.A. (1989). Hydrogen vehicles: an evaluation of fuel storage, performance, safety, environmental impact and cost. *International Journal of Hydrogen Energy* 14 (2): 81–130.

40 Kiesgen, G., Klüting, M., Bock, C. et al. (2006). The new 12-cylinder hydrogen engine in the 7 series. The H2 ICE age has begun. SAE International Paper No. 2006-01-0431.

41 Verhelst, S. and Sierens, R. (2001). Hydrogen engine—specific properties. *International Journal of Hydrogen Energy* 26: 987–90.

42 Swain, M.R., Schade, G.J., and Swain, M.N. (1996). Design and testing of a dedicated hydrogen fueled engine. SAE International Paper No. 961077.

43 Berger, E., Bock, C., Fischer, H. et al. (2006). The new BMW 12-cylinder hydrogen engine as clean efficient and powerful vehicle powertrain. Fisita World Automotive Congress Paper No. F2006P114, Yokohama, Japan.

44 Blarigan, P.V. (1996). Development of a hydrogen fueled internal combustion engine designed for single speed/power operation. SAE International Paper No. 961690.

45 Zhu, S., Akehurst, S., Andrew, L. et al. (2022). A review of the pre-chamber ignition system applied on future low-carbon spark ignition engines. *Renewable and Sustainable Energy Reviews* 154: 111872.

46 Jeelan Basha, K., Balasubramani, S., and Sivasankarlingggam, V. (2023). Effect of pre-chamber geometrical parameters and operating conditions on the combustion characteristics of the hydrogen-air mixtures in a pre-chamber spark ignition system. *International Journal of Hydrogen Energy* 48 (65): 25593–2560848.

47 Mathur, H.B. and Das, M.L.M. (1991). Performance characteristics of hydrogen-fuelled S.I. engine using timed manifold injection. *International Journal of Hydrogen Energy* 16 (2): 115–127.

48 Jing-Ding, L., Ying-ging, L., and Tian Shen, D. (1984). *Improvement on the Combustion of Hydrogen-Fuelled Engine, Hydrogen Energy Progress V*. Oxford: Pergamon Press.

49 Adt, R.R., Greenwell Jr, H., and Swain, M.R. (1974). The hydrogen and methanol fuelled air breathing automobile engine. Paper presented at the THEME Conference, University of Miami.

50 Furuhama, S., Yamane, K., and Yamaguchi, I. (1977). Combustion improvement in hydrogen-fuelled engines. *International Journal of Hydrogen Energy* 2 (3): 329–340.

51 Mogi, Y., Oikawa, M., Kichima, T. et al. (2022). Effect of high compression ratio on improving thermal efficiency and NOx formation in jet plume controlled direct-injection near-zero emission hydrogen engine. *International Journal of Hydrogen Energy* 47 (73): 31459–31467.

52 Shimomura, J. and Imanaka, T. (1985). Decomposition of carbides in 2–1/4Cr-1Mo steels during hydrogen attack. *Scripta Metallurgica* 19: 1507–1511.

53 Sakai, T., Asami, K., Kondo, N. et al. (1987). Effects of carbide forming elements on hydrogen attack and embrittlement in 2 1/4Cr-1Mo steel. *Tetsu-To-Hagane/Journal of the Iron and Steel Institute of Japan* 73 (2): 372–379.

54 Woodward, T.J. (2022). Understanding the high temperature hydrogen attack of steels utilizing novel in-situ analytical transmission electron microscopy techniques. PhD thesis submitted to The University of Manchester, Manchester, UK. https://research.manchester.ac.uk/en/studentTheses/understanding-the-high-temperature-hydrogen-attack-of-steels-util.

55 Johnson, W.H. (1875). On some remarkable change produced in iron and steel by the action of hydrogen and acids. *Proceedings of the Royal Society of London* 23: 168–179.

56 Barrera, O., Bombac, D., Chen, Y. et al. (2018). Understanding and mitigating hydrogen embrittlement of steels. *Journal of Materials Science* 53 (9): 6251–6290.

57 Dwivedi, S.K. and Vishwakarma, M. (2018). Hydrogen embrittlement in different materials: a review. *International Journal of Hydrogen Energy* 43 (46): 21603–21616.

58 Das, L.M. (1991). Safety aspects of a hydrogen-fuelled engine system development *International Journal of Hydrogen Energy* 16 (9): 619–624.

59 Meng, H., Ji, C., Xin, G. et al. (2022). Comparison of combustion, emission and abnormal combustion of hydrogen-fueled Wankel rotary engine and reciprocating piston engine. *Fuel* 318: 123660.

60 Veziroglu, T.N. (1992). *Project Hydrogen '91: Technical Proceedings*. American Academy of Science. ISBN 10: 0963163418 / ISBN 13: 9780963163417.

61 Siuru, W. (1989). R and D in the fast lane: BMW's research of alternate power. *Mechanical Engineering III.* https://www.thefreelibrary.com/R%26D+in+the+fast+lane.-a08076439.

62 Swain, M.R. (2001). Fuel leak simulation. *Proceedings of the 2001 DOE Hydrogen Program Review.* NREL/CP-570-30535.

63 Furst, S., Dub, M., Gruber, M. et al. (2005). Safety of hydrogen-fuelled vehicle with IC engines. https://h2tools.org.

64 Li, Y., Wang, Z., Shi, X. et al. (2023). Safety analysis of hydrogen leakage accident with a mobile hydrogen refueling station. *Process Safety and Environmental Protection* 171: 619–629.

8

Hydrogen in Blends with Other Fuels

If God did create the world by a word, the word would have been hydrogen.

Harlow Shapley

8.1 Introduction

The phase of research activities of using hydrogen along with gasoline is now more attractive because much work has already been done successfully on spark ignition engines using only hydrogen fuel, and even prototype vehicles initially designed to run on gasoline have been appropriately modified and road-tested. Studies and system development of existing diesel engine designs for dual fuel operation have been discussed. During this transitional phase, while infrastructure for full hydrogen operation is being established, numerous researchers are investigating the feasibility of integrating hydrogen with other fuels. This chapter focuses on the use of hydrogen along with other alternative fuels such as compressed natural gas (CNG), liquefied petroleum gas (LPG), ethanol, biogas, biodiesel, and dimethyl ether (DME).

8.2 Adding Hydrogen to CNG

Over the past several years, CNG has emerged as a prospective fuel for engines and vehicles in numerous global regions. Dedicated engines, designed to operate on gaseous fuel, have been installed in vehicles now being operated in many countries. CNG, however, with methane being its major constituent, is a carbon-based fuel with finite source reserve. Although CNG emits fewer greenhouse gases than gasoline and diesel, it still releases noxious pollutants when burned. While it offers a cleaner option in the short to medium term, its carbon emissions and finite resources make it unsustainable as a long-term solution. Use of natural gas in engines has established a gas-based fuel economy in many parts of the world and has built confidence in transitioning to hydrogen, the "cleanest burning" gaseous fuel. In this interim phase, hydrogen is added to CNG in various proportions in engines and vehicles. One essential objective of this investigation is to evaluate the feasibility of transitioning from blended to total hydrogen operation.

Degrees of blending hydrogen with CNG have varied depending on applications, influenced by the technical objectives of the research, practical feasibility, and economic considerations.

Hydrogen Engines: Design, Performance Evaluation, Combustion Analysis, and Exhaust Emissions, First Edition.
Lalit Mohan Das.

Researchers have used various terms such as HNG (hydrogen natural gas), HCNG (hydrogen + CNG), Hydrogen-added natural gas, and hythane (hydrogen + methane). Hythane is a registered trademark of a blend of hydrogen and CNG, where 20% by volume of hydrogen is blended into compressed natural gas. Table 8.1 shows some of combustion properties of hydrogen, CNG, and hythane. Flame speed, calorific value, and the stoichiometric air–fuel ratio of hythane are improved compared to CNG. Hydrogen and natural gas are miscible in all proportions, and thus blends can be prepared depending on the objectives of the investigations. Some researchers have encountered abnormal combustion in their tests on increasing hydrogen fractions in CNG above a specific level. Some studies have resulted in higher exhaust emissions that fail to meet emission standards. Each hydrogen fraction necessitates different spark timing and air–fuel ratios for optimal engine performance. Hydrogen has high flame velocity and broad flammability limits. Thus, adding hydrogen to a slow-burning fuel like CNG improves engine performance under lean burn conditions and reduces tailpipe emissions of HC, CO, and CO_2 in exhaust with higher hydrogen concentrations in CNG. However, this decrease does not proportionally match the addition of hydrogen to methane. The lean-burn combustion limit expands significantly with higher hydrogen fractions, but the maximum brake torque decreases. There is a strong correlation between hydrogen addition, excess air ratio (λ), and spark timing [1–3]. It has been a challenge to find the optimal combinations of these parameters, along with other crucial factors for optimum engine operation. Studies have demonstrated that CNG emits extremely low levels of oxides of nitrogen (NOx) compared to traditional fuels [4, 5]. On the other hand, the addition of hydrogen to CNG increases NOx emissions.

There are more than 27 million vehicles running on natural gas in Argentina, Brazil, China, India, Iran, Italy, Pakistan, Thailand, United States, and several other countries. Numbers keep rising. Years ago, road tests on hythane-fueled vehicles in the Denver Hythane Project (1992), the Montreal Hythane Bus Project (1995), and Sunline's second hythane project had shown good road performance and demonstrated drastic reductions up to 45–50% in NOx levels. In their study, Aleksejs et al. [6] analyzed the optimal blend ratio considering factors such as heat values and gas costs from a technical feasibility and economic viability point of view (Figure 8.1).

Values obtained by analyzing gas blends with varying amounts of hydrogen (0%, 1%, 5%, 10%, and 20%) are shown in Figure 8.1. A gas blend with 0% hydrogen serves as the baseline for comparison. The figure also shows the relationship between heat value losses and CO_2 emission reductions due to hydrogen's presence. CO_2 levels do not proportionately decrease the heat value of gas blends. CNG has a lower carbon content per molecule than traditional fuels like gasoline or diesel, which results in approximately 20–25% less CO_2 emissions. In fact, the absence of C—C bonds on the fuel accompanied by lean combustion strategies bring down the amount of

Table 8.1 Some combustion properties of hydrogen, CNG, and hythane.

Properties	Hydrogen	Hythane	CNG
Flame speed (cm s^{-1})	237	120	42
Diffusion coefficient (cm^2 s^{-1})	0.61	0.25	0.16
Higher calorific value (MJ Kg^{-1})	142	92	55
Stoichiometric A/F (ratio)	34.3:1	22.8:1	17.2:1
Flammability limit (vol%)	4–75	5–35	5.3–15
Minimum ignition energy (mJ)	0.02	0.21	0.28
Autoignition temperature (K)	858	825	813

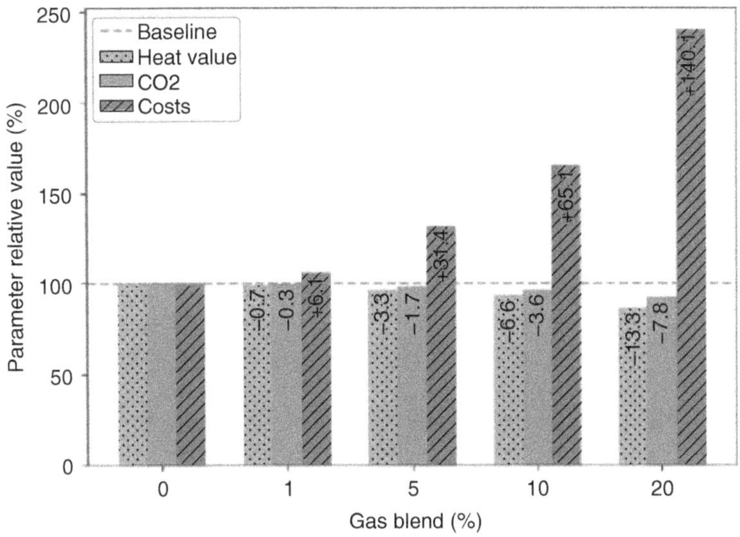

Figure 8.1 Comparison of values with different amounts of hydrogen in the gas blend [6].

particulate matter by approximately 75% [7]. CNG derived from renewables like animal waste and agricultural by-products has enhanced sustainability. Utilizing HCNG in the existing natural gas infrastructure cuts costs and several logistical hurdles. Lean combustion techniques in spark ignition engines improve efficiency and bring down pollutant emissions. This approach also decreases heat transfer loss and raises thermal efficiency [8], particularly during partial loads where increasing air dilution minimizes pumping losses. Hydrogen supplementation improves stable combustion in diluted conditions compared to CNG operation. Increasing hydrogen concentrations in CNG engines improves power output while reducing cyclic variability. Lean burn operation leads to decreased burning velocity and permits engine operation with higher compression ratios (CRs) and without knock. Recalibration of spark timing and adjustment of air/fuel ratios are necessary for stoichiometric operation in HCNG engines. Apart from performance, emission, and combustion features, studies have generated a lot of information on crucial topics such as flame stability, cycle-to-cycle variation, combustion modeling, and turbocharged engines and stratified HCNG applications, etc.

8.2.1 Significant Research Results on Combustion, Emission, and Performance

Detailed discussions of all investigators' work in this area are beyond this chapter's scope. A general description of some of these research investigations is described here highlighting some of the significant findings.

Wang et al. observed that under lean operation peak cylinder pressure and maximum rate of pressure rise increased, but cycle-by-cycle variations decreased with the increase in hydrogen fraction [9]. Salzano et al. addressed safety concerns regarding HCNG, including explosion behavior [10]. Larsen and Wallace used a blend of 85% natural gas and 15% hydrogen in volume to analyze pollutant emissions and thermal efficiency in a turbocharged SI engine, with an after-treatment system for HC and NOx emissions control [11]. Bell et al. found that blending CNG with hydrogen reduces unburned HC [12]. Molina and colleagues [13] observed that hydrogen substitution affects in-cylinder pressure and heat release rate profiles.

Molina et al. [14] analyzed methane–hydrogen combustion in an SI engine using a modeling approach. Utilizing machine-learning techniques, the combustion model estimated flame propagation based on laminar flame speed predicted by an artificial neural network (ANN). The model's reliability was validated from fundamental aspects with laminar flame speed and complex scenarios like real engine tests with different blends. Figure 8.2 shows the heat release rate, combustion duration, and combustion phasing for different fuel compositions. Modeling shows that hydrogen addition increases the burning rate, despite the adverse effects of lean mixture environments on combustion. In their analysis investigators have identified some limitations that require future refinement on the modeling. Estimating cycle-to-cycle variation is crucial for accurately predicting losses at high dilution rates due to intense misfiring. The impact of autoignition through dedicated submodels needs to be closely analyzed to predict efficiency due to severe knocking combustion. Various submodels from the literature need to be evaluated and validated under relevant engine conditions for HCNG fueling.

With the objective of assessing the effect of hydrogen addition on cycle-by-cycle variation, an experimental study was conducted in a six-cylinder throttle body injection natural gas engine using CNG and HCNG (80/20 in volume of CNG/hydrogen) as fuels [15]. Lean operation improved combustion stability, raised efficiency, and lowered emissions compared to neat CNG operation. Shortening of flame development and propagation durations and improvement in combustion phase consistency were the other dominant features of hydrogen addition.

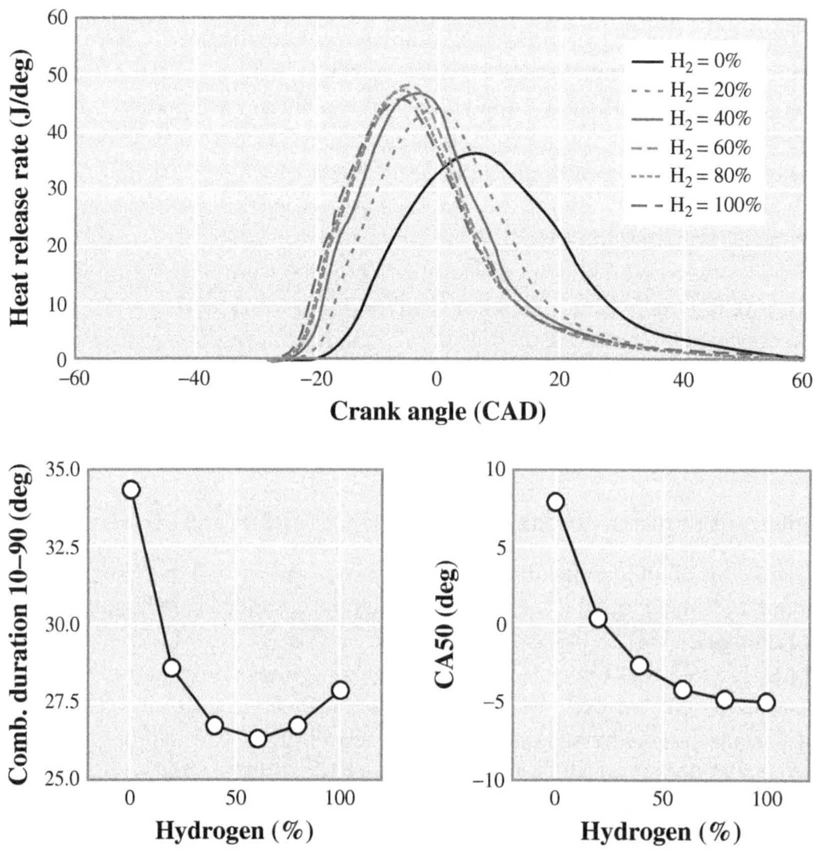

Figure 8.2 Characteristics of the combustion process: heat release rate, combustion duration, and combustion phasing for different fuel compositions [14].

It is evident in Figure 8.3 that at a given lambda and ignition timing hydrogen addition increases NOx emissions. This is expected to be due to a faster hydrogen burn rate. As evident in the figure, hydrogen addition resulted in substantial reduction in the NOx emissions level, which is primarily because of hydrogen's smaller quenching distance. Figure 8.4 gives a comparative picture of HC emission at MBT timing between CNG and HCNG fueling.

Laminar burning velocity is a crucial parameter for understanding fuels' combustion characteristics, as it directly influences speed of the flame front in combustion processes like homogeneous charge combustion. Hydrogen, because of its high laminar burning velocity, improves the combustion stability and extends the lean combustion limit when it is added to CNG. Various methods exist to analyze laminar burning velocities, including counter-flow double flames, spherically expanding flames, and flat flame burners. Among these, spherical-expanding flames technique are widely used for evaluating unstretched laminar burning velocities. In the spherical expanding flame technique, a fuel–air mixture is ignited at the center of a spherical chamber,

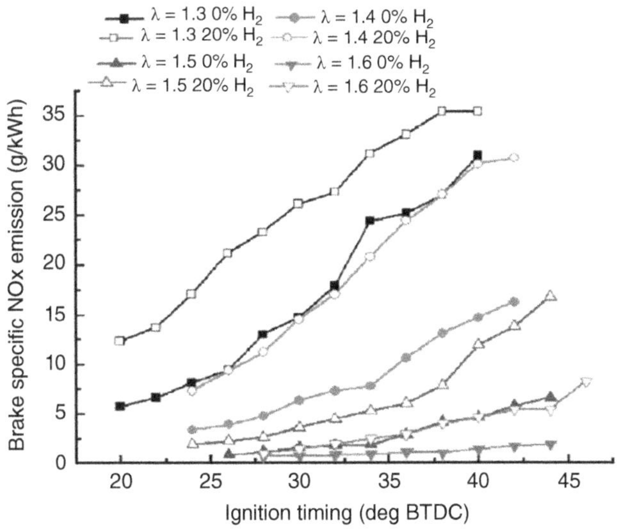

Figure 8.3 Brake specific NOx emissions at various excess air ratios at 0.55 MPa BMEP [15].

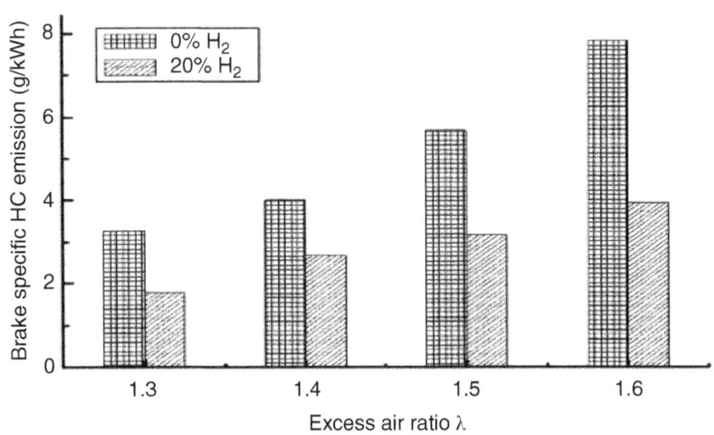

Figure 8.4 HC emission is also a significant problem in addition to NOx emission [15].

causing the flame to propagate outward uniformly. By measuring the radial expansion rate of the flame, one can determine the laminar burning velocity of the fuel under unstretched conditions [16]. Han and colleagues [17] made an exhaustive review of the uncertainty in laminar flame speed (LFS) measurement with the outwardly propagating spherical flame (OPF) method for hydrogen/air mixtures. Several studies have been done on laminar burning velocity of hydrogen–natural gas mixtures over a wide range of hydrogen fractions (0 to 100%) and under different equivalence ratios ranging from a lean to a rich mixture, $\phi = 0.6$ to 1.4 [18–23].

Huang et al. [24] studied laminar flame characteristics of natural gas–hydrogen–air flames in a constant-volume bomb under normal temperature and pressure conditions. They investigated laminar burning velocities and Markstein lengths over a range of hydrogen-to-natural gas ratios and equivalence ratios. In lean mixtures, flame radius growth slowed over time for natural gas and low hydrogen fractions, but it showed a steady, linear increase with high hydrogen fractions. Stoichiometric mixtures displayed linear flame radius growth. With hydrogen fraction, laminar burning velocities increased exponentially, while Markstein length decreased and flame instability increased. Markstein number increased with equivalence ratios, indicating improved flame stability for fixed hydrogen fractions. In this study, it was observed that flame expands spherically after ignition, and the flame radius increased rapidly in the subsequent process. For lean mixture combustion (Figure 8.5a) with $\phi = 0.6$, the flame radius increased over time, but the rate of increase slowed down as the flame expanded for natural gas and low hydrogen fraction mixtures. On the contrary, there was a linear correlation between flame radius and time for mixtures with high hydrogen fractions. In rich mixture combustion (Figure 8.5b) with $\phi = 1.3$, the flame radius exhibited a slow initial increase followed by a rapid increase during late stages of propagation for natural gas and low hydrogen fraction mixtures.

Figure 8.6 depicts the unstretched laminar burning velocity with varying hydrogen fractions increasing exponentially with increased hydrogen fraction blends. Krishnan et al. [25], in their review on use of hydrogen-enriched natural gas (HNG) for spark ignition (SI) engines, observed improved thermal efficiency, reduced fuel consumption, and lesser concentration of pollutants such as carbon monoxide (CO), carbon dioxide (CO_2), particulate matter (PM), and oxides of nitrogen (NOx). The engine showed improved combustion parameters, such as faster flame

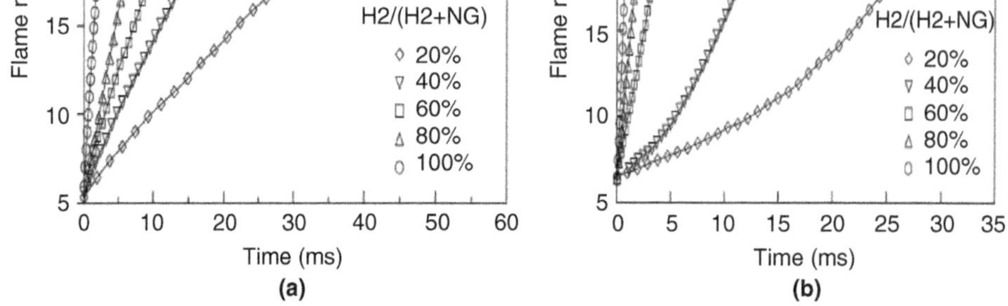

Figure 8.5 Flame radius vs. time at various hydrogen fractions: case (a) for a lean mixture and case (b) for a rich mixture [24].

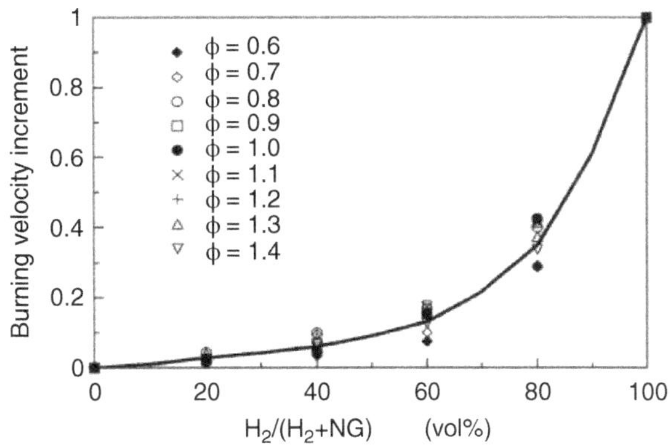

Figure 8.6 Unstretched burning velocity increment vs. hydrogen fraction at different equivalence ratios [24].

propagation, higher in-cylinder pressure, and better combustion stability, compared to natural gas operation. Higher concentrations of soot in HNG-fueled engines were found mainly due to combustion of the lubricant. Longer spark plugs reduced the flame propagation distance, leading to faster combustion. Split-injection strategies combining delayed direct injection of hydrogen with direct or port injection of methane extended the lean operating limit. Early injection of hydrogen-enriched HNG at the start of compression and with advanced spark timing resulted in reduced CO emissions. Increasing the compression ratio improved overall performance and combustion characteristics but increased NOx emissions. Nagalingam et al. [26] observed that the combustion air ratio range for hydrogen is broader than that of natural gas (NG), which facilitates unthrottled operation and reduces pumping losses. The higher burning temperature of hydrogen improved overall engine performance while simultaneously increasing NOx emissions and reducing carbon-based emissions. Strategies such as use of exhaust gas recirculation (EGR), dilution gases, hydrogen injection, inert gas induction, or water injection regulate the kinetics of specific combustion reactions that achieve controlled emissions. Park et al. [27] investigated HCNG in a vehicular engine. In a lean combustion HCNG engine, control parameters were fine-tuned to optimize combustion efficiency. In a transient mode of operation, lean combustion could satisfy the EURO-VI emission regulations even without a De-NOx catalyst. Stoichiometric operation was adopted in an oil gallery-type piston designed to withstand high temperatures, and high EGR flow rates were employed. Observations indicated that HCNG engines utilized EGR more effectively than natural gas engines. Furthermore, the stoichiometric combustion strategy in HCNG engines proved superior in addressing low efficiency and heat durability issues.

Figure 8.7 displays the thermal efficiency of CNG and HCNG under stoichiometric combustion at medium load operation and a speed of 1260 rpm and 575 Nm. There is a distinct impact on thermal efficiency of around 20% of EGR as the EGR rate increases, where HCNG offers a slight improvement over natural gas. Under various engine speeds, Figures 8.8 and 8.9 compare thermal efficiency and power characteristics of stoichiometric mixture combustion of HCNG and CNG under lean combustion. It is clear from the graphs that thermal efficiency is higher for HCNG in both low-speed and high-speed operation, with power output almost comparable.

Stable operation is achieved with hydrogen addition at higher excess air ratio conditions, leading to reduced NOx emissions. It was estimated in this study that, for the best efficiency condition, there is an approximate 67% reduction in NOx emissions with the addition of 30 vol% of hydrogen, and 40 vol% hydrogen addition brings down the NOx emissions by approximately 84% [28]. Figure 8.10 shows the effect of retarded spark timing with the addition of 30 vol% hydrogen. Apart from NOx reduction, beyond 80% both THC and CO emissions are also reduced.

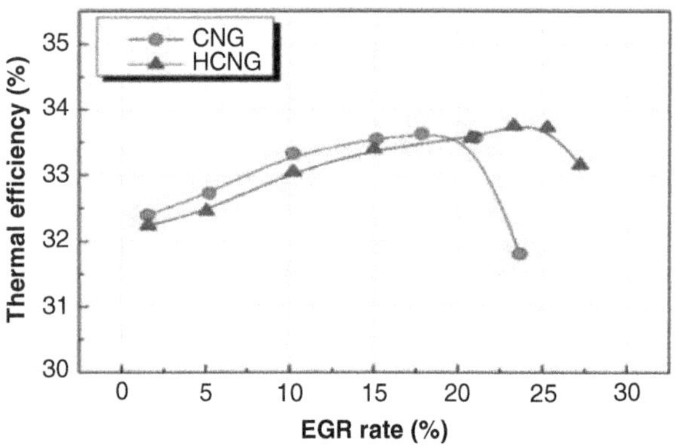

Figure 8.7 Thermal efficiency for HCNG and CNG engine with various EGR rates [27].

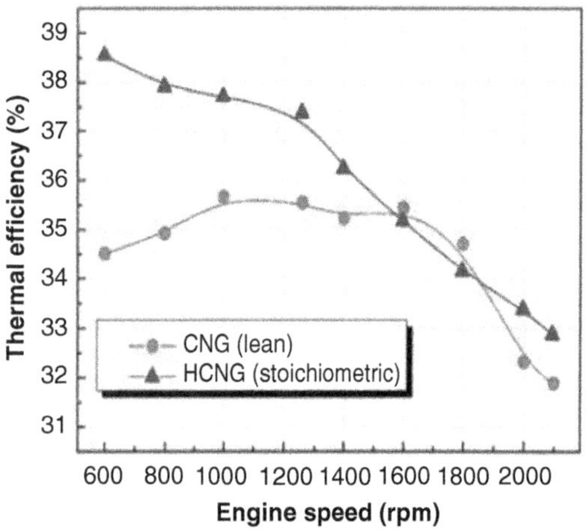

Figure 8.8 Thermal efficiency vs. speed [27].

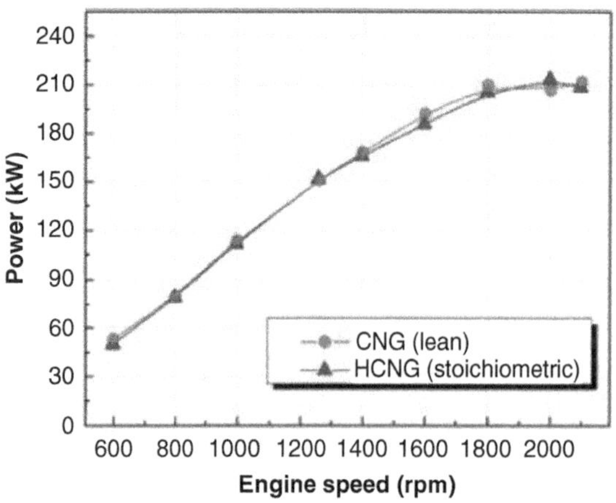

Figure 8.9 Power vs. speed [27].

Sun et al. [29] developed both 1D and 3D simulation models of NG SI engines and validated them against experimental data. Hydrogen addition led to a significant increase in NO concentration mainly inside the high-temperature combustion zone in the flame front. CO formation rates increased with hydrogen addition, and hydrogen addition also facilitated the complete oxidization rate of CO. Hydrogen addition also decreased the HC and soot emissions level.

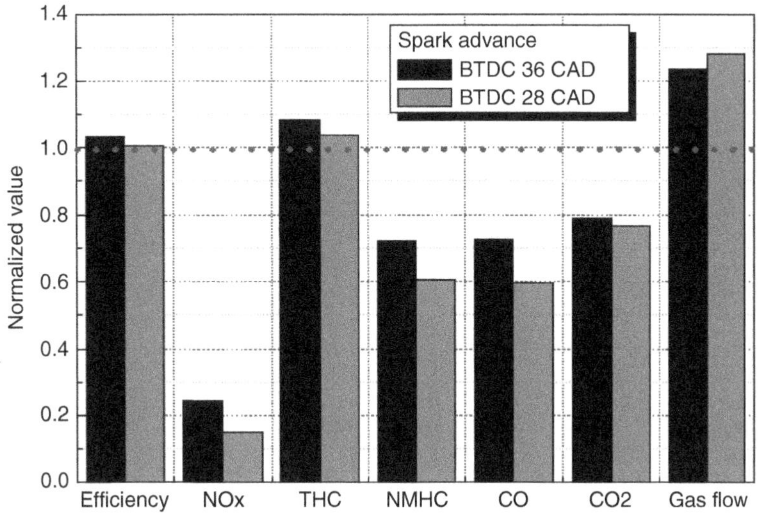

Figure 8.10 Improvement in exhaust emissions with retarded spark timing 30 vol% blending ratio of hydrogen [27].

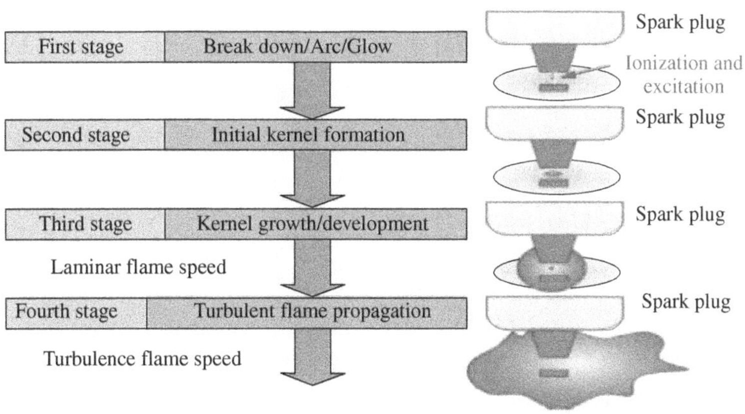

Figure 8.11 Typical spark kernel formation, growth, and flame propagation in an SI engine [29].

As shown in Figure 8.11, laminar flame speed is quite crucial to the growth and development of the initial spark kernel and its subsequent influence on turbulent flame propagation. Hydrogen has an exceptionally high laminar flame speed, approximately eight times that of natural gas. The highest laminar flame speed of methane, the basic constituent of natural gas, occurs at an equivalence ratio of 1.05 and it decreases with both leaner and richer mixtures. Hydrogen's laminar flame speed occurs at an equivalence ratio of around 1.8, which makes it possible to extend its lean-burn limit significantly. Thus, hydrogen combined with methane contributes to accelerated flame propagation with increased combustion speed, which leads to larger initial spark kernel volume for the blend.

Figures 8.12, 8.13, 8.14, and 8.15 show the experimental and simulated data for NOx, HC, CO, and soot, respectively, for natural gas as well as hydrogen-added natural gas engine operation. Higher combustion temperatures, due to hydrogen addition, increase heat transfer through the

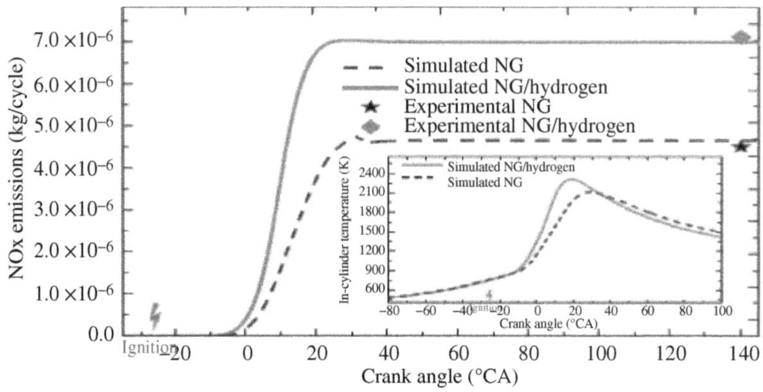

Figure 8.12 Comparison in simulated NOx emissions history and measured NOx emissions of the NG SI engine with/without hydrogen [29].

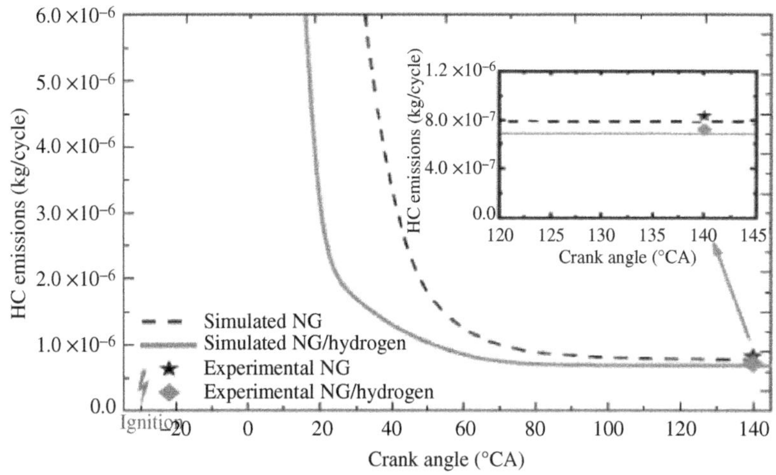

Figure 8.13 Comparison in simulated HC emissions history and measured HC emissions of an NG SI engine with/without hydrogen [29].

cylinder wall and increase the nitrogen oxide (NOx) emissions (Figure 8.12). Moreover, the reduced flame quenching distance near the cylinder wall with hydrogen addition minimized unburned mixture in crevices and near the cylinder wall, which led to reduced hydrocarbon (HC) emissions (Figure 8.13). Higher diffusivity, wider lean-burn range, and the higher flame speed of hydrogen contribute to lower HC emissions. Figure 8.14 compares simulated and measured CO emissions from the NG SI engine with and without hydrogen. With hydrogen, the CO formation rate was faster and oxidation was more complete, as evident in the enlarged area.

Hydrocarbon fuel combustion leading to solid soot particle emission involves a complex sequence of processes such as pyrolysis, precursors, cyclization, nucleation, coalescence, surface growth, and agglomeration. Several intermediated species are formed during pyrolysis, among which acetylene is crucial to soot formation. Figure 8.15 shows simulated soot emissions with and without hydrogen. As evident in the figure, hydrogen addition significantly reduced soot emissions.

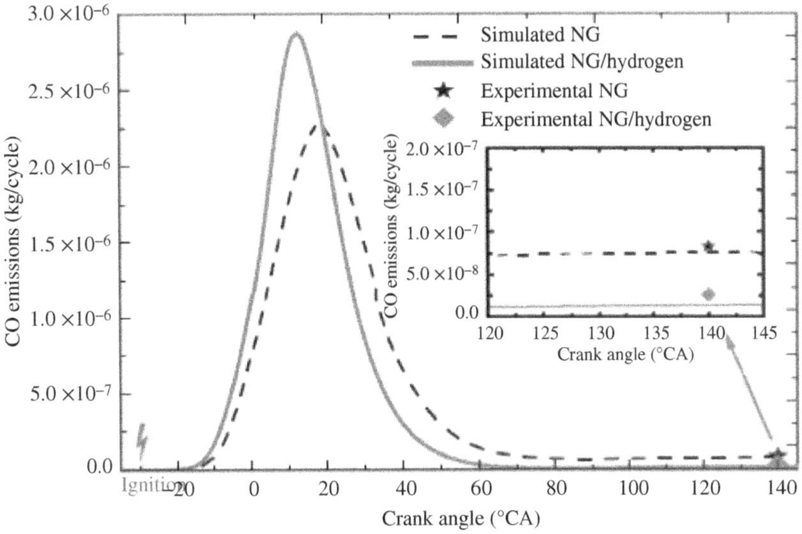

Figure 8.14 Comparison in simulated CO emission history and measured CO emission of an NG SI engine with/without hydrogen [29].

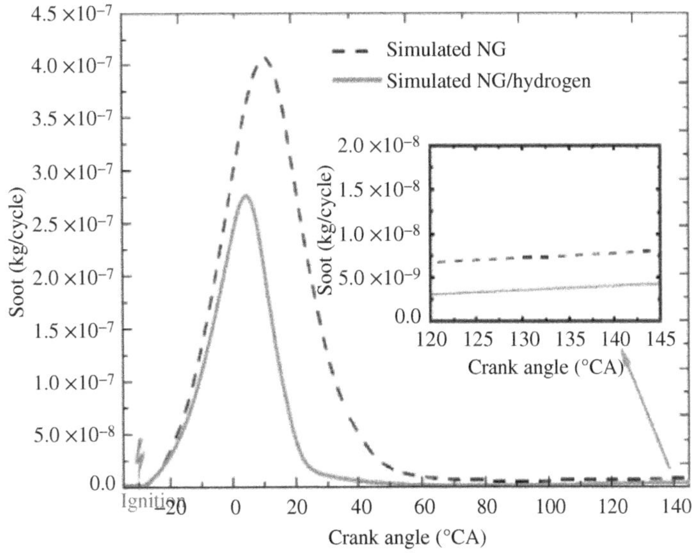

Figure 8.15 Comparison of the simulated soot emissions history of an NG SI engine with/without hydrogen [29].

Lee et al. [30] conducted experiments in a constant volume combustion chamber to study the impact of hydrogen fraction on combustion properties of HCNG blends using a single spark plug (SSP) and a double spark plug (DSP). While the DSP arrangement improved combustion efficiency and reduced pollutants like carbon monoxide (CO) and hydrocarbons (HC), it showed higher carbon dioxide (CO_2) emissions than SSP (Figure 8.16), which indicates complete combustion. NOx emissions for DSP were higher than for SSP (Figure 8.17) because better combustion caused higher temperatures.

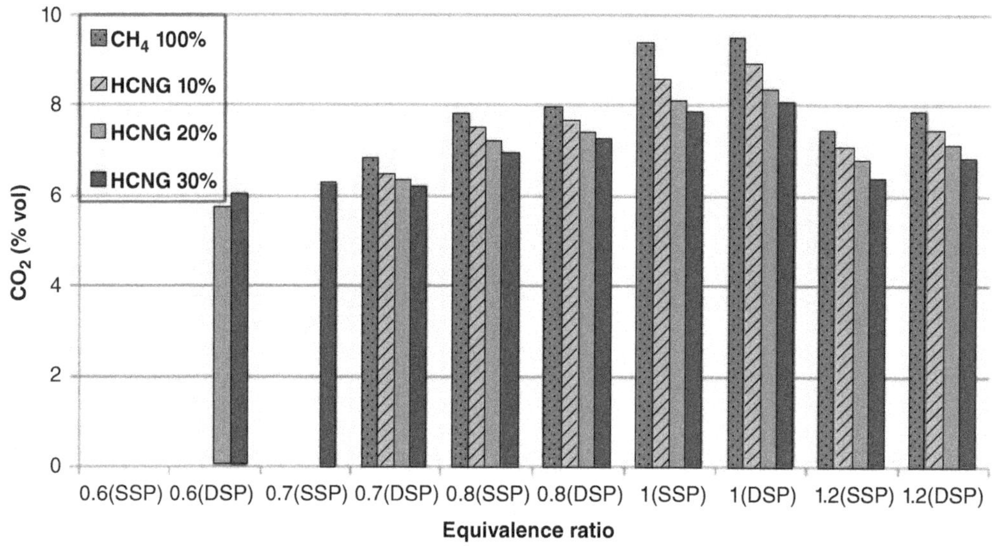

Figure 8.16 CO_2 emission for various equivalence ratios at DSP and SSP for various fractions of HCNG blends [30].

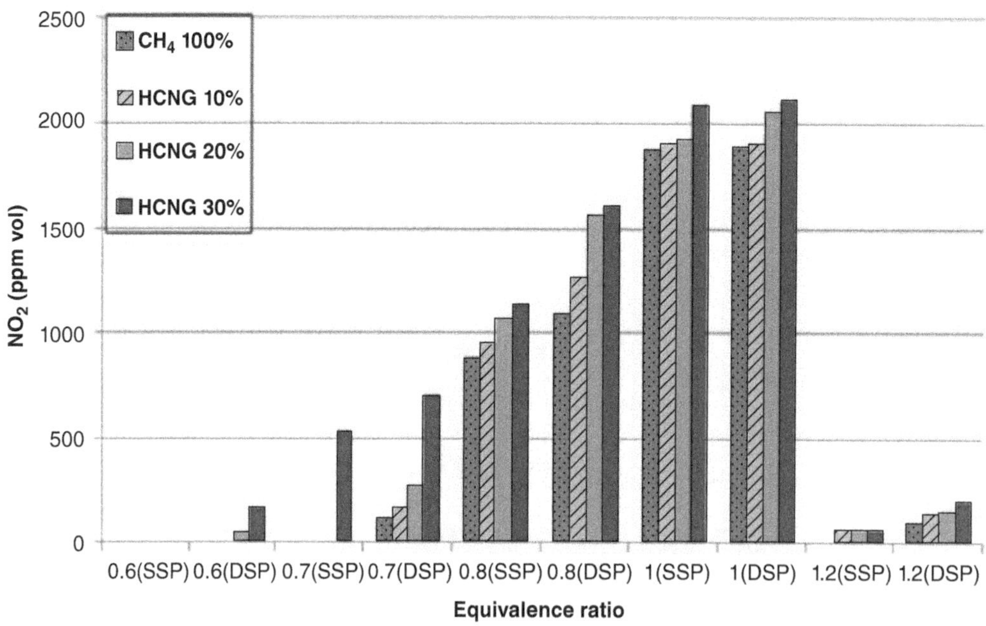

Figure 8.17 NOx emission for various equivalence ratios at DSP and SSP for various fractions of HCNG blends [30].

Excess air ratio (λ), hydrogen fraction, and dilution ratio have significant influence on combustion instability and laminar burning velocity [31,32]. Zheng et al. [33] conducted experiments to analyze the laminar burning velocity and flame stability of CO_2-diluted NG/H_2/air mixtures in a fixed-volume combustion chamber equipped with a high-speed Schlieren photography system. The unstretched laminar burning velocity of CO_2-diluted NG/H_2/air mixtures showed an increasing trend with higher hydrogen addition but deteriorated with increased CO_2 dilution ratio and initial pressure. In this paper, flame propagation of premixed CO_2 diluted natural gas/hydrogen/air mixtures under various initial pressures was studied by using a constant volume combustion bomb together with high-speed schlieren photography.

Laminar flame speed and laminar burning velocity as well as Markstein length and flame thickness were obtained for the diluted stoichiometric fuel/air mixtures with different natural gas/ hydrogen fractions and diluent ratios under normal, reduced, and elevated pressures. The results showed that both unstretched flame speed and unstretched burning velocity are reduced with the increase of diluent ratio as well as initial pressure (except when the hydrogen fraction is 80%). Hydrogen-enriched natural gas with higher hydrogen fraction can tolerate relatively more diluent gas. Ji et al. [34] studied flame propagation of premixed CO_2 diluted natural gas/hydrogen/air mixtures under various initial pressures using a constant volume combustion bomb with high-speed schlieren photography. Significant parameters such as laminar flame speed, laminar burning velocity, Markstein length, and flame thickness were evaluated for the diluted stoichiometric mixtures of natural gas/ hydrogen fractions with normal, reduced, and elevated diluent pressures. Both unstretched flame speed and unstretched burning velocity, except those with 80% hydrogen fraction, were reduced with increased diluent ratio as well as initial pressure. Tahtouh et al. [35], based on their experiments, developed a correlation for calculating the laminar burning velocity methane–hydrogen–air–nitrogen mixtures, which was verified to be in good agreement with published literature. This study showed that the laminar burning velocity showed an increasing trend with all dilution ratios. On the contrary, the burned gas Markstein length, except for high hydrogen mole fractions (>0.6), decreased with the increase in hydrogen fraction in the mixture. Nitrogen dilution has a nonlinear reducing effect on the laminar burning velocity and an increasing effect on the burned gas Markstein length. Park et al. [36] also studied the combustion of CO_2-diluted HCNG fuel mixtures and its effects in a heavy-duty SI engine. Their results show that with the increase in CO_2 thermal efficiency was reduced. The increased heat capacity of the mixture led to a lower combustion temperature, which reduced the NOx emission level. Introduction of CO caused an increase in THC emissions.

8.2.2 Exhaust Gas Recirculation

Allenby et al. [37] adopted exhaust gas recirculation (EGR) to assess the capacity of hydrogen enrichment to increase natural gas's dilution tolerance. Results of this study on a stoichiometric engine demonstrated stable combustion with significant emission reductions. A catalytic exhaust gas reforming reactor with up to 25% of EGR led to an over 80% reduction in NO emissions over the entire range of tests. However, for efficient function of the catalyst in the part-load exhaust gas temperature, it is important to assess the low temperature operation of reformers. Moreover, to ensure efficient implementation of onboard hydrogen-rich EGR, the catalyst should be able to function in extreme engine conditions, such as oxygen surges during deceleration due to fuel cutoff.

In one study, a passenger car natural gas engine was optimized for hydrogen-natural gas mixtures and high exhaust gas recirculation (EGR) rates in the part load domain [38]. Figures 8.18a and 8.18b show the fuel conversion efficiency and NOx emission, respectively, using optimum spark timing. Researchers adopted the EGR-ST model combination and verified measurements on the engine test bed to plot these graphs. The graphs were generated on the basis of the EGR-ST model combination and verified measurements on the engine test bed. An optimal combination of spark timing and EGR rate led to improved efficiency and substantial reductions in NOx emissions. Optimal rationalization of the EGR rate, hydrogen fraction in the fuel, and spark timing resulted in moderate in-cylinder peak pressure and peak combustion temperatures.

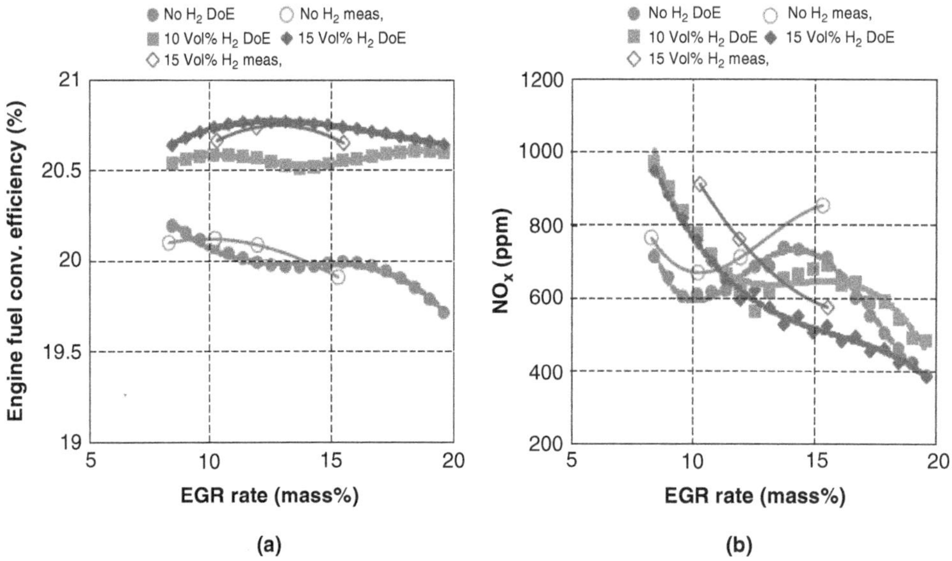

Figure 8.18 (a) Break fuel conversion efficiency vs. EGR quantity using efficiency optimal spark timing (for each point). (b) Engine-out NOx emissions vs. EGR quantity using efficiency optimal spark timing (for each point).

Liu et al. [39] investigated EGR on a stoichiometric HCNG-fueled DISI engine at 1200 rpm under wide-open throttle (WOT) conditions, with two injection timings: 300° CA BTDC (premature) and 170° CA BTDC (delayed). The results showed that BMEP decreased with increasing EGR rate, but above 15% EGR adding more hydrogen to the hydrogen–natural gas mixture stabilized combustion and improved the BMEP. Optimal brake thermal efficiency (BTE) was achieved at 5–10% EGR for all hydrogen fractions. NOx emissions decreased with higher EGR rates but increased with hydrogen enrichment. A reduction of 80% in NOx emissions was observed with 25% EGR dilution. Saanum et al. [40] researched two strategies such as lean burn operation and stoichiometric operation with EGR and a three-way catalyst (TWC) on a six-cylinder, 9.6 L, CR 11.8 engine at a constant 1200 rpm fueled by 25% by volume of hydrogen blended with 75% natural gas by volume. Experimental results showed that even with 25% by vol of hydrogen mixture, lean burn operation produced higher NOx levels compared to the EGR and catalyst approach. Mariani et al. [41] developed a numerical model to investigate the performance and emission characteristics of an HCNG engine. They analyzed the stoichiometric air–fuel ratio using a three-way catalyst (TWC) and EGR. They used optimized spark timing with each fuel blend at the maximum brake torque (MBT) point. Simulations, conducted under conditions replicating the engine cycle of a passenger car according to the New European Driving Cycle (NEDC), showed that at low loads and with a higher percentage of hydrogen in the HCNG mixture, there was an increase in brake thermal efficiency and reduction in fuel consumption over the NEDC. However, there was also an increase in NOx pollutants due to higher cylinder temperatures, although this trend was reduced using EGR (Figure 8.19).

Ma et al. [42] tested HCNG in a spark ignited natural gas engine with varying proportions of hydrogen in the blend. Their results showed that increasing hydrogen enrichment reduced the combustion duration at any excess air ratio (Figure 8.20). The tests also showed the lean limit

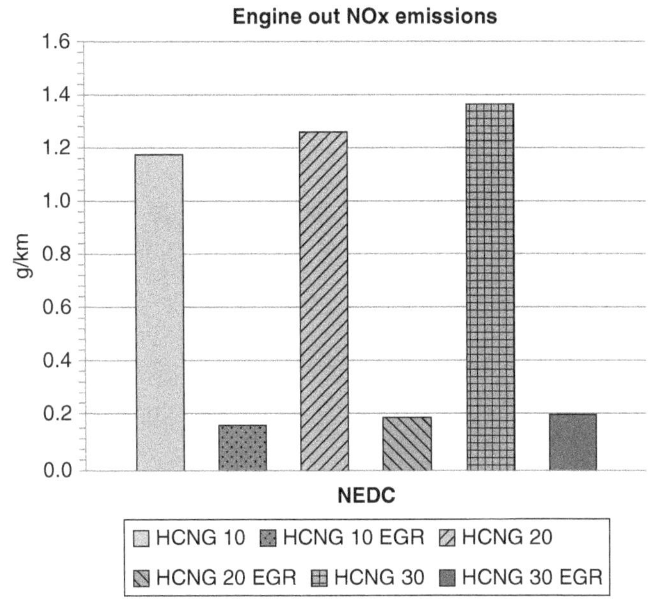

Figure 8.19 Comparison of NOx emissions over NEDC for a HCNG mixture with 10% EGR and without EGR [41].

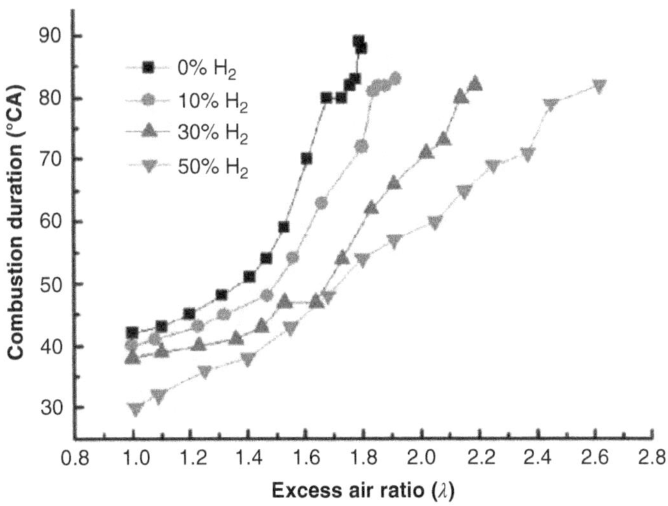

Figure 8.20 Combustion (0–98%) duration vs. excess air ratio for fuel blends of various hydrogen fractions, at engine speed 1200 rpm manifold absolute pressure (MAP) = 105 kPa, and spark advance = 30° [42].

significantly extended with higher hydrogen fractions in the fuel blend. The lean limits of blends were 1.82, 2.09, and 2.4, at 10%, 30%, and 50% hydrogen fractions, respectively, compared to a lean limit of 1.71 for natural gas alone.

Figures 8.21, 8.22, and 8.23, generated at 1200 rpm and manifold absolute pressure of 105 kPa, show the pattern of variation of CO, HC, and NOx, respectively, with respect to excess air ratio [42]. HC emissions reached a minimum when the air–fuel ratio was slightly above unity due

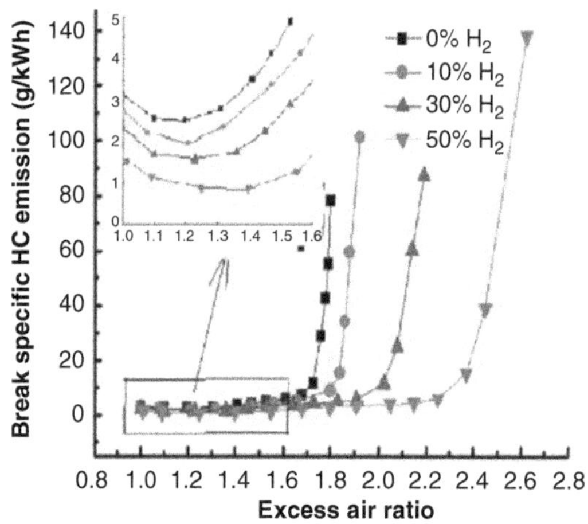

Figure 8.21 Brake specific HC emission vs. lambda for fuel blends of various hydrogen fractions, engine speed 1200 rpm, MAP =105 kPa, and spark advance 30° [42].

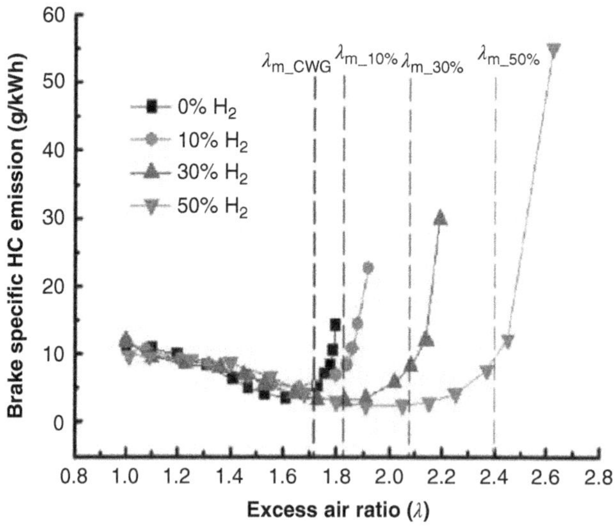

Figure 8.22 Brake specific CO emission vs. lambda for fuel blends of various hydrogen fractions, engine speed 1200 rpm, MAP =105 kPa, and spark advance 30° [42].

to sufficient air for complete combustion and high exhaust temperatures. Moreover, hydrogen's accelerated flame propagation and reduced quenching distance contributed to reduced HC emissions.

CO emissions decreased initially and rose rapidly under further lean engine conditions. Hydrogen addition reduced CO emissions significantly when the air–fuel ratio exceeded 1.7 due to hydrogen's combustion-enhancing properties. Hydrogen enrichment without optimized spark timing led to higher NOx emissions. Optimized spark timing for MBT increased engine efficiency with higher hydrogen fractions. However, at this optimized timing, NOx emissions show small variation among fuel blends with different hydrogen fractions.

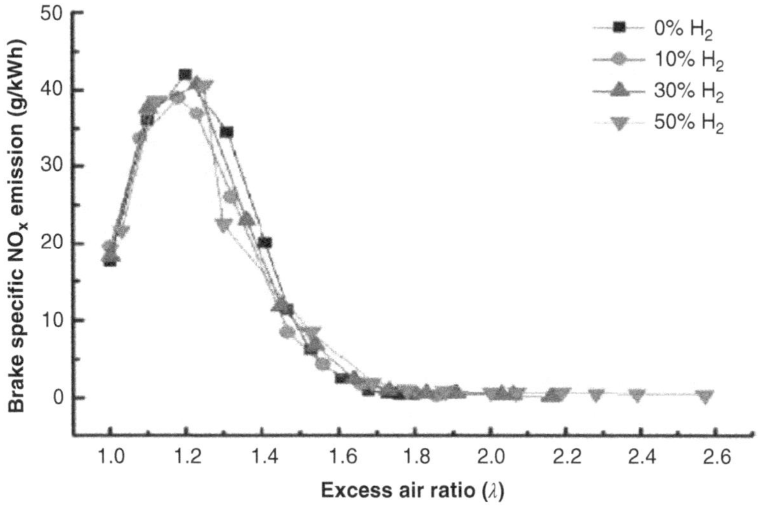

Figure 8.23 Specific NOx emission vs. lambda at MBT spark timing for fuel blends of various hydrogen fractions, at engine speed 1200 rpm and MAP = 105 kPa [42].

8.2.3 Injection

Development of the hydrogen injector for prolonged efficient operation in the thermal environment of the combustion chamber, proper synchronization of injection timing and ignition timing to ensure effective ignition, and efficient combustion of fuel–air mixtures are some of the salient features that have come out of research related to injection of gas engines using hydrogen. Huang et al. [43] studied combustion characteristics and heat release analysis of an HCNG SI engine. Figure 8.24 shows the effect of varying hydrogen fractions in natural gas–hydrogen blends on combustion characteristics under different excess air ratios. It is evident in the figure with $\lambda = 0.9$ that for rich mixtures and optimal ignition timing, increasing hydrogen fraction leads to reduction in the heat release rate, and for stoichiometric mixtures ($\lambda = 1.0$) and lean mixtures ($\lambda = 1.2$, 1.4) the blend with 10% hydrogen exhibits the highest maximum heat release rate. Adding more hydrogen lowers the maximum heat release rate because of the combined effect of increasing burning velocity and decreasing heating value of the mixture with the increasing hydrogen fraction. It was also concluded in this study that, under situations of the same excess air ratio and at optimum ignition timing, initial combustion duration and the total combustion duration become diminished as the hydrogen fraction in the natural gas–hydrogen blends is increased. In this analysis, combustion duration is divided into initial combustion duration (from ignition start to 10% accumulated heat release), rapid combustion duration (from 10% to 90% accumulated heat release) and total combustion duration (sum of initial and rapid combustion durations).

A study by Wang et al. [44] emphasized ignition timing in direct-injection HCNG engines. Shorter time intervals between fuel injection and ignition led to higher brake mean effective pressures and thermal efficiency, while combustion durations were reduced. With specific ignition timing, increasing the hydrogen fraction in natural gas improved brake mean effective pressure and thermal efficiency and combustion duration was decreased. Ignition timing advancement decreased the exhaust hydrocarbon (HC) concentration. Nitrogen oxide (NOx) concentration showed an increasing trend while carbon monoxide (CO) remained relatively unaffected (Figure 8.25).

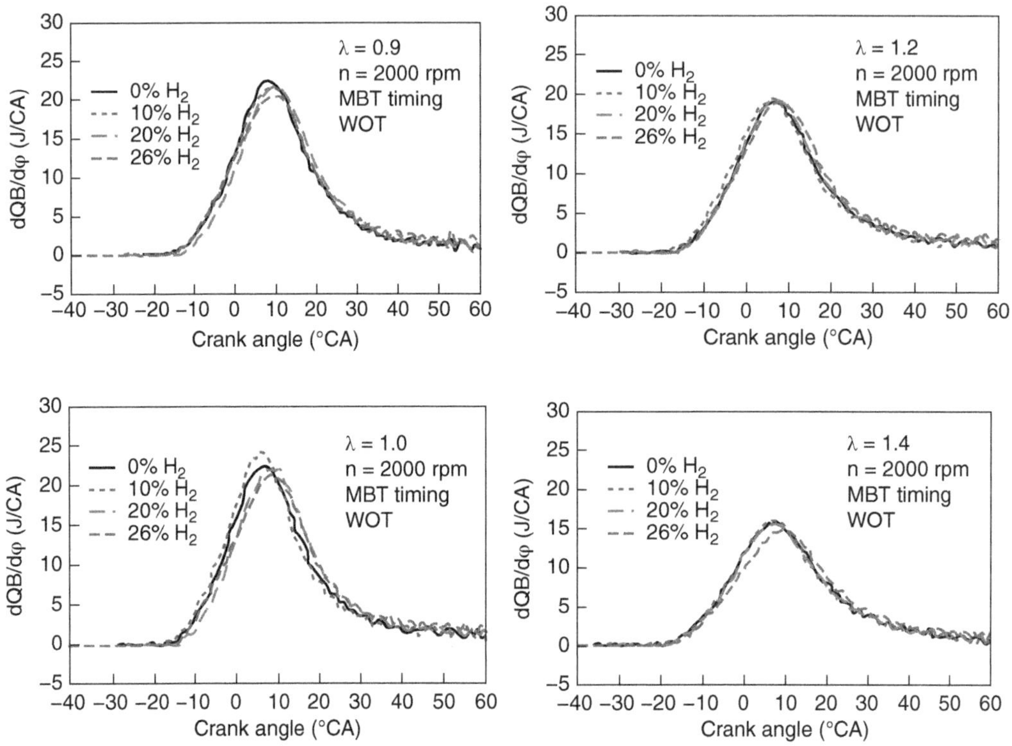

Figure 8.24 Heat release rate of natural gas–hydrogen blends [43].

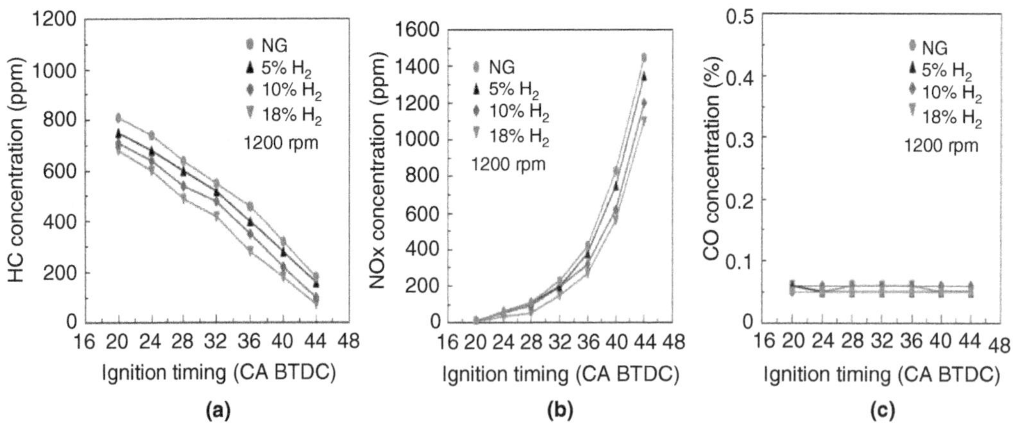

Figure 8.25 Variation of HC, NOx, and CO with ignition timing [44].

Sandhu et al. [45] studied single-cylinder motorbike engines with HCNG as fuel. A timed manifold injection system through an electronic fuel injection system was used, and the engine was fueled with 5, 10, 15, and 20% hydrogen energy in the HCNG mixture. BTE increased with increased hydrogen percentage, and this increase was more pronounced in leaner regions. In general, cylinder pressure and heat release show increasing trend with hydrogen addition. Figures 8.26 and 8.27 represent typical conditions of variation of cylinder pressure and heat release, respectively, with respect

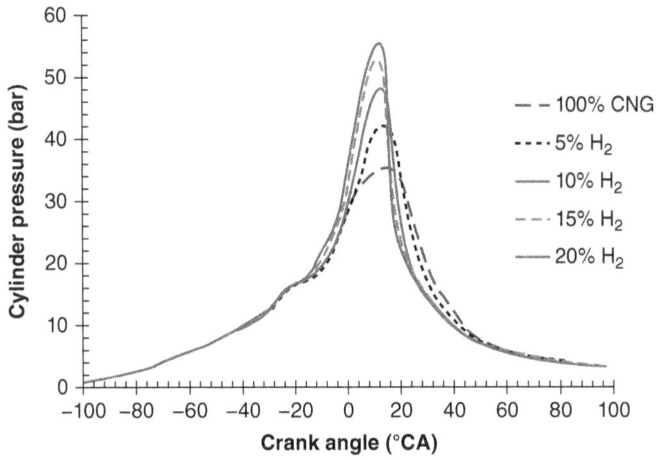

Figure 8.26 Variation of cylinder pressure with crank angle [45].

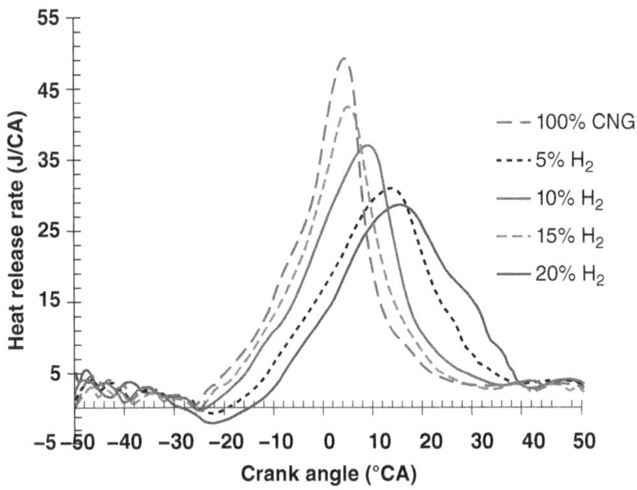

Figure 8.27 Variation of heat release with crank angle [45].

to crank angle at 2800 rpm and equivalence ratio of 0.96. Emissions of HC and CO decreased with increasing hydrogen percentages while increasing NOx. Moreover, combustion analysis showed that hydrogen supplementation shifted the peak pressure toward TDC and increased heat release rate [46,47].

Biffiger et al. [48] compared blends of premixed methane and hydrogen with split injections of methane (port injected) and small amounts of hydrogen (direct injected). While split hydrogen injection and methane did not provide significant advantages, split methane injection alone showed promising results. Injecting methane directly into the spark plug had a similar impact on combustion as adding hydrogen. Engine operation on PFI with hydrogen addition to methane improved engine efficiency, achieved the lean limit, and resulted in lower NOx emissions. It also decreased combustion durations and ignition delay. Both CO_2 and UHC also were reduced. An engine system with hydrogen (DFI) and methane (PFI) facilitated a higher range of lean operation compared to a premixed combustion of pure methane, but it was less with respect to premixed hydrogen-enriched methane. Compared to the premixed combustion, UHC emission were higher, which led to a fall in efficiency. Engine operation with hydrogen (directly injected) and methane (port fuel injected) demonstrated an increase in the lean limit compared to pure methane, although to a lesser extent than observed with hydrogen-enriched methane. Improve kinetics proved more effective. Split injection proved an effective approach as it allowed precise control to optimize an engine and avoided the stringent condition of direct injection of the entire amount of fuel. Charge stratification near the spark plug formed by direct injection acts as a "virtual prechamber" [48].

A number of studies have been carried out at the Indian Institute of Technology, Delhi, using various proportions of hydrogen blends in CNG [49–52]. Apart from performance, exhaust emission and combustion, long-term sustained engine operation and lubrication tests have been reported on a modified gasoline genset engine. Over a 60-hour test, HCNG showed lower BSFC, CO, and HC emissions but higher NOx. HCNG led to higher iron deposits on spark plugs and cylinder liners in comparison to CNG. Lubricant properties like viscosity and total base number (TBN) decreased significantly with HCNG. Higher levels of wear metals (iron and copper) were also found in used oil. Das et al. [49] conducted three series of experiments on a gasoline-powered single-cylinder SI engine. Neat gasoline was the initial fuel. In the next series, modifications were made, including installing a gas injector and positioning the engine intake manifold. In the final stage, the test rig was further modified to use a mixture of hydrogen and CNG as fuel. Results showed that blending 10% and 15% hydrogen to CNG significantly improved performance and emission behavior. HCNG blends consistently yielded higher power output versus CNG due to hydrogen's combustion properties. BTE increased with ϕ up until a certain limit, then showed a decreasing trend due to poor combustion stability. Results obtained by different research activities in several industries on different design of vehicles showed that existing CNG engines could run efficiently with 18% hydrogen by volume without major alteration. Keeping that in view, extensive tests were conducted at the Indian Institute of Technology, Delhi, to investigate performance, emission, and combustion studies with HCNG fuelling (18% by volume of hydrogen blended to CNG) on a gasoline genset engine that was retrofitted and appropriately modified (Figure 8.28).

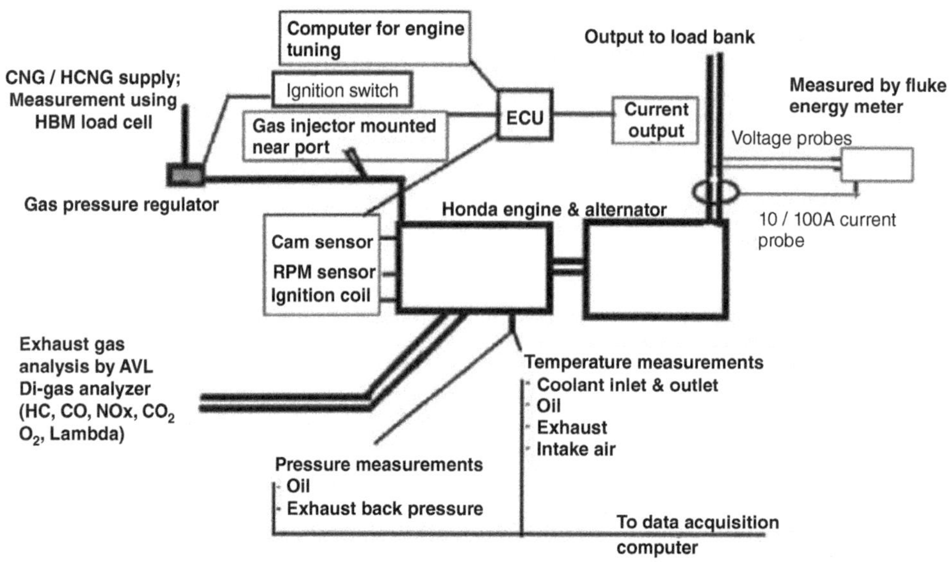

Figure 8.28 Schematic of experimental set-up [50, 51].

Another goal was to analyze deposits and lubricant conditions within the combustion chamber after prolonged operation.

In an effort to evolve an optimized engine configuration, initial engine tests were conducted with varying compression ratios (of 8.5, 9.3, and 10.0) and spark timing. A CR of 9.3 at 24° BTDC spark timing showed maximum power and improved brake thermal efficiency. Subsequent tests varied injection timing from 24° BTDC to 18° and 21° BTDC to optimize for HCNG fuel. However, spark timing at 24° BTDC remained optimal for best HCNG engine performance. Long-duration tests were conducted under these optimal conditions of 9.3 compression ratio and 24° BTDC ignition timing for both CNG and HCNG fuels.

8.2.4 Performance and Emission Characteristics During Long-Term Endurance Tests

Endurance tests were conducted using CNG and HCNG fuels. Before each test, the power section (rings and pistons) of the engine was replaced, and it was overhauled according to the manual. Engine run-in was performed using the respective test fuel, followed by draining and refilling with fresh SAE 15W40 oil.

Performance tests were conducted at varying loads, maintaining a constant engine speed. Endurance tests were run at constant speed of 3000 rpm and load of 2.8 kW (BMEP: 3.1 bar) for 60 hours without oil top-up. Engine performance and emissions were evaluated every 10 hours, with used oil analysis conducted every 15 hours. The engine was stopped and oil samples were drawn for analysis. After each test, the engine was dismantled and deposits on components were assessed. Performance results indicated a marginal increase in brake specific fuel consumption (BSFC) by 2% for HCNG operation over CNG fueling at the end of tests (Figure 8.29). The results suggest consistent performance without abnormal deposit formation or engine component wear.

During the entire range of endurance tests, HCNG demonstrated, on average, 8.3% lower BSFC relative to CNG. This was because of higher calorific value and improved combustion efficiency due to a higher hydrogen content. Emission results depicted in Figure 8.30 show that HCNG operation yielded lower CO and HC emissions but higher NOx emissions relative to CNG. The increase in hydrogen content in HCNG (2% by weight or 18% by volume) reduced the C/H ratio and resulted in better combustion degrees, primarily due to hydrogen's higher flame speed. Moreover, the quenching gap of hydrogen is 0.6 mm compared to 1.9 mm for CNG. This leads to complete combustion of the fuel–air mixture adjacent to cylinder walls and crevice volumes, further reducing HC emissions. A decrease in HC and CO emissions of 89.4% and 42.4%, respectively, relative to CNG was

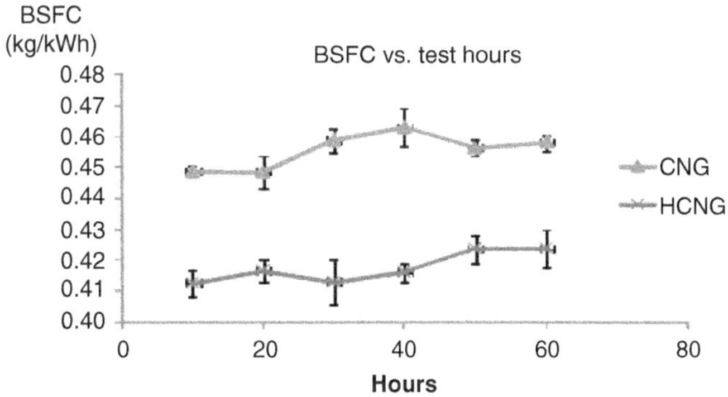

Figure 8.29 Variation in BSFC during endurance tests of 60 hours [50].

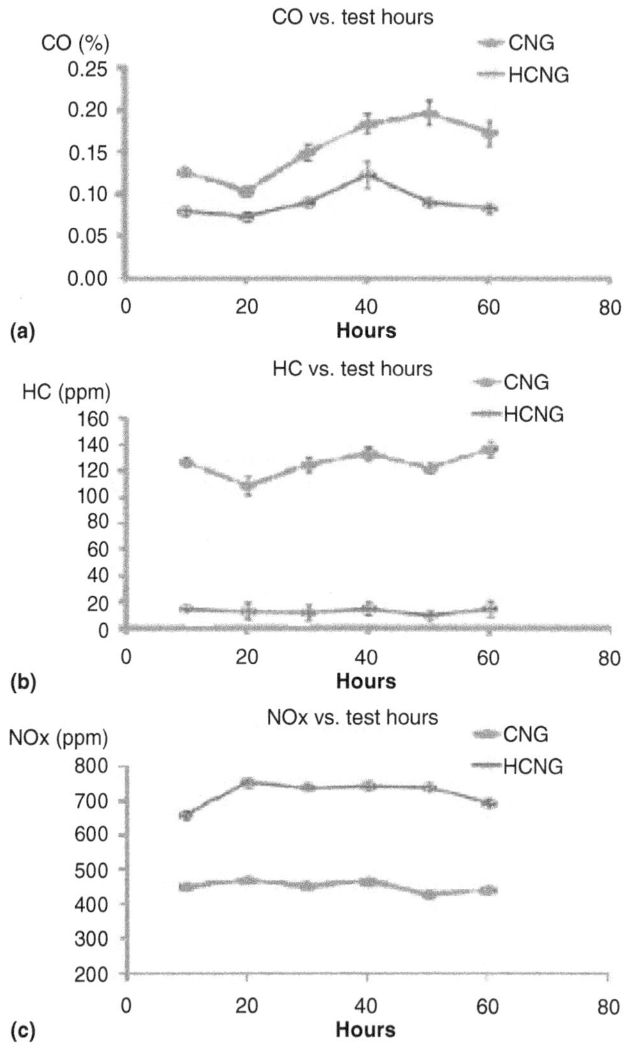

Figure 8.30 Variation of CO, HC, and NOx emissions during endurance tests. (a) CO variation of CNG and HCNG endurance test for 80 hours. (b) HC variation of CNG and HCNG endurance test for 80 hours. (c) NOx variation of CNG and HCNG endurance test for 80 hours [50].

observed. However, the accelerated HCNG combustion process increased thermal NOx formation, resulting in a 59.1% increase in NOx emissions vis-à-vis CNG due to factors like increased oxygen concentration, peak temperature, and residence time at peak temperature.

8.3 Oil Analysis

The variation of kinematic viscosity of lube oil was measured at 100 °C using CANNON CAV 2000 series equipment, following ASTM D445, with CNG and HCNG as illustrated in Figure 8.31.

Kinematic viscosity indicates oil thickening, increased contamination levels, or insoluble presence. A reduction in viscosity suggests oil dilution or shearing of viscosity index improvers. Initially, SAE 15W40 oil had a viscosity of 14.18 cSt. With CNG, viscosity dropped marginally by

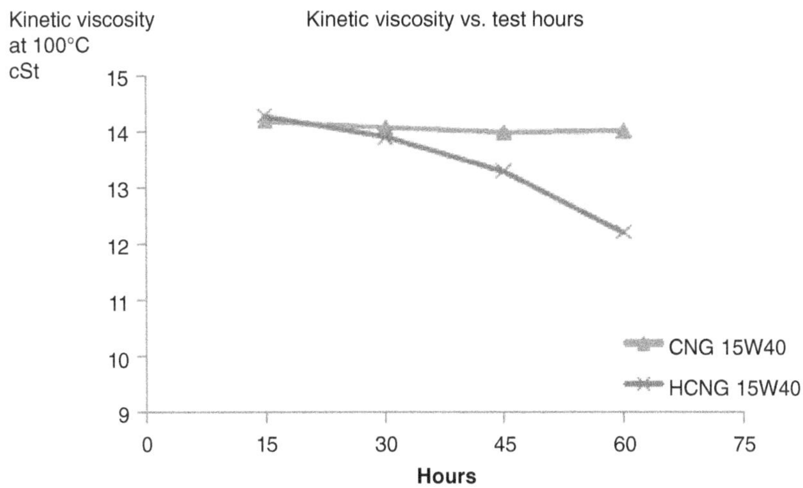

Figure 8.31 Variation in kinematic viscosity during an endurance test at 100 °C [50].

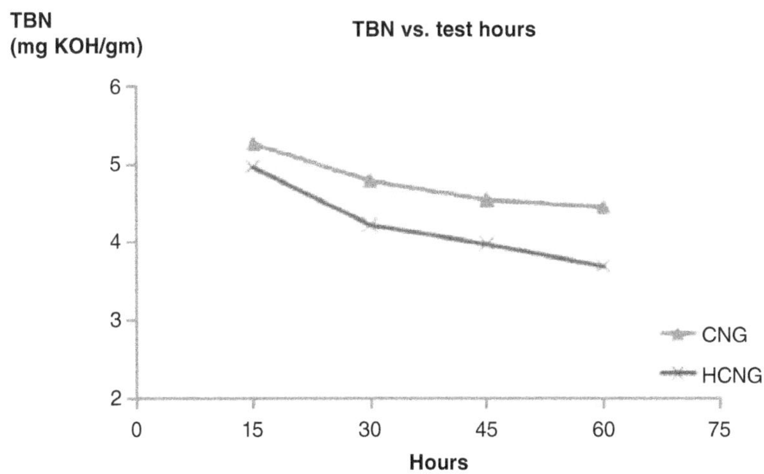

Figure 8.32 Variation in total base number during endurance tests [50].

1.3%, within limits due to viscosity index improved shearing. However, with HCNG, viscosity dropped by 14%, indicating permanent shearing of viscosity index polymers, likely due to high combustion temperatures compared to CNG.

8.3.1 Total Base Number (TBN)

TBN indicates reserve alkalinity in engine oil assessed per ASTM D2896 using Metrohm equipment. TBN reflects the ability of detergent/dispersant to neutralize acids formed during combustion. Figure 8.32 shows TBN variation in engine oil runs with CNG and HCNG. At the end of tests, the TBN drop for CNG and HCNG with SAE 15W40 oil was 16% and 30%, respectively, exceeding a 50% rejection limit from the initial value. Higher TBN reduction with HCNG suggests acidic component formation that might lead to corrosive wear.

8.3.2 Total Acid Number (TAN)

TAN measures acid value in engine oil evaluated according to ASTM D664 using Metrohm equipment.

A high TAN indicates nitration, oxidation, and contamination in the oil. It was observed that TAN increased by 20% and 48.6% using 15W40 oil with CNG and HCNG, respectively (Figure 8.33). The higher TAN increase with HCNG suggests the formation of acidic components, possibly due to elevated combustion temperatures, presence of water, and nitrogen oxides. However, further investigation is needed for confirmation.

8.3.3 Spectrochemical Analysis for Wear Metals

Spectrochemical analysis conducted according to ASTM D6595 measures wear metal concentrations in used oil, providing insight into wear rate of engine components. Multi-element Oil Analyser II Plus equipment was used for these measurements, typically analyzing wear particles in the 5–6 μm range. Elements analyzed include Fe, Cu, Al, Cr, Pb, and Sn. Significant variations in Fe and Cu concentrations were observed, particularly in SAE 15W40 oil used with HCNG (Figure 8.34).

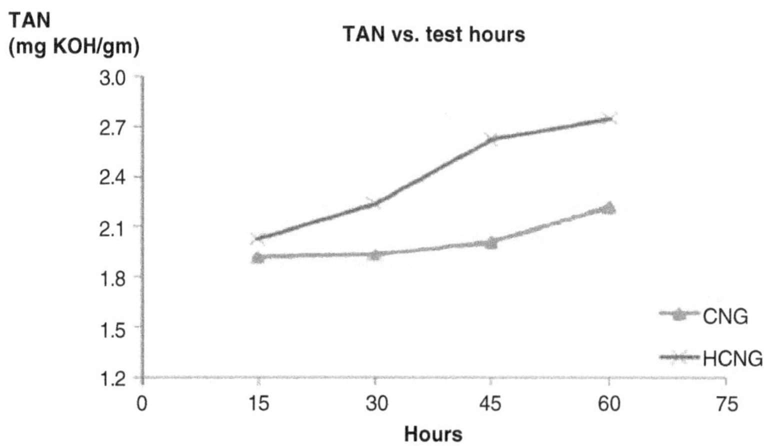

Figure 8.33 Variation of total acid number during endurance tests [50].

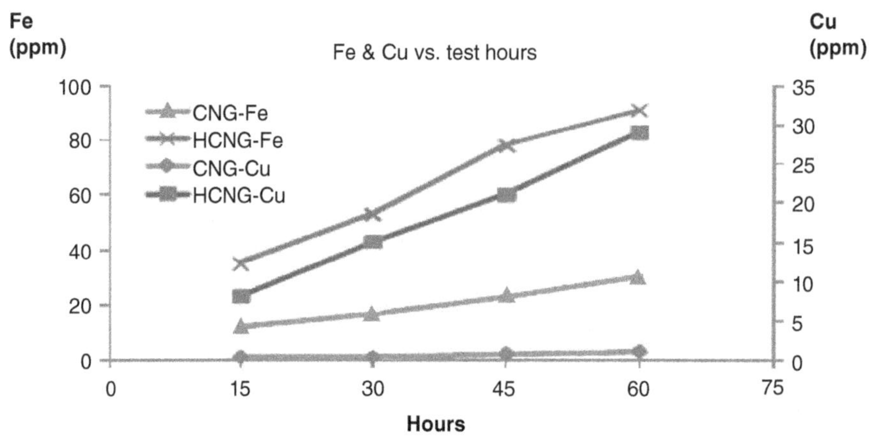

Figure 8.34 Variation of iron and copper in used oil during endurance tests [50].

Higher wear rate associated with HCNG operation is mainly attributed to viscosity reduction, resulting in thinner oil film between piston rings and liner. Additionally, formation of acidic compounds, leading to a lower total base number (TBN), can increase corrosive wear on bearings. Polar additives, such as anti-wear agents, serve critical roles in lubricating oils by forming protective layers on metal surfaces to reduce friction and prevent wear. Extreme pressure agents are specifically designed to offer additional protection under high-load conditions by forming a barrier that prevents metal-to-metal contact and subsequent damage. It was observed that wear particles depleted the oil of polar additives and accelerated wear in engine components, reducing oil's protective capabilities.

8.3.4 Nitration

Natural gas engine oils, nitration stemming from interaction between oil and nitrogen oxides (NOx) in combustion, stand as critical concern for oil performance. The nitration rate depends on various factors like air–fuel ratio, load, and oil temperature. At IIT Delhi study [50, 51], a Perkin Elmer IR spectrometer, following ASTM D7418, was used to measure nitration. It involves carbon chains reacting with nitrogen oxides (NOx) from combustion, thickening the oil prematurely, forming varnish and carbon deposits. Engine manufacturers like Waukesha and Caterpillar typically set a condemning limit at 2.5 abs/mm. As shown in Figure 8.35, nitration with HCNG increased mildly from 0.02 to 0.09 abs/mm, staying within acceptable limits.

Nitration rates depend on various factors like air–fuel ratio, load, and oil temperature. In the tests, oil temperatures ranged from 97 to 106 °C, with engine load at around 40% of capacity, mitigating severe nitration.

8.3.5 Deposits on Cylinder Head and Piston

After HCNG engine tests, piston deposits were assessed with the CRC rating manual [53]. For base CNG fuel, no coloration or deposits were observed (Figure 8.36) except for the first ring land. With HCNG, no carbon deposits were found in the first and second ring grooves, but lacquer was present. The third ring groove and land were deposit-free. Both pistons had carbon deposits on the first ring land due to combustion exposure.

However, HCNG pistons had higher deposits, possibly due to lubricant degradation at high temperatures. Presence of lacquer on the HCNG piston land indicated lubricant thermal stability, formed from prolonged exposure to combustion gases and thermal stress. Cylinder head deposits

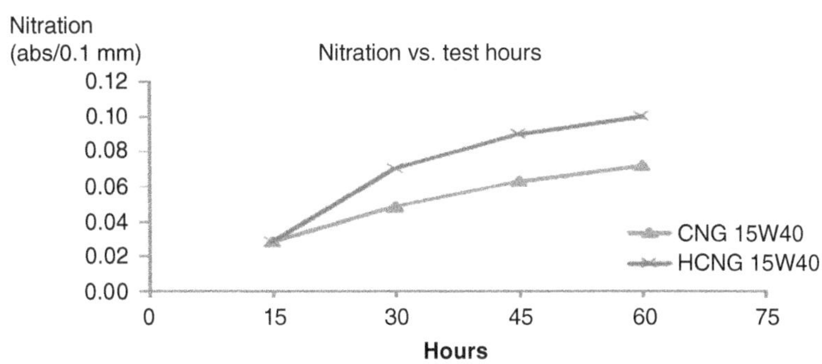

Figure 8.35 Nitration in used oil during endurance tests [50].

Piston - CNG   Piston - HCNG

Figure 8.36 Deposits on a piston after tests with CNG (left) and HCNG (right) [50, 51] / with permission of Elsevier.

Cylinder head - CNG   Cylinder head - HCNG

Figure 8.37 Cylinder head deposits with CNG (left) and HCNG (right) operation [50, 51] / with permission of Elsevier.

Spark plug - CNG   Spark plug - HCNG

Figure 8.38 Spark plug after tests with CNG and HCNG [50, 51] / with permission of Elsevier.

after CNG tests were grayish, whereas HCNG tests had more grayish- and reddish-color deposits (Figure 8.37).

Greyish color deposits were observed on spark plugs and cylinder bore after CNG testing, while grayish-reddish deposits were found after HCNG testing (Figures 8.38 and 8.39).

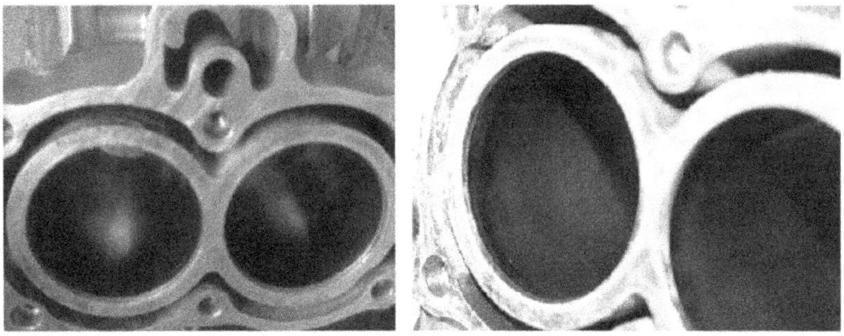

Figure 8.39 Cylinder bore after test with CNG (left) and HCNG (right) [50, 51] / with permission of Elsevier. Please visit the companion website to view a color version of this image.

Scanning electron microscope analysis of a nonmetallic spark plug part showed iron content of 1.1% for CNG and significantly higher percentages (12.9 and 17.1%) for HCNG. High temperatures and corrosive wear of engine components might have resulted in decreased total base number (TBN) and higher iron content in used oil analysis. Water and iron interaction at high temperatures and pressure could also be a factor.

8.4 Hydrogen Added to Biogas, Biodiesel

The primary combustible element in biogas is methane, which is also the major constituent of natural gas. Researchers are also working on different aspects of blending hydrogen with biogas for engine applications [54–64]. Porpatham et al. [54] experimented with the addition of hydrogen in small amounts (5, 10, and 15% on an energy basis) to biogas and evaluated the performance, combustion, and emission characteristics of a spark ignition engine at constant speed and different equivalence ratios. Hydrogen improved the combustion rate and extended the lean limit of biogas combustion, which led to improved brake thermal efficiency and brake power. Hydrogen operation in a rich mixture was not an attractive option as it required retarding ignition timing to control knock, which reduced thermal efficiency and power output. There was a significant reduction in hydrocarbon emissions with 10% hydrogen addition because of the fast combustion rate associated with extended lean operation. Addition of hydrogen to biogas did not show an appreciable increase in nitric oxide emissions. This was due to the combined effect of retarded ignition timing to avoid knock along with the presence of carbon dioxide, which acted as a diluent. Considering performance, emission, and smooth combustion, 10% hydrogen addition was observed to be most suitable for overall engine application in this study.

Figures 8.40 (a–e) show a pattern of variations of thermal efficiency and major pollutants with equivalence ratio. NO emission is almost negligible up to an equivalence ratio of 0.6, then it goes up until it peaks at close to the stoichiometric condition. CO level is extremely low in lean mixtures, but takes up an increasing trend in the rich zone, primarily due to incomplete combustion. Figures 8.41, 8.42, and 8.43 show results of combustion-related investigations. Figure 8.41 shows that peak pressure was significantly reduced because of increased hydrogen added for retarded ignition timing. The heat release rate shown in Figure 8.42 corresponds to an equivalence ratio of 0.9. The heat release rate was observed to decrease at the point (15% hydrogen addition) at which the spark timing had been retarded to control knock. Figure 8.43 illustrates the variations in peak pressure of 100 consecutive cycles at different equivalence ratios with 10% hydrogen addition to biogas. Cyclic variations were observed to be high at an equivalence ratio of 0.86 where knock occurred. It has also been found that retarding the timing at high equivalence ratio, adopted to avoid knock, also resulted in low peak pressure variation.

Pandey and a team of researchers [55] analyzed performance and emissions of an IC engine using biogas blended with 5, 10, and 15% hydrogen. Increasing the hydrogen content improved engine performance, power, and reduced carbon monoxide (CO) and total hydrocarbons (THCs), though nitrogen oxides (NOx) increased due to higher combustion temperatures. For 15% hydrogen-enriched biogas, there was a 17.5% increase in brake power, a 17% increase in brake thermal efficiency (BTE), a 50% decrease in CO, a 68% decrease in unburned hydrocarbons (UHC), and a 71% increase in NOx at an equivalence ratio of 0.6. Results of this study are shown in Figures 8.44 to 8.48.

Ali et al. [56] studied the combustion and emission characteristics in a thermal barrier coating (TBC) diesel engine fueled with CH_4–CO_2–H_2 mixtures. Decreasing heat transfer from the combustion chamber to the cooling system improved BTE. Ceramic coatings with lower thermal

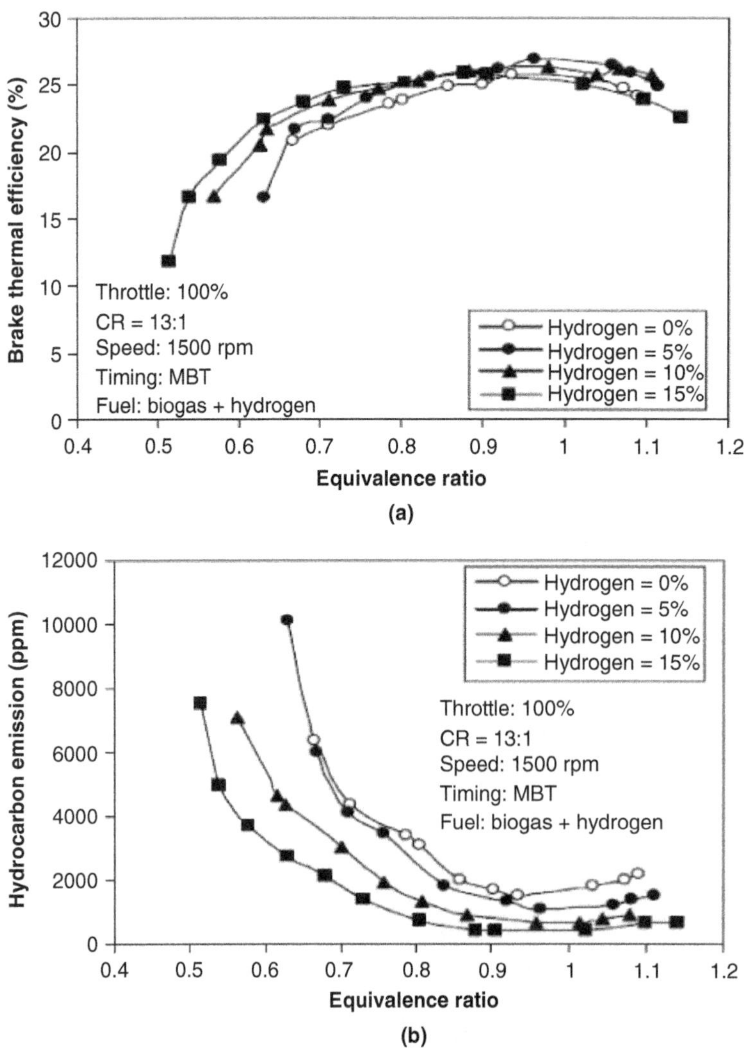

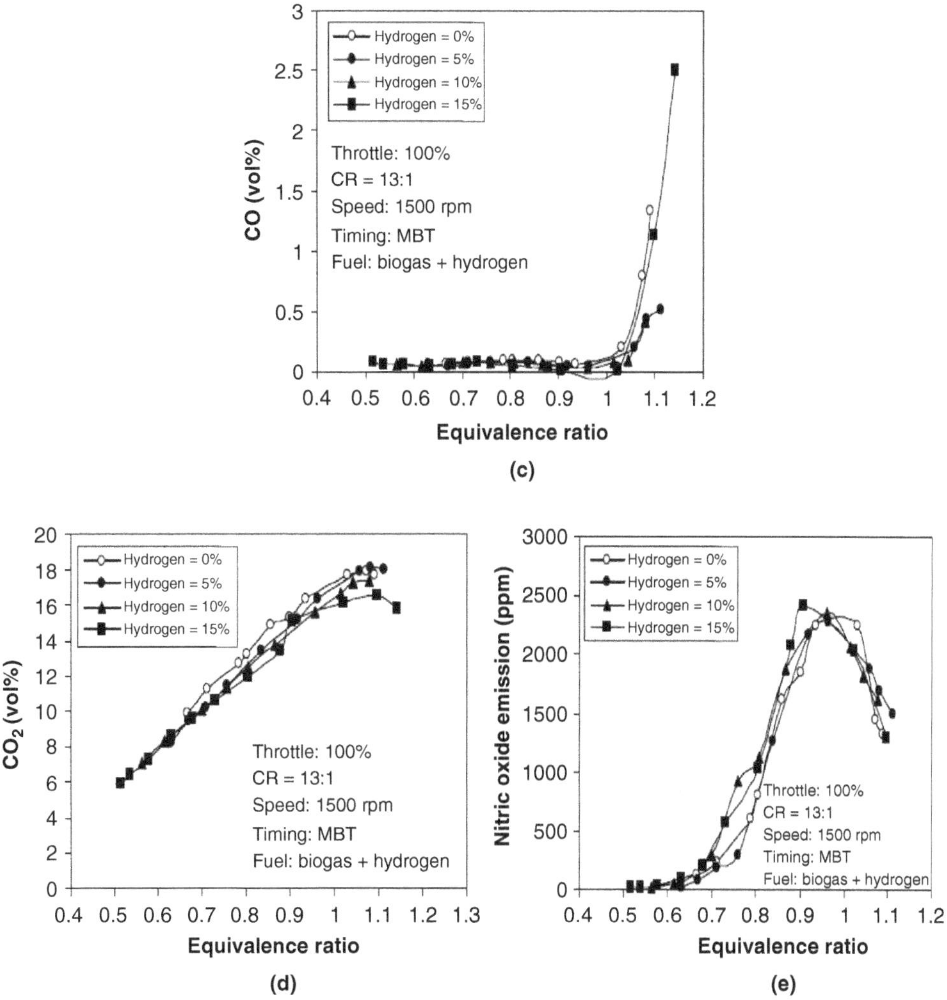

Figure 8.40 Variation of (a) thermal efficiency and emissions of (b) HC, (c) CO, (d) CO_2, and (e) nitric oxide with equivalence ratio [54].

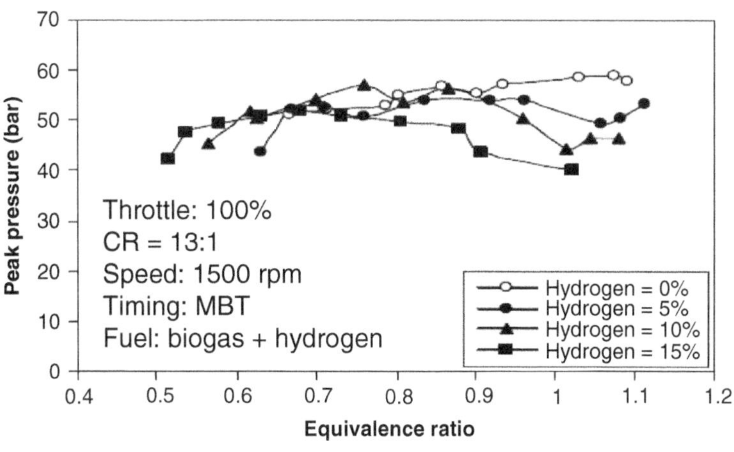

Figure 8.41 Peak pressure vs. equivalence ratio [54].

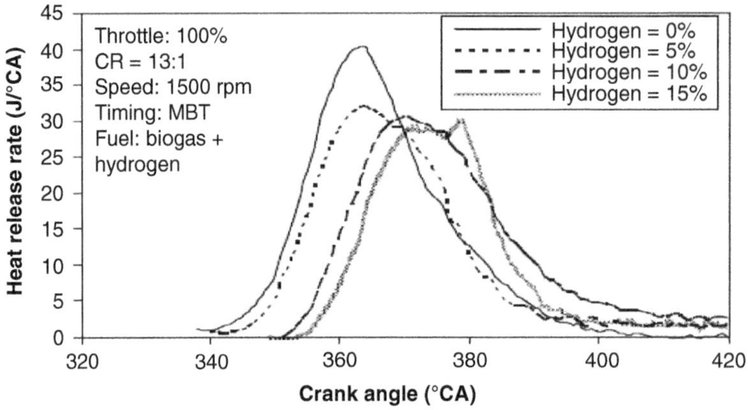

Figure 8.42 Heat release rate vs. crank angle [54].

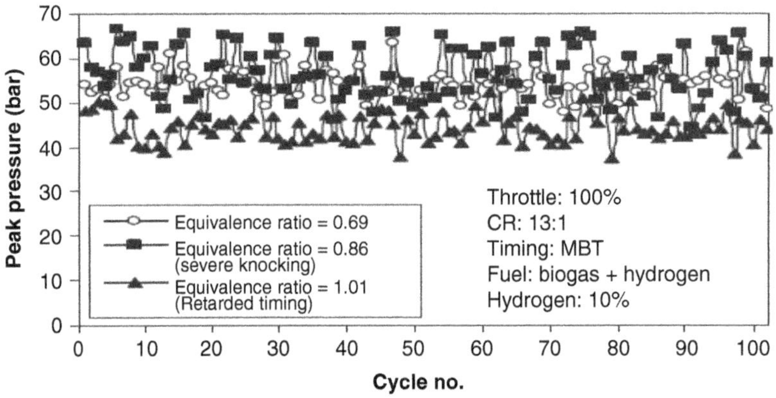

Figure 8.43 Cyclic variations of peak pressure at different equivalence ratios [54].

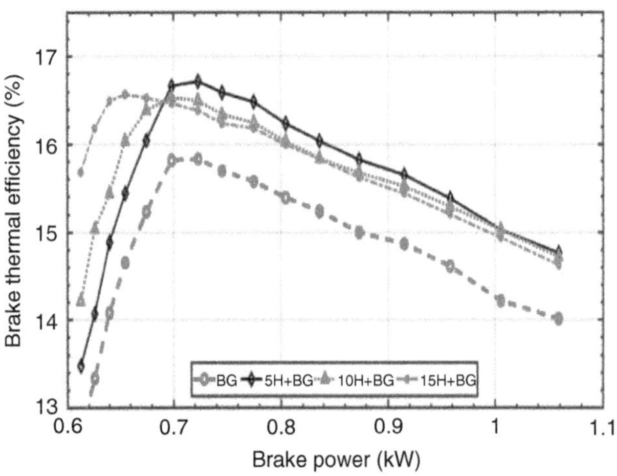

Figure 8.44 Thermal efficiency of a hydrogen-blended biogas engine [55].

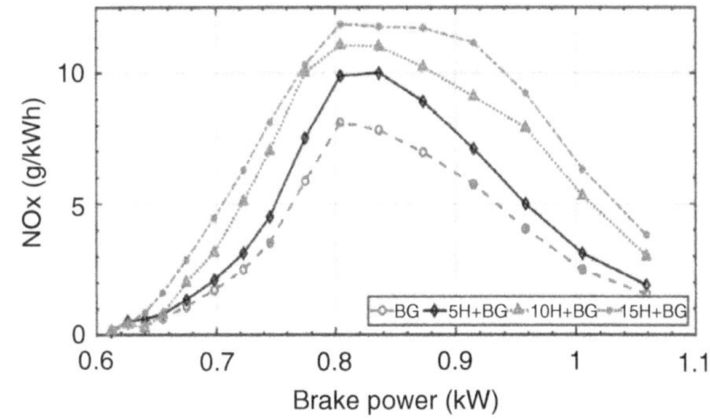

Figure 8.45 Variation of brake specific NOx emission with engine power [55].

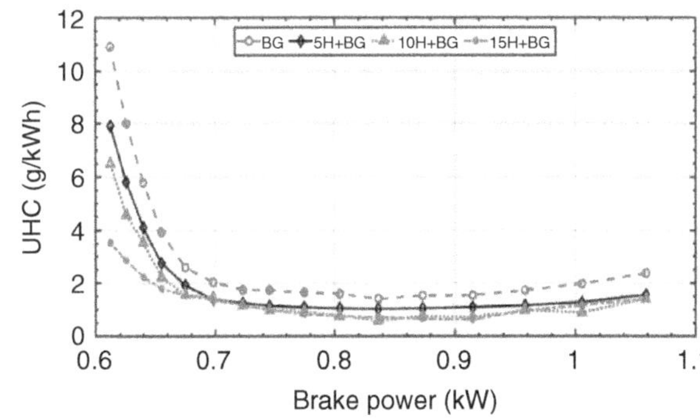

Figure 8.46 Unburned hydrocarbon (UHC) variation with brake power [55].

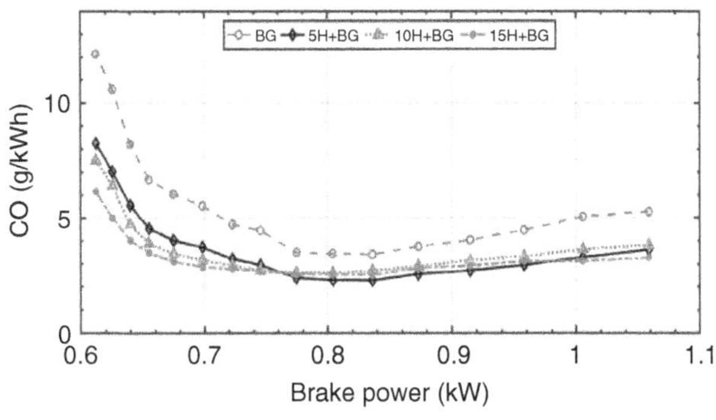

Figure 8.47 CO variation with brake power [55].

conductivity and higher heat resistance functioned as thermal barrier coatings (TBC). This led to reduced heat transfer in chamber walls and also protected them from thermal shock. Engine pistons were coated with zirconia and Ni-Al bond coats by the plasma spray method. With a small amount of pilot diesel, port fueled methane (100% CH_4), synthetic biogas (80% CH_4 and 20% CO_2), and a mixture consisting of 80% methane (CH_4), 10% carbon dioxide (CO_2), and 10% hydrogen (H2) were used as the main fuel. These fuels were tested at different loads (50, 75, and 100 Nm) and at a constant speed of 1750 rpm. Comparative evaluation of combustion, fuel economy, and

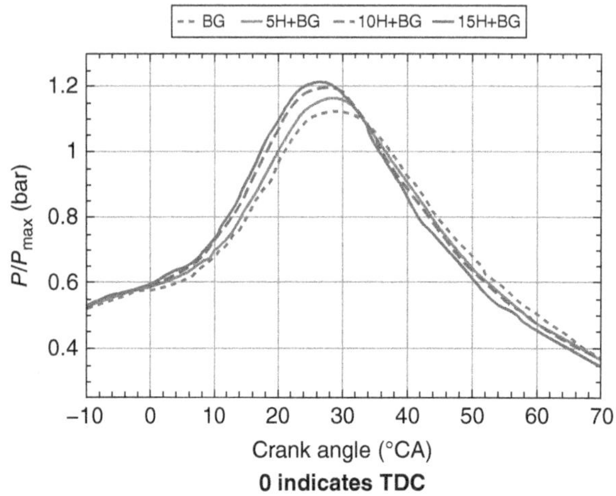

Figure 8.48 Cylinder average pressure variation with respect to crank angle at $\phi = 0.8$ [55].

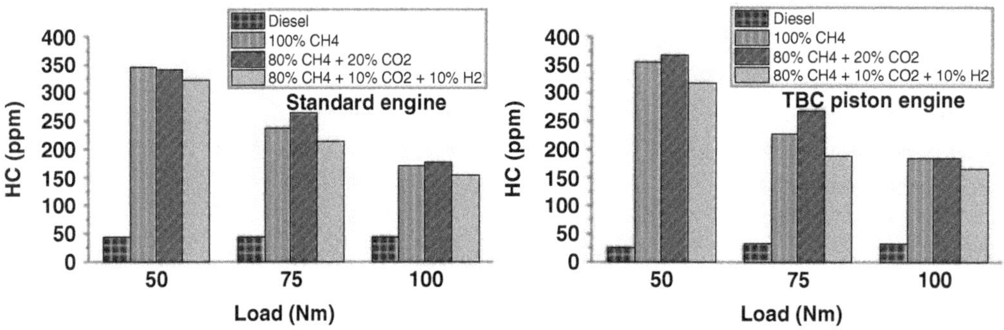

Figure 8.49 HC emissions of the fuel combinations for all conditions [56]. Please visit the companion website to view a color version of this image.

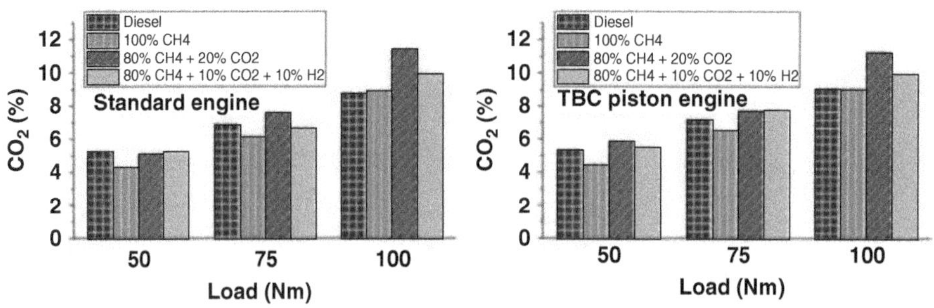

Figure 8.50 CO_2 emissions of the fuel combinations for all conditions [56]. Please visit the companion website to view a color version of this image.

emissions was done with and without piston coating. Coating-enhanced combustion characteristics slightly affected BSFC and emissions.

Figures 8.49 through 8.52 indicate the pattern of variation of various pollutants (HC, CO, NOx, and smoke) of a standard engine with a thermal barrier coating (TBC). TBC engines showed higher peak cylinder pressure, pressure release rate (PRR), and cumulative heat release (CHR) at all loads.

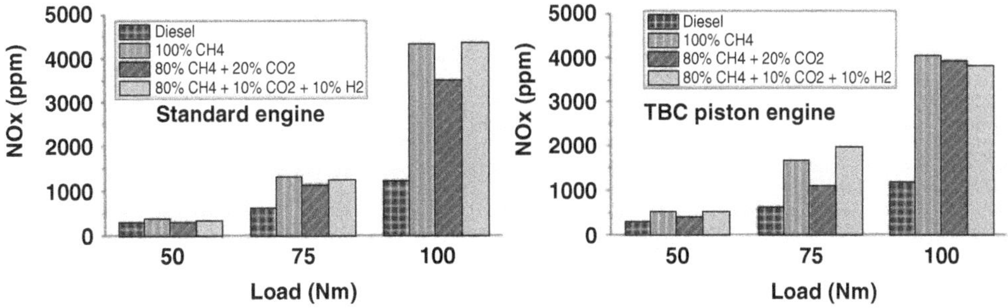

Figure 8.51 NOx emissions of the fuel combinations for all conditions [56]. Please visit the companion website to view a color version of this image.

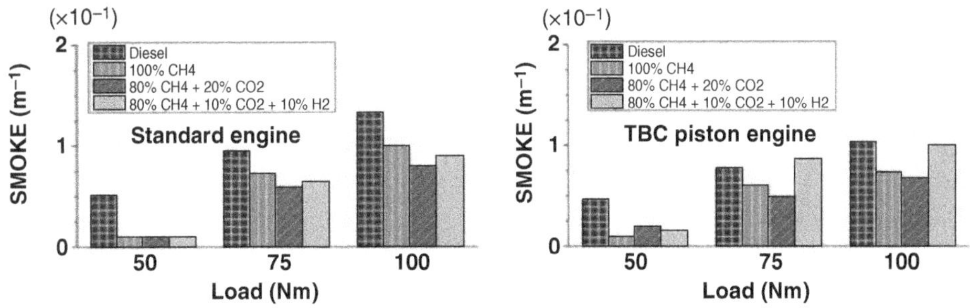

Figure 8.52 Smoke emissions of the fuel combinations for all conditions [56]. Please visit the companion website to view a color version of this image.

Sabrina et al. [57] investigated the combustion effects of blending hydrogen with biogas in air under various conditions. They used a GRI Mech 3.0 mechanism and carried out simulation studies to analyze laminar flame speed and ignition delay. Thr results of this analysis were in good agreement with experimental data. Even with a small amount of hydrogen, there was a significant improvement in low-temperature combustion. Ignition delay was reduced by five to ten times compared to a pure biogas–air mixture with 10–50% (v/v) addition of hydrogen with biogas. Similarly, high-temperature combustion characteristics (laminar flame speed) demonstrated a two to three times increase in laminar flame with 10–50% (v/v) of H_2 addition in the biogas. The 1D simulation studies with more than 30% hydrogen in the biogas/hydrogen–air mixture showed a similar laminar flame speed range comparable to hydrocarbon fuel–air mixtures under similar conditions. The researchers also concluded that the increase in the laminar flame speed, heat release rate, and adiabatic flame temperature with an increase in initial temperature are due to the large production of radicals accelerating the combustion process, which increases laminar flame speed, heat release rate, and adiabatic flame temperature with an increase in initial temperature.

Elhawary et al. [58] analyzed the effect of hydrogen addition on the detonation performance of pulse detonation engines (PDE) using biogas as fuel using NASA's Chemical Equilibrium with Applications (CEA) software. The equivalence ratio was varied from 0.7 to 1.4. Hydrogen-enriched biogas effects on PDE operation were investigated, adding hydrogen from 5 to 20%. CO_2 was found to act as a diluent in biogas and reduced flame speed, temperature, and pressure compared to pure methane. Optimal hydrogen addition to biogas was 15%, which improved detonation pressure by 23% at stoichiometric condition.

Khatri and Kishore [59] carried out experimental studies on a four-stroke, single-cylinder compression ignition (CI) engine, analyzing the impact of hydrogen addition on performance and emissions of biogas–diesel dual fuel operation. The optimum blend of hydrogen–biogas–diesel demonstrated improved brake thermal efficiency with a 3.09% increase at an optimum load of 12 kg. In this study the injection timing for the selected hydrogen–biogas–diesel fuel blend was optimized for brake thermal efficiency and brake specific consumption by retarding the injection to 21° CA BTDC. Typical BSEC value for hydrogen–biogas–diesel at 12 kg load was estimated to be 14076kJ/kWh. Some results from the paper [59] are given in Table 8.2 with reference to the corresponding value for neat diesel at a load of 12 kg.

Overall performance and emission characteristics have demonstrated the potential of hydrogen–biogas–diesel blends as a sustainable solution for CI engines. Further research needs to aim at optimization of parameters like injection pressure and compression ratio, along with studying the effect of varying hydrogen flow rates.

In their experimental study on enriched biogas with hydrogen, Rekia et al. [60] observed improvement in BSEC. In addition, there was a significant reduction in the emissions of UHC and CO compared to pure biogas. As shown in Figure 8.53a, specific NOx levels in dual fuel

Table 8.2 Some performance and emission parameters of hydrogen–biogas–diesel, biogas–diesel, and hydrogen–diesel.

Performance/emission parameters	Hydrogen–biogas-diesel	Biogas–diesel	Hydrogen–diesel
Brake specific diesel consumption (BSDC)	Reduced by 71.05%	Reduced by 23.68%	Reduced by 7.89%
Brake specific energy consumption (BSEC)	Decreased by 12.13%	Increased by 32.58%	Increased by 3.14%
Carbon monoxide (CO)	Decreased by 88.10%	Decreased by 57.14%	Decreased by 52.38%
Unburned hydrocarbon	Decreased by 5.68%	Increased by 5.68% for	Decreased by 62.50%
Oxides of nitrogen (NOx)	Decreased by 83.01%	Decreased by 61.39%	Decreased by 51.74%

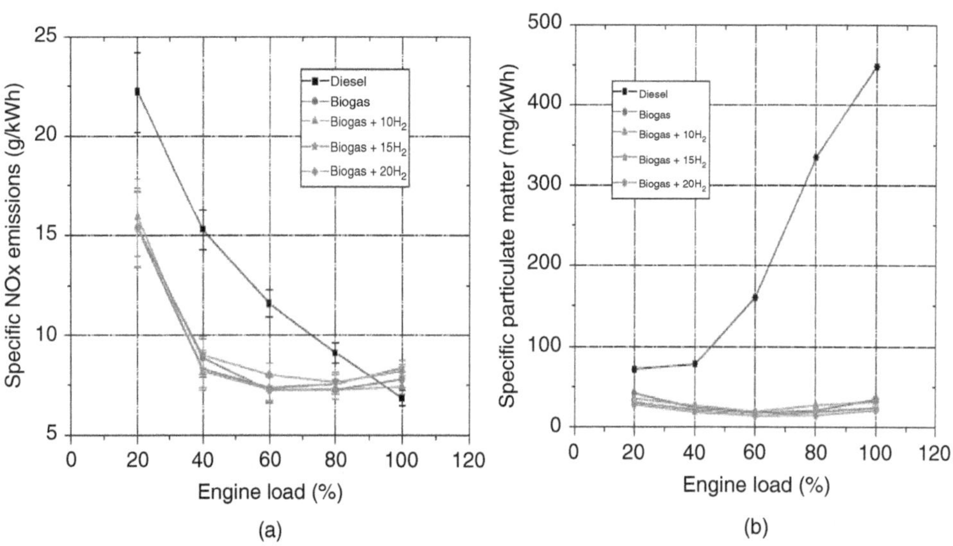

Figure 8.53 (a) Specific NOx and (b) particulate matter emissions variation with engine load [60].

mode with either pure biogas or biogas–hydrogen blends are significantly lower, from 20% to 80% of full engine loads compared to diesel operation. This is due to the combined effect of low combustion temperatures and inadequate oxygen in the mixture. High temperature pyrolysis of heavy hydrocarbons in diesel fuel is responsible for higher particulate matter (PM) formation. Transition to dual fuel mode, where hydrogen addition has almost no effect, led to a remarkable decrease in particulate matter emissions, up to 95% less than those of diesel mode (Figure 8.53b). Performance and emission were improved in dual fuel mode over conventional diesel operation.

Karagöz [61] studied the impact of gasoline, biogas, and biogas–hydrogen fuels on emissions and vehicle performance on the Worldwide Harmonized Light Duty Test Cycle (WLTC) and New European Driving Cycle (NEDC). Engine maps for a six-cylinder engine were prepared using an empirically validated engine model and were initially verified with AVL Boost software. Subsequently, fuel consumption and emissions (CO, HC, and NOx) data were gathered and compared for the NEDC and WLTC cycles using Matlab and Simulink.

Increase in fuel consumption was observed when biogas was used. When running on biogas, fuel consumption increased by 15.6% under the WLTC and by 16.3% under the NEDC compared to gasoline. However, when a blend of biogas with 5% hydrogen (molar) was used, the increase in fuel consumption was mitigated, with the rise reducing to 11.4% under the WLTC and 12.1% under the NEDC compared to gasoline. This reduction in the percentage increase in fuel consumption with hydrogen addition indicates an improvement in fuel economy. Thus, addition of 5% hydrogen demonstrated significant recovery in fuel efficiency under both test cycles.

Important results of this investigation are given in Table 8.3. It was clear from this study that increase in fuel consumption could be restricted, adding hydrogen. Hydrogen decreased the increase in THC emissions even though it caused certain levels of NOx emissions to increase. With hydrogen added, NOx emissions could be reduced significantly by creating a natural EGR effect due to the CO_2 in biogas.

Table 8.3 Operating conditions of the WLTC and NEDC cycles.

Operating condition	WLTC cycle	NEDC cycle
When biogas was used compared to gasoline fuel	Increase of 15.6% in fuel consumption	Increase of 16.3% in fuel consumption
When biogas–hydrogen fuel was used compared to gasoline fuel	Increase of 11.4% in fuel consumption	Increase of 12.1% in fuel consumption
When biogas fuel was used compared to gasoline fuel	25% increase in CO emission	21.6% increase in CO emission
When biogas–hydrogen was used compared to gasoline fuel	14.5% increase in CO emission	11.7% increase in CO emission
When biogas was used compared to gasoline fuel	34.4% increase in THC emission	33.5% increase in THC emission
When biogas–hydrogen was used compared to gasoline fuel	24.6% increase in THC emission	24.0% increase in THC emission
When biogas fuel was used compared to gasoline fuel	8.0% decrease in NOx emission	7.4% decrease in NOx emission
When biogas–hydrogen was used compared to gasoline fuel	3.8% increase in NOx emission	4.1% increase in NOx emission

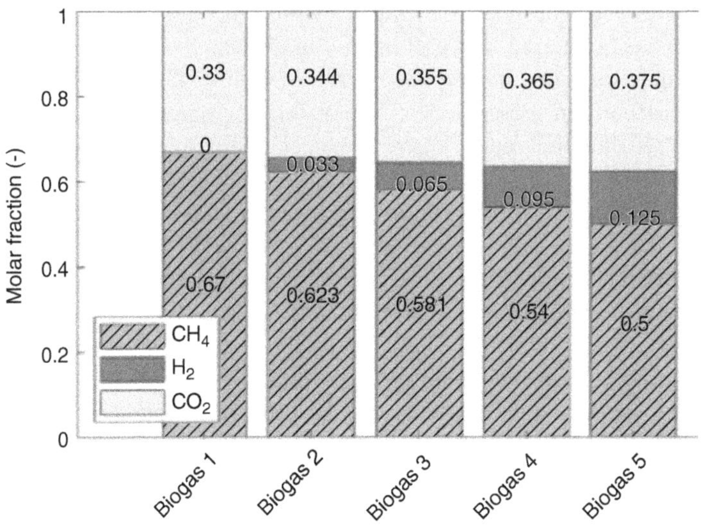

Figure 8.54 Different biogas compositions with varying amounts of hydrogen [62].

Antonio et al. [62] made a numerical study of performance and emissions characteristics of a controlled auto-ignition (CAI) engine fueled by biogas of different compositions. The biogas used was a conventional type (Biogas 1) and four innovative types containing varying amounts of hydrogen as shown in Figure 8.54.

Innovative biogas, in this study, consists of CH_4, CO_2, and H_2 with H_2/CH_4 ratios ranging from 0.05 (Biogas 2) to 0.25 (Biogas 5). The composition of the fuels is shown in Figure 8.54. One of the major objectives of this study was to find the optimal fuel composition to reduce oxides of nitrogen. Autoignition was avoided by controlling in-cylinder temperature through exhaust gas recirculation (EGR). With increasing hydrogen content in biogas, EGR requirements decreased, from 9% for Biogas 1 to 6.9% for Biogas 5. The optimal composition of biogas resulted in a 32% reduction in NOx as compared to conventional biogas. An optimal biogas composition was identified, allowing a 32% reduction in NOx emission compared to that for conventional biogas.

Juan et al. [63] converted a diesel engine to an SI engine and used gaseous fuel to improve the output power and efficiency of an original diesel engine. The compression ratio of the original diesel engine was not changed. Biogas was used as the primary fuel to boost the knocking resistance, and biogas was blended with other gaseous fuels such as propane, hydrogen, and natural gas to improve combustion properties. Composition of fuel blends used in the investigation and their properties such as low heating value (LHV), low wobble index (LWI), and methane number (MN) are given in Table 8.4 [63].

As evident in Figure 8.55, specific fuel consumption with 100NG was the lowest, as the fuel did not have any diluent (CO_2). Tests were carried out keeping the knocking peak pressure between 0.3 and 0.5 bar, and the coefficient of variation (COV) indicated mean effective pressure (IMEP) below 4% over 200 consecutive cycles as a benchmark. Maximum output power in gas engines was independent of throttle percentages and instead depended on knocking threshold and cyclic variations.

In the entire series of tests, 50B50M (a mix of 50% biogas and 50% natural gas) appeared to be most desirable option from a performance and combustion point of view. It developed the highest power output, which was 8.25% more than that for a neat diesel engine with 100% diesel. In

Table 8.4 Compositions and properties of several designated fuels [63].

Fuel designation	Fuel composition	Fuel properties			
		LHV MJ/m^3 fuel	LWI MJ/m^3 fuel	Energy density MJ/m^3 air	MN
100NG	100% NG	34.55	45.13	3.58	87.2
100B	100% biogas	20.35	20.99	3.44	140.0
50B50M	50% biogas + 50% methane	27.14	31.40	3.57	120.0
57B38M5H	57% biogas + 38% methane + 5% hydrogen	25.00	29.01	3.58	105.3
54B36M10H	54% biogas + 36% methane + 10% hydrogen	24.22	28.80	3.59	96.5
83B17P	83% biogas + 17% propane	31.98	31.38	3.64	65.8
79B16P5H	79% biogas + 16% propane + 5% hydrogen	30.89	31.04	3.65	65.2
75B15P10H	75% biogas + 15% propane + 10% hydrogen	30.71	30.71	3.77	63.8

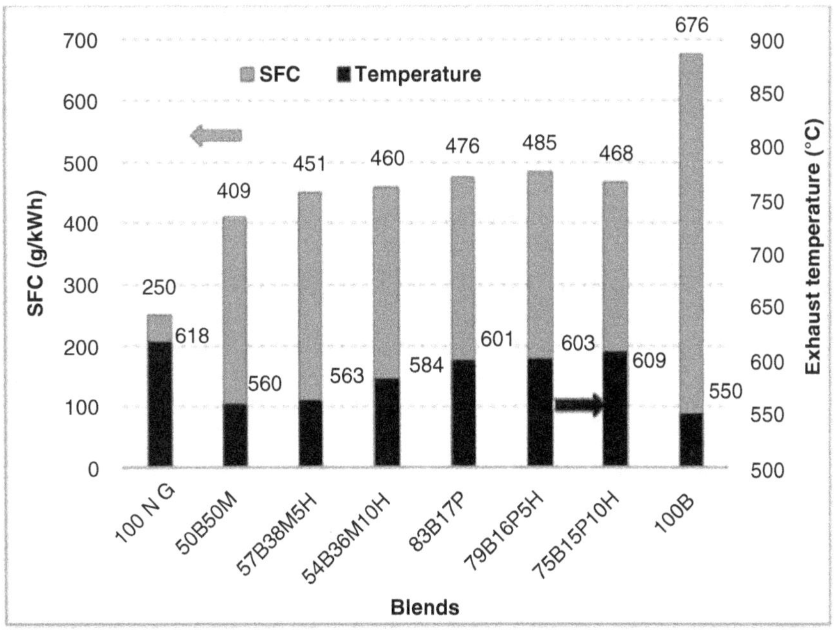

Figure 8.55 Specific fuel consumption and exhaust temperature [63].

addition, it provided a rational balance between severe knocking resistance, flame speed, and energy density. 100B developed an output power of only 7.5 kW. This de-rating of 6.25% compared to the diesel engine was because of its low energy density.

Xin et al. [64] conducted an experimental study on a port injection spark ignited single-cylinder engine fueled by biogas with 15, 29, 25, 30, and 35% (in volume) hydrogen. When hydrogen (H_2)

was added to fuel blends within the range of 15 to 35% volume fraction, the engine's maximum pressure increased, leading to higher engine power output. Increasing the volume fraction of carbon dioxide (CO_2) in biogas gave rise to a higher coefficient of the cyclic variation of pressure (IMEP). Mixing biogas with hydrogen in the range of 15 to 35% established an improved correlation between peak pressure (P_{max}) and IMEP. This range also accelerated the heat release rate and led to stable combustion in the early phase. Further study need be done to integrate engine power, cyclic variation, and emissions (NOx, CO_2) to propose optimal operating conditions and corresponding hydrogen concentrations.

The literature presents promising prospects for biodiesel–biogas-based dual fuel engines owing to their renewable nature; yet addressing challenges like reduced efficiency and increased emissions requires further research. Tests conducted at IIT Delhi [65, 66] investigated hydrogen enrichment in biogas as a potential solution, with a focus on optimizing injection parameters for improved performance and safety using a simplified simulation methodology. In addition, the study applied exergy analysis and irreversibility quantification to assess various engine processes. The experimental set-up involved modifying a single-cylinder diesel engine to operate in dual fuel mode, utilizing *Jatropha curcas* biodiesel as the pilot fuel and biogas as the main fuel, with an in-house electronic control unit developed to investigate hydrogen enrichment levels. Computational fluid dynamics-based simulations were employed to optimize injection parameters; for instance, an injection angle of 60° and an injection pressure of 150 kPa were found to offer good mixture formation. Subsequently, experimental evaluations of engine performance and emissions characteristics were conducted, including analysis of hydrocarbon, carbon monoxide, oxides of nitrogen, and smoke emissions. The results highlighted the critical role of injection angle and pressure in mixture formation, with late injection timing proving effective in preventing backfire. Optimal conditions were identified, leading to significant improvements in exergy efficiencies. For instance, at 20% hydrogen enrichment, exergy efficiencies of a dual fuel engine increased from 8.4% to 10.1% at low load and from 23.3% to 25.5% at high load. Moreover, the combination of biogas, biodiesel, and hydrogen offers potential benefits such as enhanced combustion efficiency, increased power output, improved knock resistance, and reduced engine wear.

Verma et al. [67] performed a comparative analysis of exergetic performance and emissions of a pilot diesel-ignited dual fuel (DF) engine using biogas (BG), compressed natural gas (CNG), and hydrogen as primary fuels. Optimal injection timing required for the highest efficiency or lowest emissions varies with main fuels and engine loads. Variation in maximum diesel substitution depends on the type of gaseous fuel used, engine load, and injection timing. From a second-law perspective, hydrogen–dual fuel combustion outperformed other diesel–dual fuel operations. However, its utilization was limited due to low diesel substitution and high NOx emissions at full load. Hydrogen–dual fuel operation showed higher thermal efficiencies than diesel operation at low load, whereas at full load, efficiency was generally higher for dual fuel operations except for biogas–dual fuel.

Figure 8.56 displays the exergetic performance coefficient (EPC) for diesel and dual fuel (DF) operations at their optimized injection timings. EPC, which defines ratio of work transferred exergy to exergy destroyed, is an extremely significant thermodynamic parameter to assess overall engine performance. It is evident in the figure that even at low load conditions, the EPC of hydrogen dual fuel operation is same as with neat diesel. Moreover, H_2-DF operation shows the highest EPC of 0.98 at full load.

A team of researchers [68] analyzed the impact of varying hydrogen injection timing and duration on combustion and emissions of a compression ignition (CI) engine, utilizing a

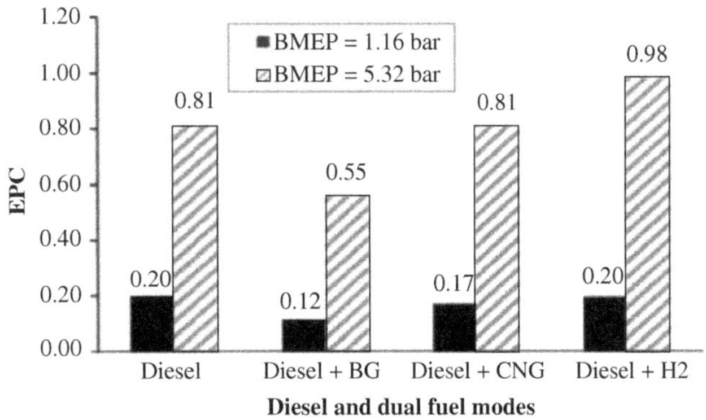

Figure 8.56 Exergetic performance coefficient (EPC) for diesel and dual fuel operations at their optimum injection timings [67].

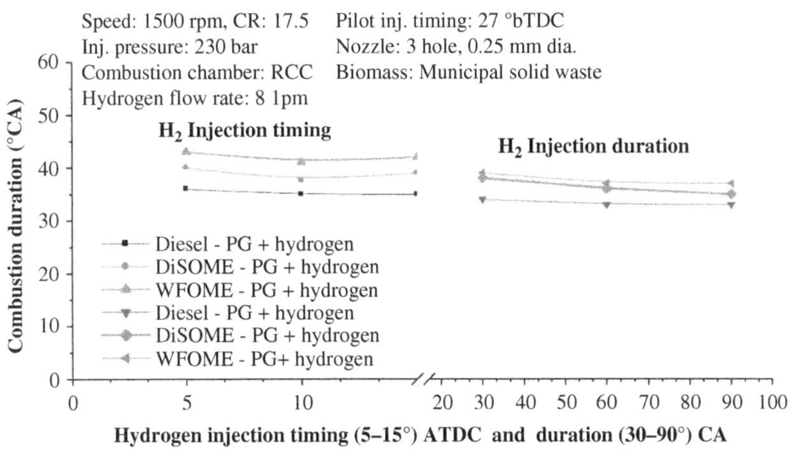

Figure 8.57 Comparison of combustion modes of hydrogen induction and injection [68].

combination of diesel/dairy scum oil methyl ester (DiSOME)/waste frying oil methyl ester (WFOME) with producer gas (PG). Hydrogen was injected at a rate of 8 lpm into the air–producer gas mixture in the inlet manifold using a gas injector. Injection timing was varied from top dead center (TDC) to 15° after top dead center (ATDC) in 5° increments, while the injection duration ranged from 30° CA to 90° CA in steps of 3°. The study found that optimal injection timing and duration were 10° ATDC and 60° CA, respectively (Figure 8.57). Optimum injection parameters led to significant improvements in engine performance and emission characteristics. Results showed that the diesel–producer gas combination improved brake thermal efficiency by 6.7% and 12.4%, respectively, compared to DiSOME and WFOME operation with hydrogen addition. It was observed that, with similar condition of hydrogen supply system, the biodiesel–PG combination demonstrated smooth engine operation with lower knocking tendency than dual fuel diesel engine operation. Similarly, under optimized injection characteristics (timing and duration), diesel-supported dual fuel operation showed better results than engines using biodiesel.

8.5 Hydrogen–Ethanol Blend

Al-Baghdadi [69] performed tests on a four-stroke spark ignition engine using varying percentages (2, 4, 6, 8, 10, and 12% by mass) of hydrogen–ethanol blends with compression ratios (CRs) varying from 7 to 12 using gasoline as a reference fuel. Hydrogen addition extended the combustibility range of ethanol such that the engine could operate over a wider range of fuel mixtures. Addition of hydrogen resulted in increased engine power output and a reduction in specific fuel consumption leading to improved fuel efficiency. Ethanol fuel's high heat of vaporization resulted in reduction of the peak temperature inside the cylinder, lowering nitrogen oxide (NOx) emissions. Even though NOx emissions increased with higher percentages of hydrogen, still at a CR of 12 and with hydrogen addition (up to 3.5% by mass), NOx emissions remained lower than those observed with gasoline fuel at a CR of 7.

In another paper [70] it has been reported that using 8 % of hydrogen by mass blended with 30 vol% of an ethyl alcohol–gasoline engine operating at a compression ratio of 9 and 1500 rpm demonstrated 48.5% reduction in CO emission, 31.1% reduction in NOx emission, and 58.5% reduction in specific fuel consumption. Moreover, thermal efficiency and output power increased by 10.1 and 4.72%, respectively [70]. Studies have been made to identify the safe zone of operation for an engine fueled with hydrogen and ethyl alcohol. In this context, the safe zone develops lower fuel consumption with higher power output and lower NOx and CO emission [71]. Figure 8.58 displays the safe zone that satisfies the above conditions. Shaded areas in the figure indicate the safe blending of the two fuels with gasoline to restrict NOx emission levels [71].The researchers developed an equation from experimental data to evaluate the least amount of ethyl alcohol that need to be added to gasoline to achieve the specific blending ratio when combined with hydrogen fuel to restrict NOx emission.

In their studies on an ethanol-fueled SI engine with hydrogen, Marwaha et al. [72] evaluated regulated emissions such as carbon monoxide (CO), non-methane hydrocarbons (NMHC), nitric oxide (NO), nitrogen dioxide (NO_2), and unregulated emissions such as nitrous oxide (N_2O), ethene, acetaldehyde, etc. Tests were carried out at a constant torque of 5.5 Nm and at varying speeds of 3500, 3800, and 4000 rpm. Ethanol and hydrogen were separately introduced into the engine's intake manifold during the suction stroke. The optimal proportion of hydrogen energy share, over all speed ranges, was observed to be 15%, based on improved combustion stability compared to using only ethanol, and several significant parameters were evaluated at 15% hydrogen energy share (HES).

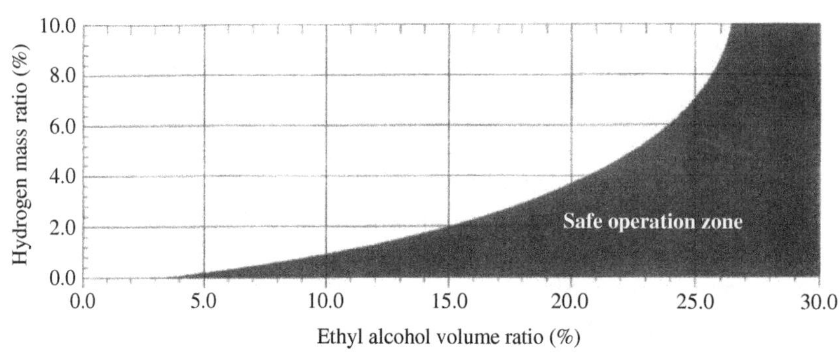

Figure 8.58 Safe operation zone of the engine [71].

Tests showed that 15% HES with ethanol (E100) improved combustion stability compared to using ethanol alone. Brake thermal efficiency showed an increasing trend at 3800 and 4000 rpm with 15% HES, but decreased beyond 15% HES. Peak pressure increased with higher HES, advancing the start of occurrence due to higher heat release rates compared to ethanol. Hydrogen addition increased emissions of NO, NO_2, and N_2O, whereas CO emissions were nearly zero at all speeds. Non-methane hydrocarbon (NMHC) emissions decreased significantly, by 41.1 and 31.9% at 3800 and 4000 rpm, respectively. Unburned ethanol emissions also reduced by 47.2 and 43.4% at 3800 and 4000 rpm with 15% HES. Ethene emissions decreased with higher HES due to increased combustion temperature and better oxidation. Acetaldehyde emissions dropped with 15% HES at engine speeds of 3800 and 4000 rpm, while methanol emissions decreased at all speeds. Further increase in HES to 20% increased methanol emissions. Some relevant graphs, especially on nonregulated emissions, are shown in Figure 8.59.

Massalami et al. [73] performed computer simulation to study the combined effect of hydrogen addition to ethanol. Software was used to conduct a parametric analysis, varying the levels of hydrogen addition and excess air ratio. Combustion and performance parameters within the

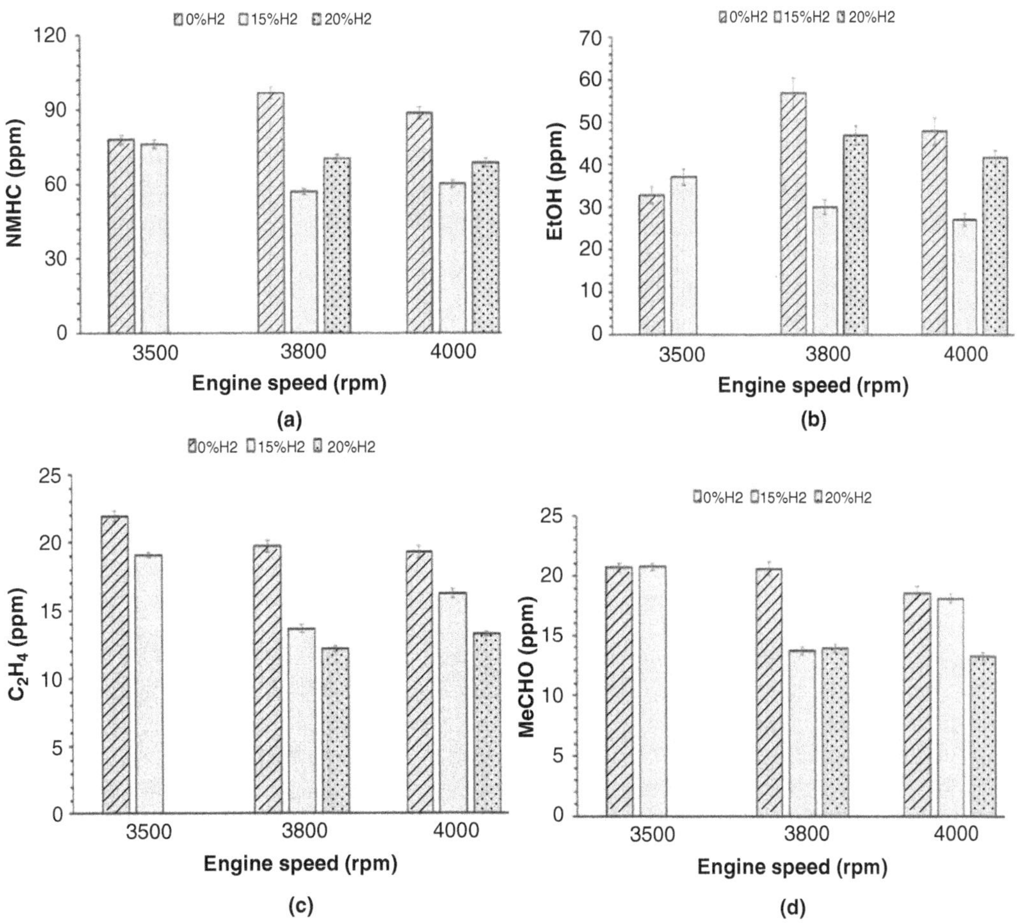

Figure 8.59 Averaged emissions values of (a) NMHC, (b) unburned ethanol, (c) ethene, and (d) acetaldehyde [72].

software were calibrated against experimental data and fine-tuned to optimize engine operation. The results of this analysis showed that lean operation up to 20% of excess air, addition of hydrogen up to 6%, did not affect the performance and combustion characteristics. A minimum of 6% hydrogen addition was essential for lean operation with 40% air excess to achieve good performance and smooth combustion. This study concluded that a rational combination of turbocharging, hydrogen enrichment, and lean operation can ensure high power output with improved efficiency and low emission.

Greenwood et al. [74] studied hydrogen enrichment of ethanol under ultra-lean operating conditions, using a modified 0.745 L two-cylinder SI engine. The study examined part-throttle, fixed rpm operation with varying levels of hydrogen (0, 15, and 30%) mixed with ethanol. Various performance and emission parameters such as power, exhaust gas temperature, thermal efficiency, volumetric efficiency, brake specific fuel consumption, NO and HC emissions, and Wiebe burn fraction curves were measured. Results demonstrated that hydrogen enrichment of ethanol led to a reduction in NOx emissions and improved combustion stability and acceleration. Using 15 and 30% hydrogen by volume, near the lean operating limit (LOL), decreased NOx emission by more than 95% as compared to stoichiometric operation with gasoline. However, this work reports that hydrogen addition brought down the specific NOx emission level. Moreover, there was no effect on the thermal efficiency due to hydrogen addition. Such observations are inconsistent with published literature. This difference, even though very small, could have been due to operation at part throttle, engine geometry, or the diffusion of hydrogen into crevices and needs to be further investigated [74].

García-Morales et al. [75] designed an adaptive control feed system to regulate a hydrogen-enriched ethanol–gasoline blend in an internal combustion engine. The objective of this control scheme was to swiftly regulate the overall air–fuel ratio in the engine operating with hydrogen-enriched ethanol-gasoline blend E10. The control scheme had two distinct control laws: a feedback control law for regulating hydrogen and an adaptive nonlinear control law for managing the injection of E10 mass flow. The adaptive controller resulted in smaller overshoot magnitudes in the air–fuel ratio (AFR) control compared to a proportional-integral (PI) controller. Rapid control action of the adaptive controller and its swift response can improve the combustion process in the engine at any operating point. A distinctive feature of this work was the consolidation of controllers, in contrast to conventional strategies in which separate controllers are utilized for controlling E10 mass flow injection.

8.6 Hydrogen–DME

Jeon and Bae [76] carried out a comparative assessment of dimethyl ether (DME) partially premixed charge compression ignition (PCCI) and hydrogen–DME PCCI compression ignition combustions in a single-cylinder direct injection compression ignition engine. Intake manifold injection of hydrogen was scheduled at –210° ATDC at a pressure of 0.5 MPa in the DME PCCI operation.

Figure 8.60 shows the experimental set-up. To simulate homogeneous charge in DME PCCI engine operation, DME, at the first stage, was injected directly into the cylinder at –120° CA after top dead center at a pressure of 30 MPa. In the second stage, DME was injected with varying injection timing. Timing of the second DME injection had a significant impact on engine performance and emissions.

Figure 8.61 shows the heat release rate comparison between PCCI combustion with single DME injection and DME PCCI combustion with double injection. The solid line in the figure represents

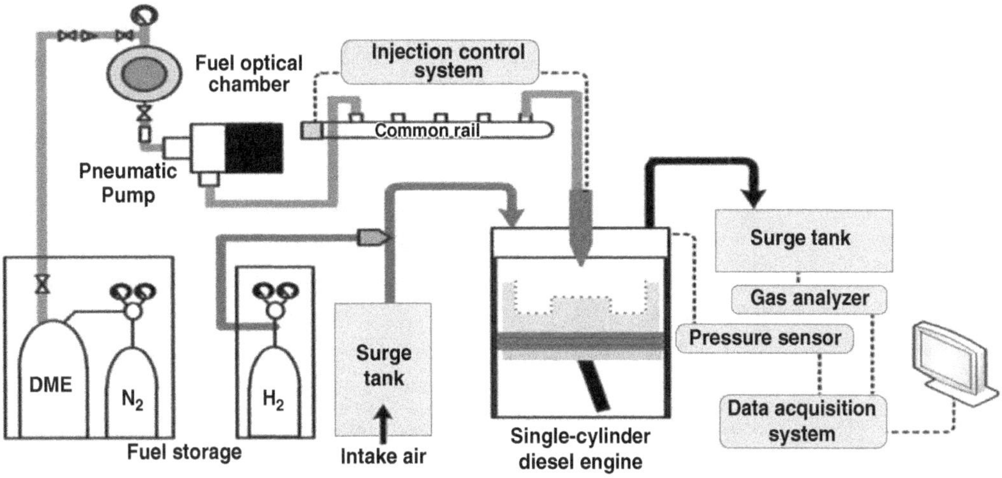

Figure 8.60 Experimental set-up [76]. Please visit the companion website to view a color version of this image.

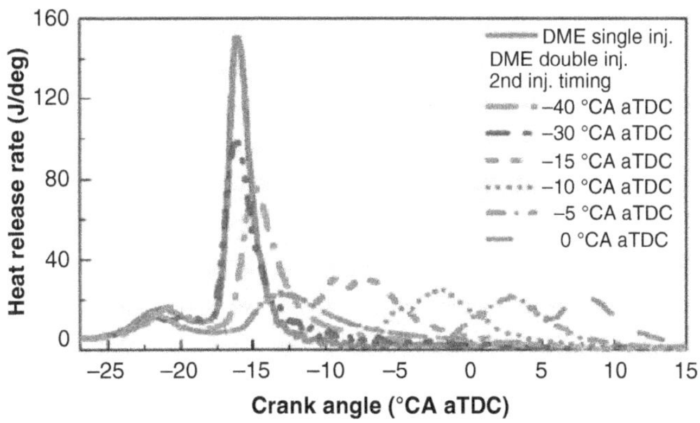

Figure 8.61 Heat release rate of DME single injection and DME double injection [76]. (Engine speed, 1200 rpm; coolant temperature, 353K; intake temperature, 298K; injected energy, 400 J; and hydrogen quantity, 35.5%.). Please visit the companion website to view a color version of this image.

the single injection strategy whereas the dashed lines stand for double injection. The single injection strategy shows a higher heat release rate because of precise timing and fuel distribution.

Figure 8.62 illustrates the impact of DME injection timing on exhaust emission characteristics at different injection timings. The pattern of variation of each pollutant with respect to the various injection strategies is clearly shown in the figure. In Figure 8.62a, it is evident that CO_2 emissions were lower with hydrogen–DME PCCI compared to DME PCCI. HC emissions decreased nearly to zero, with the trend of CO emissions shown in Figure 8.62b. The dual fuel injection strategy demonstrated a dramatic reduction in CO emissions as DME injection was delayed, while DME single-fuel injection maintained higher levels across all timing points. Delayed injection allowed for DME stratification, reducing HC and CO emissions (Figure 8.62c). This is in close agreement with earlier research highlighting incomplete combustion in lean, homogeneous mixtures at low-load conditions [77–79]. Because of reduced carbon in the fuel, increase in hydrogen addition resulted in a decrease of CO_2, CO, and HC emission. Addition of hydrogen had no effect on NOx

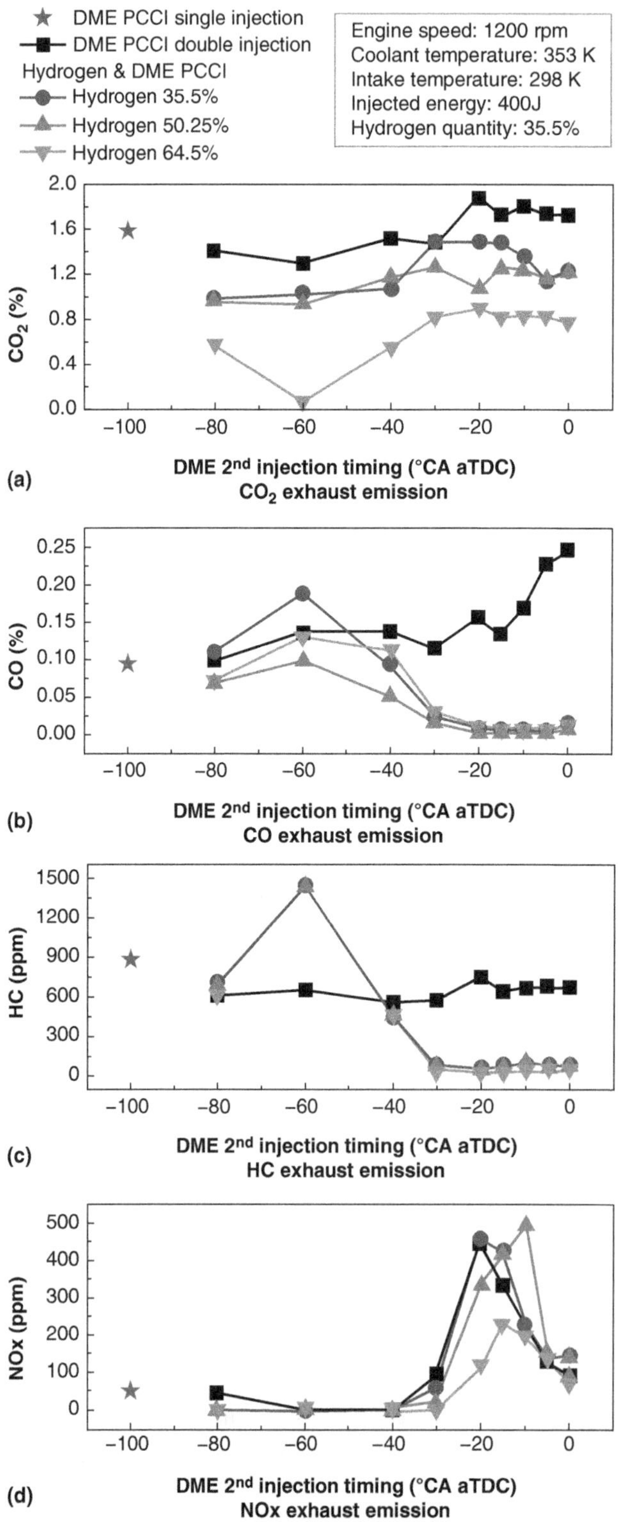

Figure 8.62 (a) CO_2, (b) CO, (c) HC, and (d) NOx emissions depending on DME injection timing of each strategy [76].

emission, but NOx emission increased with retarded injection timing in all cases. DME injection timing and quantity play a significant role in controlling the hydrogen ignition delay and ensuring stable engine combustion when high amounts of hydrogen are added. In the series of tests, the highest amount of hydrogen addition was 64.5%. Some symptoms of misfire were encountered, and additional DME was injected to ensure stable combustion.

Shudo and Yamada [80] studied the effect of hydrogen in ignition control for HCCI combustion of DME. Adding hydrogen to a HCCI engine slows down initial heat release and delays the temperature rise during the low-temperature oxidation of DME. This delayed temperature rise pushes back the crank angle where the second heat release is delayed without significantly affecting the starting temperature of the high-temperature oxidation. By preventing premature ignition, hydrogen addition ensures higher load operation in HCCI engines. Hydrogen abstraction from DME produces CH_3OCH_2 mainly from OH reactions during DME's low-temperature oxidation. Introduction of hydrogen to the low-temperature oxidation of DME brings down OH concentration and retards CH_3OCH_2 consumption. Both hydrogen and carbon monoxide retard H abstraction from DME by OH. Hydrogen reacts faster with OH than CO, which significantly retards DME autoignition. Reaction of H and O_2 plays a dominant role in hydrogen combustion producing OH and O. Even though less reactive, HO_2 is still dominant in the temperature range of the low-temperature oxidation and so inhibits progress during low-temperature DME oxidation.

Cong et al. [81] studied spark ignition combustion and emission using hydrogen–dimethyl ether under various dimethyl ether fractions (αDME). Initially, after DME blending, flame development and propagation periods were longer. Increases of both periods were 1.7° and 2.0° CA when αDME varied from 1.4% to 3.0%, respectively. However, at αDME of 3.4%, due to spontaneous high-temperature oxidation of a few ends unburned mixtures, the flame propagation period experienced a 1.7° CA reduction. Power output improved with DME addition, with a maximum 18.3% increase in indicated mean effective pressure. Although cyclic variation increased when αDME was lower than 3.0%, the variations were within 0.13%. Because of DME addition, there was an improvement in DME output leading to a maximum 18.3% increase in IMEP. Even though cyclic variation showed an increasing trend when αDME was lower than 3.0%, still the variations were within 0.13%. Addition of DME decreased emissions of nitrogen oxides substantially decreased: up to 49.9% compared to neat hydrogen operation, whereas both hydrocarbon and carbon monoxide emissions increased. Specifically, increase in hydrocarbon emissions was about one-third compared to that of carbon monoxide emission.

Faingold et al. [82] performed a numerical investigation on the utilization of dimethyl ether (DME) and hydrogen in a reactivity controlled compression ignition (RCCI) engine, using advanced numerical simulations with Reynolds averaged Navier–Stokes (RANS) as well as large eddy simulation (LES) methodology. They carried out a modeling analysis of DME combustion on the basis of its chemical characteristics and interactions within the combustion chamber. Experimental data from premixed DME jet flames and an optically accessible engine with direct hydrogen injection were gathered and validated against numerical models. Comparison between RANS and LES from the point of view of chemistry of combustion process and turbulence showed the better predictive capability of the latter.

A study by Shere et al. [83] focused on improving energy efficiency and reducing emissions in a dimethyl ether (DME) fueled HCCI engine by adopting hydrogen and exhaust gas recirculation (EGR). Figure 8.63 shows challenges normally encountered in DME HCCI combustion and methods adopted in the present study to overcome the difficulties. Initially, in neat DME mode, the maximum load limit was 24% with controlled autoignition (CAI), reaching a BMEP of

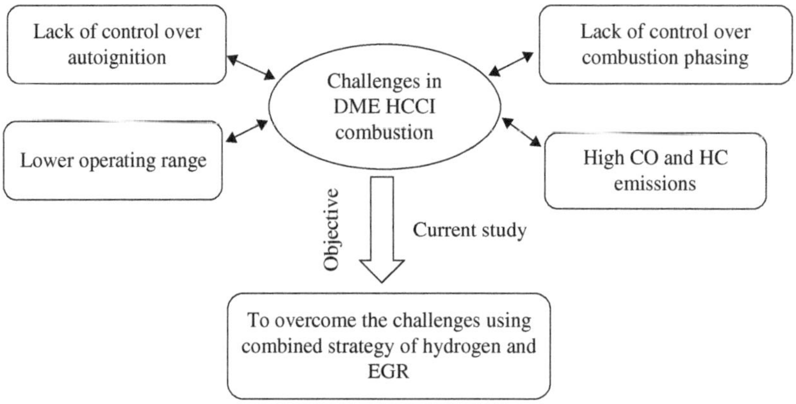

Figure 8.63 Challenges in DME HCCI combustion [83].

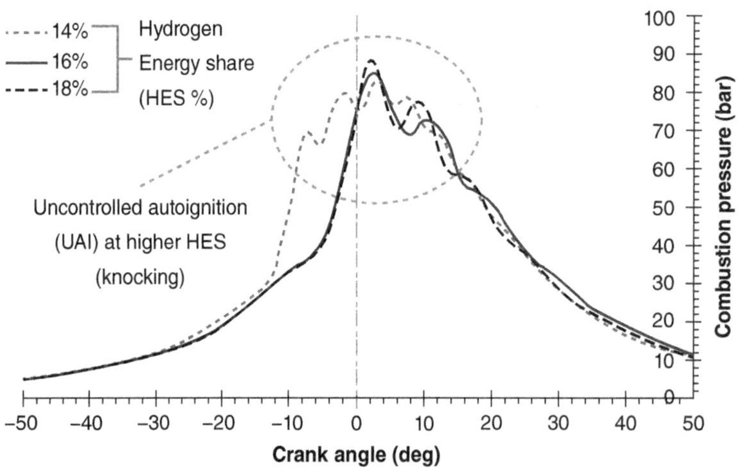

Figure 8.64 Combustion pressure with hydrogen addition in DME HCCI engine at 1400 rpm [83].

1.67 bar, and beyond this point, knocking occurred leading to uncontrolled auto-ignition (UAI) combustion. Excessive EGR led to unstable combustion, and 35% EGR was experimentally evaluated to be the optimum to avoid abnormal combustion symptoms. The combined strategy of introducing hydrogen with DME at an optimized EGR rate had many significant features. It retarded the combustion phase and eliminated the low-temperature reaction (LTR) region and improved thermal efficiency. There was practically no smoke and NOx emission and lower CO_2 emissions up to 12% hydrogen energy share (HES).

Figure 8.64 illustrates combustion pressure with hydrogen addition in a DME HCCI engine at an optimized EGR rate (35%) under DME-fueled HCCI mode. Without hydrogen addition, the peak combustion pressure was 71.21 bar. With 12% hydrogen energy share (HES), peak pressure increased to 80.07 bar with peak pressure of 88.9 bar at 18% HES, causing high-intensity knocking. Figures 8.65 and 8.66 show NOx, smoke, CO, and HC emission levels of DME HCCI engines with and without EGR mode, respectively.

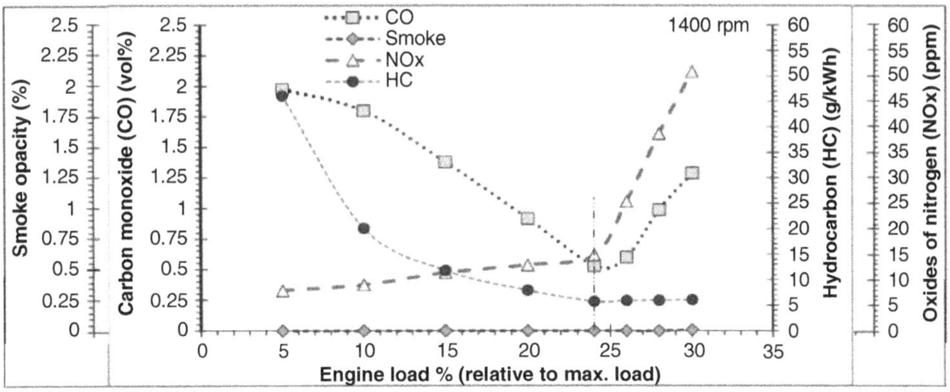

Figure 8.65 NOx, smoke, CO, and HC in a DME-fueled HCCI engine without EGR [83].

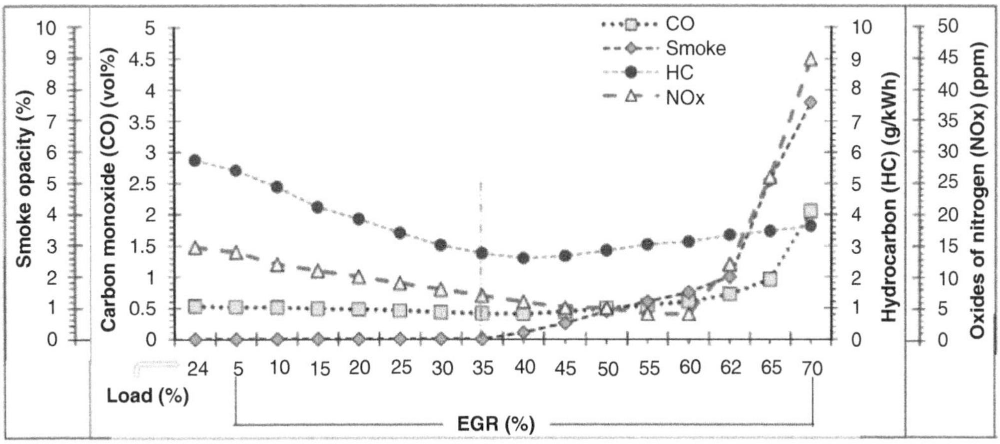

Figure 8.66 NOx, smoke, CO, and HC emission under EGR mode in a DME-fueled HCCI engine [83].

Table 8.5 gives a comparative picture of hydrogen addition on the performance, emission, and combustion features of a DME HCCI engine with and without EGR [83].

Kakoee et al. [84] carried out a numerical investigation on the impact of hydrogen addition on a reactivity controlled compression ignition (RCCI) natural gas–dimethyl ether engine, with six varying hydrogen percentages from 0 to 50% in steps of 10%. The study aimed at controlling low temperature combustion (LTC) strategies as well as emissions using DME and natural gas (CH_4). With 50% hydrogen addition, total gross work increased up to 19.7% and cumulative heat release (CHR) also increased by 15.9%. As the DME decomposition process occurs, more than 10% of it was decomposed to its products, particularly CH_4, which affected the fuel mixture cetane number and in-cylinder heat release. Some of the other significant conclusions from this study with 50% hydrogen addition shows that because of the high heating value, flame speed and flammability of hydrogen in-cylinder temperature increased from 1455 K to 1800 K and caused rise in the pressure of about 20 bar showing an increase by 16.3% and 25.6%, respectively. UHC and CO decreased by about 90% and 80%, respectively, while NO emissions increased by 95%.

Liu et al. [85] investigated hydrogen addition to a DME–CH_4 dual fuel RCCI engine using three-dimensional calculations and chemical kinetics. The analysis was based on a novel reduced

Table 8.5 Hydrogen addition in a DME HCCI engine with and without EGR [83].

S. No.	Parameters	DME HCCI without EGR	DME HCCI With increasing H_2 energy share at optimized EGR
1	Energy efficiency	Increased to 17.5%	Increased to 26.90%
2	Indicated thermal efficiency	Increased to 44.08%	Increased to 52.62%
3	Smoke emission	Zero	Zero
4	NO_x, emission	Ultra-low (10–15 ppm)	Ultra-low (10–20 ppm)
5	CO emission	Decreased to 0.53% vol.	Ultra-low (0.06% vol.)
6	CO_2 emission	Increased to 5.2% vol.	Decreased to (1.8% vol.)
7	HC emission	Decreased to 5.75 g/kWh	Ultra-low (0.66 g/kWh)
8	LTR max	Increased	Decreased
9	HTR max	Increased	Increased
10	LTR start timing	Advanced	Retarded
11	HTR start timing	Advanced	Retarded
12	Combustion duration	Decreased	Increased with EGR and decreased slightly with H_2 addition
13	Equivalence ratio	0.55 at 24% load	1.06 at 12% HES

DME oxidation mechanism that was proposed for simulating the ignition and combustion process of DME. Hydrogen addition required advanced ignition time, and it led to an increase in peak cylinder pressure. Combustion had a more dominant effect in the initial phase than in latter stages. CH_4 emission, even though decreased, still remained high because CH_4 flame did not propagate throughout the charge. CO emissions declined, mostly produced by premixed CH_4 combustion. Hydrogen addition increased NO emission levels. The simulation study showed that injection strategy and pilot fuel quantity during dual fuel operation had a significant effect on NOx emission.

Pan et al. [86] adopted the CHEMKIN package along with chemical kinetic models and simulated the impact of hydrogen on ignition characteristics of dimethyl ether (DME) when mixed with air under conditions relevant to engine operation. In the analysis to validate previously measured ignition delay times from the literature, they used four models. Among all the models NUIG Mech_56.54 demonstrated the closest agreement with published literature and so was used for subsequent simulations and chemical kinetic analyses. Simulation studies with the NUIG Mech_56.54 model on DME–H_2–air mixtures using a wide range of blending ratios from 0 to 100% showed that addition of hydrogen delayed ignition timing of DME. With increasing hydrogen-blending ratio, this delaying effect showed a nonlinear trend. The study also involved radical mole fraction analysis to interpret this nonlinear behavior. Sensitivity and reaction pathway analysis were also conducted to have a more clear picture on the ignition characteristics of DME–H_2–air blends. Huang et al. [87] used a constant volume combustion bomb to investigate the laminar combustion characteristics of DME–hydrogen–air mixtures using various equivalence ratios, ratio of hydrogen to DME plus hydrogen, and initial pressures. Increasing the hydrogen ratio to DME plus hydrogen had a number of well-pronounced effects. It led to higher flame speed, laminar burning velocity, and mass burning rate, and also advanced the timing of peak pressure and shortened the combustion duration.

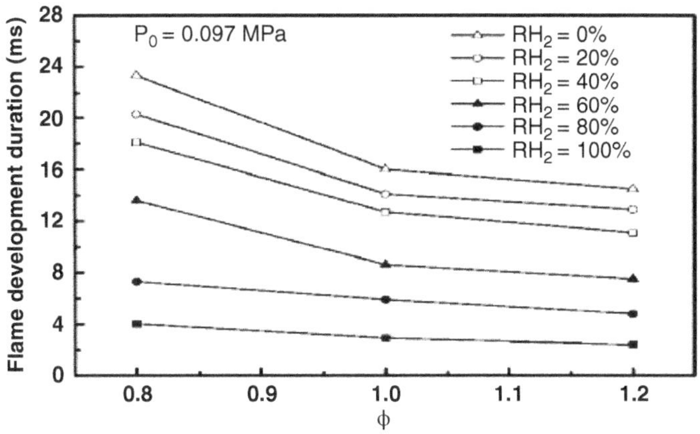

Figure 8.67 Flame development duration vs. equivalence ratio at different ratios of hydrogen to DME plus hydrogen [87].

For small hydrogen additions, lean mixtures exhibited better flame front stability, decreased Markstein length, and increased equivalence ratio. Increased Markstein length with increase of equivalence ratio indicated better flame front stability in rich mixtures. Figure 8.67 shows the flame development duration decreased with increased hydrogen addition to a specified equivalence ratio. RH_2 in this graph stands for the hydrogen fraction; for example, 20% refers to the case of 20% hydrogen volume fraction and 80% DME volume fraction at initial pressure of 0.097 MPa and initial temperature of 293 K. This occurs due to improved ignition and faster flame speed caused due to increased hydrogen content. Similarly, for a fixed DME–hydrogen–air premixed mixture, decrease in flame development with increasing equivalence ratio indicates better mixture ignitability and rapid flame propagation, particularly in a rich mixture.

8.7 Hydrogen with LPG (Liquefied Petroleum Gas) and Propane

LPG, or liquefied petroleum gas, consists primarily of propane (C_3H_8) with butane (C_4H_{10}) and other light hydrocarbons. Since propane is LPG's major component, "propane" and "LPG" are often used interchangeably, even though they are not the same. The discussions here are related to engine operation on hydrogen with propane and LPG.

Choi et al. [88] conducted tests with hydrogen-enriched LPG on a heavy-duty variable compression ratio single-cylinder engine at 1400 rpm with a compression ratio of 8 and MBT (minimum spark advance for best torque) spark timing conditions (Figure 8.68). High-pressure hydrogen stored at 200 bar was controlled through a pressure controller and then measured via a mass flow meter. A solenoid valve, operated by a duty drive, controlled both LPG and hydrogen, and the fuels were mixed in the desired proportions. Before entering the combustion chamber, the mixture passed through a flame arrestor for safety. Thus, accurate fuel control and safety were ensured before delivering the desired fuel mixture for engine combustion. Figure 8.69 shows a slight decrease in thermal efficiency with addition of hydrogen, which is due to rapid burning of hydrogen. Figure 8.70 shows the fuel consumption with the relative air–fuel ratio with varying percentages of hydrogen (0, 10, and 20%). With increased hydrogen additions, fuel consumption decreased compared to LPG combustion, due to the fast flame propagation velocity of hydrogen.

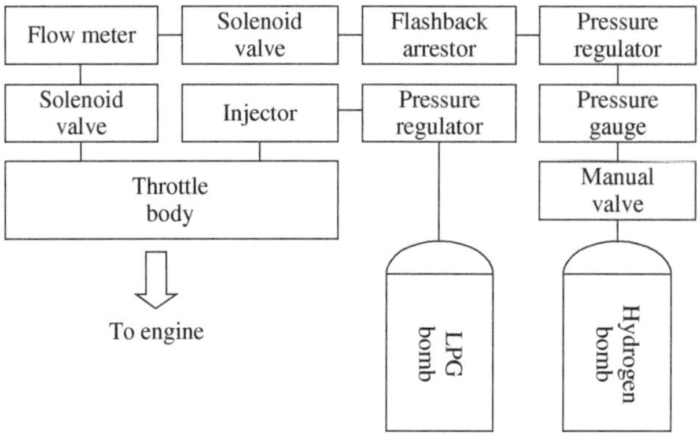

Figure 8.68 Fuel supply system [88].

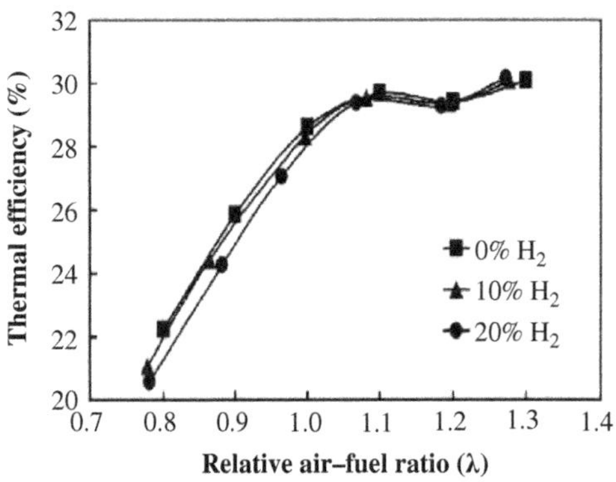

Figure 8.69 Variation of thermal efficiency [88].

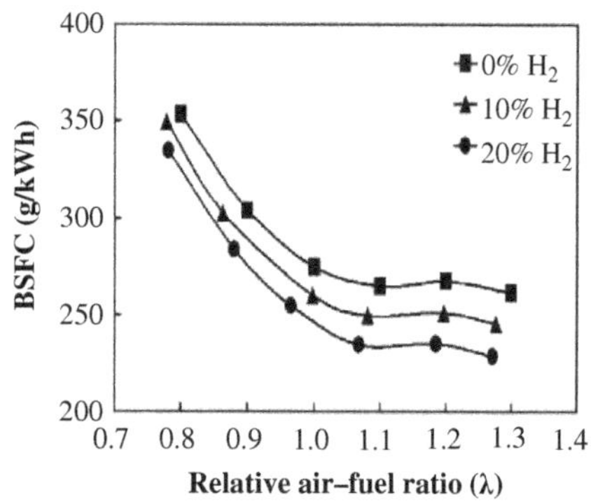

Figure 8.70 Variation of BSFC with λ [88].

Figures 8.71, 8.72, and 8.73 illustrate CO, HC, and NOx emissions, respectively, relative to the air-fuel ratio (λ) at 1400 rpm and a compression ratio of 8, with varying percentages of hydrogen (0, 10, and 20%). CO emissions decreased as λ increased from 0.76 to 1.3. There was no significant difference in CO emissions with hydrogen addition. THC emissions show a trend similar to CO emissions as λ increases from 0.76 to 1.3 (Figure 8.72). THC emissions, which originate from quench regions within engines, are expelled during exhaust processes. In the large quench zones of lean combustion, much of the THC emissions are destroyed during exhaust mixing and combustion becomes erratic. Figure 8.73 shows that addition of 20% hydrogen resulted in an approximately 20% increase in NOx emissions compared to pure LPG combustion. Hydrogen combustion occurs faster than LPG and results in higher maximum temperatures and pressures within the cylinder than with LPG fuel. Figure 8.74 shows heat release as a function of crank angle. In this analysis, the heat release rate was calculated from average pressure versus crank angle variation for 250 cycles.

Karagöz and Köten [89] investigated emissions of blending LPG and hydrogen at varying energy contents (0, 15, 30, and 45%) when operating a diesel engine with full load at a constant torque of 70 Nm and speed of 1200 rpm (Figure 8.75). Regulator, shut-off valve, check valve, relief valve, and flowmeters were installed at appropriate locations to avoid pressure and pressure fluctuations, ensure the desired quantity of fuel is delivered to the engine, and prevent undesirable combustion phenomena.

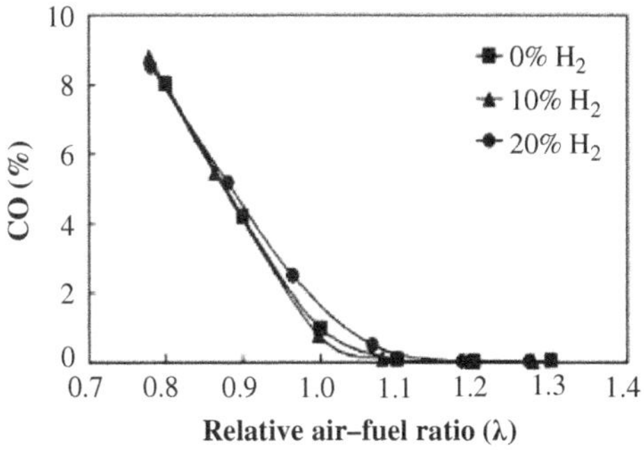

Figure 8.71 CO vs. λ [88].

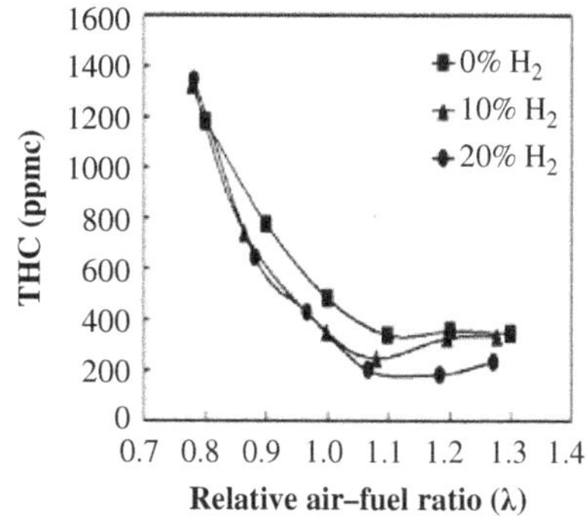

Figure 8.72 Total hydrocarbon vs. λ [88].

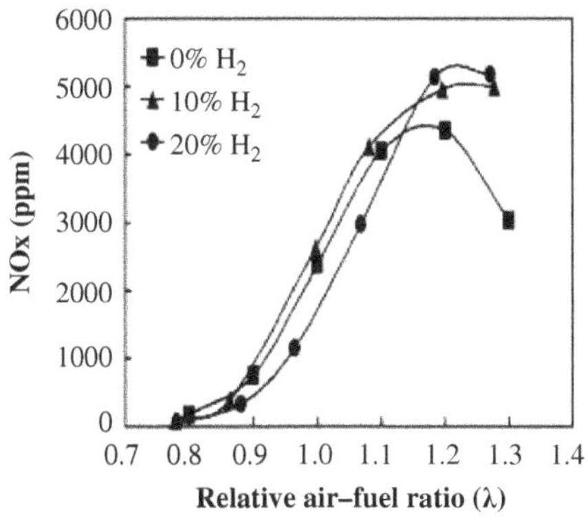

Figure 8.73 NOx vs. λ [88].

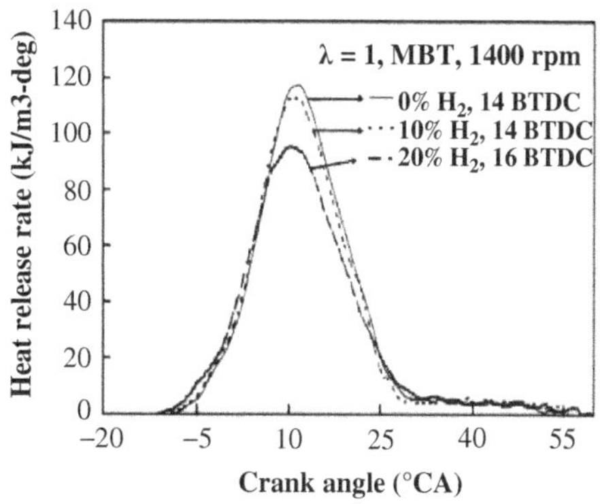

Figure 8.74 Heat release vs. crank angle [88].

It is evident in Figure 8.76 that brake thermal efficiency decreased by 0.8, 2.1, and 2.9% with 15, 30, and 45% LPG + hydrogen addition, respectively, compared to neat diesel operation. This result indicates that there is further scope to optimize the LPG injection process. However, high flame, speed, and high diffusivity of hydrogen were useful in preventing further decline. Both CO (Figure 8.77) and THC (Figure 8.78) emissions show the impact of adding gaseous fuel (LPG + H2) with different level of energy contents. Compared to neat diesel, CO emissions increased by 4, 12, and 16%, respectively, with 15, 30, and 45% LPG. This was mainly due to reduced oxygen availability in the combustion chamber. Moreover, the crevices in the combustion chamber also contributed to further CO emission while using gaseous fuel. **Increase in THC level** with 15, 30, and 45% LPG by 8.3, 12.5, and 25%, respectively, indicates appropriate modification in the system is essential to respond to the combustion characteristics of LPG and hydrogen. Similarly, reduced NOx emission reflected in Figure 8.79 is due to decreased oxygen availability in the combustion chamber. In the absence of oxygen, gaseous fuel effectively carries out the role of exhaust gas recirculation (EGR). Reduction of smoke emission as shown in Figure 8.80 was chiefly due to the high diffusion coefficient of gaseous fuels. The high flame speed of gaseous fuels also contributed to mitigating smoke emissions.

Kacem et al. [90] studied hydrogen enrichment on an LPG-fueled engine through numerical simulations and experiments. Besides engine performance and exhaust emissions, the researchers also analyzed how the intake system's geometry influenced mixture flow behavior within the cylinder. Both simulations and experimental investigations done in this study show specific features of hydrogen enrichment on in-cylinder flow characteristics. In order to analyze the intake flow

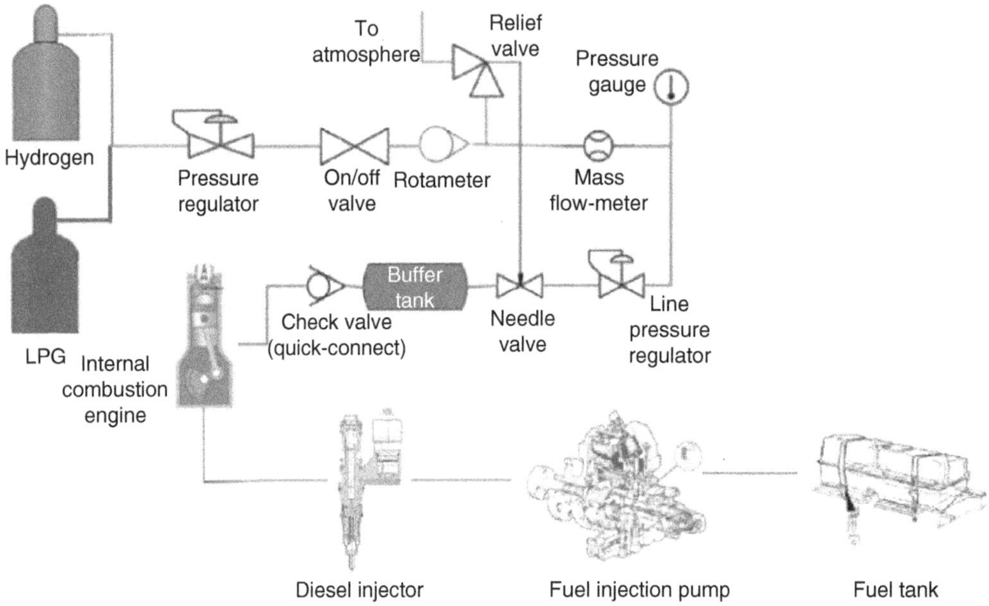

Figure 8.75 LPG and hydrogen with diesel fuel system [89]. Please visit the companion website to view a color version of this image.

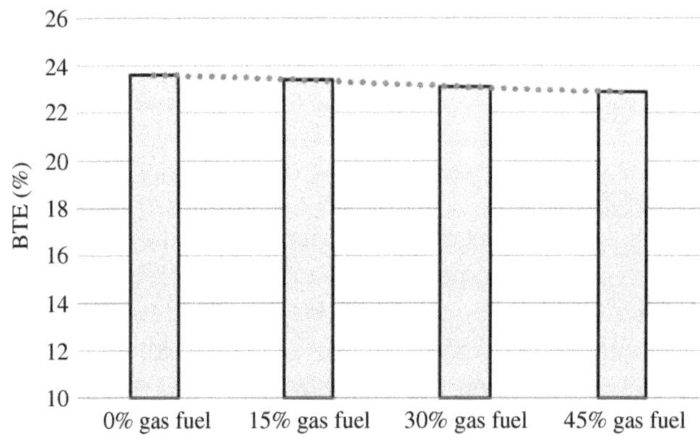

Figure 8.76 Various levels of LPG + H_2 addition on brake thermal efficiency [89].

behavior, in-cylinder flow velocity, turbulence characteristics, and hydrogen mass fraction were determined during intake stroke using a 3D computational fluid dynamics (CFD) code. Tests were carried out to evaluate variations in brake torque (BT), brake specific fuel consumption (BSFC), NOx, and CO emissions with various engine speeds for gasoline, LPG, and LPG–hydrogen fueled engines. Engine performance was evaluated across various fractions of LPG–hydrogen blends (0, 10, and 20% H_2). Numerical analysis of in-cylinder flow characteristics showed a 2.4% average increase in admission velocity and 8.12% average increase in turbulent kinetic energy.

Gaseous fuels (LPG and hydrogen) have an inherent limitation due to their low density, and this effect is reflected in Figure 8.81, where torque of gaseous fuels (LPG, hydrogen) is relatively less compared to that of gasoline (liquid) at low engine speeds. However, at higher speed ranges, a LPG–hydrogen blend with a 20% hydrogen fraction exhibits a 12 and 20% increase in brake torque compared to gasoline and pure LPG operations, respectively.

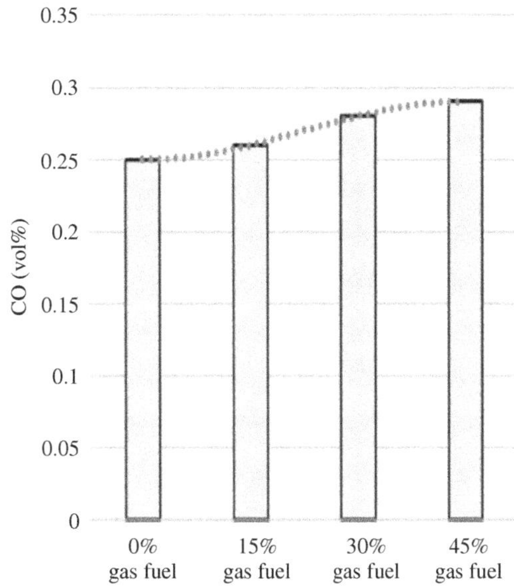

Figure 8.77 LPG + H_2 addition on CO emissions [89].

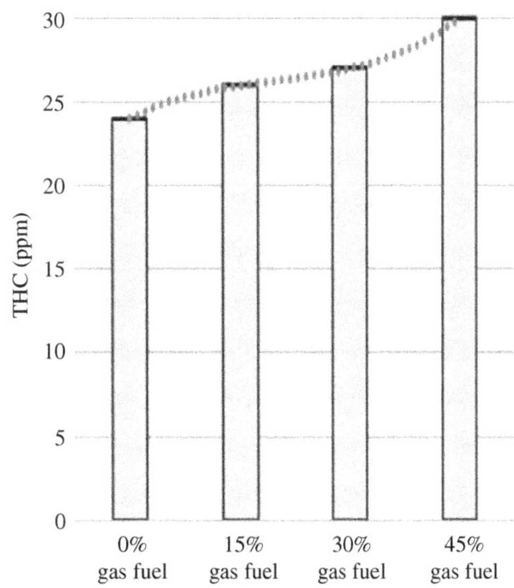

Figure 8.78 LPG + H_2 addition on THC emissions [89].

It is evident in Figure 8.82 that gasoline exhibited the lowest specific consumption between 1000 and 1800 rpm, while a LPG–20% hydrogen blend showed it between 2000 and 3000 rpm. Adding 20% hydrogen to LPG led to a 35.37% decrease in specific fuel consumption compared to the gasoline operation.

CO emission of the gaseous-fueled engine was substantially less than gasoline operation (Figure 8.83). CO emissions of the gasoline engine shows an increasing trend as engine speed increased, whereas for all gas-fueled operation, it remains constant for all engine speeds.

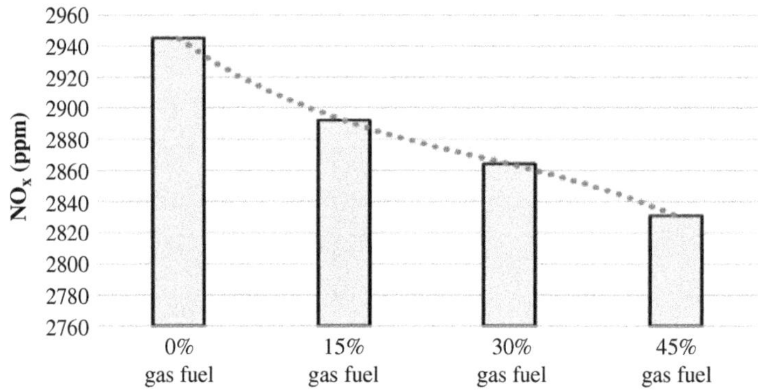

Figure 8.79 Effect of various levels of LPG + H_2 addition on NOx emissions [89].

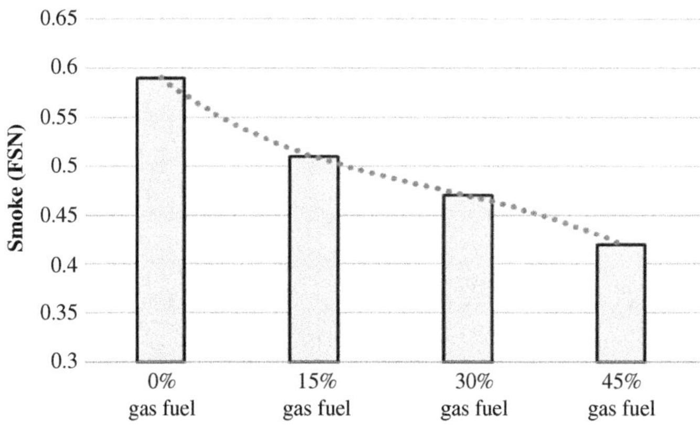

Figure 8.80 Effect of various levels of LPG + H_2 addition on smoke emissions [89].

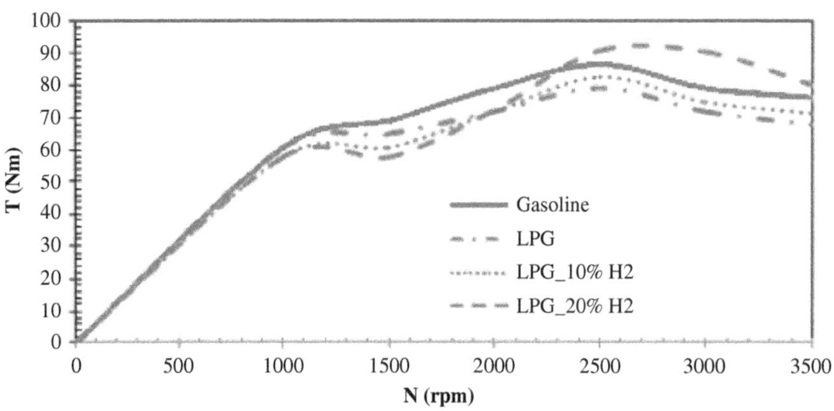

Figure 8.81 Torque vs. speed [90].

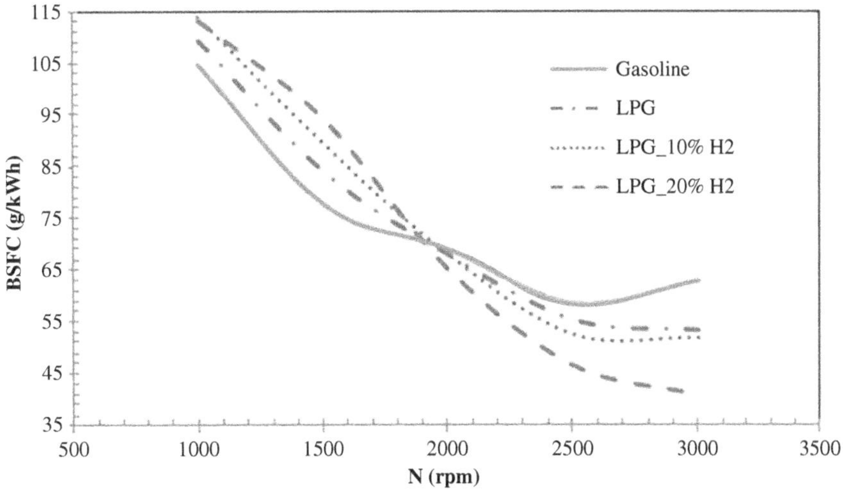

Figure 8.82 BSFC vs. speed [90].

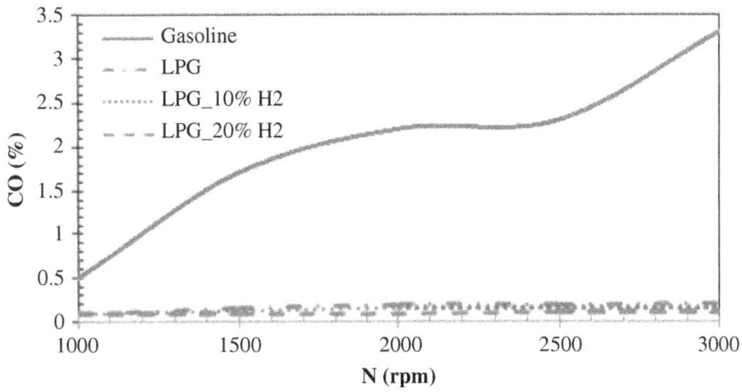

Figure 8.83 CO emissions of various fuels with speed [90].

Figure 8.84 shows a decreasing trend in NOx emissions for LPG–hydrogen operation as engine speed rises. There is hardly any difference between LPG and LPG + 10% H_2 operations. At starting speed, the LPG + 20% H_2 mixture emits more NOx due to its leaner nature. At higher engine speeds, NOx emissions from LPG–hydrogen (20% H_2) are lower than gasoline and pure LPG. This is due to the faster combustion and higher pressure and temperature of hydrogen than of gasoline and LPG.

Ravi et al. [91] modified a single-cylinder compression ignition (CI) engine to operate on liquefied petroleum gas (LPG) in a lean burn spark ignition (SI) engine mode. Tests were conducted at 1500 rpm and 20% throttle opening at compression ratios of 9:1, 10:1, 10.5:1, and 11:1, with varying equivalence ratios. Figures 8.85 and 8.86 show that increasing the compression ratio improves brake thermal efficiency and reduces specific fuel consumption. Lean mixture led to low CO emission levels (Figure 8.87) and at higher equivalence ratios, a slight increase in CO was observed due to partial combustion. Interestingly, the compression ratio of 9:1 produced the lowest CO levels on the rich side, possibly due to higher temperatures leading to CO_2 dissociation. It was

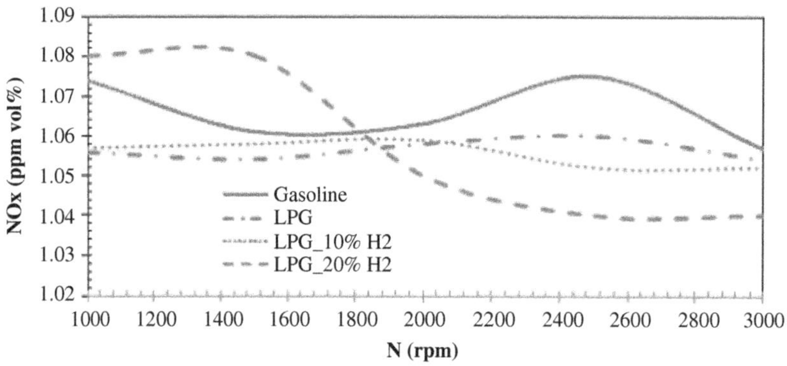

Figure 8.84 NOx emission of various fuels with speed [90].

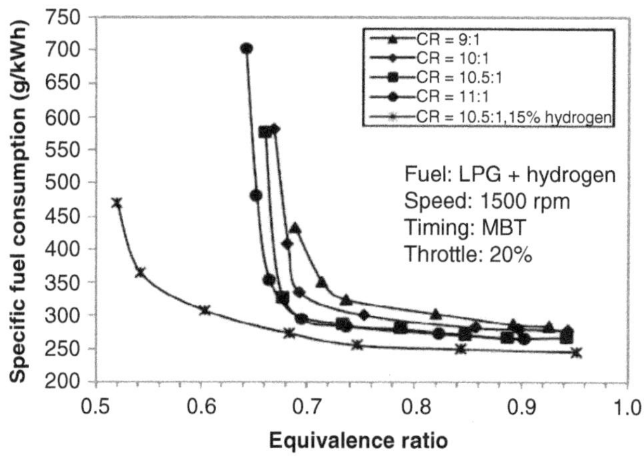

Figure 8.85 Specific fuel consumption vs. equivalence ratio [91].

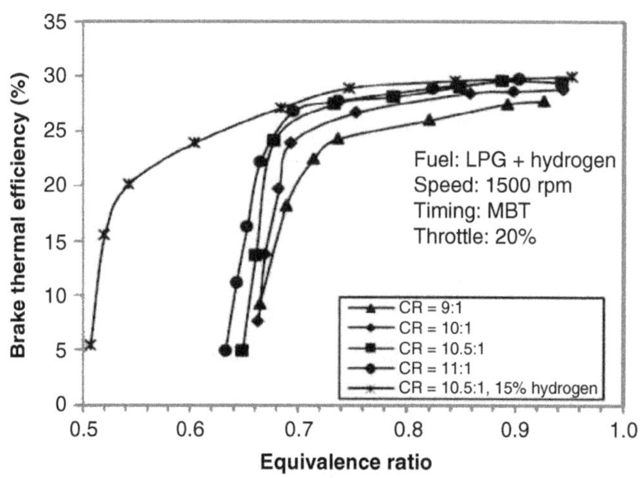

Figure 8.86 Brake specific fuel consumption vs. equivalence ratio [91].

also reported in this study that maximum power output was achieved with 40% reduction in CO_2 levels with 15% hydrogen addition.

Figure 8.88, which uses a split Y-axis for clarity, shows that hydrocarbon emissions rise at lean equivalence ratios. Increasing the compression ratio raises hydrocarbon emissions due to more charge trapped in crevices and increased post-oxidation in the exhaust manifold. As the equivalence ratio drops to a certain point, hydrocarbon (HC) levels spike due to misfiring. At an equivalence ratio of 0.9, with 15% hydrogen substitution, HC emissions are 0.14 g/kWh, compared to 0.31 g/kWh with pure LPG at a 10.5:1 compression ratio [91].

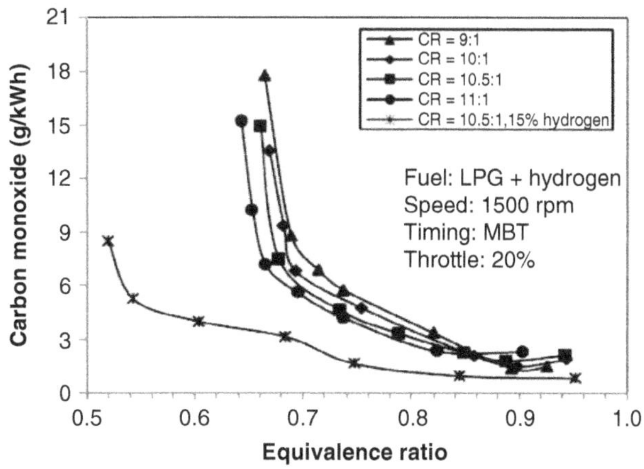

Figure 8.87 CO emission vs. equivalence ratio [91].

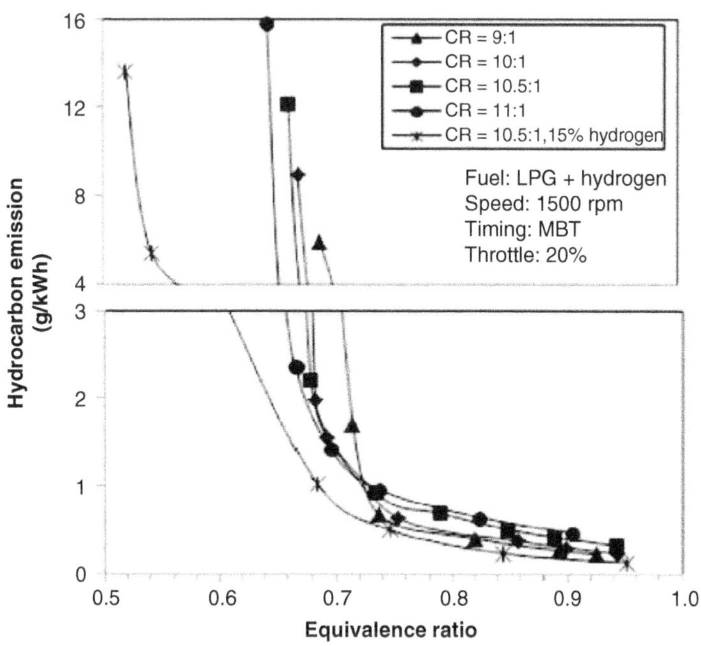

Figure 8.88 HC emission vs. equivalence ratio [91].

A higher compression ratio increases the peak combustion temperature and consequently leads to higher emissions of both nitric oxide (NO) and nitrogen dioxide (Figure 8.89).

(MBT) timing was more advanced as the mixture became leaner. This was necessary to compensate for the prolonged flame development and reduced flame speed. Figure 8.90 shows the flame development period against the equivalence ratio. At the same time, a higher compression ratio led to accelerated combustion rates, and the situation made it necessary to retard the MBT. Moreover, the addition of hydrogen improved flammability and flame speed, requiring reduced flame development time. Tests showed that with 15% hydrogen-added LPG, optimal spark timing needed to be retarded by at least 6° for peak performance. Increasing the compression ratio raises the heat release rate. Moreover, addition of hydrogen further amplifies the heat release rate, even though the pressure does not rise as quickly. The cumulative heat release in Figure 8.91 indicates that this significant increase in heat release was due to hydrogen addition [91].

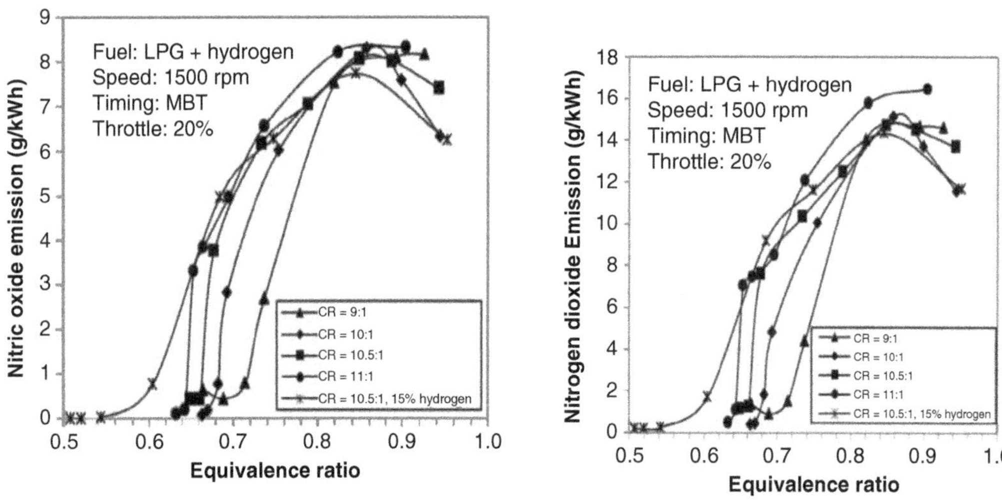

Figure 8.89 Nitric oxide emission vs. equivalence ratio [91].

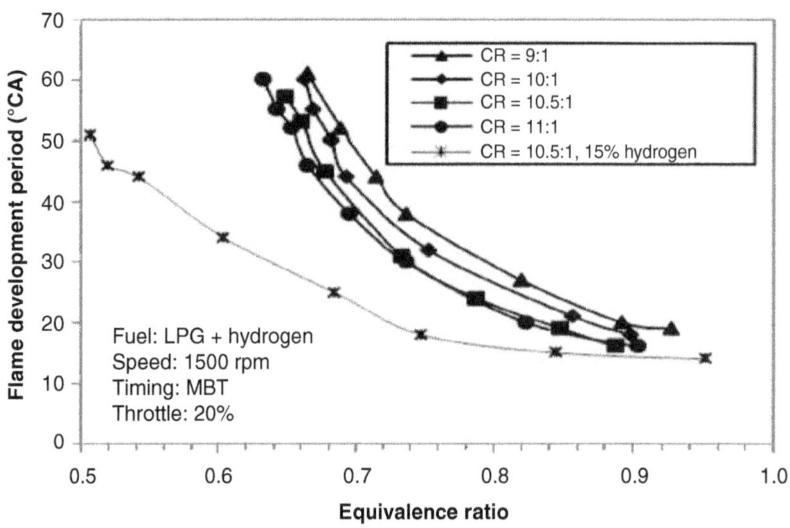

Figure 8.90 Flame development period vs. equivalence ratio [91].

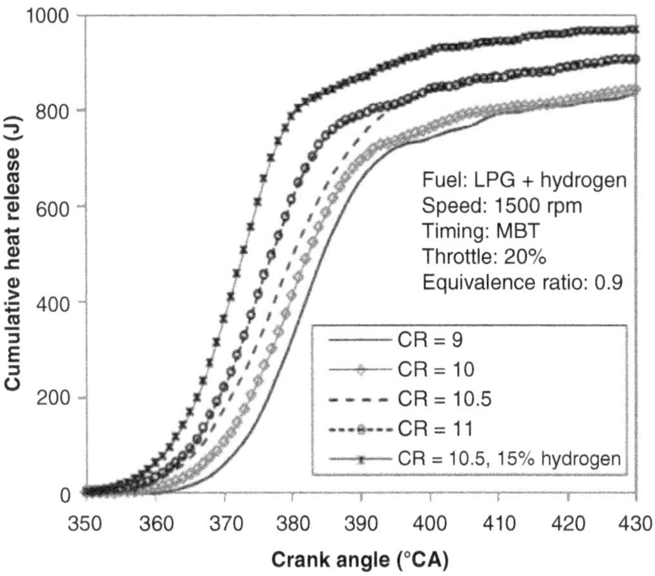

Figure 8.91 Cumulative heat release vs. equivalence ratio [91].

On a four-cylinder turbocharged and intercooled 62.5 kW diesel engine, Lata and co-researchers [92] investigated hydrogen, liquefied petroleum gas (LPG), and a mixture of LPG and hydrogen as secondary fuels. Performance and exhaust emissions characteristics of the engine were evaluated under various operating conditions. Brake thermal efficiency showed improvements with a secondary fuel addition, particularly at higher load conditions. Efficiency improved when hydrogen or LPG was used individually as a secondary fuel, though they exhibited certain limitations. Very encouraging results were achieved when a mixture of LPG and hydrogen was used as a secondary fuel, particularly under high load conditions. With a 40% substitution rate (in a 70:30 ratio of LPG to hydrogen), a substantial 27% increase in efficiency was achieved, accompanied by a significant 68% reduction in HC emissions. Figures 8.92, 8.93, 8.94, and 8.95 show variation in CO, HC, NOx, and smoke emission with respect to gaseous fuel substitution at loads of 10% in corresponding (a) and 80% in corresponding (b) in all of the figures. Even though the units chosen for the graphs in the two studies are different, the logic behind the analysis and the broad conclusions on these pollutants agree with results of Karagöz and Köten [89].

Studies have also identified severe knock during dual fuel operation with high substitution rates of hydrogen or LPG at higher loads, so it is desirable that a threshold for safe substitution levels should be determined. Lata and co-workers investigated the combustion parameters of the four-cylinder, turbocharged and intercooled 62.5 kW gen-set dual fuel diesel engine utilizing hydrogen and LPG as secondary fuels [93]. They analyzed several significant combustion characteristics such as maximum pressure rise rate, peak cylinder pressure, heat release rate in the first phase of combustion, and combustion over a wide range of load conditions with different fuel substitutions. With 30% hydrogen as a secondary fuel, there was an increase in maximum rate of pressure rise by 0.82 bar/° CA compared to pure diesel operation. Moreover, peak cylinder pressure and combustion duration increased by 8.44 bar and 5° CA, respectively. Similarly, with 30% LPG substitution, maximum pressure rise rate, peak cylinder pressure, and combustion duration increased by 1.37 bar/° CA, 6.95 bar, and 5° CA, respectively. The heat release rate in the first phase of combustion decreased at all conditions of loading in tests with both substitutions

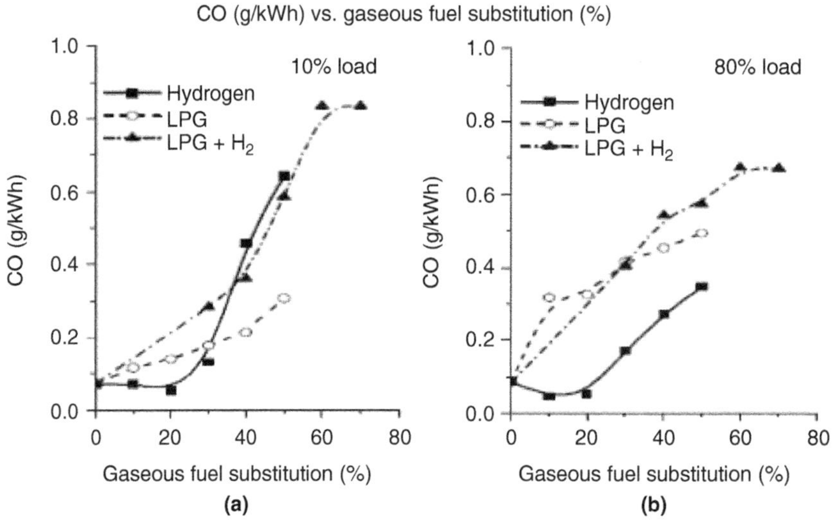

Figure 8.92 Carbon monoxide (g/kWh) vs. gaseous fuels substitution at (a) 10% load and (b) 80% load [92].

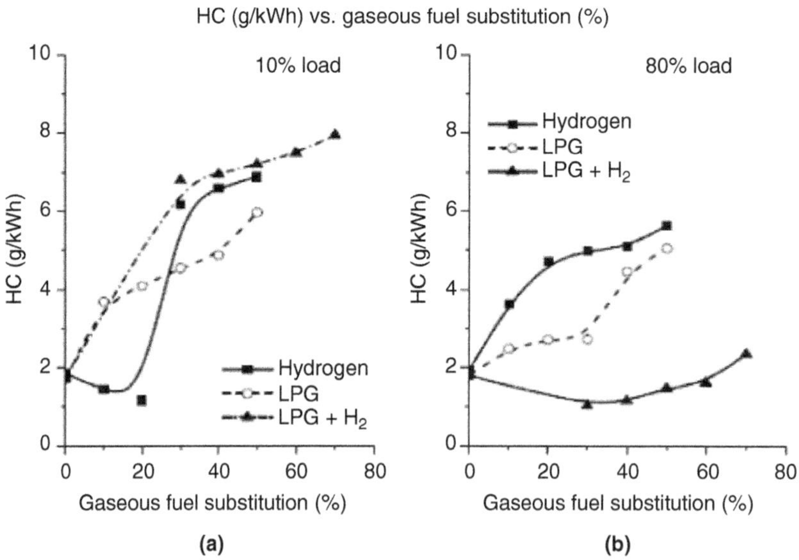

Figure 8.93 Hydrocarbon (g/kWh) vs. gaseous fuels substitution at (a) 10% load and (b) 80% load [92].

compared to pure diesel. A significant improvement in engine performance occurred when a mixture of LPG and hydrogen was used as secondary fuel. In a mixture of LPG and hydrogen (40% in the ratio LPG:hydrogen = 70:30) used as secondary fuel, both the maximum pressure rise rate (by 0.88 bar/° CA) and combustion duration reduced by 4° CA, but peak cylinder pressure and heat release rate were increased by 5.25 bar and 35.24 J/° CA, respectively.

Lata and co-researchers have also worked on mathematical models on dual fuel diesel engines operating on hydrogen, LPG, and a mixture of LPG and hydrogen as secondary fuels. A host of performance and critical combustion parameters such as brake thermal efficiency, net heat

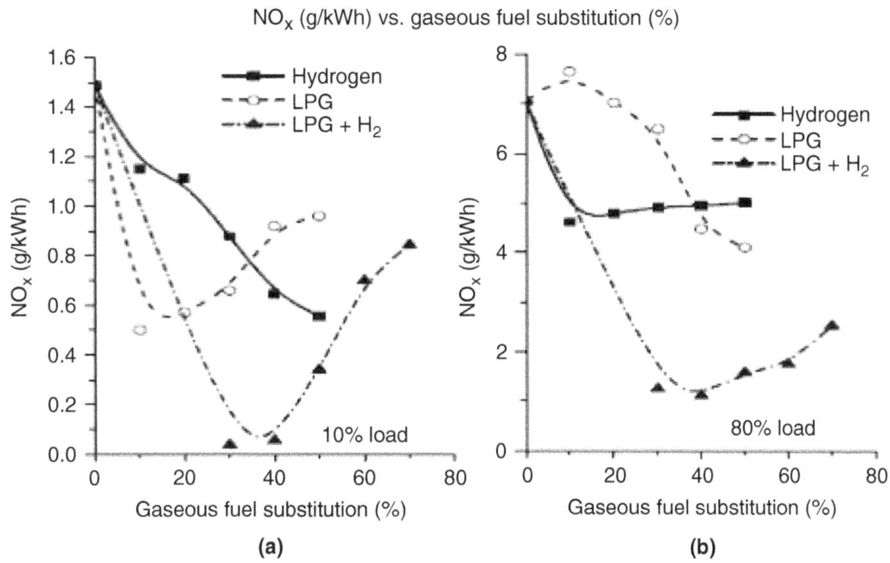

Figure 8.94 Oxides of nitrogen (g/kWh) vs. gaseous fuels substitution at (a) 10% load and (b) 80% load [92].

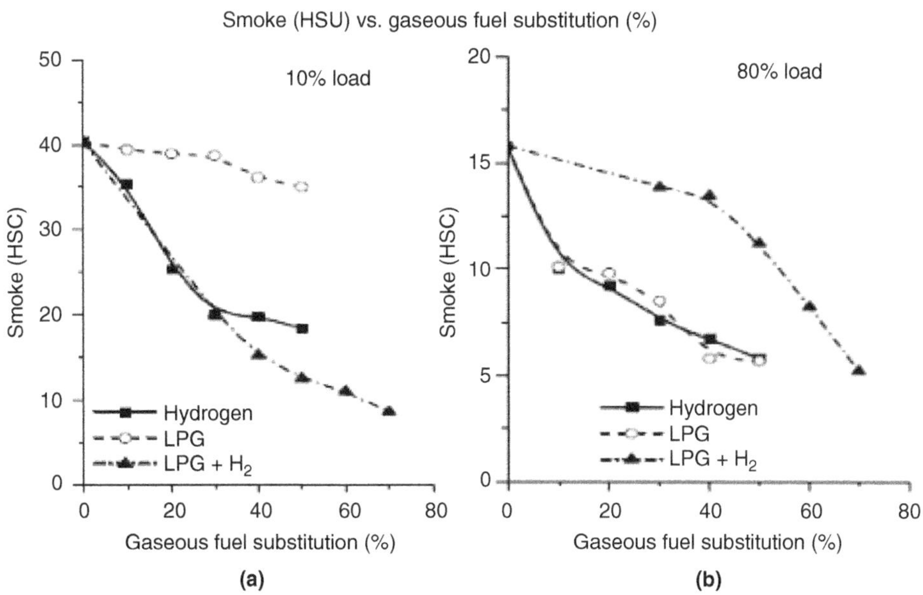

Figure 8.95 Smoke (HSU) vs. gaseous fuels substitution at (a) 10% load and (b) 80% load [92].

release rate, spray mixing characteristics, flame propagation, equilibrium combustion products, and in-cylinder processes were analyzed and used empirical equations for computation and were subsequently compared against experimental data [94]. The comparative analysis of predicted results and experimental data mostly agreed with quantitative as well as qualitative accuracy related to in-cylinder processes and engine performance during combustion. However, the predicted brake thermal efficiency was 3–4% lower than experimentally measured values,

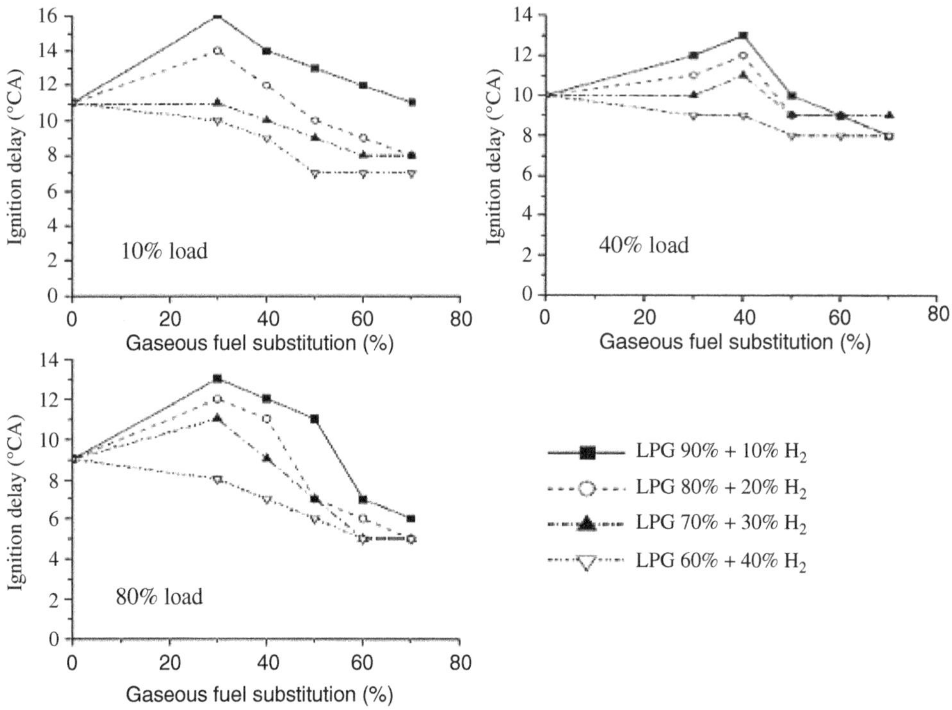

Figure 8.96 Ignition delay (° CA) vs. gaseous fuels substitution [95].

mostly due to quasi-static processes. Similarly, predicted maximum pressure rise and net heat release rate fell short by 6–7% compared to experimental results. This discrepancy was mainly because of inaccurate assessment of several factors such as charge/air velocity, turbulence, and heat transfer to cylinder walls and combustion chambers. Researchers also investigated ignition delay periods under different load conditions and varying levels of diesel substitution and compared with existing ignition delay correlations for diesel and dual-fuel diesel engines by other researchers [95]. Ignition delay was influenced not only by the type and concentration of gaseous fuels but also by factors such as charge temperature, pressure, and oxygen concentration. Ignition delay in dual fuel diesel engines varied with the type and concentration of gaseous fuels in the cylinder charge. Figure 8.96 shows ignition delay with respect to various degrees of hydrogen substitution with LPG. Ignition delay in dual fuel engines using hydrogen, LPG, and their combination could deviate by up to 33.3% compared to that in conventional diesel engines. These variations might be because of the preignition reaction rates and energy release rates of different gaseous fuels.

Concluding Remarks and Perspectives

Among several fuels adopted for use in engines along with hydrogen, compressed natural gas (CNG) has received the most attention. Hydrogen-blended compressed natural gas (HCNG) engines have been a subject of significant research mainly with the objective of introducing hydrogen through the well-established infrastructure of CNG. Several researchers have considered various fractions of hydrogen added to CNG and studied many aspects of performance, emission, combustion, long-term endurance, and tribological studies. It has been widely tested that 18 to 20% by volume is the optimum blend of hydrogen for use in engines for increased brake thermal efficiency (BTE) and decreased brake specific energy consumption (BSEC). Hydrogen enrichment ensures faster flame propagation, higher in-cylinder pressure, and improved combustion stability. HCNG engines lead to lower emissions CO, CO_2, particulate matter (PM), and NOx. Sometimes higher soot concentrations occur due to lubricants burning in the combustion chamber. Several strategies including exhaust gas recirculation (EGR) have proven effective in lowering the NOx emission level.

Ethanol possesses some desirable combustion characteristics, such as a high octane rating and oxygen content, which can enhance engine efficiency and power output when blended with hydrogen. Additionally, hydrogen supplementation can further optimize combustion by promoting faster flame propagation and better combustion stability. Blending ethanol with hydrogen results in greater engine power output because of ethanol's higher octane rating and hydrogen's combustion-enhancing properties. Ethanol's high vaporization heat contributes to lowering peak temperatures inside the cylinder, reducing nitrogen oxide (NOx) emissions. Alternatively, hydrogen combustion can increase NOx emissions. Thus, the overall emissions profile of ethanol–hydrogen blends can be optimized, especially for higher compression ratios. While ethanol alone may exhibit poor specific fuel consumption (SFC) compared to gasoline, blending it with hydrogen leads to significant SFC improvements. Although adequate research activities have not been carried out on blending ethanol with hydrogen for engine applications, the limited work done shows promise in both improved performance and emission concerns.

From whatever limited research has been reported in the literature, it is clear that blending dimethyl ether (DME) with hydrogen for engine applications presents both challenges and opportunities, but there are opportunities for improvement. Blending DME with hydrogen can enhance power output, with improvements in IMEP observed compared to neat hydrogen fuel-supply modes. However, overall thermal efficiency (ITE) might decrease due to prolonged heat release duration, especially at higher DME concentrations. As far as emission aspects are concerned, blending DME with hydrogen results in significant NOx emission reductions, although emission of hydrocarbon (HC) and carbon monoxide (CO) increase. Precise injection timing control needs optimization, and the quantity of DME used with hydrogen needs critical investigation depending on engine configurations. For example, DME exhibits a shorter ignition delay than hydrogen, leading to advanced combustion phases and lower indicated mean effective pressure (IMEP) in DME partially premixed charge compression ignition (PCCI) compared to hydrogen–DME PCCI combustion. Therefore, adjustments in DME injection quantity are required to ensure stable combustion, particularly at higher hydrogen addition levels.

Integration of biogas with hydrogen for engine applications involves an intricate interaction between combustion characteristics, emission profile, and overall engine performance. Advanced injection timing or higher compression ratio improves brake thermal efficiency and reduces fuel consumption. Combustion parameters such as peak pressure and heat release rate can be significantly improved by supplying biogas and hydrogen into the cylinder with advanced injection timing and higher compression ratio. LPG or propane is the other alternative fuel, which has been widely investigated with hydrogen addition in engines. Typical combustion features such as high-octane LPG rating along with wide flammability range and hydrogen's high flame speed together can improve combustion kinetics and raise engine efficiency and power output. Several researchers have brought out test condition in which in-cylinder flow characteristics can lead to increased velocity and kinetic energy to facilitate better mixing of air fuel.

References

1 Shioji, M., Kitazaki, M., Mohammadi, A. et al. (2004). Knock characteristics and performance in an SI engine with hydrogen and natural-gas blended fuels. SAE Technical Paper No. 2004-01-1929.

2 Das, L.M. and Polly, M. (2005). Experimental evaluation of a hydrogen added natural gas (HANG) operated SI engine. SAE Technical Paper No. 2005-26-029.

3 Pede, G., Rossi, E., Chiesa, M. et al. (2007). Test of blends of hydrogen and natural gas in a light duty vehicle. SAE Technical Paper No. 2007-01-2045.

4 Hoekstra, R.L., Van Blarigan, P., and Mulligan, N. (1996). NOx emissions and efficiency of hydrogen, natural gas, and hydrogen/natural gas blended fuels. SAE Technical Paper No. 961103.

5 Wang, D., Mao, X., Wang, J. et al. (2009). Experimental study on the operating range restrictions of a lean-burn turbocharged SI natural gas engine. *Energy Fuels* 23: 3054–3062.

6 Zacepins, A., Kotovs, D., Komasilovs, V. et al (2024). Economic modelling of mixing hydrogen with natural gas. *Processes* 12: 262. https://doi.org/10.3390/pr12020262.

7 McTaggart-Cowan, G.P., Reynolds, C.C., and Bushe, W.K. (2006). Natural gas fuelling for heavy duty on-road use: Current trends and future direction. *International Journal of Environmental Studies* 63 (4): 421–440. https://doi.org/10.1080/00207230600802056.

8 Germane, G.J., Wood, C.G., and Hess, C.C. (1983). Lean combustion in spark-ignited internal combustion engines–a review. *SAE Technical Papers*. https://doi.org/ 10.4271/831694.

9 Wang, J., Chen, H., Liu, B. et al. (2008). Study of cycle-by-cycle variations of a spark ignition engine fueled with natural gas-hydrogen blends. *International Journal of Hydrogen Energy* 33 (18): 4876–4883. https://doi.org/10.1016/j.ijhydene.2008.06.062.

10 Salzano, E., Cammarota, F., and Benedetto, A. (2012). Explosion behavior of hydrogen–methane/air mixtures. *Journal of Loss Prevention in the Process Industries* 25 (3): 443–447. https://doi.org/10.1016/J.JLP.2011.11.010.

11 Larsen, J.F. and Wallace, J.S. (1997). Comparison of emissions and efficiency of a turbocharged lean-burn natural gas and hythane-fueled engine. *Journal of Engineering for Gas Turbines and Power* 119 (1): 218–226. https://doi.org/10.1115/1.2815553, 10.1115/1.2815553.

12 Bell, S.R. and Gupta, M. (1997). Extension of the lean operating limit for natural gas fueling of a spark ignited engine using hydrogen blending. *Combustion Science and Technology* 123 (1–6): 23–48. https://doi.org/10.1080/00102209708935620.

13 Molina, S., Ruiz, S., Gomez-Soriano, J. et al. (2023). Impact of hydrogen substitution for stable lean operation on spark ignition engines fueled by compressed natural gas. *Results in Engineering* 17: 100799.

14 Molina, S., Novella, R., Gomez-Soriano, J. et al. (2021). New combustion modelling approach for methane-hydrogen fueled engines using machine learning and engine virtualization. *Energies* 14 (20): 6732. https://doi.org/10.3390/ en14206732.

15 Ma, F., Wang, Y., Liu, H. et al. (2008). Effects of hydrogen addition on cycle-by-cycle variations in a lean burn natural gas spark-ignition engine. *International Journal of Hydrogen Energy* 33 (2): 823–831. https://doi.org/10.1016/J. IJHYDENE.2007.10.043.

16 Mehra, R.K., Duan, H., Juknelevičius, R. et al. (2017). Progress in hydrogen enriched compressed natural gas (HCNG) internal combustion engines–a comprehensive review. *Renewable and Sustainable Energy Reviews* 80: 1458–1498.

17 Han, W., Dai, P., Gou, X. et al. (2020). A review of laminar flame speeds of hydrogen and syngas measured from propagating spherical flames. *Applications in Energy and Combustion Science* 1–4: 100008.

18 Milton, B.E. and Keck, J.C. (1984). Laminar burning velocities in stoichiometric hydrogen and hydrogen hydrocarbon gas mixtures. *Combustion and Flame* 58: 13–22.

19 Yu, G., Law, C.K., and Wu, C.K. (1986). Laminar flame speeds of hydrocarbon + air mixtures with hydrogen addition. *Combustion and Flame* 63: 339–47.

20 Liu, Y., Lenze, B., and Leuckel, W. (1991). Investigation of the laminar and turbulent burning velocities of premixed lean and rich flames of CH 4-H 2-air mixtures. *Dynamics of Deflagrations and Reactive Systems: Flames* 131: 259–274.

21 Halter, F., Chauveau, C., Djebaili-Chaumeix, N. et al. (2005). Characterization of the effects of pressure and hydrogen concentration on laminar burning velocities of methane–hydrogen–air mixtures. *Proceedings of the Combustion Institute* 30: 201–208.

22 Ilbas, M., Crayford, A.P., Yılmaz, I. et al. (2006). Laminar-burning velocities of hydrogen–air and hydrogen–methane–air mixtures: an experimental study. *International Journal of Hydrogen Energy* 31: 1768–1779.

23 Coppens, F.H.V., De Ruyck, J., and Konnov, A.A. (2007). Effects of hydrogen enrichment on adiabatic burning velocity and NO formation in methane + air flames. *Experimental Thermal and Fluid Science* 31: 437–44.

24 Huang, Z., Zhang, Y., Zeng, K. et al. (2006). Measurements of laminar burning velocities for natural gas–hydrogen–air mixtures. *Combustion and Flame* 146: 302–11.

25 Krishnan, B.N., Chelladorai, P., Teoh, Y.H. et al. (2022). A review on hydrogen-natural gas engines for sustainable mobility. *Materials Today: Proceedings* 68: A8–A25.

26 Nagalingam, B., Duebel, F., and Schmillen, K. (1983). Performance study using natural gas, hydrogen-supplemented natural gas and hydrogen in AVL research engine. *International Journal of Hydrogen Energy* 8 (9): 715–720. https://doi.org/10.1016/0360-3199(83)9018.

27 Park, C.W., Kim, C.G., Choi, Y. et al. (2017). Development of hydrogen-compressed natural gas blend engine for heavy duty vehicles. *International Journal of Automotive Technology* 18 (6): 1061—1066.

28 Park, C., Kim, C., Choi, Y. et al. (2011).The influences of hydrogen on the performance and emission characteristics of a heavy duty natural gas engine. *International Journal of Hydrogen Energy* 36 (5): 3739—3374.

29 Sun, X., Liu, H., Duan, X. et al (2022). Effect of hydrogen enrichment on the flame propagation, emissions formation and energy balance of the natural gas spark ignition engine. *Fuel* 307: 121843. https://doi.org/10.1016/j.fuel.2021.121843

30 Lee, S.-W., Choi, W., and Cho, Y.-S. (2012). Characterization of HCNG combustion and emission characteristics in a constant volume chamber with a single and a dual spark plug configuration. *International Journal of Hydrogen Energy* 37: 682–90.

31 Li, H., Karim, G.A., and Sohrabi. A. (2003). Knock and combustion characteristics of CH_4, CO, H_2 and their binary mixtures. SAE Technical Paper No. 2003-01-3088.

32 Miao, H., Ji, M., Jiao, Q. et al. (2009). Laminar burning velocity and Markstein length of nitrogen diluted natural gas/hydrogen/air mixtures at normal, reduced and elevated pressures. *International Journal of Hydrogen Energy* 34: 3145–3155.

33 Zheng, S., Zhang, X., Xu, J. et al. (2012). Effects of initial pressure and hydrogen concentration on laminar combustion characteristics of diluted natural gas–hydrogen–air mixture. *International Journal of Hydrogen Energy* 37: 12852–12859.

34 Ji, M., Miao, H., Jiao, Q. et al. (2009). Flame propagation speed of CO2 diluted hydrogen-enriched natural gas and air mixtures. *Energy Fuels* 23: 4957–4965.

35 Tahtouh, T., Halter, F., Samson, E. et al. (2009). Effects of hydrogen addition and nitrogen dilution on the laminar flame characteristics of premixed methane–air flames. *International Journal of Hydrogen Energy* 34: 8329–38.

36 Park, C., Won, S., Kim, C. et al. (2012). Effect of mixing CO2 with natural gas-hydrogen blends on combustion in heavy-duty spark ignition engine. *Fuel* 102: 299–304.

37 Allenby, S., Chang, W.C., Megaritis, A. et al. (2001). Hydrogen enrichment: a way to maintain combustion stability in a natural gas fuelled engine with exhaust gas recirculation, the potential of fuel reforming. *Proceedings of the Institution of Mechanical Engineers, Part D, Journal of Automobile Engineering* 215: 405–418.

38 Dimopoulos, P., Rechsteiner, C., Soltic, P. et al. (2007). Increase of passenger car engine efficiency with low engine-out emissions using hydrogen–natural gas mixtures: a thermodynamic analysis. *International Journal of Hydrogen Energy* 32: 3073–3083.

39 Liu, Y.F., Liu, B., Zeng, K. et al. (2012). Performance and emission characteristics of a hydrogen-enriched compressed-natural-gas direct-injection spark ignition engine diluted with exhaust gas recirculation. *Proceedings of the Institution of Mechanical Engineers, Part D, Journal of Automobile Engineering* 226: 123–132.

40 Saanum, I., Bysveen, M., Tunestål, P. et al. (2007). Lean burn versus stoichiometric operation with EGR and 3-way catalyst of an engine fueled with natural gas and hydrogen enriched natural gas. SAE Technical Paper No. 2007-01-0015.

41 Mariani, A., Morrone, B., and Unich, A. (2012). Numerical evaluation of internal combustion spark ignition engines performance fuelled with hydrogen–natural gas blends. *International Journal of Hydrogen Energy* 37: 2644–2654.

42 Ma, F., Wang, Y., Liu, H. et al. (2007). Experimental study on thermal efficiency and emission characteristics of a lean burn hydrogen enriched natural gas engine. *International Journal of Hydrogen Energy* 32: 5067–5075.

43 Huang, Z., Liu, B., Zeng, K. et al. (2007). Combustion characteristics and heat release analysis of a spark-ignited engine fueled with natural gas-hydrogen blends. *Energy & Fuels* 21 (5): 2594–2599.

44 Huang, Z., Wang, J., Liu, B. et al. (2007). Combustion characteristics of a direct-injection engine fueled with natural gas–hydrogen blends under different ignition timings. *Fuel* 86: 381–387.

45 Sandhu, S.S., Babu, M.K.G., and Das, L.M. (2011). Performance and combustion characteristics of a typical motor bike engine operated on blends of CNG and hydrogen using electronically controlled solenoid actuated injection system. *ARPN Journal of Engineering and Applied Sciences* 6 (2): ISSN 1819-6608.

46 Sandhu, S.S., Babu, M.K.G., and Das, L.M. (2012). Effect of hydrogen supplementation on the performance, combustion and emission characteristics of a natural gas fuelled S.I. engine. *International Journal of Alternative Propulsion* 2 (2): 181–196.

47 Sandhu, S.S. Babu, M.K.G., and Das, L.M. (2013). Investigations of emission characteristics and thermal efficiency in a spark-ignition engine fuelled with natural gas–hydrogen blends. *International Journal of Low-Carbon Technologies* 8 (1): 7–13. https://doi.org/10.1093/ijlct/ctr038.

48 Biffiger, H. and Soltic, P. (2015). Effects of split port/direct injection of methane and hydrogen in a spark ignition engine. *International Journal of Hydrogen Energy* 40: 1994–2003.

49 Das, L.M., Swain, R.K., and Polly, M. (2004). Performance improvement of a cng-fuelled spark ignited engine by hydrogen addition. SAE Technical Paper No. 2004-28-0033. https://doi.org/10.4271/2004-28-0033.

50 Mathai, R., Malhotra, R.K., Subramanian, K.A. et al. (2012). Comparative evaluation of performance, emission, lubricant and deposit characteristics of spark ignition engine fuelled with CNG and 18% hydrogen-CNG. *International Journal of Hydrogen Energy* 37: 6893–6900.

51 Mathai, R. (2012). Hydrogen and CNG as fuel for IC engines durability studies of a spark ignition engine for hydrogen blended CNG fuel. PhD thesis. Indian Institute of Technology, Delhi, India.

52 Athawe, M. (2013). Experimental investigation on utilization of hydrogen and CNG in a port injected SI engine. PhD thesis. Indian Institute of Technology, Delhi, India.

53 Coordinating Research Council Inc. (2001). *CRC Manual No. 20, CRC Deposit Rating Manual.* Coordinating Research Council Inc. https://crcao.org.

54 Porpatham, E., Ramesh, E., Nagalingam, B. (2007). Effect of hydrogen addition on the performance of a biogas fuelled spark ignition engine. *International Journal of Hydrogen Energy* 32 (12): 2057–2065.

55 Pandey, V., Shahapurkar, K.H., Guluwadi, S. et al. (2023). Studies on the performance of engines powered with hydrogen-enriched biogas. *Energies* 16: 4349.

56 Şanlı, A., Yılmaz, İ.T., and Gümüş, M. (2021). Investigation of combustion and emission characteristics in a TBC diesel engine fuelled with CH_4–CO_2–H_2 mixtures. *International Journal of Hydrogen Energy* 46 (47): 24395–24409.

57 Benaissa, S., Adouane, B., Ali, S.M. et al. (2021). Effect of hydrogen addition on the combustion characteristics of premixed biogas/hydrogen-air mixtures. *International Journal of Hydrogen Energy* 46 (35): 18661–18677.

58 Elhawary, S., Saat, A., Wahid, M.A. (2020). Experimental study of using biogas in pulse detonation engine with hydrogen enrichment. *International Journal of Hydrogen Energy* 45 (30): 15414–15424.

59 Khatri, N. and Khatri, K.K. (2020). Hydrogen enrichment on diesel engine with biogas in dual fuel mode. *International Journal of Hydrogen Energy* 45 (11): 7128–7140.

60 Bouguessa, R., Tarabet, L., Loubar, K. et al. (2020). Experimental investigation on biogas enrichment with hydrogen for improving the combustion in diesel engine operating under dual fuel mode. *International Journal of Hydrogen Energy* 45 (15): 9052–9063.

61 Karagöz, Y. (2019). Analysis of the impact of gasoline, biogas and biogas + hydrogen fuels on emissions and vehicle performance in the WLTC and NEDC. *International Journal of Hydrogen Energy* 44 (59): 31621–3163.

62 Mariani, A., Minale, M., and Unich, A. (2021). Use of biogas containing CH4, H2 and C02 in controlled auto-ignition engines to reduce NOx emission. *Fuel* 301 (11): 1209–1225.

63 Montoya, J.P.G., Amell, A.A., Olsen, D.B. et al. (2018). Strategies to improve the performance of a spark ignition engine using fuel blends of biogas with natural gas, propane and hydrogen. *International Journal of Hydrogen Energy* 43 (46): 21592–21602.

64 Xin, Z., Jian, X., and Shizhuo, Z. (2013). The experimental study on cyclic variation in a spark ignited engine fueled with biogas and hydrogen blends. *International Journal of Hydrogen Energy* 38 (25): 11164–11168.

65 Verma, S., Kumar, K., Das, L.M. et al. (2021). Effect of hydrogen enrichment strategy on performance and emission features of biodiesel–biogas dual fuel engine using simulation and experimental analyses. *Journal of Energy Resources Technology* 143 (9): 1–18.

66 Verma, S. (2019). Comparative assessment of an IC engine using alternative fuels based on energy and exergy. PhD thesis. Indian Institute of Technology, Delhi, India.

67 Verma, S., Das, L.M., Bhatti S.S. et al. (2017). A comparative exergetic performance and emission analysis of pilot diesel dual-fuel engine with biogas, CNG and hydrogen as main fuels. *Energy Conversion and Management* 151: 764–777.

68 Lalsangi, S., Yaliwal, V.S., Banapurmath, N.R. et al. (2023). Influence of hydrogen injection timing and duration on the combustion and emission characteristics of diesel engine operating on dual fuel mode using biodiesel of dairy scum oil and producer gas. *International Journal of Hydrogen Energy* 48 (55): 21313–21330.

69 Al-Baghdadi, M.A.S. (2003). Hydrogen–ethanol blending as an alternative fuel of spark ignition engines. *Renewable Energy* 28 (9): 1471–1478.

70 Al-Baghdadi, M.A.-R.S. (2000). Performance study of a four-stroke spark ignition engine working with both of hydrogen and ethyl alcohol as supplementary fuel. *International Journal of Hydrogen Energy* 25 (10): 1005–1009.

71 Al-Baghdadi, M.A.S. (2001). The safe operation zone of the spark ignition engine working with dual renewable supplemented fuels (hydrogen+ethyl alcohol). *Renewable Energy* 22 (4): 579–583.

72 Marwaha, A. and Subramanian, K.A. (2023). Performance enhancement and emissions reduction of ethanol-fueled spark ignition engine with hydrogen. *Biofuels, Bioproducts, and Biorefining* 13(3). https://doi.org/10.1002/bbb.253.

73 Ayad, S.M.M.E., Belchior, C.R.P., and Sodré, J.R. (2024). Hydrogen addition to ethanol-fuelled engine in lean operation to improve fuel conversion efficiency and emissions. *International Journal of Hydrogen Energy* 49 Part B: 744–752.

74 Greenwood, J.B., Erickson, P.A., Hwang, J. et al. (2014). Experimental results of hydrogen enrichment of ethanol in an ultra-lean internal combustion engine. *International Journal of Hydrogen Energy* 39 (24): 12980–12990.

75 García-Morales, J., Escobar-Jiménez, R.F., Ramos-Negrón, O.J. et al. (2022). Design of an adaptive control to feed hydrogen-enriched ethanol-gasoline blend to an internal combustion engine. *International Journal of Chemical Engineering* 2022: Article ID 7413554, 12 pp. https://doi.org/10.1155/2022/7413554.

76 Jeon, J. and Bae, C. (2013). The effects of hydrogen addition on engine power and emission in DME premixed charge compression ignition engine. *International Journal of Hydrogen Energy* 38 (1): 265–273.

77 Mariott, C.D. and Reitz. D. (2002). Experimental investigation of direct injection-gasoline for premixed compression ignition combustion phasing control. SAE Technical Paper No. 2002-01-0418. http://dx.doi.org/10.4271/2002-01-0418.

78 Mariott, C.D., Kong, S., and Reitz, R.D. (2002). Investigation of hydrocarbon emission from a direct injection-gasoline premixed charge compression ignited engine. SAE Technical Paper No. 2002-01-0419. http://dx.doi.org/10.4271/2002-01-0419.

79 Kuboyama, T., Moriyoshi, Y., Hatamura, K. et al. (2011). Extension of operating range of a multi-cylinder gasoline HCCI engine using the blowdown supercharging system. SAE Technical Paper No. 2011-01-0896. http://dx.doi.org/10.4271/2011-01-0896.

80 Shudo, T. and Yamada, H. (2007). Hydrogen as an ignition controlling agent for HCCI combustion engine by suppressing the low-temperature oxidation. *International Journal of Hydrogen Energy* 32: 3066–3072.

81 Cong, X., Ji, C., and Wang, S. (2021). Investigation into engine performance of a hydrogen-dimethyl ether spark-ignition engine under various dimethyl ether fractions. *Fuel* 306: 121429.

82 Faingold, G., Tartakovsky, L., and Franke, S.H. (2018). Numerical study of a direct injection internal combustion engine burning a blend of hydrogen and dimethyl ether. *Drones* 2 (23): https://doi.org/10.3390/drones2030023www.mdpi.com/journal/drones.

83 Shere, A. and Subramanian, K.A. (2024). Effects of hydrogen and EGR on energy efficiency improvement with ultra low emissions in a common rail direct injection compression ignition engine fueled with dimethyl ether (DME) under HCCI mode. *International Journal of Hydrogen Energy* 52: 1447–1474.

84 Kakoee, A., Bakhshan, Y., Gharehghani, A. et al. (2019). Numerical comparative study of hydrogen addition on combustion and emission characteristics of a natural-gas/dimethyl-ether RCCI engine with pre-chamber. *Energy* 186: 115878.

85 Liu, J., Yang, F., Wang, H. et al. (2012). Numerical study of hydrogen addition to DME/CH4 dual fuel RCCI engine. *International Journal of Hydrogen Energy* 37 (10): 8688–8697.

86 Pan, L., Hu, E., Meng, X. et al. (2015). Kinetic modeling study of hydrogen addition effects on ignition characteristics of dimethyl ether at engine-relevant conditions. *International Journal of Hydrogen Energy* 40 (15): 5221–5235.

87 Huang, Z., Chen, G., Chen, C. et al. (2008). Experimental study on premixed combustion of dimethyl ether–hydrogen–air mixtures. *Energy & Fuels* 22 (2): 967–971.

88 Choi, G.H., Chung, Y.J., and Han, S.B. (2005). Performance and emissions characteristics of a hydrogen enriched LPG internal combustion engine at 1400 rpm. *International Journal of Hydrogen Energy* 30 (1): 77–82.

89 Karagöz, Y. and Köten, H. (2019). Effect of different levels of hydrogen + LPG addition on emissions and performance of a compression ignition engine. *Journal of Thermal Engineering* 5 (2, Special Issue 9): 58–69. Yildiz Technical University Press, Istanbul, Turkey.

90 Kacem, S.H., Jemni, M.A., Driss, Z. et al. (2016). The effect of H2 enrichment on in-cylinder flow behavior, engine performances and exhaust emissions: case of LPG-hydrogen engine. *Applied Energy* 179: 961–971.

91 Ravi, K., Bhasker J.P., and Porpatham, E. (2017). Effect of compression ratio and hydrogen addition on part throttle performance of a LPG fuelled lean burn spark ignition engine. *Fuel* 205: 71–79.

92 Lata, D.B., Misra, A., and Medhekar, S. (2012). Effect of hydrogen and LPG addition on the efficiency and emissions of a dual fuel diesel engine. *International Journal of Hydrogen Energy* 37: 6084–6096.

93 Lata, D.B., Misra, A., and Medhekar, S. (2011). Investigations on the combustion parameters of a dual fuel diesel engine with hydrogen and LPG as secondary fuels. *International Journal of Hydrogen Energy* 36: 13808–13819.

94 Lata, D.B. and Misra, A. (2010). Theoretical and experimental investigations on the performance of dual fuel diesel engine with hydrogen and LPG as secondary fuels. *International Journal of Hydrogen Energy* 35: 11918–11931.

95 Lata, D.B. and Misra, A. (2011). Analysis of ignition delay period of a dual fuel diesel engine with hydrogen and LPG as secondary fuels. *International Journal of Hydrogen Energy* 36: 3746–3756.

9

Some Upgraded Strategies for Hydrogen Engines

The only thing that interferes with my learning is my education.

Albert Einstein

9.1 Introduction

Existing designs of both spark and compression ignition engines have been appropriately modified to operate on hydrogen and in some cases, practical prototype systems have been developed for use in vehicles, gensets, pump sets, lawn mowers, etc. Research activities are being carried out to incorporate advanced concepts in systems and strategies to further improve these engines. This chapter highlights some of such upgraded and augmented hydrogen engines involving supercharging, turbocharging, HCCI combustion and stratification, etc. Alternative configurations such as in rotary hydrogen engines are briefly discussed.

9.2 Supercharging and Turbocharging

The low volumetric energy density of hydrogen–air mixtures has been responsible for low power output. Turbochargers and superchargers are used to increase the amount of oxygen entering the engine, which improves combustion and increases engine power output through different mechanisms. Turbochargers utilize exhaust gases to drive a turbine connected to a compressor, while superchargers are mechanically driven by the engine's crankshaft. Despite this operational difference, both share the common purpose of boosting the horsepower of a normally aspirated engine.

Nguyen et al. [1] investigated boosting methods on a 2.4 L supercharged port fuel injection engine running at 2000 rpm and 3000 rpm with torque at low-load settings of 50 Nm. The excess air ratio (lambda) was varied from stoichiometric to 2.8, and spark advance timing was set at maximum brake torque conditions.

One of the major objectives of this study was to investigate effectiveness of a supercharger system (S/C) in a hydrogen engine under low-load conditions. Turbocharging (T/C) can improve power by extending high-load conditions, but high exhaust pressure limits performance, especially at high speeds and heavy loads. To ensure a comparative evaluation between the supercharger system and a previously tested turbocharger system [2], the amount of injected hydrogen and throttle settings were kept consistent with the same as were adopted in turbocharger experiments by Lee et al. [2]. Results

Hydrogen Engines: Design, Performance Evaluation, Combustion Analysis, and Exhaust Emissions, First Edition.
Lalit Mohan Das.

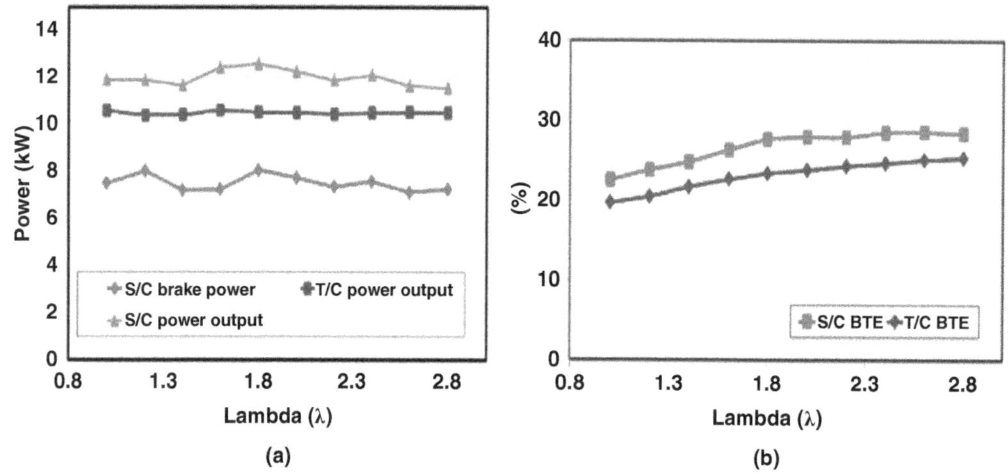

Figure 9.1 (a) Net power output and (b) brake thermal efficiency at 2000 rpm under various air excess ratios for supercharging (S/C) and turbocharging (T/C) in engines [1].

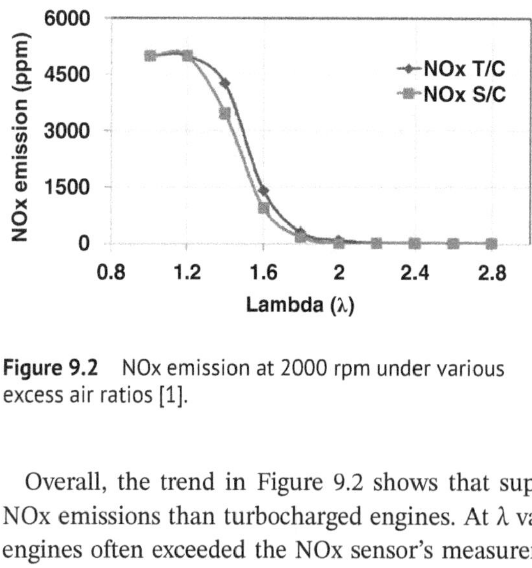

Figure 9.2 NOx emission at 2000 rpm under various excess air ratios [1].

showed that supercharging reduced pumping losses and increased indicated mean pressure over turbocharging under low-load conditions. The power output of a hydrogen engine with a supercharger system showed an average improvement of 10% compared to a turbocharger-boosted engine (Figure 9.1a). At its highest, the power output reached 28% greater with the supercharger system. Figure 9.1b shows the brake thermal efficiency of a supercharged and turbocharged engine. The measured NOx emissions using the supercharger system were lower than those recorded earlier using the turbocharger.

Overall, the trend in Figure 9.2 shows that supercharged hydrogen engines exhibited lower NOx emissions than turbocharged engines. At λ values below 1.4, emissions from supercharged engines often exceeded the NOx sensor's measurement range of 5000 ppm, even though values were lower than for turbocharged engines. At λ values of 1.4, 1.6, and 1.8, NOx emissions from a turbocharged to a supercharged engine dropped by 20%, 33%, and 44%, respectively. This is because the air induced by the supercharger was higher than for the turbocharged engine particularly at lambda 1.4 and 1.8 and increased air volume increased oxygen concentration. Verhelst et al. [3] studied single-cylinder hydrogen engines equipped with a supercharger and exhaust gas recirculation (EGR) system. The value of λ was restricted to a minimum of 1.3 to prevent preignition and backfire. A three-way catalyst (TWC) was also used for NOx control. The EGR enabled stoichiometric supercharged operation, and maximum power output was raised up to 30% higher than for gasoline engines assuming 95% TWC conversion efficiency. Nakagawa et al. [4] developed supercharged, lean burn, hydrogen fueled, direct injection engines and achieved large output power, high thermal efficiency, and near-zero NOx emissions. In these tests, cooling loss decreased

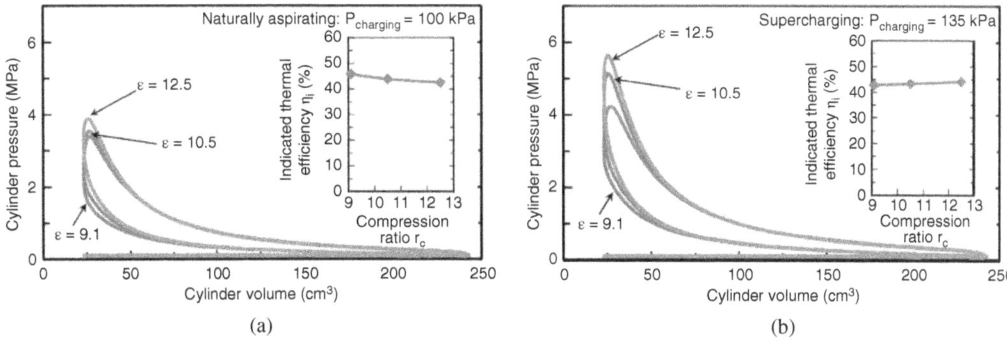

Figure 9.3 Cylinder pressure variation for (a) naturally aspirated and (b) supercharged operation [4].

because of lean combustion. Thus, it was possible to achieve a brake thermal efficiency of 34% at the highest intake pressure of 200 kPa.

Figure 9.3 shows two scenarios at compression ratios (ε) of 9.1, 10.5, and 12.5: one with natural aspiration (Figure 9.3a) and the other with a supercharging pressure of 135 kPa (Figure 9.3b). In the case of naturally aspirated engines, with increasing compression ratio, the work in the compression stroke becomes higher than the work in the expansion stroke. On the contrary, in the case of the supercharging operation, with the increase in compression ratio, work in the compression stroke is less than the work in the expansion stroke [4].

9.2.1 Performance Features

Rogaliskiet al. [5] investigated a supercharged spark ignition hydrogen engine in which hydrogen was delivered in two different methods: with injector and mixer installed in the intake port. In-cylinder pressure during hydrogen combustion was analyzed with various coefficients of stoichiometry and boost pressures, which were constrained by the occurrence of knock since. hydrogen's tendency to knock increases due to boosting pressure. Higher boosting pressure led to shorter combustion duration, and shifting the spark timing closer to TDC kept the knock intensity within the permissible limit. Supercharging the hydrogen engine proved an effective strategy for enhancing engine performance without knock, particularly when operating at a reasonable compression ratio of 8.6. A mixture with a λ of 1.3 ensured stable combustion without backfire. With a temperature increase from 20 to 70 °C and boost pressure of 1.9 bar, rise in knock intensity was only marginal. This insignificant knock intensity was primarily due to the relatively low compression ratio of 8.6.

Work by Ling-zhi et al. [6] on a direct injection (DI) turbocharged hydrogen engine indicated superior power and torque density compared to other fuels, as shown in Figure 9.4 [6].

After simulation, calculation, and reconfiguration, several parameters were evaluated for optimized performance and emission characteristics with smooth combustion. The power of the turbocharged engine increased by 123%, reaching a maximum value of 120 kW at 4400 rpm (Figure 9.5). Engine operation at λ = of 1.91 at 2000 rpm achieved the maximum brake thermal efficiency (BTE) of 42.6%, and 40.4% BTE was obtained with λ of 2.47 at 3000 rpm. The engine could achieve high NOx emissions conversion efficiency of over 99.5% at low speeds (below 2000 rpm) and 90% at 4400 rpm with an NH3-SCR system installed for after-treatment (Figure 9.6). Port fuel injection (PFI) systems lose about 30% of power due to hydrogen's ultra-low density. DI engines, which avoid such problems, are the preferred mode. In direct injection (DI) operation, ongoing

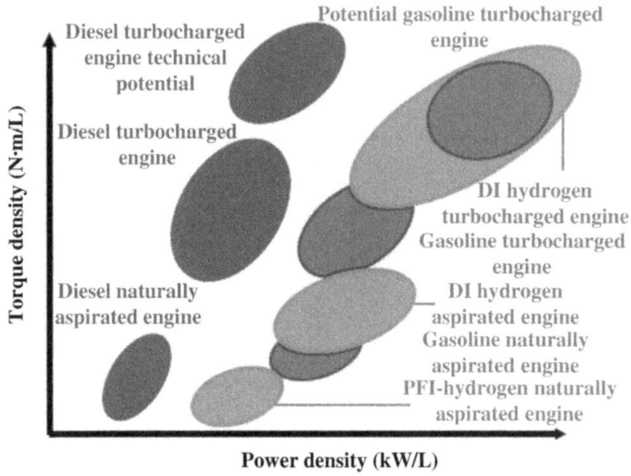

Figure 9.4 Power and torque potential of different hydrogen engines with respect to gasoline and diesel engines [6]. Please visit the companion website to view a color version of this image.

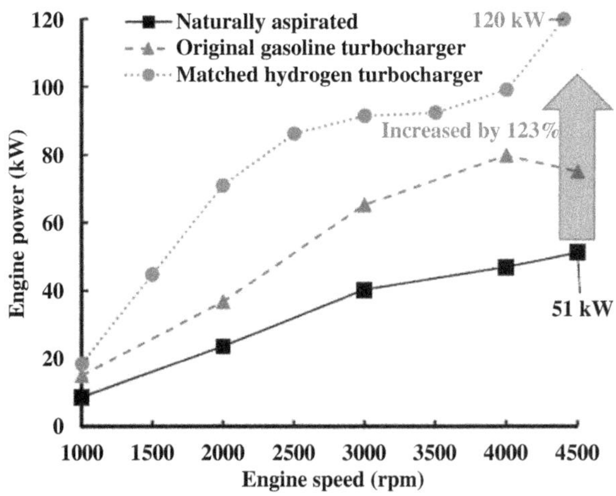

Figure 9.5 Comparison of engine power with different intake types [6].

research should focus on implementing a late injection strategy to achieve near-ideal combustion and reduce efficiency losses such as compression work, improper combustion, and heat transfer to the coolant.

Advantages of direct injection (DI) over port fuel injection (PFI) have been elaborately discussed in Chapters 3 and 4. Hybrid fueling systems integrating both direct injection and port fuel injection technologies offer the distinctive advantage of improved efficiency and reduced emissions [7]. Direct injection increases a mixture's calorific value by 41% compared to premixed hydrogen engines, which improves performance and efficiency [8]. A combined approach of direct injection with $\lambda = 1$ and turbocharging makes it possible to increase load density through turbocharging, while direct hydrogen injection increases calorific value of the mixture [8]. Wang and co-researchers [9] studied the energy and exergy distribution of a 2.3 L turbocharged hydrogen engine through characteristic mapping experiments. The energy loss estimated during fuel energy conversion is given in Table 9.1, which shows that depending upon the situation, the effective work is within the

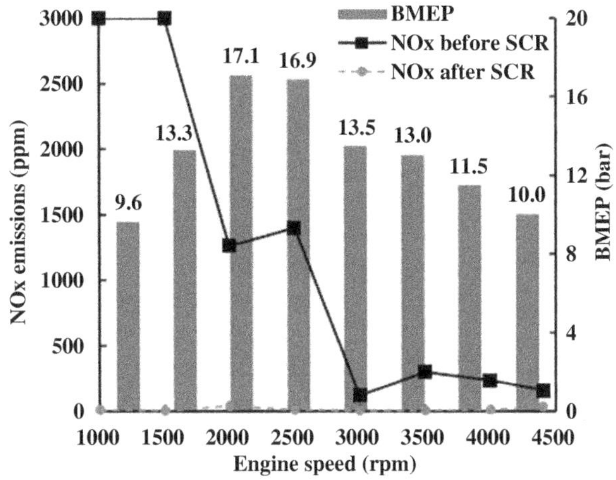

Figure 9.6 NOx emission before and after the SCR [6].

Table 9.1 Range of energy loss during fuel-energy conversion [9].

Energy	% loss
Exhaust energy	23.5–34.7
Cooling medium (coolant an oil) energy	21.3–34.8
Intercooler energy	0.5–3.6
Unaccounted-for energy	5.8–14.1

range from 25.7% to 35.1%. Cooling medium energy has two shares: coolant energy and oil energy. Between the two, coolant energy is much larger than oil energy. Combined total energy of both is more affected by changes in the engine speed, which means that as the engine speed changes, the sum of both parts changes significantly. However, the proportion of each of these two parts is more sensitive to load, which implies that as the load changes, the individual contributions of coolant and oil energy to the total energy vary more noticeably. This trend of proportions decreasing with increasing load has been reported to be 34.8% at 1500 rpm, 0.3 MPa in the analysis. Broadly speaking, all kinds of energy reacted to engine speed and showed an increasing trend, but they were insensitive to load. However, the proportions of the different energy types varied. A turbocharger improved brake thermal efficiency by up to 4.8%. Exergy efficiency of coolant energy did not exceed 5%, and exergy efficiency of exhaust energy could be raised to 23%. The total hydrogen fuel thermal efficiency theoretically exceeded 59% [9].

9.2.2 Combustion Characteristics: Peak Pressure, Pressure Rise Rate, Heat Release Rate, Combustion Duration, Preignition, and Knock

Duan et al. [10] carried out tests on a turbocharged direct injection hydrogen engine, over a wide range of operating conditions with the focus on combustion. It has been adequately reported in the literature that equivalence ratio, spark timing, injection timing and duration, etc. have a large influence on combustion characteristics. Tests were designed to assess the influence of

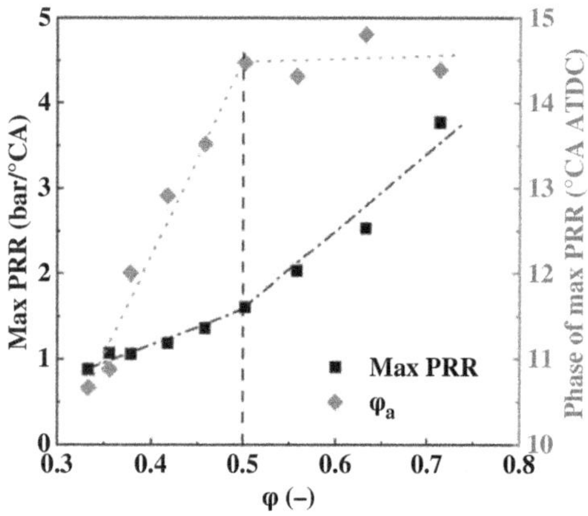

Figure 9.7 Maximum pressure rise rate and phase of maximum PRR with φ at 2500 rpm [10].

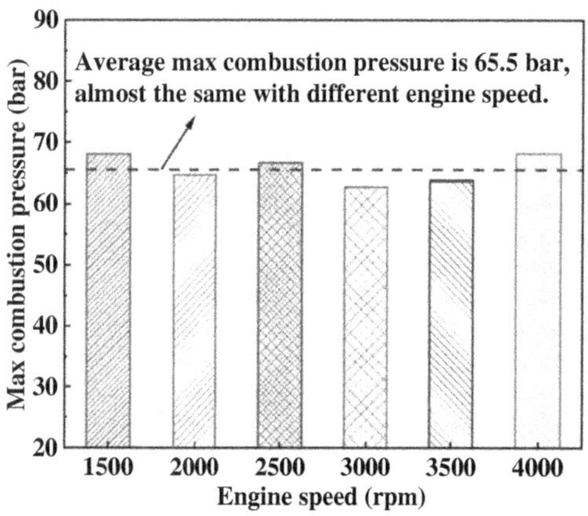

Figure 9.8 Maximum combustion pressure with engine speed at BMEP = 10 bar [10].

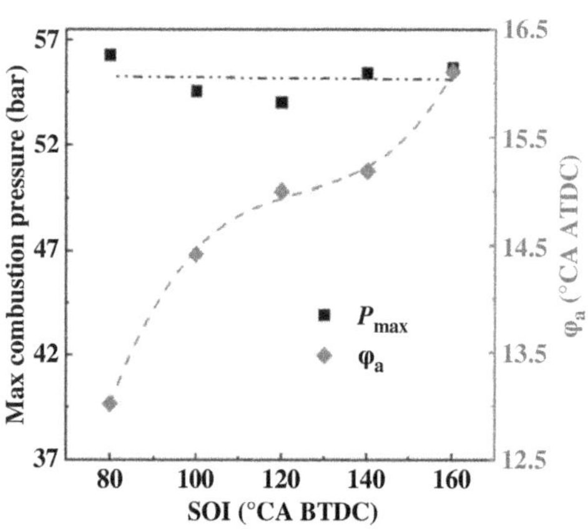

Figure 9.9 Maximum combustion pressure and corresponding phase wth SOI [10].

these combustion parameters. Speeds varied from 1000 to 4000 rpm with loads within the range of 3.7 to 10.6 bar. Test were restricted to an equivalence ratio between 0.33 and 0.71. Spark timing was adjusted within 21° of the crank angle, and the start of injection varied within 80° of the crank angle.

Pressure rise rate (PRR) is a significant parameter as it quantifies combustion, and maximum PRR directly indicates the level of combustion noise. Maximum PRR reached 3.85 bar/°CA at 1500 rpm with wide-open throttle (WOT), higher than turbocharged port fuel injection hydrogen engines (PFI-HICE). In the analysis, the relationship between ϕ and maximum PRR was divided into two stages on the basis of a value of $\phi = 0.5$ as the reference threshold value. In the first stage, increasing ϕ from 0.33 to 0.50 led to an 83% increase in the maximum PRR: from 0.88 bar/°CA to 1.61 bar/°CA. In the second stage, an increase in ϕ from 0.50 to 0.71 resulted in a significant 134% increase in maximum PRR from 1.61 bar/°CA to 3.77 bar/°CA (Figure 9.7).

Figure 9.8 shows the value of BMEP at 10 bar. As evident in the figure, there is an extremely small difference in the maximum combustion pressure with engine speed, ranging from 62.8 to 68.5 bar. Maximum combustion pressure remains almost constant despite changes in SOI. This is so because the amount of hydrogen injected and air mass flow per cycle are consistent. However, in the case of SOI getting shifted from 160 °CA to 80 °CA before TDC, the phase of maximum combustion pressure shifts from 16 °CA to 13 °CA before TDC (Figure 9.9). In such a case, as SOI is delayed, the proximity of the injector to spark results in increasing the rich mixture close to the spark plug.

During their investigation of a 2.0 L turbocharged direct injection hydrogen engine, Wang et al. [11] observed that optimal turbocharger selection requires proper charge density, λ, and exhaust back pressure. Some of the observations during the study, as given below, provide significant guidelines on combustion. Adjustments of a variable geometry turbocharger (VGT) with 50% at 2500 rpm and 80%

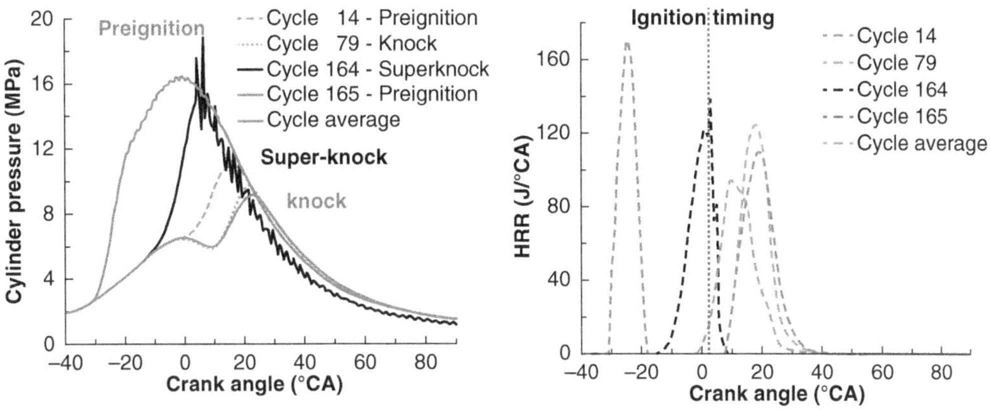

Figure 9.10 Abnormal combustion under high loads (BMEP = 1.561 MPa) at 2500 rpm [11]. Please visit the companion website to view a color version of this image.

at 4500 rpm cause the minimum pumping loss, and a maximum exhaust back pressure is below 270 kPa. Injection timing plays a crucial role. Too early injection of SOI = 240 °CA BTDC increased preignition, and late injection at SOI caused unstable combustion. Optimal injection timing, to get rid of these undesirable combustion phenomena, occurred after the closure of the intake valve and was in the range of 60–80 °CA after BTDC. Compared to load, engine speed, λ, and ignition timing had a bigger impact on burn duration. Burn duration (BD) extended from 11 °CA to 29 °CA as speed increased from 1500 to 4500 rpm and λ from 1.7 to 2.9. The minimum λ limited by knock (KLλ) was established at λ = 1.39 at 2500 rpm. The minimum value of λ limited by knock (KLλ) is considered a significant primary parameter to control knock. When λ decreased, it increased the laminar flame velocity leading to a higher possibility of knock. Frequent preignition and knock at λ = 1.39 and 2500 rpm triggered conditions of superknock. Figure 9.10 shows the close relationship between preignition and knock. Initially sporadic preignition (cycle 14) occurred near top dead center (TDC) with increased cylinder temperature, which subsequently led to knock (cycle 79). As a result, cylinder pressure and instantaneous heat release rate (HRR) went far beyond the typical cycle levels. Frequent preignition and intense knock resulted in a superknock cycle (164) and the peak pressure reached 18.8 MPa. In the next cycle (cycle 165) knock occurred relatively earlier and showed severe preignition, with a maximum HRR of 165 J/° CA.

Wang et al. [12] developed a computation model of the direct injection (DI) turbocharged H2ICE using one-dimensional (1D) performance modeling tools and engine test bench. Performance ceiling was restricted to avoid knock and backfire conditions, keeping operational safety and reliability in mind. Part of this investigation was to determine the performance boundary by simulating supercharging. Optimization steps including adjustments to valve phase, hydrogen injection angle, compression ratio, and turbocharging, collectively improved the engine's performance. Consequently, the engine achieved peak power output of 133 kW at 4500 rpm, maximum torque of 339 N·m at 2000 rpm, and high thermal efficiency of 43.06%.

Luo et al. [13] investigated combustion characteristics and NOx emission of a turbocharged hydrogen engine. The combustion duration (the crank angle of 10–90% fuel burned) at both 1500 rpm and 2000 rpm was found to be equal and significantly longer than other loads, specifically at a BMEP of 0.27 MPa. This may be due to the turbocharger's influence on gas exchange, which influences the combustion process. Over the entire range of operating conditions, the peak pressure rise rate remained below 0.25 MPa/CA. NOx emissions varied from 300 ppm to 1200 ppm with increasing engine speed, peaking at 7000 ppm at 2500 rpm with an equivalence ratio of 0.88 (Figure 9.11).

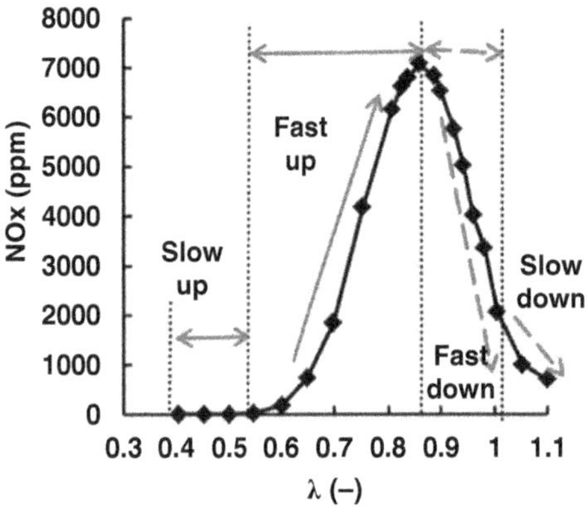

Figure 9.11 Variation of NOx emission with λ [13].

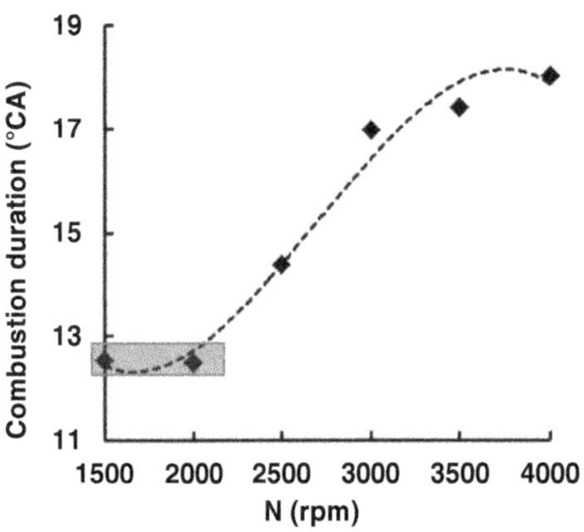

Figure 9.12 Variation of combustion duration changes with engine speeds at WOT [13].

It is evident from Figure 9.12 that as engine speed increases from 1500 rpm to 4000 rpm, combustion duration (the crank angle of 10–90% fuel burned) extends from 12.5 °CA to 18.1 °CA. Increase of engine speed decreases time corresponding to the crank angle per degree and so leads to increased combustion duration. Increase in speed also increases the mass of air flow into the cylinder, especially after turbocharging. Combustion duration at 1500 rpm and 2000 rpm is the same. This is because the turbine resists the exhaust gas and increases in residual gas fraction and subsequent reduction in combustion velocity.

Bao et al. [14] focused their study on achieving improved thermal efficiency and higher power output in a 2.0 L turbocharged direct injection hydrogen engine with near-zero emission (below 20 ppm). The study involved lean combustion and delayed ignition. At 2500 rpm, adjusting λ from 2 to 2.4 reduced NOx emissions from 2500 ppm to 130 ppm. Since maximum brake torque (MBT) timing changes significantly with different engine speeds and loads, CA50 (the crank angle at which 50% of the fuel has burned) was kept within the range of 0 to 20 degrees of crank angle to ensure smooth knock-free combustion.

Figure 9.13 shows the effect of ignition timing on the power, efficiency, and emission. Advancing ignition by bringing forward CA50 from 7 °CA to 0 °CA resulted in a significant increase in the NOx emission level. Both BMEP and BTE are slightly decreased. Delayed ignition was a highly effective in bringing down the NOx level to near-zero emission. Retarding CA50 from 7 °CA to 20 °CA causes a drop in BTE from 41.9% to 39.3% at 2500 rpm, but there was only a 0.07 MPa loss in BMEP. Thus, NOx reduction by delayed ignition is a very effective technique. Even the little energy loss by retardation can be recycled from the high-temperature exhaust gases of the turbine.

Figure 9.14 shows the average cylinder temperature over 200 cycles. In lean combustion, the temperature starts rising slowly after 10 °CA and peaks at 1401 K. For delayed ignition, the temperature rise occurs later, with a lower peak temperature of 1388 K. In another paper on a turbocharged direct-injection engine, researchers have reported [15] effects of the coefficient of excess air (λ), start of injection (SOI) timing, and injection pressure within an engine speed range of 1000–4000 rpm. Figure 9.15 shows part of the important results in that study. There are three load characteristic curves with each curve indicating its boundary. The black dotted line indicates maximum λ that could be achievable at full throttle with IMEP's low coefficient of vibration. Conditions below this line require throttle control to maintain combustion stability. Red blocks signify the NOx level, which indicate the need for exhaust post-treatment. Green blocks, which

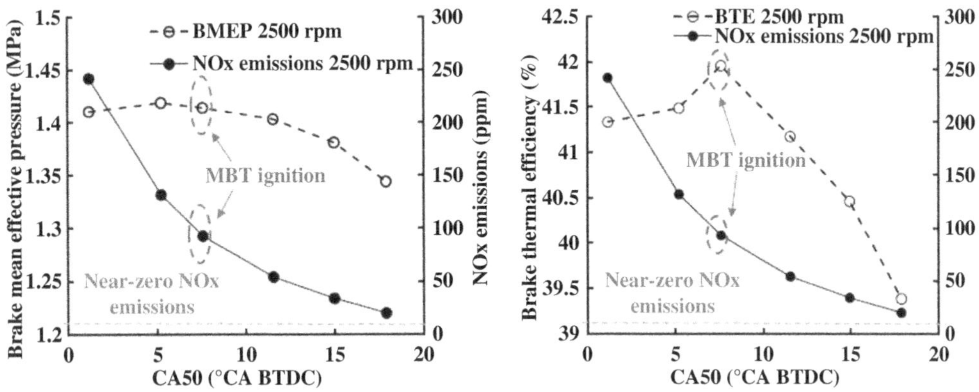

Figure 9.13 Effect of ignition timing on BTE, BMEP, and NOx emissions [14].

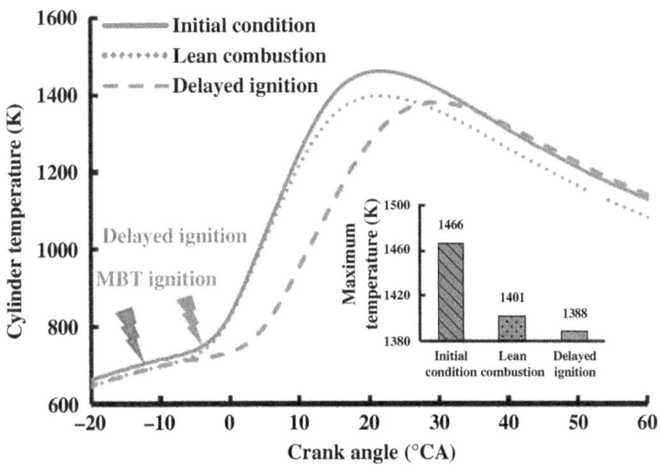

Figure 9.14 Average cylinder temperature of lean combustion and delayed ignition timing [14].

cover approximately two-thirds of the total space, indicate that the majority of conditions can attain near-zero emissions after optimization. In the near-zero emission region, maximum BMEP reaches 13.3 bar at 3000 rpm (equivalent to 212.8 N·m), maximum power is 71 kW at 3500 rpm, and the highest BTE is 40.4% at 2000 rpm. Figure 9.16 shows the correlation between λ and near-zero NOx emissions. For naturally aspirated engines, λ exceeding 2.53 ensured nearly zero emissions while a λ of 2.62 was required for turbocharged engines at 2000 rpm. However, near-zero emissions could be attained at a λ of 3.29 only at 3000 rpm and leaner combustion.

Hong and co-researchers [16] examined thermal efficiency and NOx emissions in a three-cylinder turbocharged DI H2-ICE using a single-hole injector over a wide range of high injection pressures and timings. Results indicated both turbocharging and lean burn strategies had a strong influence on thermal efficiency. Injection strategy is chiefly dependent on wall heat loss. Wall heat loss decreased by about 2% in late injection due to strong mixture stratification, and thus late injection achieves higher efficiency than early injection over various load conditions. However, late injection always leads to a high rise in NOx emission. Thus, a moderate optimization strategy between late injection and NOx emission was optimized at an end-of-injection (EOI) of about 40° CA.

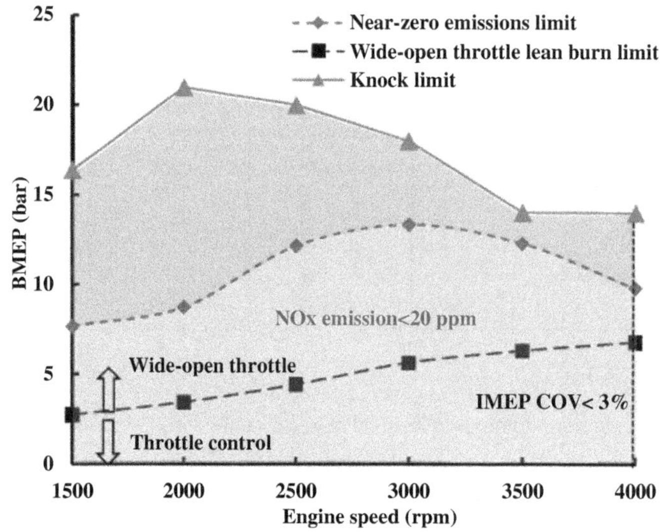

Figure 9.15 Load characteristics of the turbocharged direction injection engine [15]. Please visit the companion website to view a color version of this image.

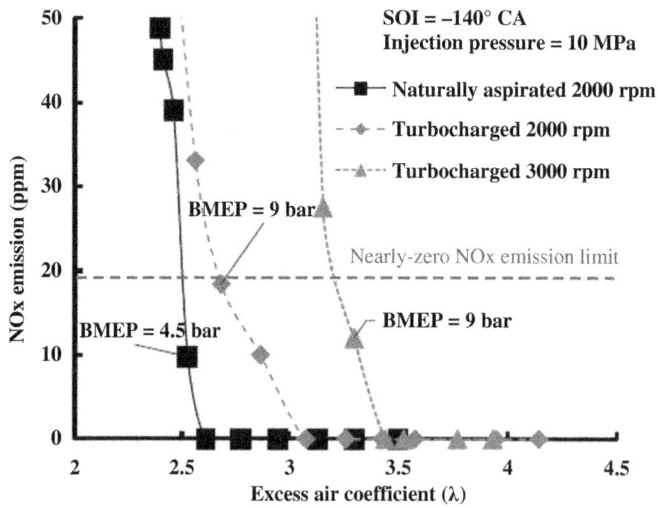

Figure 9.16 Effects of λ on the near-zero emission characteristics for various engine speeds [15].

Figure 9.17 shows the effects of injection pressure on brake thermal efficiency (BTE) and nitrogen oxide (NOx) emissions at brake mean effective pressure (BMEP) of 10 bar. It indicates that while the start of injection (SOI) had a similar influence on BTE regardless of injection pressure, BTE was more sensitive to SOI under higher injection pressures. Specifically, optimizing SOI timing leads to 0.5% improvement in BTE at 5 MPa injection pressure, increasing to about 2.3% at 15 MPa injection pressure. The researchers also observed that hydrogen injection duration was reduced approximately threefold when the injection pressure was increased from 5 to 15 MPa. Moreover, at a given start of injection (SOI), the highest BTE at an injection pressure of 5 MPa was about 1.5% higher than that at an injection pressure of 15 MPa.This shows the potential of the lower injection pressures with large-flow injectors.

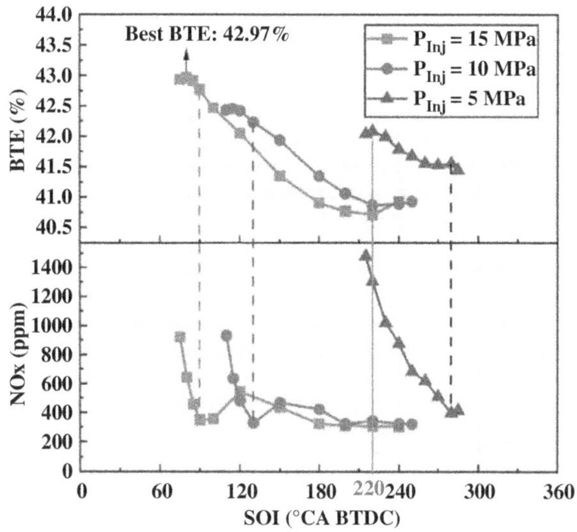

Figure 9.17 BTE and NOx emissions at variable SOI under different injection pressures [16].

Qing-he and team [17] observed that optimizing the combustion process with rich mixtures significantly improved power in turbocharged hydrogen engines at low engine speeds. At 2000 rpm, power output increased from 6.8 kW to 21 kW at 0.9 equivalence ratio and at 1500 rpm, it increased from 6.4 kW to 16.5 kW at an equivalence ratio of 0.8. The study also agreed with many report in the literature that increasing the equivalence ratio improved the brake thermal efficiency (BTE). Turbocharged pressure and volumetric efficiency played important roles in deciding the optimal equivalence ratio for maximum power and for increasing brake thermal efficiency in certain ranges of the equivalence ratio (ϕ in range of 0.6 to 0.8). Increasing the equivalence ratio resulted in speeding up the combustion velocity and improved gas exchange processes at low engine speeds. NOx emission decreased significantly and became almost negligible with equivalence ratios higher than 0.95. NOx emission decreases from 877 ppm to 0 ppm at 1500 rpm and from 1259 ppm to 17 ppm at 2000 rpm. This indicates that reduction efficiency of hydrogen and a three-way catalyst could go beyond 99%.

9.3 Stratification and Injection Strategy

In a stratified charge hydrogen engine, a richer fuel mixture is created near the spark plug, with leaner mixture further away. This rich mixture ignites easily, initiating combustion that spreads to the leaner mixture. Low minimum ignition energy and high hydrogen flame speed ensure stable combustion, which makes it possible for engines to operate with an overall leaner air–fuel mixture. A leaner mixture leads to improved efficiency and reduced NOx emissions. It is important that appropriate mixture formation at the desired locations needs to be done to ensure effective stratification. Zhendong et al. [18] investigated the effects of five different mixture formation methods on efficiency, combustion, emissions, and energy distribution in a hydrogen engine. The methods, shown in Figure 9.18, include homogeneous charge (HC), in-cylinder mildly stratified charge (IMSC), in-cylinder severely stratified charge (ISSC), spark plug mildly stratified charge (SMSC) and spark plug severely stratified charge (SSSC).

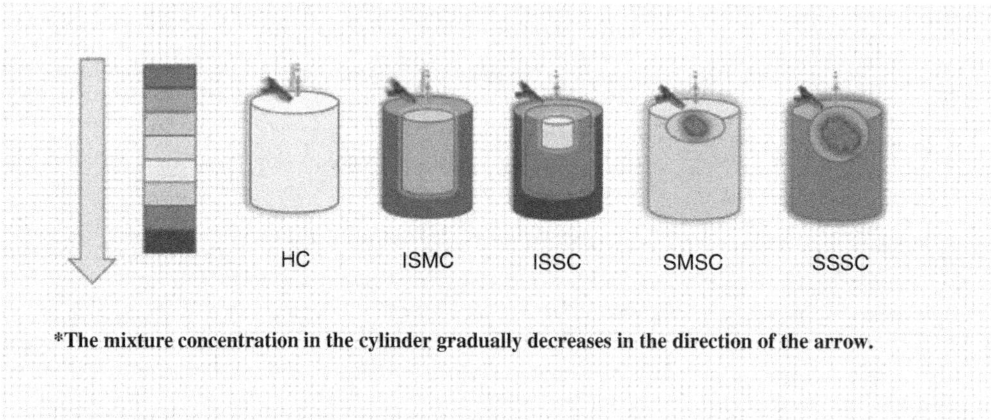

Figure 9.18 Schematic diagram of in-cylinder mixture distribution for different methods [18]. Please visit the companion website to view a color version of this image.

Table 9.2 Mixture formation and effects.

Mixture formation method	Description of the injection process	Effects
Homogeneous charge (HC)	Single injection at 180 °CA BTDC to create a uniform mixture throughout the cylinder.	Higher intake efficiency, but slightly decreased mixture homogeneity.
In-cylinder mildly stratified charge (IMSC)	Injection at 110 °CA BTDC to form stratified mixture slightly richer mixture near the spark plug and leaner elsewhere.	Improved combustion efficiency and reduced emissions due to controlled and localized combustion.
In-cylinder severely stratified charge (ISSC)	Injection at 50 °CA BTDC to create a much higher degree of stratification–a significantly richer mixture near the spark plug and leaner elsewhere.	Enhanced combustion efficiency and emissions control by concentrating fuel near the spark plug for easier ignition.
Spark-plug mildly stratified charge (SMSC)	Double injection: first injection timing same as homogeneous charge (HC) method, second at 50 °CA BTDC to enrich mixture near the spark plug.	More precise control over mixture formation, improving combustion stability and reducing emissions.
Spark plug severely stratified charge (SSSC)	Double injection: first at HC timing, second at 50 °CA BTDC with higher fuel proportion for the second injection. This was similar to the SMSC method, but the higher injection proportion for the second injection resulted in a more pronounced stratification.	Improved ignition and combustion stability with a higher equivalence ratio near the spark plug while maintaining efficiency and emissions control.

Table 9.2 gives a comparative picture of the mixture formation method, injection process, and results of each of these methods. HC timing is significantly earlier than in conventional gasoline engines, where injection timing is usually set at 300 °CA BTDC. However, because hydrogen has an extremely high diffusion rate in air, it is acceptable to delay injection timing to ensure a homogeneous mixture. Delaying injection timing also contributes to higher intake efficiency [19].

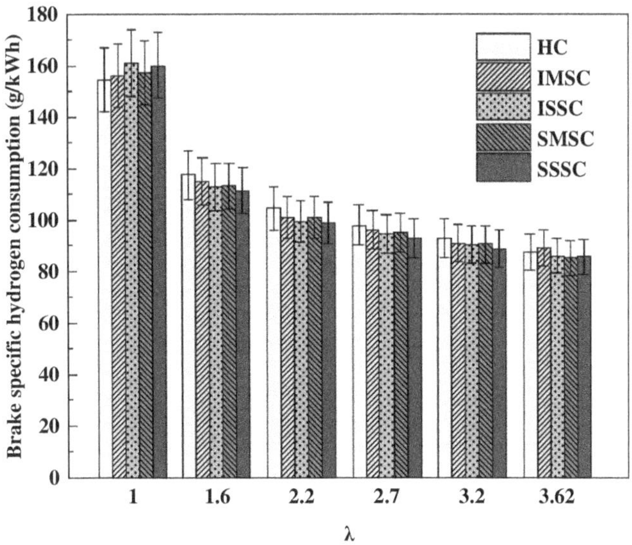

Figure 9.19 Hydrogen consumption vs. λ [18]. Please visit the companion website to view a color version of this image.

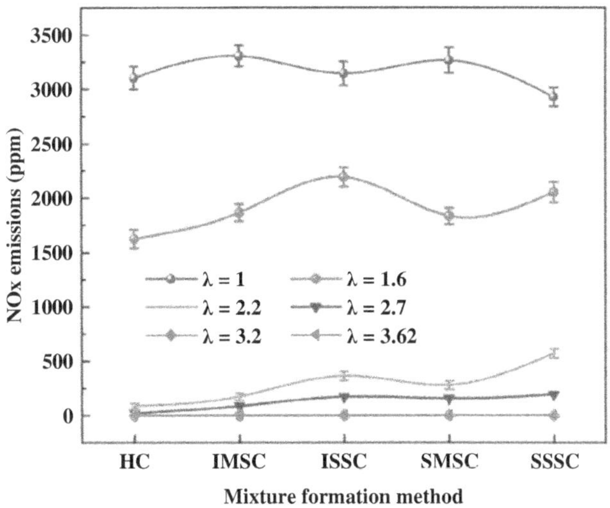

Figure 9.20 NOx variation for mixture formation methods [18]. Please visit the companion website to view a color version of this image.

Computational fluid dynamics (CFD) analysis and planar laser-induced fluorescence (PLIF) have shown that delayed injection timing might decrease mixture homogeneity to some extent [20]. The SSSC approach was in agreement with earlier studies [21] that a double-injection strategy with hydrogen could generate a higher concentration of fuel near the spark plug while still ensuring adequate homogeneity in other areas of the cylinder.

Figures 9.19, 9.20, and 9.21 show the fuel consumption, NOx emissions, and combustion duration, respectively, of the various mixture formation techniques. The study identified the SMSC method as the optimal mixture formation method for achieving efficient combustion and reducing NOx emissions. At $\lambda = 3.62$, the SMSC method resulted in NOx emissions of less than 10 ppm,

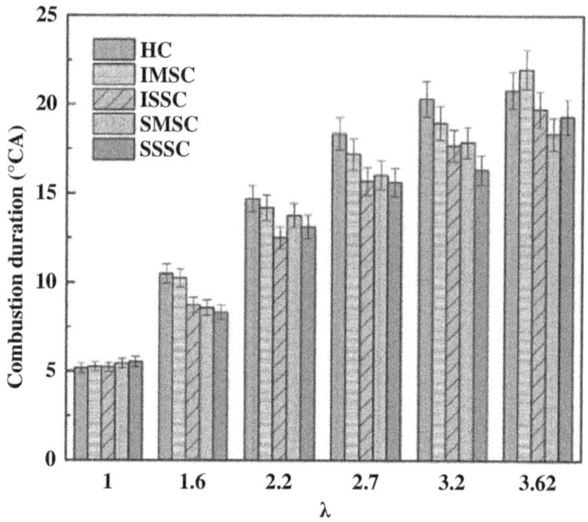

Figure 9.21 Combustion duration vs. λ for mixture formation method [18]. Please visit the companion website to view a color version of this image.

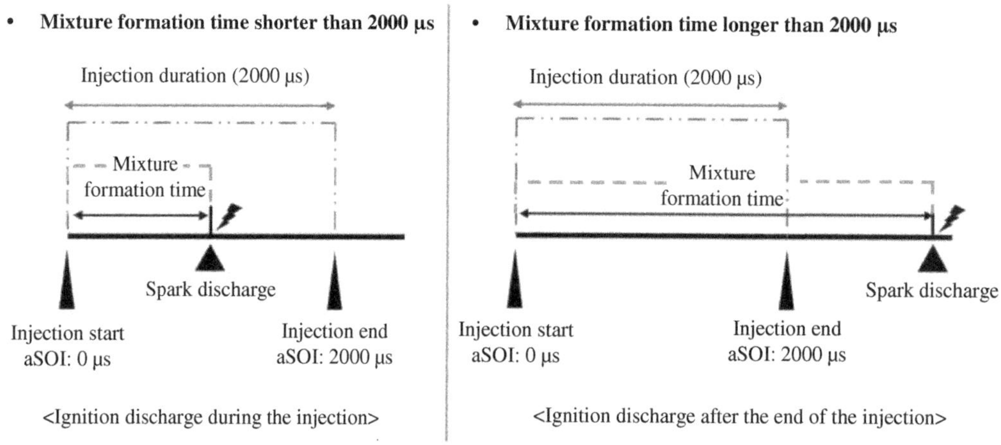

Figure 9.22 Definition of mixture formation time (left: ignition before end of injection; right: ignition after end of injection) [22].

achieving near-zero emissions. The SMSC method at $\lambda = 3.62$ significantly reduced the coefficient of variation of P_{max} ($COVP_{max}$) by about 7% compared to the homogeneous charge (HC) method. Moreover, at $\lambda = 3.62$, the SMSC method achieved reduction of 75.8 g/(kW×h) in hydrogen consumption over the ISSC method at $\lambda = 1$. NOx emissions still exceeded 100 ppm at $\lambda = 2.7$, despite effective reductions over other methods.

Sanguk et al. [22] investigated hydrogen stratified charge combustion (SCC) in a constant-volume chamber. The effects of mixture formation time (MFT) and ambient density on hydrogen SCC were analyzed. MFT, as applied to the present analysis, is represented in Figure 9.22. The researchers observed that short MFTs resulted in high burning velocities, while long MFTs led to lower burning velocities due to fuel stratification. High ambient density caused nonlinear combustion characteristics due to collapsed hydrogen jet structure. Optical combustion techniques like direct flame, schlieren, and OH* chemiluminescence imaging were used to study hydrogen SCC characteristics.

The local equivalence ratio near the spark plug significantly influenced combustion. Turbulent intensity from injection increased the flame propagation speed when ignition occurred during injection. Short MFT resulted in quick flame development and rise and produced a wrinkled flame boundary surface. In contrast, long MFT led to slower combustion. It produced a circular flame boundary surface and improved mixture homogeneity but lower burning velocity. Ignition timing during injection affected flame trajectory, with top-to-bottom movement for ignition during injection and bottom-to-top movement for ignition after injection.

In this study, during combustion analysis, the researchers conducted pressure-based heat release analysis with MATLAB post-processing and calculated the representative heat release rate (RHRR), representative total heat release (RTHR), and normalized cumulative heat release (NCHR) from the following equations. The representative total heat release (RTHR) was calculated from

$$\text{RTHR} = \int_{t_{ignition}}^{t_{end}} \text{RHRR}\, dt$$

where

$t_{ignition}$ is the time the ignition starts, and

t_{end} is the time the combustion ends.

The representative heat release rate (RHRR) was calculated from the following equation:

$$\text{RRHR} = \frac{dP_{in-chamber}}{dt}$$

where

$P_{in-chamber}$ is the measured pressure in-chamber, and

t is the time after ignition.

The normalized cumulative heat release (NCHR) was evaluated from the equation

$$\text{NCHR} = \frac{\int_{t_{ignition}}^{t} \text{RHRR}\, dt}{\text{RTHR}}$$

Figure 9.23 shows RHCR and NCHR. Flame development time (FDT) is the duration between the time of ignition discharge and 10% total cumulative heat release. Flame rise time (FRT), in the present context, is defined as the time between 10 and 90% total cumulative heat release.

Short mixture formation times (MFT) in hydrogen stratified charge combustion (SCC) resulted in visible flame emissions, with intensity decreasing as MFT increased. A red-colored flame, indicating hydrogen-rich combustion, was consistently emitted, while hydroxyl radical extinction

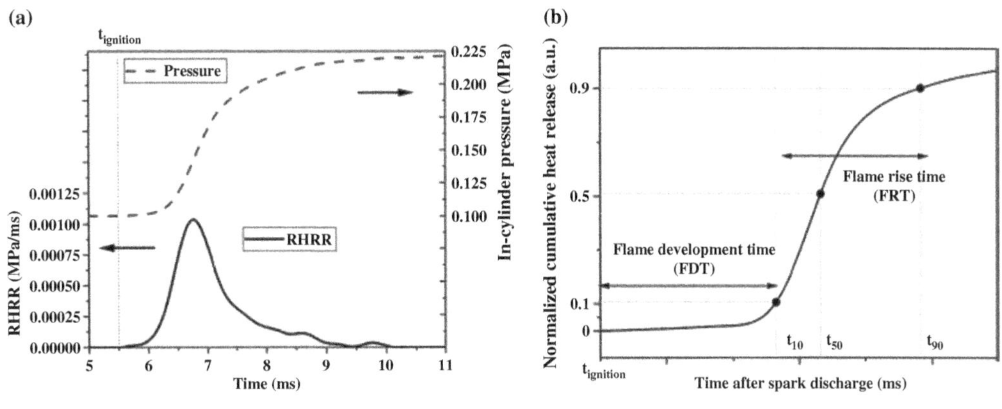

Figure 9.23 Schematic of RHRR and NCHR (a) pressure curve and RHRR (b) NCHR.

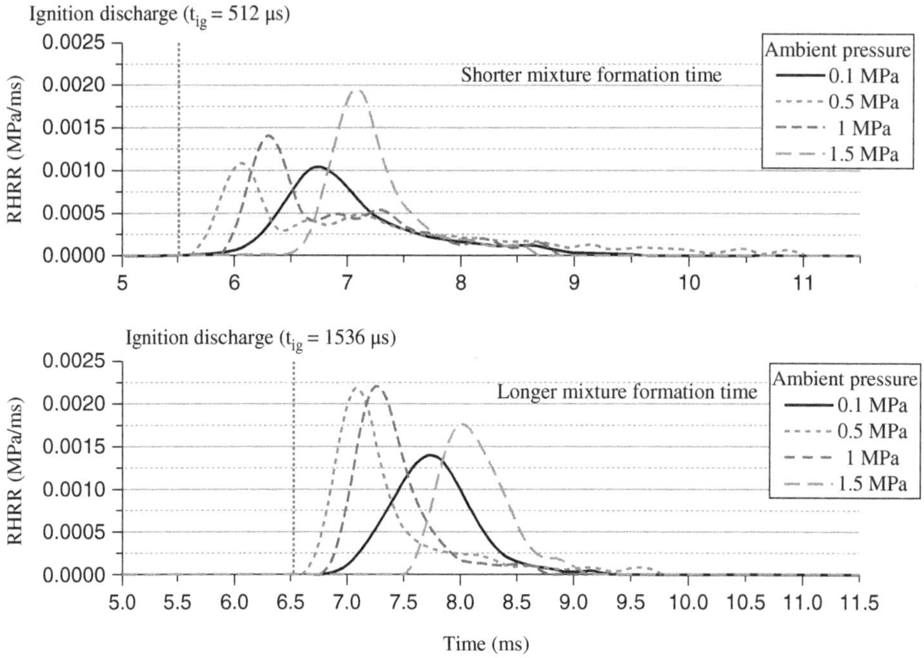

Figure 9.24 Representative heat release rate at different ambient pressures (mixture formation time). Top: 512 μs; bottom: 1536 μs [22].

led to a bluish flame. Representative total heat release (RTHR) improved, particularly with short MFTs. Higher ambient pressure produced a rich mixture, but excessive pressure led to a prolonged flame development time (FDT) because of jet contraction.

The dashed lines in Figure 9.24 indicate two different combustion speed peak locations for both cases of MFT. This variation suggests two distinct combustion mechanisms. With short MFT, early injected hydrogen forms a premixed state with a relatively high equivalence ratio and when ignited, the premixed portion burns.The remaining injected hydrogen participates in combustion diffusion and raises the high equivalence ratio due to the short MFT. As a result, the initial premixed mixture results in a relatively high combustion speed, which gets reduced on subsequent hydrogen injection after spark discharge. At low engine loads, stratification increases NOx emissions compared to early injection methods [23–26]. Stratifying the air–hydrogen mixture offers advantages in minimizing heat losses during combustion. Late direct injection of fuel during the compression stroke increases the combustion rate significantly through stratification. At the same time it minimizes fuel wetting and reduces the associated heat losses [27–29]. Proper nozzle design and injection strategies are crucial to balance the benefits and challenges associated with hydrogen combustion in direct injection engines. Optimizing this stratification process depends on nozzle design, which controls the spray layering at different concentrations in desired levels. Sometimes the Coanda effect, based on wall geometries, poses challenges for both homogenization and stratification because it is difficult to mix hydrogen uniformly and control the layers of hydrogen concentration. Higher hydrogen diffusivity relative to gasoline is advantageous for homogenizing mixtures but poses challenges for controlling stratification and spread toward the cylinder liner and piston crevices [27–29].

Lee et al. [30] investigated the impact of injection and ignition timing on hydrogen-lean stratified charge combustion in a single-cylinder engine. Injection timing control is mild and moderate

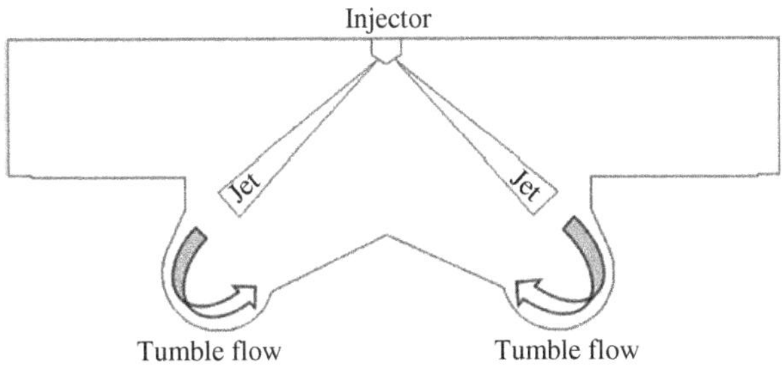

Figure 9.25 Schematic diagram of jet-room coordination [35].

and hard combustion, while combustion stability was regulated through ignition timing control. NOx emissions, which primarily occurred due to high-temperature and oxygen-rich conditions, was brought down by adjusting the dual injection strategy to form locally too-rich or too-lean mixtures. Biffiger et al. [31] and Li et al. [32, 33] studied the effects of the ratio and timing of dual injection and observed that dual injection could create a stratified mixture, with central area richer. Such a situation accelerated the combustion speed with lean mixture near the cylinder wall so that wall heat loss got reduced and and knocking tendency was suppressed. Early injection can lead to premixed flame propagation, whereas late injection results in mixture stratification and partially premixed combustion.

Hamzehloo et al. [34] applied a fully implicit computational methodology, coupled with the Reynolds-averaged Navier–Stokes (RANS) approach to investigate mixture formation and combustion in a DISI hydrogen engine. Hydrogen was injected into the combustion chamber through a six-hole side-mounted injector. The study focused on examining the effects of two injection strategies: single- and double-pulse injections per cycle. The researchers observed that the speed of flame and its direction of travel were influenced by the velocity field at ignition timing. The effect of velocity field was more dominant than the impact of fuel stratification. Li et al. [35] conducted a numerical analysis to investigate the impact of various injection parameters on the formation of hydrogen–air mixtures. They found that achieving a uniform mixture was possible by setting the start of injection (SOI) at 88 °CA before top dead center (BTDC) along with the jet-room coordination to increase tumble strength. In the context of injection timing, "jet-room coordination" refers to precise timing and placement of hydrogen injection to align the fuel jet with the u-pit in the piston (Figure 9.25).This alignment strengthened tumble flow and accelerated hydrogen–air mixing and flame propagation. The technique involved optimizing injection, which was crucial for implementing split-injection strategies in creating stratified mixtures. Delaying SOI facilitated formation of stratified mixtures within the cylinder, resulting in reduced heat transfer losses and improved thermal efficiency.

Sanguk et al. [36] investigated three mixture formation modes—homogeneous charge (HC), lean-homogeneous charge (LHC), and lean-stratified charge (LSC)—using direct injection in a single-cylinder research engine. Due to hydrogen's high heat transfer loss characteristics, optimized combustion-phasing angle was retarded by approximately 5–10 CAD compared to conventional engines. LSC mode, despite emitting high NOx emissions (6.68 g/kWh at low load and 10.85 g/kWh at high load) due to locally rich hydrogen mixtures demonstrated the highest indicated thermal efficiency (ITE) of 34.09%, particularly under low-load conditions compared to the HC mode. Further research could utilize optical diagnostics with chemiluminescence to better understand combustion mechanisms. In the LSC mode, improvements in heat transfer loss could be analyzed through proper image data and quantitative heat transfer analysis using heat balance.

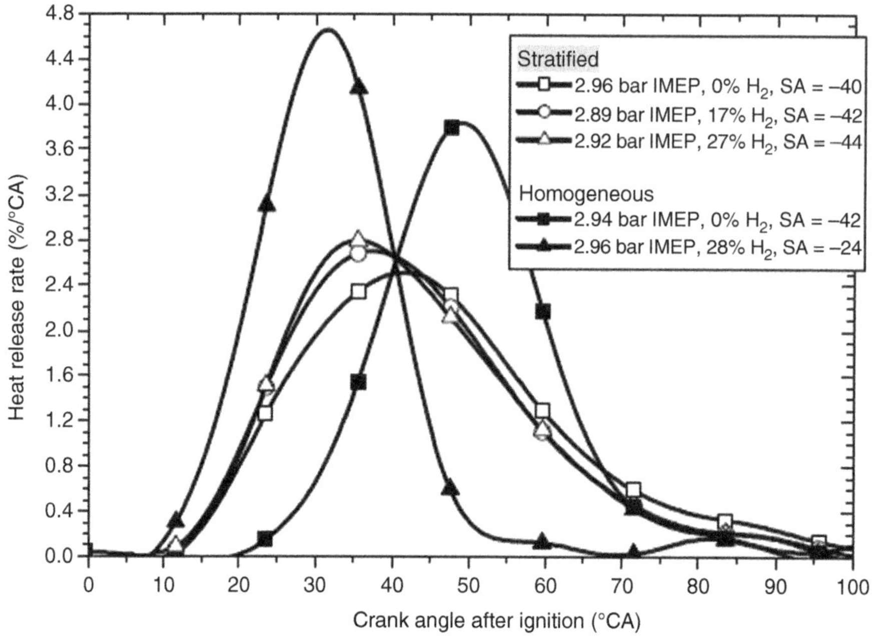

Figure 9.26 Heat release rate for stratified and homogeneous operation [37].

To achieve high efficiency with low emissions, lean boosting or exhaust gas recirculation (EGR) should be considered even with locally hydrogen-rich mixtures.

Conte and Boulouchos [37] investigated stratified combustion of hydrogen-enhanced gasoline in a gasoline direct injection (GDI) engine. By injecting up to 27% hydrogen (on an energy basis) in the intake port, a reactive homogeneous mixture was generated for direct gasoline injection. This method allowed the engine to use high exhaust gas recirculation (EGR), which proved useful in lowering NOx and HC emissions. In addition to this, steps to delay ignition under certain situations compared to GDI gave "unexpected results" on ignition delay. In practice, the hydrogen-containing reformer gas used onboard would have been the ideal for vehicular application, but safety concerns prevented the use of carbon monoxide in the tests. Several combinations of stoichiometries at the spark plug location had different impacts on the ignitability and NOx formation on the H_2 fraction, EGR rate, and spark delay. Spark delay alone was not a successful technique to compensate for the higher NOx production at lower load ranges where EGR was significantly reducd NOx. Under low load condition, with 30% EGR, NOx concentration was reduced by a factor of 3 compared to pure gasoline operation. On the other hand, large spark delay by hydrogen enrichment alone substantially reduced the NOx level and EGR provided further improvement. Hydrogen addition brought down HC emissions consistently. As shown in Figure 9.26, in the case of homogeneous combustion, hydrogen addition leads to quicker combustion compared to pure gasoline. On the contrary, in stratified combustion, the initial heat release rate remains similar irrespective of the extent of hydrogen enrichment. This is because stratified combustion is constrained by evaporation of gasoline and subsequent ignitable mixture formation.

Shi et al. [38] installed new hydrogen and gasoline injection systems controlled by an electronic control unit on an original gasoline engine and studied the impact of spark timing on combustion and emissions in a hydrogen–gasoline stratified engine. Optimal stratified combustion with improved thermal efficiency, better fuel economy, and low emission characteristics was achieved with hydrogen injection timing at 110 °CA BTDC. Hydrogen direct injection after intake valve closure increased volumetric efficiency. Some of the significant observations from this study are discussed below, particularly with respect to the influence of spark timing. Increasing the spark advance angle initially increased brake thermal efficiency, but subsequently the efficiency decreased. There was an improvement in brake thermal efficiency when hydrogen was added to gasoline. The maximum instantaneous heat release rate showed a decreasing trend with increasing spark advance angle, accompanied by decrease in crank angle. For 9% hydrogen volume fraction, 10 °CA BTDC was found to be the optimum spark advance angle for maximum brake thermal efficiency. Addition of hydrogen to gasoline increased the peak pressure at the same spark timing. Also, hydrogen addition to gasoline accelerated the combustion process, and the combustion rate was improved by increasing the hydrogen volume fraction. Both larger spark advance angle and higher hydrogen volume fraction contributed to the rise in peak cylinder pressure. With a specified hydrogen fraction, NOx, HC, and CO emissions heightened with increased spark advance. NOx surged rapidly, whereas HC and CO increased slowly.

In another study on a gasoline/hydrogen spark ignition (SI) engine conducted by Yu et al. [39], the effects of homogeneous or stratified hydrogen on combustion and emissions were investigated. Compared to neat gasoline, hydrogen addition improved brake thermal efficiency by 1.96%, decreased flame development duration by 6.6 °CA, reduced rapid combustion duration by 3.7 °CA, and decreased the coefficient of variation (COV) by 21.1% on average. Both HC and CO emissions were decreased by 32.0% and by 0.91%, respectively, but there was an excessive rise in NOx emission by 85.9% on average. Stratified hydrogen, particularly under lean burn conditions, showed an improvement in combustion rate and brake thermal efficiency. Gasoline with stratified hydrogen raised brake thermal efficiency by 0.55% and maximum pressure by 3.3%. The position of maximum pressure advanced by 2.4 °CA. Duration of flame development and rapid combustion duration were decreased by 1.0 °CA and 1.3 °CA, respectively. Emissions from stratified hydrogen were higher compared to homogeneous hydrogen. In the case of gasoline and stratified hydrogen operation, both HC and NOx emissions were increased by 12.8% and 14.3%, compared to homogeneous hydrogen [39].

Typical combustion characteristics of hydrogen like higher burning velocity and shorter quenching distance lead to increased heat transfer to combustion chamber walls, causing cooling loss, which affects an engine's thermal efficiency. However, decreasing cooling losses does not always necessarilty improve thermal efficiency because decreasing cooling loss can sometimes coincide with increase in exhaust heat loss. Therefore, an integrated approach, considering both cooling loss reduction and exhaust heat management, is essential for optimum efficiency. Shudo [40] conducted experiments in constant volume combustion vessels and designed a fuel injection pattern to achieve a stratified mixture to reduce the heat transfer to chamber walls. Some significant observations showed that lower compression ratios necessitate higher degrees of constant volume to compensate for cooling loss reduction and improve efficiency. Retarding spark ignition timing is an effective way to reduce the cooling losses by lowering the maximum temperature. Quite often, retarding spark timing tends to decrease the degree of constant volume, and the improvement in thermal efficiency depends upon how close the actual combustion process is to the ideal constant volume combustion. In direct injection, stratified charge is generated, which brings down the temperature of burned gas near the chamber walls and minimizes cooling losses. Thus, stratified charge increases the apparent heat release fraction in hydrogen combustion, which improves thermal efficiency by reducing cooling loss without compromising the degree of constant volume. Shudo and Oba [41] used laser-induced breakdown spectroscopy (LIBS) to measure hydrogen concentration distribution and evaluate instantaneous local equivalence ratios in the direct injection stratified charge engine to have a deeper insight on mixture formation characteristics. The instantaneous local equivalence ratio was measured using intensity ratios of atomic emissions of hydrogen and nitrogen at specific wavelengths (656 nm and 871 nm) for accurate characterization of hydrogen distribution within the stratified charge. It was observed that hydrogen injection showed faster diffusion compared to methane in both radial and axial direction. Tests with two-stage split injection were conducted to assess control over hydrogen diffusion patterns for split injection. Split injection with a 7:3 fuel ratio caused hydrogen diffusion in awider region, but in a 3:7 fuel ratio, split injection weakened jet penetration and radial diffusion simultaneously. Optimization of the split injection process proved useful to strategically identify zones of higher equivalence ratio so that desired stratification could be achieved with rich mixtures close to the spark plug to have improved ignition with efficient stable combustion.

9.4 Homogeneous Charge Compression Ignition (HCCI)

Homogeneous charge compression ignition (HCCI) blends features of traditional SI and CI engines and typically operates on principles of the diesel engine, using high compression ratios to ignite a homogeneous air–fuel mixture without a spark plug. HCCI is receiving a lot of attention because of its potential to bring down NOx and particulate emissions and still maintain high thermal efficiency. Numerous terms have been used in the literature to describe the process of charge formation. These are controlled autoignition (CAI), lean homogeneous combustion (LHC), compression ignited homogeneous charge combustion (CICH), active radical combustion (ARC), homogeneous charge compression ignition diesel combustion (HCDC), diesel fumigation, multiple staged diesel combustion (MULDIC), premixed compression ignited combustion (PCIC), active thermo-atmosphere combustion (ATAC), premixed charge compression ignition (PCCI), premixed lean diesel combustion (PREDIC), and compression-ignited homogeneous charge (CIHC) [42, 43]. Strictly speaking, even though most of these technologies share similarities with HCCI, sometimes some of them have their distinct features. However, HCCI is the most widely used term and is the one used here. Figure 9.27 shows the mechanism of HCCI combustion along with spark ignition

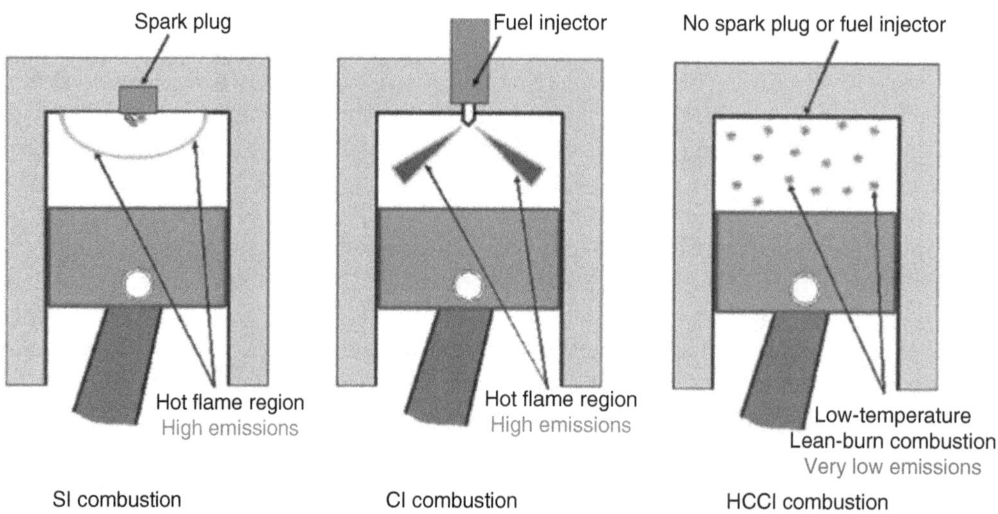

Figure 9.27 HCCI combustion mechanism [43].

and compression ignition engines. In the HCCI process, a premixed fuel–air charge is introduced into the cylinder within equivalence ratios ranging from 0.15 to stoichiometric.

Gopinathan and Ibrahim [44] investigated parameters such as compression ratio equivalence ratio and intake charge temperature on hydrogen-fueled HCCI engine performance using a 0D multizone numerical simulation to assess their influence on engine performance. The optimal equivalence ratio range was observed to be within 0.1 to 0.3, with intake charge temperature playing a significant role in promoting autoignition and accelerating the combustion process. At an equivalence ratio of 0.4 and an intake charge temperature of 393 K, higher indicated thermal efficiency was achieved. However, the range of equivalence ratio was restricted by knocking due to a high rate of pressure rise. A significant aspect of this study also highlights charge dilution as a method to extend the range of equivalence ratio in HCCI mode [44]. As the charge undergoes compression within the cylinder, its temperature rises until it reaches ignition temperature. Ignition occurs simultaneously at multiple points around the cylinder, leading to rapid combustion, and all heat is released within a short duration with a minimum of 4 °CA [45, 46]. The combination of fast heat release in HCCI mode and lean cylinder charge bring the combustion close to constant volume combustion. This leads to reduced peak gas temperatures and decreased heat transfer losses through the cylinder walls. Such characteristics contribute to high indicated thermal efficiencies in HCCI engines. However, since HCCI combustion depends on autoignition, it needs the use of high compression ratios. In simulation studies the optimal compression ratio has been found suitable in the ranges between 18:1 and 25:1, in the absence of charge heating [42].

9.4.1 Significant Features of HCCI Combustion

In their investigation, Gomes and a team of researchers [42] injected hydrogen close to the inlet valve port using a fast-response solenoid valve and injection controller to ensure precise control over the timing and quantity of hydrogen injected into the cylinder. The engine, with a compression ratio of 17:1, was appropriately instrumented to avoid backfire. Figure 9.28 shows the diagram of the experimental set-up. Maximum hydrogen was fed at a temperature of approximately 20 °C with injection pressure of 6.0 bar. In an effort to ensure autoignition of the hydrogen charge, the air inlet temperature was increased using a 2800 W electric heater controlled by a proportional–integral–derivative (PID) temperature controller. Tests were conducted at an engine speed of 2200 rpm, air inlet temperature at 93 °C, and a hydrogen flow rate of 90 dm^3/min to ensure stable and

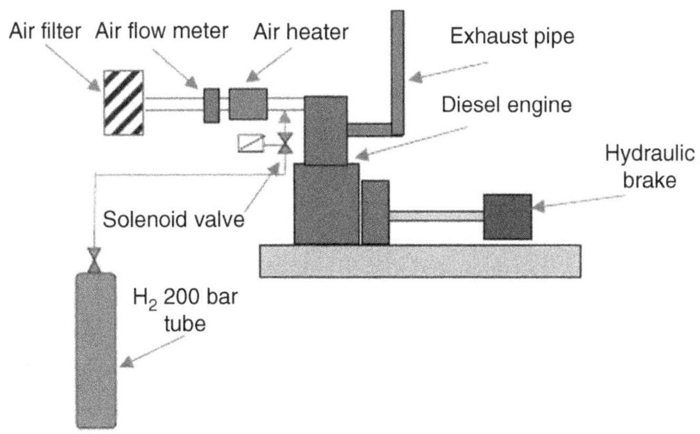

Figure 9.28 Hydrogen HCCI engine test set-up [42].

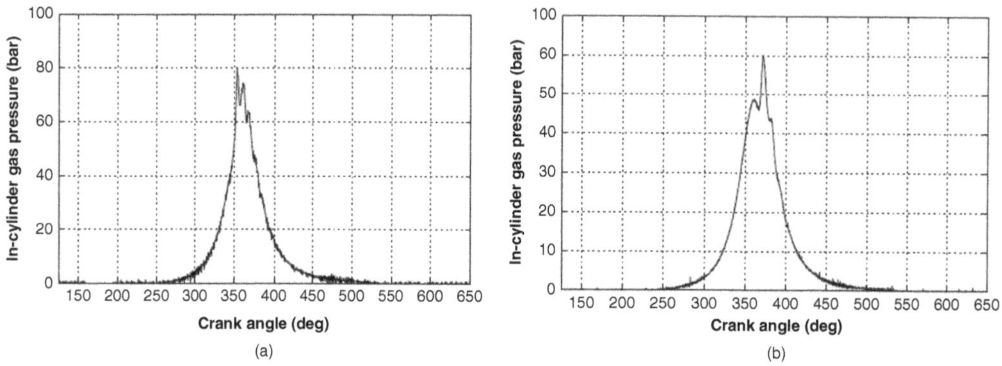

Figure 9.29 In cylinder pressure traces for (a) HCCI and for (b) conventional diesel operation [42].

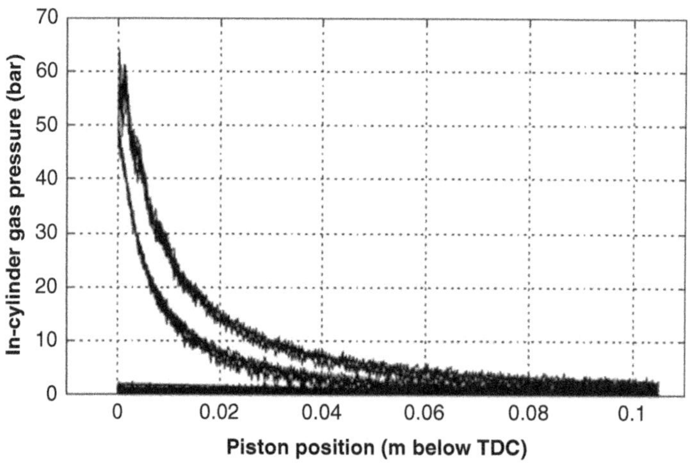

Figure 9.30 Cylinder pressure plots in HCCI operation [42].

consistent operation of the engine in HCCI mode. It was observed that peak in-cylinder pressures and pressure rise rates were higher in the HCCI hydrogen engine than in conventional diesel operation (Figure 9.29). HCCI hydrogen engines demonstrated significantly higher fuel efficiency even with very lean cylinder charges. However, it was necessary to heat the inlet air to facilitate autoignition. Controlling the inlet air temperature was most effective for regulating ignition timing.

The presence of hydrogen in exhaust gases of engines is referred to as hydrogen slip. Hydrogen slip occurs during the valve overlap period and the nonoptimized hydrogen injection valve period. During the valve overlap period, both the intake and exhaust valves are partially open simultaneously. This allows some unburned hydrogen fuel to escape into the exhaust manifold without undergoing combustion. Similarly, during the nonoptimized hydrogen injection valve period, where hydrogen injection timing may not be perfectly synchronized with the engine's combustion cycle, some hydrogen also escapes. Figure 9.30 shows the in-cylinder gas pressure for 10 consecutive engine cycles when the engine was operating at a constant speed of 2200 rpm with an air excess ratio of 3. The little fluctuations that were observed in the maximum pressure between consecutive cycles were primarily due to the nature of HCCI combustion. Even despite such variations, the engine's brake thermal efficiency was close to 45%.

Persson et al. [47] studied early flame development in spark-assisted HCCI combustion. To achieve autoignition with SI compression, some fresh charge was replaced with hot residuals. In this study, the engine was operated with negative valve overlap (NVO) to trap hot residuals in the cylinder. By increasing the negative valve overlap (NVO), the initial charge temperature was raised, which made it possible to study and analyze closely the intermediate combustion zone between spark ignition (SI) and homogeneous charge compression ignition (HCCI). The researchers also investigated the interaction between the growing SI flame and subsequent HCCI combustion as the amount of residuals increased. Tests were carried out on a single-cylinder SI engine with optical access with in-cylinder pressure measurement and high-speed chemiluminescence imaging to monitor the combustion process. It was observed that even with significant NVO and high residual fractions, it was primarily the SI flame that governed the HCCI combustion process. Spark timing played a crucial role in phasing combustion timing, even when the majority of the released heat came from HCCI combustion. The flame expansion speed showed a decreasing trend with higher NVO, but it continued to exist even when there were high fractions of residual gases. To some extent, higher spark advance compensated for slower flame expansion caused by higher NVO.

9.4.2 Modeling and Simulation Studies

Namar et al. [48] developed a single-zone thermo-kinetic model (TKM) to study the effects of engine speed, inlet pressure and temperature, equivalence ratio, and exhaust gas recirculation on the performance of a hydrogen-fueled HCCI engine. The study involved detail energy and exergy analysis. Among the several parameters analyzed, it was seen that engine IVC pressure and equivalence ratio had the most significant impact on irreversibity and exergy. Increasing the equivalence ratio in the lean burn zone led to increase in both power and irreversibility. Higher inlet temperature decreased both engine power and chemical exergy. Engine speed had practically no effect on irreversibity production. Increase in EGR rate reduced chemical exergy while increasing thermochemical exergy. Effect of cooled EGR results in decreasing the volumetric efficiency, but it is not the same as hot EGR. Caton and Pruitt [49] investigated hydrogen HCCI using a single-zone model on a single-cylinder CFR diesel engine and made a comparative evaluation with conventional diesel combustion at the same compression ratio and load. Simulation studies indicated that peak efficiency with HCCI hydrogen operation could be achieved at compression ratios around 18 with minimum intake heating to 100 °C and at equivalence ratios in the range of 0.2–0.5. Experimental results showed that HCCI could be accomplished at an equivalence ratio range between 0.15 and 0.37 with intake heating to 80 °C and 100 °C. However, hydrogen HCCI's efficiency was lower than for the corresponding diesel cycle mainly due to higher thermal and chemical enthalpy losses with lower combustion efficiency. As far as emissions are concerned, in HCCI combustion around 0.5–1.0% of the hydrogen remains unburned after combustion. This unburned hydrogen indicates that the engine's combustion efficiency is between 90 and 95%. Nitric oxide emissions were below 1 ppm, significantly lower than for diesel combustion.

Nöda and Foster [50] developed a simulation code with multizone chemical kinetics and studied combustion in HCCI hydrogen engines. To achieve sufficient combustion duration, which is caused due to the self-accelerating nature of the chemical reaction, the simulation included different temperature and fuel distributions within the combustion chamber. Calculations considered various conditions over a wide range starting from perfectly homogeneous cases to cases with varying degrees of inhomogeneity. As far as the effect on ignition timing of HCCI is concerned, the intake gas temperature proved to be a more dominant factor compared to the equivalence ratio.The simulation considered several conditions, ranging from a perfectly homogeneous scenario to cases with varying degrees of inhomogeneity. Gas temperature had a stronger impact on ignition timing than equivalence ratio, and gas temperature distribution proved more effective for controlling combustion duration. Direct injection technology facilitates easier implementation of fuel inhomogeneity. Gas temperature inhomogeneity and mean gas temperature should be regulated for optimal combustion duration. The study observed that a temperature inhomogeneity in the in-cylinder mixture can control the HCCI combustion.

Maurya and Akhil [51] used 15 hydrogen mechanisms to simulate HCCI combustion using a stochastic reactor model over a range of engine operating conditions such as intake temperature, pressure, speed, and air–fuel ratios. Various combustion parameters like start of combustion (CA10), maximum cylinder pressure (P_{max}) and maximum heat release rate (HRR_{max}) were evaluated to assess the accuracy of different mechanisms. The Maurya-2017 mechanism showed best accuracy for cylinder pressure and heat release rate, whereas the Oconaire-2004, Maurya-2017, NUIG-NGM-2010, USC-II-2007, and CRECK-2014 mechanisms predicted experimental ignition delay accurately. Deviations in combustion parameters (P_{max}, HRR_{max}, CA10) for each mechanism were analyzed relative to the Maurya-2017. The Oconaire-2004 mechanism closely agreed with the predicted values for Maurya-2017. The NUIG-NGM-2010 and CRECK-2014 mechanisms performed well at higher intake temperatures, pressures, and engine loads. The Burke-2012 mechanism accurately predicted start of combustion but not P_{max} and HRR_{max}. Liu and Karim [52] developed a turbulent transient 3D predictive computational model with detailed chemical kinetics specific to hydrogen oxidation and applied it to simulations of HCCI hydrogen engines. The study made some significant observations on the influence of changes in swirl velocity of intake mixture on combustion processes. Circumferential velocity of the mixture within the engine cylinder increased with the increase in the initial intake swirl ratio. A higher initial intake swirl ratio resulted in a more vigorous swirl and subsequently intensified turbulence in the combustion chamber. The increased turbulence delayed the onset of autoignition and prolonged the combustion duration. NOx emission levels, maximum and minimum cylinder pressure, temperature, and knocking tendency were significantly decreased, but heat loss increased. To obtain high thermal efficiencies, low NOx emission, and reduced knock tendency, it was necessary to optimize the initial intake swirl ratio and engine speed for certain values of mixture intake temperature, compression ratio, equivalence ratio, and initial intake pressure.

Research on hydrogen HCCI engines by Stenlåås et al. [53] show that the heat release rate was extremely high and the burn angle was narrow and decreased with increasing intake manifold temperature. Ignition temperature decreased as λ decreased. Changes in engine speed or compression ratio did not have any significant effect on the start of combustion. λ also has hardly any impact on the start of combustion range. It has been reported in the literature that combustion in hydrogen HCCI was possible even at very lean values of $\lambda = 6$. Hydrogen HCCI indicated higher efficiency than hydrogen SI in the range at which it is possible to operate an engine in HCCI mode. NOx emissions in HCCI mode were so low in higher λ that it did not even require a TWC. Rosati and Aleiferis [54] studied hydrogen SI and HCCI combustion in a direct injection optical single-cylinder research engine equipped with both port fuel injection (PFI) and direct injection (DI) systems. The engine was run at constant speed of 1000 rpm. Optical access to in-cylinder phenomena was provided by extended piston and optical crown. A high-speed camera with a resolution of 1 or 2 °CA was used to capture the combustion images for multiple engine cycles. Even at a low compression ratio, it was possible to achieve stable HCCI combustion. Both the PFI and DI operation developed a spherical flame growing across a range of λ between 1.2 and 2. DFI did not encounter any symptoms of backfire and demonstrated more rapid flame propagation and peak pressure compared to PFI. With an internal EGR rate adopted via modified valve timing (negative valve overlap) it was possible for a HCCI engine to operate at a higher injection pressure (100 bar) and at an intake temperature of 200–400 °C and up to $\lambda = 3.0$ [54].

9.4.3 Exhaust Emission in a HCCI Hydrogen Engine

During the valve overlap period, when both the intake and exhaust valves are partially open simultaneously, a situation might occur in which some unburned hydrogen fuel might escape into the exhaust manifold. Similarly, it is quite likely that some hydroen might escape during the nonoptimized hydrogen injection valve period if hydrogen injection timing is not synchronized with the engine's combustion cycle. Ducduy and James [55] investigated the effect of nitric oxide (NO) on the performance and emissions of a hydrogen HCCI engine.They experimentally evaluated the operating range of hydrogen port fuel injection in an HCCI engine with and without NO addition. The engine operated at air–fuel ratios of $\lambda = 2.7$ to 4.0 at 150 °C inlet air temperature without NO. Addition of NO caused autoignition and extended the operating range to ultra lean zone ($\lambda = 5$) at relatively lower inlet air temperatures (90 °C). However, overall NOx emissions increased as about 50 ppm of NO got converted to NO_2 during combustion.

Boretti investigated the operation of a hydrogen truck engine in HCCI mode [56]. The simulation studies utilized detailed chemical kinetics data from GT-POWER and established mathematical Wiebe models, which accurately predict combustion processes for both diesel and gasoline engines. In this analysis four modes of injection and combustion were investigated on a diesel truck engine equipped with a power turbine design, converted to hydrogen with a "direct fuel injector and a jet ignition pre-chamber." The four modes of combustion were as follows: (i) Diffusion with jet ignition with prechamber injection before main chamber injection; the combustion process resembles a traditional diesel engine, where fuel ignites due to compression and heat generated during compression stroke. (ii) Mixed diffusion/premixed with jet ignition, in which there is gasoline-like operation with prechamber injection after main chamber injection. Injection takes place in the jet ignition prechamber after the main chamber fuel is injected and mixed with air. (iii) Mixed diffusion/premixed diesel/gasoline in which injection occurs in the jet ignition prechamber after only part of the main chamber fuel is injected and mixed with air. (iv) Premixed without jet ignition where no injection occurs in the jet ignition prechamber and the engine operates like an HCCI engine. It was theoretically computed that adding power turbine in exhaust flow can recover up to 20% of the 20–25% of fuel energy exhausted in modern heavy-duty diesel engines. However, it is not practically achievable due to added exhaust back-pressure, increased pumping losses, and temperature level required to ensure efficient efficient after treatment. Further constraints on this technique are due to the use of cooled EGR.Simulations on a 12.8 L in-line six-cylinder turbocharged diesel engine indicated that a rise in efficiency up to 2% could be attained at high loads and speeds. Turbo compound improved fuel conversion efficiency at high engine loads but had negative impacts at light loads. Low engine load brings down the efficiency as the converted energy available beyond the turbine is not adequate to counterbalance the back-pressure losses. Simulation results for a diesel truck engine converted to hydrogen operation demonstrated better brake efficiencies than diesel at the same BMEP outputs. Over a wide range of speed and load, a hydrogen engine in diesel-like mode operation had distinct advantages over conventional diesel. It is possible to achieve even better fuel efficiency in HCCI/gasoline-like operation depending upon the permissible limit of peak pressure and λ. This work suggests that a mixed diesel/gasoline-like mode of operation, if properly modulated, offers the potential to show much improved performance.

Hydrogen HCCI is a promising and emerging technique. Several researchers have also investigated hydrogen HCCI combustion along with other alternative fuels such as CNG, methane, DME, LPG, etc. [57–62].

9.5 Reactivity Controlled Compression Ignition (RCCI)

In the domain of HCCI engines, two control techniques have emerged: temperature control and fuel reactivity control. The latter is usually referred to in the literature as reactivity-controlled compression ignition or RCCI [63–68]. Temperature control methods focus on regulating temperature, but RCCI (also known as dual-fuel HCCI) is an advanced combustion strategy in which two fuels with different reactivities are injected into the combustion chamber. The low-reactivity fuel and air are compressed but not up to the point of autoignition. Later, still during the compression stroke, high-reactivity fuel is injected to form a local mixture of low- and high-reactivity fuel. A close correlation between the fuel injection event and combustion event is essential to monitor the combustion process precisely from cycle to cycle [69].

Many researchers have investigated the potential of several aspects of PCCI and HCCI strategy in different categories of engines with emphasis on a performance, emission, and combustion point of view using gasoline, diesel, ethanol, etc. [70–76]. However, a detailed discussion on these topics lies outside the scope, objectives, and primary focus of the present discussion. The research community in engine technology has put a lot of emphasis on premixed charge compression ignition (PCCI) combustion, which is characteristically a hybrid approach combining the features of both HCCI and conventional direct injection compression ignition (DICI) combustion. Synchronization between initiation of fuel injection and onset of combustion in PCCI combustion plays a crucial role, so fuel injection is scheduled early in the compression stroke to ensure thorough mixing of fuel and air before ignition. A number of studies have discussed the potential of early injection in reducing NOx and soot emission in both HCCI and PCCI combustion.

Loyte et al. [77] did a numerical study on a 2.44 L heavy-duty engine using CONVERGE CFD code and analyzed the in-cylinder combustion phenomenon of RCCI engines with hydrogen-enriched natural gas (HENG). The fuel mixture consisted of 50% each of HENG and diesel. Port fuel injection was adopted for HENG and direct injection for diesel. Three different load (low, medium, and high) conditions were analyzed by varying the diesel injection timing from −11° to −21° after top dead center (ATDC). Results showed a faster burning rate with hydrogen addition and complete combustion in the RCCI mode with higher combustion efficiency. Optimum injection timings were 21°, 15°, and 11° BTDC for maximum output under low, mid, and high load conditions, respectively. This strategy enabled low-temperature combustion at high load leading to relatively less NOx emission. NOx emissions were reduced at mid-load range with retarded injection timing. Maximum IMEP was achieved at advanced injection timing of −21° ATDC under low-load conditions, but optimum injection timing had to be retarded with increasing load.

A study by Watgave and colleagues [78] aimed at comparing the performance of hydrogen (H_2) and hydrogen-blended compressed natural gas (HCNG) as primary fuels, along with diesel and Simarouba biodiesel as secondary fuels. They investigated compression ignition (CI) engines operating in homogeneous charge compression ignition (HCCI) and reactivity-controlled compression ignition (RCCI) modes. Compressed natural gas (CNG) was blended with 20% hydrogen. Experiments were conducted on a single-cylinder CI engine running at 1500 rpm at 75% load for

RCCI mode and 50% load for HCCI mode, varying the gaseous energy share. Results reported that RCCI engines, operated with H_2 and HCNG, exhibited higher brake thermal efficiency (BTE) than HCCI engines using the same gaseous fuel. At 40% gaseous fuel energy share (GFES), there was substantial increase in BTE along with reduced emissions. Nitric oxide and smoke emissions were significantly reduced in the RCCI mode compared to the HCCI combustion mode. Beyond 40% GFES, there was a penalty for lower thermal efficiency and higher emissions. Among the primary fuels tested, hydrogen yielded better results than HCNG in both HCCI and RCCI combustion modes.

Rahimi et al. [79] investigated the increase in hydrogen energy share under RCCI combustion in a heavy-duty diesel engine fueled with diesel and natural gas enriched with hydrogen or syngas. The syngas used consisted of hydrogen and carbon monoxide mixed together in a volumetric ratio of 80:20. The engine was operated at 9.4 bar gross indicated mean effective pressure (mid-load). A fixed amount of diesel oil was injected per cycle into the combustion chamber. Hydrogen or syngas mixed with exhaust gas recirculation (EGR) was gradually introduced along with natural gas, leading to a low-temperature combustion process. Simulation of RCCI engine operation was conducted using commercial software coupled with a chemical kinetics solver. Simulation results indicated that the addition of hydrogen to natural gas showed an increase in hydrogen energy share up to 40.43% with hardly 1% reduction in engine power output and without diesel knock. Furthermore, addition of syngas to natural gas enabled an increase in hydrogen energy share up to 27.05%, improving engine power output by more than 4%. Implementing the two strategies, overall hydrocarbon fuel consumption per cycle could be reduced by up to 46.60% and 33.86%, respectively. Moreover, RCCI engines demonstrated a gross indicated efficiency of well over 50% with significant reductions in engine emissions than RCCI combustion fueled solely with natural gas and diesel oil.

9.6 Hydrogen Rotary Engine

Despite the advantages of high power density and light weight, some serious drawbacks such as excessive fuel consumption and severe maintenance have stood in the way of widespread adoption of rotary engines. However, recently there is a spate of interest in both industry and academia to utilize the distinctive features of rotary engines in future propulsion systems. The rotary engine has already been used as a power plant for unmanned aerial vehicles owing to its high power density, light weight, and less vibration [80–82]. Mazda's interest in hydrogen rotary power has grown persistently since debut of its HR-X hydrogen concept car at the 1991 Tokyo Motor Show. A series of other hydrogen efforts by Mazda over the years, including HR-X2, MX-5, and Capella Cargo, are all powered by hydrogen rotary engines. The RX8 hydrogen rotary engine in 2003 used both hydrogen and gasoline as fuel sources [83]. The Mazda RX-8 RE integrates Mazda's Renesis hydrogen rotary engine in the RX-8 sports car. Even when running on conventional gasoline, the new Renesis model features significant environmental improvements over previous generation rotary engines with better fuel economy and reduced emissions. The hydrogen-fueled rotary engine (HRE) has demonstrated comparable thermal efficiency and power density [84) with the reciprocating gasoline engine.

Meng et al. [85] made a comparative assessment between combustion and emission characteristics of a hydrogen Wankel rotary engine (HWRE) with a hydrogen reciprocating piston engine (HRPE). The fast burning velocity of hydrogen makes it possible for hydrogen flame to overcome the difficulties such as long flame propagation path and reverse unidirectional flow, etc.

[86] usually posed by the rotary engine.Thus, the hydrogen flame can propagate to the rear of the combustion chamber, overcoming challenges posed by the HWRE's long flame propagation path and reverse unidirectional flow [86]. Numerical simulation of hydrogen combustion in a laser-ignited rotary engine has indicated that combustion and ignition is "highly transient" [87] and so ignition timing has to be very precise to avoid any abnormal combustion. Ignition efficiency and hydrogen burning rate were simulated and compared for single and dual laser ignition systems. It was found that a H_2-fueled rotary engine with the dual laser ignition system is a promising technique with high efficiency and near-zero emission. A critical analysis of the phases of combustion like flame development and flame propagation period are extremely important parameters for hydrogen-enriched Wankel rotary engines. Wang et al. have analyzed five machine learning (ML) models, such as linear regression (LR), regression tree (TR), ensembles of trees (EnTR), support vector machine (SVM), and Gaussian process regression (GPR), to predict combustion phases and control mechanism [88]. Meng and co-researchers studied knock characteristics and analyzed the correlation among various parameters under knocking conditions in a WRE. The study evaluated the "lognormal distribution in four zones: knock-free zone, main knock zone, strong knock zone, and danger knock zone" [89]. Knock intensity (KI), knock duration (KD), and maximum pressure rising rate (MPRR) were evaluated, and it was observed that a logarithmic relationship between the KI and KD were linked with a logarithmic relationship, whereas a linear relation existed between KD and MPRR.

It has been reported that HWREs achieve higher power per displacement vis-a-vis both hydrogen and gasoline reciprocating piston engines (RPEs). HWRE has higher power per displacement (PPD) than both the hydrogen reciprocating piston engine (HRPE) and gasoline reciprocating piston engine (GRPE). At 2500 r/min, HWRE has a maximum PPD of 22.51 kW/L, which is about 1.66 times the maximum PPD of HRPE and 1.23 times that of GRPE respectively [85]. As λ is reduced, the NO emission of HWRE increases to about three times the NO emission of HRPE in stoichiometric ratio combustion. HWREs exhibit slightly lower brake thermal efficiency, with reductions of 4.96% and 3.06% lower than HRPE and GRPE, respectively. Figure 9.31 shows the variation of PPD and BTE at different speeds.WRE with hydrogen shows higher PPD compared to corresponding values of RPE with gasoline fuel.

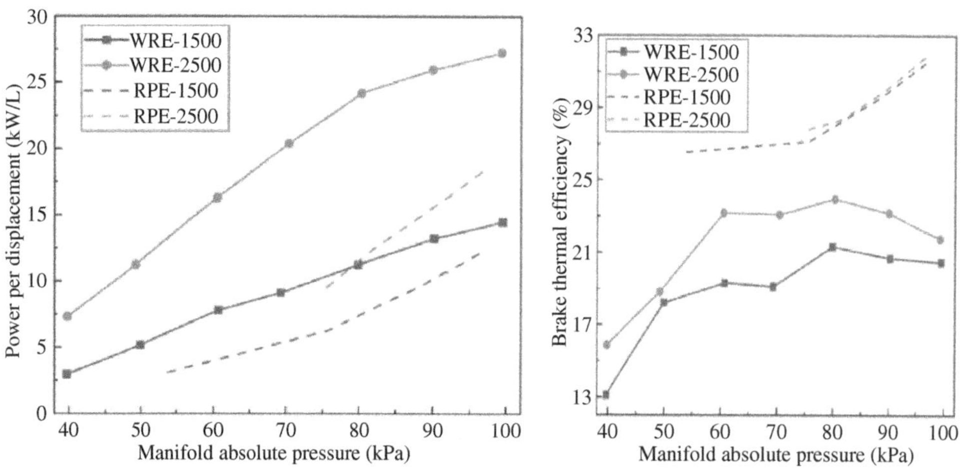

Figure 9.31 The PPD (left) and BTE (right) of GWRE and GRPE at different engine speeds [85].

BTE of GWRE was observed lower than that of GRPE, particularly at high-load or wide-open throttle conditions.

Hydrogen rotary engines offer several technical advantages. Physical separation of combustion phases in different chambers within the Wankel engine design increases resistance to abnormal combustion events like knocking or detonation. Hydrogen's wide range of flammability enables lean operation of the rotary engine with improved efficiency and lesser fuel consumption while maintaining engine power. It also provides more stable engine operation, and lower engine noise and vibrations. Similarly, properties such as high calorific value, flame speed, small quenching distance, and diffusivity, contribute to better-premixed fuel and more complete combustion within the rotary engine. The dual fuel mode of operation of hydrogen in a Wankel engine is very suitable for a wider driving range in vehicles [90].

There are significant benefits of qualitative control in hydrogen reciprocating engines to achieve higher thermal efficiency and reduce NOx emissions. However, compared to HRPE, HWRE has some drawbacks such as low BTE, uneven thermal load, high tendency to knock. and high NOx emissions [91]. There are problems of qualitative control and excessive air ratio leading to significant cyclic variation and combustion instability. Thus, identifying an appropriate load control strategy is crucial in HWRE. In their tests with a HWRE, Meng et al. [91] adopted a combination of quantitative and qualitative control strategy to evaluate efficient and stable operation of a hydrogen-fueled WRE. The combined control strategy offered the flexible option of matching logic between qualitative control and quantitative control tailored to specific needs such as emissions, efficiency, stability, and power. Combined load control led to improved thermal efficiency, increased volumetric efficiency, substantial reduction in maximum pressure rising rate, and reduced NO emission in most ranges of engine operation. The increased cyclic variation was also less than 4.5%. However, post-treatment methods are required to control high NO emissions under stoichiometric combustion. Further studies on knock and NO emission need to be carried out for HWRE. Amrouche et al. [92] worked on a modified Wankel rotary engine using a hydrogen–gasoline blend. Various hydrogen energy fractions (0%, 2%, 4%, 5%, 7%, and 10%) were introduced into the engine intake. The engine operated at 3000 rpm under high-speed and wide-open throttle conditions with original ignition timing. The brake thermal efficiency of the GWRE was improved by about 28% over the baseline gasoline operation by adding 10% energy fraction of hydrogen to the gasoline fuel mixture. Brake specific fuel consumption and the exhaust temperature of the engine were decreased. Brake specific NOx emissions increased by 137% as the energy fraction of hydrogen increased from 0 to 10%. However, brake specific HC emissions were reduced by 85%, CO by 64%, and CO_2 by 36% with increasing hydrogen fraction.

Ji et al. [93] investigated stratified mixture distribution in a hydrogen-enriched compound-intake gasoline Wankel rotary engine (WRE) by utilizing the difference in gas velocity from different intake ports. Compound intake method refers to various strategies for introducing both hydrogen and gasoline into the engine's intake system by utilizing differences in gas velocity from various intake ports. A three-dimensional computational fluid dynamics (CFD) WRE model was developed for mixture formation and combustion under five intake modes. Thus, the way in which the binary fuel entered the combustion chamber through different intake ports (such as side intake port, peripheral intake port, or both ports) was considered. In this analysis, a geometric model and a CFD simulation model were respectively established with CATIA and CONVERGE software.

The S-HGP intake mode (mixtures of H_2, gasoline, and air in the peripheral intake port and pure air in the side intake port) produced the most suitable distribution of fuel equivalence ratio.

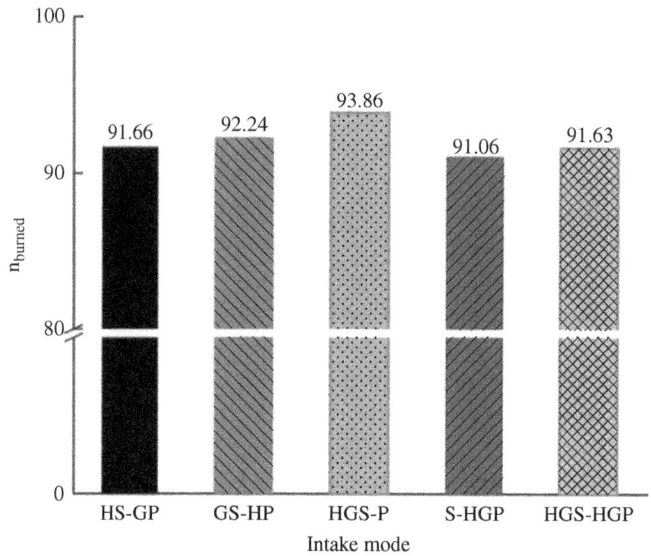

Figure 9.32 Volume fraction of the burned zone at the opening time of the exhaust valve in each case [93]. Please visit the companion website to view a color version of this image.

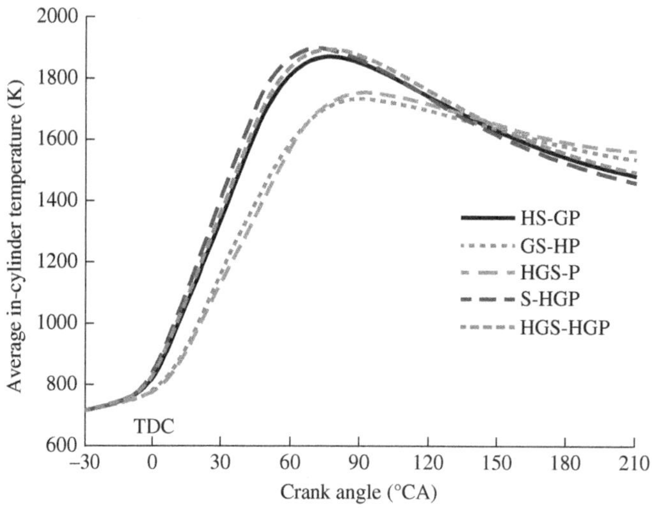

Figure 9.33 Average in-cylinder temperature for five cases [93]. Please visit the companion website to view a color version of this image.

Thus, the mixture distribution with local equivalence ratio is dependent on intake mode. Fuel from the side intake port accumulates in the rear zone of the combustion chamber, whereas the fuel from the peripherical intake port remains in the front region. Figure 9.32 shows the volume fraction of the burned zone in the total combustion chamber at the exhaust valve timing, which clearly shows that the burned areas of GS-HP and HGS-P are larger than the other three. Proper fuel distribution led to moderate combustion with low CO emission with S-HGP near the exhaust valve opening timing (210 °CA ATDC).

Figures 9.33 and 9.34 show the in-cylinder temperature and NO emission in all five cases. NO formation was closely correlated to combustion temperature. The high temperature zone

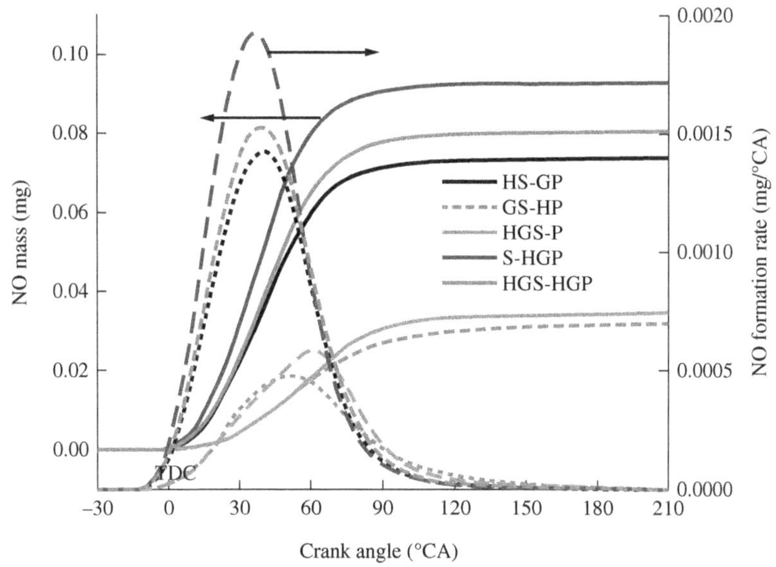

Figure 9.34 Mass and formation rate of NO in the combustion chamber under [93]. Please visit the companion website to view a color version of this image.

remained mainly from TDC to 90 °CA ATDC. NO formation of GS-HP and HGS-P was each slightly behind the other three. The higher temperature inside the cylinder with S-HGP resulted in slightly higher NO emissions. This study shows compound intake, combined with intake mode control, is an effective approach for optimizing fuel distribution, with the benefits due to direct fuel injection.

9.6.1 Leakage Issues in the Hydrogen Rotary Engine

Lower combustion efficiency [94] and higher levels of HC and CO emissions are two major shortcomings of rotary engines because of gas leakage problems [95]. Small molecular weight and the fast flame speed of hydrogen lead to greater risk of leakage. Apex seals in a Wankel rotary engine, which perform a similar function as piston rings in reciprocating internal combustion piston engines, have often faced the problem of leakage in the actual working process of a rotary engine.

Spark plug cavity and corner seal gap are major sources of leakage in hydrogen rotary engines. Some studies have been done in which a 3D dynamic simulation model was developed for an engine fueled with natural gas/hydrogen blended fuel [96, 97]. On this 3D model, action of the apex seal leakage (ASL), diffusion and combustion processes of natural gas–hydrogen blended fuel in the cylinder with different hydrogen injection angle (HIA), and hydrogen injection timing (HIT) were calculated. The results indicated that the hydrogen jet flow first affected the rotor wall after entering the cylinder with apex seal leakage. Yang et al. [98], using CONVERGE software, studied leakage characteristics in a HWRE, optimizing the location of the leading spark plug. There was only one leakage process in the trailing spark plug cavity referred to as residual gas leakage. On the other hand, two leakage processes occurred at the leading spark plug such as residual gas and fresh mixture leakage.

Figure 9.35 shows that the leakage paths of the leading spark plug cavity involve direct leakage between adjacent chambers and leakage through the spark plug cavity itself. During the residual gas leakage stage, as depicted in Figure 9.35a, most residual gas leaks from leading spark plug cavity into chamber I are via path B, while a smaller amount leaks directly from chamber II to

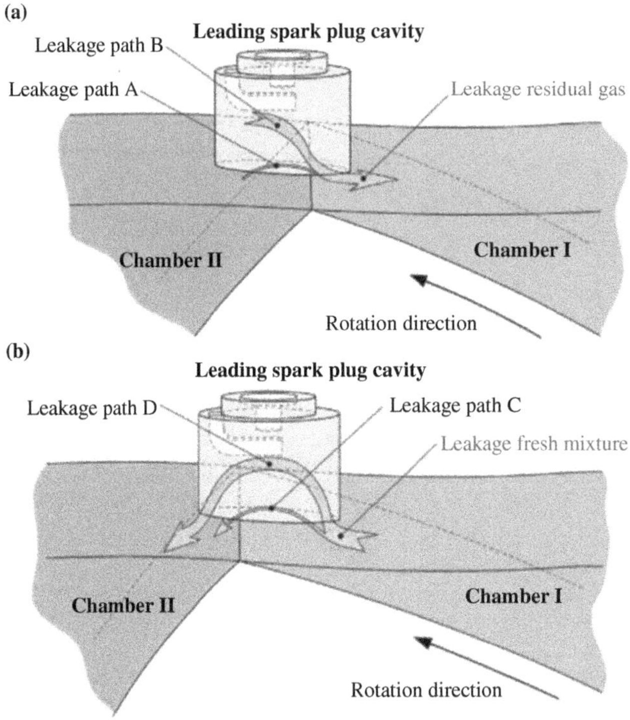

Figure 9.35 Leakage paths of leading spark plug cavity: (a) the leakage path of residual gas, (b) the leakage path of the fresh mixture [98]. Please visit the companion website to view a color version of this image.

chamber I via path A. During the fresh mixture leakage stage, as shown in Figure 9.35b, most residual gas leaks occurred from chamber I to chamber II through the leading spark plug cavity via path D, with a smaller amount directly leaking from chamber I to chamber II via path C. Thus, the spark plug's position significantly influenced these leakages and affected the HWRE's performance. Changing position of the leading spark plug affected the flow characteristics and combustion behavior of the HWRE. Residual gas leakage might generate new vortices in the primary flow region, whereas fresh mixture leakage diminished existing vortices in the same area. Moving the spark plug closer to recess midpoint reduces hydrogen leakage, which improves flame diffusion and combustion.

Research and development activities in this area are still being pursued. A variety of materials ranging from cast iron to fiber-reinforced ceramics have been tried in rotary engines with varying degrees of success. It is essential to identify the appropriate material needed to balance wear resistance with optimum combustion chamber sealing in these engines.

Concluding Remarks and Perspectives

Technologies such as supercharging, turbocharging, HCCI combustion, stratification, and rotary engine integration into hydrogen engines play a crucial role in the transition to a cleaner and more sustainable future. Even though these are still in an early stage of research and development, some definite qualitative pictures have emerged about prospects of these technologies and how best they could be implemented. Supercharging and turbocharging are well-proven in increasing the power output of hydrogen engines operating in lean mode without undesirable combustion phenomena. An optimized DI hydrogen engine can achieve brake thermal efficiency of about 42.6% and near-zero NOx emissions of less than 20 ppm. At low speeds of less than 2000 rpm, high NOx emission conversion efficiency of over 99.5% can be achieved with NH3-SCR. Multiple research results on several governing parameters need to be integrated to evolve an optimum configuration.

The HCCI mode makes it possible for hydrogen to utilize some specific benefits of the CI engine with a homogeneous mixture. It is possible to achieve stable engine operation with an excess air ratio of 6. Some studies have shown that maximum fuel efficiency can be achieved with an excess air ratio of 3. In these engines, the heat release rate is very high, requiring accurate ignition timing control. Cycle-to-cycle variations are observed in the absence of accurate control of ignition timing. Some studies have demonstrated satisfactory monitoring of ignition timing using heating inlet air. Inlet air heating to ensure autoignition in HCCI mode reduces the air mass flow through the engine, which leads to modest maximum power. Hydrogen HCCI engines can operate in very lean mode leading to low NOx emissions. It can control the reactivity of air–fuel mixtures through use of dual injection systems and optimize combustion stability. From the extensive research results showing both advantages and shortcomings, the positive features can be identified and implemented to maximize potential benefits from these research results. Mixture strength and distribution are critical to stratification of charge in hydrogen engines, which can be accomplished through several routes such as premixed hydrogen mixture distribution (PHMD), split hydrogen direct injection (SHDI) with partially premixed hydrogen mixture distribution (PPHMD). Among these, PPHMD has demonstrated the fastest combustion and highest brake thermal efficiency. Of course, its emissions are a little higher than for the PHMD. Thus, by selecting the appropriate injection strategy, better control over the combustion process can be achieved, resulting in improved fuel efficiency and reduced emissions. This critical area needs to be further investigated.

Integration of the rotary engine into a hydrogen propulsion system offers several advantages, including compactness, simplicity, and smooth operation. Disadvantages stem primarily from high hydrocarbon emissions, low combustion efficiency, and gas leakage. The rotary engine's long sealing line makes sealing more difficult than in piston engines. HREs offer advantages such as higher power density and reduced backfire and exhibit knock problems due to their extended combustion chamber. Ignition timing, spark plug number and location, excess air ratio, and engine speed are crucial factors in designing a hydrogen-specific rotary engine. Apex seal, combustion chamber profile, rotor profile, and port designs need special attention to ensure optimal functionality and efficiency.

References

1 Nguyen, D., Choi, Y., Park, C. et al. (2021). Effect of supercharger system on power enhancement of hydrogen-fueled spark-ignition engine under low-load condition. *International Journal of Hydrogen Energy* 46 (9): 6928–6936.

2 Duan, J., Liu, F., and Sun, B. (2014). Backfire control and power enhancement of a hydrogen internal combustion engine. *International Journal of Hydrogen Energy* 39 (9): 4581–4589.

3 Verhelst, S., Maesschalck, P., Rombaut, N. et al. (2009). Increasing the power output of hydrogen internal combustion engines by means of supercharging and exhaust gas recirculation. *International Journal of Hydrogen Energy* 34 (10): 4406–4412.

4 Nakagawa, K., Yamane, K., and Ohira, T. (2012). Potential of large output power, high thermal efficiency,near-zero nox emission, supercharged, lean-burn, hydrogen-fuelled, direct injection engines, world hydrogen energy conference 2012. *Energy Procedia* 29: 455–462.

5 Szwaja, S. and Grab-Rogalinski, K. (2012). Hydrogen combustion in supercharged SI engine. *Journal of KONES Powertrain and Transport* 19 (3): 149–155.

6 Bao, L.-Z., Sun, B.-G., Luo, Q.-H. et al. (2022). Development of a turbocharged direct-injection hydrogen engine to achieve clean, efficient, and high-power performance. Fuel 324: 124713.

7 Shinde, B.J. and Karunamurthy, K. (2022). Recent progress in hydrogen fuelled internal combustion engine (H2ICE)–a comprehensive outlook. *Materials Today: Proceedings* 51: 1568–1579. https://doi.org/10.1016/j.matpr.2021.10.378.

8 Kiesgen, G., Klüting, M., Bock, C. et al. (2006). The new 12-cylinder hydrogen engine in the 7 series: The H2 ICE age has begun. SAE Technical Paper 2006-01-0431. https://doi.org/10.4271/2006-01-0431.

9 Wang, X., Sun, B.-G., and Luo, Q.-H. (2019). Energy and exergy analysis of a turbocharged hydrogen internal combustion engine. *International Journal of Hydrogen Energy* 44 (11): 5551–5563.

10 Duan, Y.-H., Sun, B.-G., Li, Q. et al. (2023). Combustion characteristics of a turbocharged direct-injection hydrogen engine. *Energy Conversion and Management* 291: 117267.

11 Wang, K.-D., Zhang, Z.-F, Sun, B.-G. et al. (2024). Experimental investigation of the working boundary limited by abnormal combustion and the combustion characteristics of a turbocharged direct injection hydrogen engine. *Energy Conversion and Management* 299: 117861.

12 Wang, K.-D., Sun, B.-G., Luo, Q.-H. et al. (2023). Performance optimization design of direct injection turbocharged hydrogen internal combustion engine. *Applications in Energy and Combustion Science* 16: 100204.

13 Luo, Q.-H., Hu, J.-B., Sun, B.-G. et al. (2019). Experimental investigation of combustion characteristics and NOx emission of a turbocharged hydrogen internal combustion engine. *International Journal of Hydrogen Energy* 44: 5573–5584.

14 Bao, L.-Z., Sun, B.-G., and Luo, Q.-H. (2022). Optimal control strategy of the turbocharged direct-injection hydrogen engine to achieve near-zero emissions with large power and high brake thermal efficiency. *Fuel* 325: 124913.

15 Bao, L.-Z., Sun, B.-G., Luo, Q.-H. (2022). Experimental investigation of the achieving methods and the working characteristics of a near-zero NOx emission turbocharged direct-injection hydrogen engine. *Fuel* 319: 123746.

16 Hong, W., Hu, Z., Ma, J. et al. (2023). Experimental study of thermal efficiency and NOx emission of turbocharged direct injection hydrogen engine based on a high injection pressure. *International Journal of Hydrogen Energy* 48: 12905–12916.

17 Luo, Q.-H., Hu, J.-B., Sun, B.-G. et al. (2019). Effect of equivalence ratios on the power, combustion stability and NOx controlling strategy for the turbocharged hydrogen engine at low engine speeds. *International Journal of Hydrogen Energy* 44: 17095–17102.

18 Liang, Z., Xie, F, Jiang, B. et al. (2024). Evaluating the potential of mixture formation methods to achieve efficient combustion and near-zero emissions on a hydrogen direct injection engine. *Journal of Cleaner Production* 439: 140930.

19 Lee, S., Kim, G., and Bae, C. (2022). Effect of mixture formation mode on the combustion and emission characteristics in a hydrogen direct-injection engine under different load conditions. *Applied Thermal Engineering* 209: 118276. https://doi.org/10.1016/j.applthermaleng.2022.118276.

20 Baratta, M. and Rapetto, N. (2015). Mixture formation analysis in a direct-injection NG SI engineunder different injection timings. *Fuel* 159: 675–688. https://doi.org/ 10.1016/ j.fuel.2015.07.027.

21 Li, G., Yu, X., Jin, Z. et al. (2020). Study on effects of split injection proportion on hydrogen mixture distribution, combustion and emissions of a gasoline/hydrogen SI engine with split hydrogen direct injection under lean burn condition. *Fuel* 270: 117488. https://doi.org/10.1016/ j.fuel.2020.117488.

22 Sanguk, L., Gyeonggon, K., and Choongsik, B. (2021). Lean combustion of stratified hydrogen in a constant volume chamber. *Fuel* 301: 121045.

23 Korn, T., Nobile, R.-F., and Grassinger, D. (2021). Zero-emission, maximum performance–the latest generation of hydrogen combustion engines. *Proceedings of the 42nd International Vienna Motor Symposium*, Vienna, Austria (29–30 April 2021).

24 Shudo, T., Omori, K., and Hiyama, O. (2008). NOx reduction and NO2 emission characteristics in rich-lean combustion of hydrogen. *International Journal of Hydrogen Energy* 33: 4689–4693.

25 Salazar, V. and Kaiser, S. (2011). Interaction of intake-induced flow and injection jet in a direct-injection hydrogen-fueled engine measured by PIV. SAE Technical Paper No. 2011-01-0673.

26 Boretti, A., Watson, H., and Tempia, A. (2010). Computational analysis of the lean-burn direct-injection jet ignition hydrogen engine. *Proceedings of the Institution of Mechanical Engineers, Part D: Journal of Automobile Engineering* 224: 261–269.

27 Wallner, T., Ciatti, S., and Bihari, B. (2007). Investigation of injection parameters in a hydrogen DI engine using an endoscopic access to the combustion chamber, SAE Technical Paper No. 2007-01-1464, 2007, https://doi.org/10.4271/2007-01-1464.

28 Wallner, T. (2004). Entwicklung von brennverfahrenskonzepten für einen pkw–motor mit wasserstoffbetrieb (Development of combustion concepts for a hydrogen-powered internal combustion engine). PhD dissertation. TU Graz, Graz, Austria.

29 Dreisbach, R., Arnberger, A., Zukancic, A. et al. (2021). The heavy-duty hydrogen engine and its realization until 2025. *Proceedings of the 42nd International Vienna Motor Symposium*, Vienna, Austria (29–30 April 2021).

30 Lee, S., Kim, G., and Bae, C. (2022). Effect of injection and ignition timing on a hydrogen-lean stratified charge combustion engine. *International Journal of Engine Research* 23 (5):816—829.

31 Biffiger, H. and Soltic, P. (2015). Effects of split port/direct injection of methane and hydrogen in a spark ignition engine. *International Journal of Hydrogen Energy* 40 (4): 1994—2003.

32 Li, G., Yu, X., Shi, W. et al. (2019). Effects of split injection proportion and the second injection timings on the combustion and emissions of a dual fuel SI engine with split hydrogen direct injection. *International Journal of Hydrogen Energy* 44 (21):11194—11204.

33 Li, G., Yu, X., Jin, Z. et al. (2020). Study on effects of split injection proportion on hydrogen mixture distribution, combustion an emissions of a gasoline/hydrogen SI engine with split hydrogen direct injection under lean burn condition. *Fuel* 270: 117488.

34 Hamzehloo, A. and Aleiferis, P. (2014). Numerical modelling of mixture formation and combustion in disi hydrogen engines with various injection strategies. SAE Technical Paper No. 2014-01-2577.

35 Li, Y., Gao, W., Zhang, P. et al. (2019). Effects study of injection strategies on hydrogen-air formation and performance of hydrogen direct injection internal combustion engine. *International Journal of Hydrogen Energy 44* (47):26000—26011.

36 Lee, S., Kim, G., and Bae, C. (2022). Effect of mixture formation mode on the combustion and emission characteristics in a hydrogen direct-injection engine under different load conditions. *Applied Thermal Engineering* 209: 118276.

37 Conte, E. and Boulouchos, K. (2006). Hydrogen-enhanced gasoline stratified combustion in SI-DI engines. *Proceedings of the ASME 2006 Internal Combustion Engine Division Spring Technical Conference* (ICES2006), Aachen, Germany (7–10 May), pp. 429–436. https://doi.org/10.1115/ICES2006-1399.

38 Shi, W., Yu, X., Zhang, H. et al. (2017). Effect of spark timing on combustion and emissions of a hydrogen direct injection stratified gasoline engine. *International Journal of Hydrogen Energy* 42: 5619–5626.

39 Yu, X., Li, G., Du, Y. et al. (2019). A comparative study on effects of homogeneous or stratified hydrogen on combustion and emissions of a gasoline/hydrogen SI engine. *International Journal of Hydrogen Energy* 44: 25974–25984.

40 Shudo, T. (2007). Improving thermal efficiency by reducing cooling losses in hydrogen combustion engines. *International Journal of Hydrogen Energy* 32: 4285–4293.

41 Shudo, T. and Oba, S. (2009). Mixture distribution measurement using laser induced breakdown spectroscopy in hydrogen direct injection stratified charge. *International Journal of Hydrogen Energy* 34:2488 –2493.

42 Gomes Antunes, J.M., Mikalsen, R., and Roskilly, A.P. (2008). An investigation of hydrogen-fuelled HCCI engine performance and operation. *International Journal of Hydrogen Energy* 33: 5823–5828.

43 Alexandros, G. (2013). Homogeneous charge compression ignition (HCCI) engines. *Advances in Internal Combustion Engines and Fuel Technologies*. InTech. https://doi.org/10.5772/55807.

44 Gopinathan, R. L. and Ibrahim, M. M. (2024). Hydrogen driven HCCI engine combustion and performance indices: A numerical investigation. *Australian Journal of Mechanical Engineering*, 1–18. https://doi.org/10.1080/14484846.2024.24325.

45 Tsujimura, T., Mikami, S., Achiha, N. et al. (2003). A study of direct injection diesel engine fuelled with hydrogen. SAE Technical Paper No. 2003-01-0761.

46 Naber, J.D. and Siebers, D.L. (1998). Hydrogen combustion under diesel engine conditions. *International Journal of Hydrogen Energy* 23: 363–371.

47 Persson, H., Hultqvist, A., Johansson, B. et al. (2007). Investigation of the early flame development in spark assisted HCCI combustion using high speed chemiluminescence imaging. SAE Technical Paper No. 2007-01-0212.

48 Namar, M.M. and Jahanian, O. (2019). Energy and exergy analysis of a hydrogen-fueled HCCI engine. *Springer Journal of Thermal Analysis and Calorimetry* 137: 205.

49 Caton, P.A. and Pruitt, J.T. (2009). Homogeneous charge compression ignition of hydrogen in a single-cylinder diesel engine. *International Journal of Engine Research* 10 (1): 45–63. https://doi.org/10.1243/14680874 JER0220.

50 Nöda, T. and Foster, D.E. (2001). A numerical study to control combustion duration hydrogen-fueled HCCI by using multi-zone chemical kinetics simulation. *SAE Transactions: Journal of Engines* 110 Section 3: 186–191.

51 Maurya, R.K. and Akhil, N. (2017). Comparative study of the simulation ability of various recent hydrogen combustion mechanisms in HCCI engines using stochastic reactor model. *International Journal of Hydrogen Energy* 42 (16): 11911–11925.

52 Liu, C. and Ghazi, A.K. (2008). A simulation of the combustion of hydrogen in HCCI engines using a 3D model with detailed chemical kinetics. *International Journal of Hydrogen Energy* 33: 3863–3875.

53 Stenlåås, O., Christensen, M., Egnel, l R. et al. (2004). Hydrogen as homogeneous charge compression ignition engine fuel. *SAE Transactions, 2004: Journal of Fuels and Lubricants* 113 Section 4: 1317–132.

54 Rosati, M. and Aleiferis, P. (2009). Hydrogen SI and HCCI combustion in a direct-injection optical engine. *SAE International Journal of Engines* 2 (1): 1710–1736. https://doi.org/10.4271/2009-01-1921
55 Nguyen, D. and Turner, J.W.G. (2023). Effect of nitric oxide on the performance and emissions of a hydrogen-fueled HCCI engine. *Fuel* 349: 128611.
56 Alberto, A. (2011). Boretti diesel-like and HCCI-like operation of a truck engine converted to hydrogen. *International Journal of Hydrogen Energy* 36: 15382–15391.
57 Narioka, Y., Yokoyama, T., Iio, S. et al. (2006). HCCI combustion characteristics of hydrogen and hydrogen-rich natural gas reformate supported by dme supplement. SAE Technical Paper No. 2006-01-0628. https://doi.org/10.4271/2006-01-0628.
58 Hairuddin, A.A., Yusuf, T., and Wandel, A.P. (2014). A review of hydrogen and natural gas addition in diesel HCCI engines. *Renewable and Sustainable Energy Reviews* 32: 739–761.
59 Raitanapaibule, K. and Aung, K. (2005). Performance predictions of a hydrogen-enhanced natural gas HCCI engine. *Proceedings of the ASME 2005 International Mechanical Engineering Congress and Exposition, Energy Conversion and Resources*, Orlando, Florida, USA (5–11 November), pp. 289–294. https://doi.org/10.1115/I MECE2005-80963.
60 Hu, E., Chen, Y., Cheng, Y. et al. (2015). Study on the effect of hydrogen addition to dimethyl ether homogeneous charge compression ignition combustion engine. *Journal of Renewable and Sustainable Energy* 7: 063121. https://doi.org/10.10 63/1.4937139.
61 Bae, C. (2009). Dual-fuel HCCI operation with DME/LPG/gasoline/hydrogen. *Proceedings of IEA Combustion 2009: IEA 31*. Task Leaders Meeting on Energy Conservation and Emissions Reduction in Combustion.
62 Wang, Z.-H., Wang, C.-M., Tang, H.-X. et al. (2009). Effect of hydrogen addition on methane HCCI engine ignition timing and emissions using a multi-zone model. *Chinese Journal of Chemical Physics* 22 (3):article 290–296.
63 Flowers, D., Aceves, S., Westbrook, C.K. et al. (2001). Detailed chemical kinetic simulation of natural gas HCCI combustion: gas composition effects and investigation of control strategies. *Journal of Engineering for Gas Turbines and Power* 123 (2): 433e9. http://dx.doi.org/10.1115/1.136452.
64 Kokjohn, S.L., Hanson, R.M., Splitter, D.A. et al. (2010). Fuel reactivity controlled compression ignition (RCCI): a pathway to controlled high-efficiency clean combustion. *International Journal of Engine Research* 12 (3): 209e26. http://dx.doi.org/10.1177/1468087411401548.
65 Papagiannakis, R.G. and Hountalas, D.T. (2004). Combustion and exhaust emission characteristicsof a dual fuel compression ignition engine operated with pilot diesel fuel and natural gas. *Energy Conversion and Management* 45 (18e19): 2971e87. http://dx.doi.org/10.1016/j.enconman.2004.01.013.
66 Stanglmaier, R., Ryan, T., and Souder, J. (2001). HCCI operation of a dual-fuel natural gas engine for improved fuel efficiency and ultra-low NOx emissions at low to moderate engine loads. SAE Technical Paper No. 2001-01-1897. http://dx.doi.org/10.4271/ 2001-01-1897.
67 Kaddatz, J., Andrie, M., Reitz, R. et al. (2012). Light-duty reactivity controlled compression ignition combustion using a cetane improver. SAE Technical Paper No. 2012-01- 1110. http://dx.doi.org/10.4271/2012-01-1110.
68 Aldawood, A., Mosbach, S., and Kraft, M. (2012). HCCI combustion control using dual-fuel approach: experimental and modeling investigations. SAE Technical Paper No. 2012-01-1117. http://dx.doi.org/10.4271/2012-01-1117.

69 Dempsey, A., Walker, N., Gingrich, E. et al. (2014). Comparison of low temperature combustion strategies for advanced compression ignition engines with a focus on controllability. *Combustion Science and Technology* 86 (2): 210e41.

70 Manente, V., Tunestal, P., and Johansson, B. (2009). Partially premixed combustion at high load using gasoline and ethanol, a comparison with diesel. SAE Technical Paper No. 2009-01-0944.

71 Splitter, D.A., Hanson, R.M., Kokjohn, S.L. et al. (2010). Improving engine performance by optimizing fuel reactivity with a dual fuel PCCI strategy. THIESEL 2010, Conference on Thermo and Fluid Dynamic Processes in Diesel Engines, Valencia, Spain. https://www.researchgate.net/publication/267927564_Improving_engine_performance_by_optimizing_fuel_reactivity_with_a_dual_fuel_PCCI_strategy.

72 Kokjohn, S.L., Splitter, D.A., Hanson, R.M. et al. (2010). Experiments and modeling of dual fuel HCCI and PCCI combustion using in-cylinder fuel blending. *SAE International Journal of Engines* 2 (2): 24e39.

73 Manente, V., Johansson, B., Tunestal, P. et al. (2010). Effects of ethanol and different type of gasoline fuels on partially premixed combustion from low to high load. SAE Technical Paper No. 2010-01-0871.

74 Hasegawa, R. and Yanagihara, H. (2003). HCCI combustion in DI diesel engine. SAE Technical Paper No. 2003-01-0745.

75 Hardy, W.L. and Reitz, R.D. (2006). A study of the effects of high EGR, high equivalence ratio, and mixing time on emissions levels in a heavy-duty diesel engine for PCCI combustion. SAE Technical Paper No. 2006-01-0026.

76 Opat, R., Ra, Y., Gonzalez, M.A. et al. (2007). Investigation of mixing and temperature effects on HC/CO emissions for highly dilute low temperature combustion in a light duty diesel engine. SAE Technical Paper No. 2007-01-0193.

77 Loyte, A., Suryawanshi, J., Bhiogade, G. et al. (2023). Influence of injection strategies on ignition patterns of RCCI combustion engine fuelled with hydrogen enriched natural gas. *Environmental Research* 234: 116537.

78 Watgave, S., Banapurmath, N., and Harari, P. (2021). Comparative study on effect of hydrogen and hydrogen blended compressed natural gas on compression ignition engine operated under homogeneous charge compression ignition and reactivity controlled compression ignition mode of combustion. SAE Technical Paper No. 2021-28-0010. https://doi.org/10.4271/2021-28-0010.

79 Rahimi, H.M., Zazayeri, S.A., and Ebrahimi, M. (2020). Hydrogen energy share enhancement in a heavy duty diesel engine under RCCI combustion fuelled with natural gas and diesel oil. *International Journal of Hydrogen Energy* 45 (35): 7975–17991.

80 Kweon, C.-B.M. (2011). *A Review of Heavy-Fueled Rotary Engine Combustion Technologies.* Adelphi, MD, USA: Army Research Lab.

81 Drbal, M.A. (2020). Method of rotary engine performance prediction. *Scientific Journal of Silesian University of Technology. Series Transport 2020* 108: 37–43.

82 Qin, Z., Jia, M., and Yang, H. (2020). Study on vortex characteristics and velocity distribution in small rotary engine. *Energy* 206: 118065.

83 Heller, D.A. (2005). International strategic alliances and technology strategy: the case of rotary-engine development at Mazda. *Economics Review* 52: 31–56.

84 Meng, H., Ji, C., Wang, D. et al. (2022). Research on the load control of hydrogen-fueled Wankel rotary engine. *International Journal of Hydrogen Energy* 47: 16665–16675.

85 Meng, H., Ji, C., Xin, G. et al. (2022). Comparison of combustion, emission and abnormal combustion of hydrogen-fueled Wankel rotary engine and reciprocating piston engine. *Fuel* 318: 123675.

86 Zambalov, S.D., Yakovlev, I.A., and Maznoy, A.S. (2020). Effect of multiple fuel injection strategies on mixture formation and combustion in a hydrogen-fueled rotary range extender for battery electric vehicles. *Energy Conversion and Management* 220: 113097. https:// doi.org/10.1016/ j.enconman.2020.113097.

87 Zambalov, S.D., Yakovlev, I.A., and Skripnyak, V.A. (2017). Numerical simulation of hydrogen combustion process in rotary engine with laser ignition system. *International Journal of Hydrogen Energy* 42 (27): 17251–17259. https://doi.org/10.1016/j.ijhydene.2017.05.142.

88 Wang, H., Ji, C., Su, T. et al. (2022). Comparison and implementation of machine learning models for predicting the combustion phases of hydrogen-enriched Wankel rotary engines. *Fuel* 310: 122371. https://doi.org/10.1016/j.fuel.2021.122371.

89 Meng, H., Ji, C., Su, T. et al. (2022). Analyzing characteristics of knock in a hydrogen-fueled Wankel rotary engine. *Energy* 250: 123828.

90 Ozcanli, M., Bas, O., Akar, M.A. et al. (2018). Recent studies on hydrogen usage in Wankel SI engine. *International Journal of Hydrogen Energy* 43: (38): 18037–1804591.

91 Meng, H., Ji, C., Wang, D. et al. (2022). Research on the load control of hydrogen-fueled Wankel rotary engine. *International Journal of Hydrogen Energy* 47 (37): 16665–16675.

92 Amrouche, F., Erickson, P., Park, J. et al. (2014). An experimental investigation of hydrogen enriched gasoline in a Wankel rotary engine. *International Journal of Hydrogen Energy* 39 (16): 8525–8534.

93 Ji, C., Meng, H., Wang, S. et al. (2020). Realizing stratified mixtures distribution in a hydrogen-enriched gasoline Wankel engine by different compound intake methods. *Energy Conversion and Management* 203: 112230.

94 Chen, W., Yu, S., Zuo, Q. et al. (2022). Combined effect of air intake method and hydrogen injection timing on airflow movement and mixture formation in a hydrogen direct injection rotary engine. *International Journal of Hydrogen Energy* 47: 12739–12758.

95 Yang, J., Meng, H., Ji, C. et al. (2022). Comparatively investigating the leading and trailing spark plug on the hydrogen rotary engine. *Fuel* 308: 122005.

96 Fan, B., Zeng, Y., Pan, J. (2022). Evaluation and analysis of injection strategy in a peripheral ported rotary engine fueled with natural gas/hydrogen blends under the action of apex seal leakage. *Fuel* 310: 122315.

97 Fan, B., Zeng, Y., Zhang, Y. et al. (2021). Research on the hydrogen injection strategy of a direct injection natural gas/hydrogen rotary engine considering apex seal leakage. *International Journal of Hydrogen Energy* 46 (13): 9234–9251.

98 Yamg, Z., Ji, C., Wang, H. et al. (2023).The optimization of leading spark plug location and its influences on combustion and leakage in a hydrogen-fueled Wankel rotary engine. *International Journal of Hydrogen Energy* 48 (53): 20465–204.

10

The Path Forward

The future belongs to those who believe in the beauty of their dreams.

Eleanor Roosevelt

10.1 Introduction

The transport sector, which is built on internal combustion engines, has been badly affected due to fossil fuel depletion. Several alternative emerging technologies are still developing, and it is uncertain which will be adopted globally or regionally. Despite their significant potential, hydrogen vehicles are still in the early stages of commercialization and do not have substantial market penetration. This is due to several factors, including a limited number of refueling stations, technological adaptation challenges, and the lack of infrastructure for economically viable production. Recently, it has been in the news that BMW, after a production run of nine years and more than 250,000 units sold, is ceasing production of EVs (the BMW i3) and is switching to hydrogen with fuel cells [1]. Buses and hydrogen-powered trains (with fuel cells) have already been demonstrated. Fuel cell and hydrogen IC engines are not rivals; instead, they are complementary technologies. Both pursue the same goal through distinct methods, working together to achieve a common objective. The circumstances will dictate which technologies are widely adopted and when and where they are most effectively implemented depending on local resources, available infrastructure, and the price of green energy.

10.2 Hydrogen Engines for Land and Marine Transport

ICE technology will continue to be vital for many decades. In view of hydrogen's infinite source potential and clean-burning characteristics, introducing hydrogen to replace petroleum-based fuels will ensure its continued dominance on IC engines. Transport is the most carbon-intense sector in urban areas of most countries. Reducing carbon emissions in this sector can be achieved by switching to clean-burning alternative fuels such as hydrogen. Hydrogen-fueled vehicles can revolutionize transportation by using fuel cells or modified internal combustion engines tailored to hydrogen's specific combustion properties. Each approach can provide versatile pathways to a sustainable future. In this context, it is relevant to mention that internal combustion engines (ICEs) and fuel cells are not competing technologies but rather complementary solutions, both contributing to

Hydrogen Engines: Design, Performance Evaluation, Combustion Analysis, and Exhaust Emissions, First Edition.
Lalit Mohan Das.

the same goal of zero emissions. Development of one supports the other, as both have common hydrogen production, transportation, and distribution infrastructure. In a vehicle, both involve the same vehicle storage tanks. Of course, both have different maturity levels. Internal combustion engines, with decades of widespread application, have an extensive and well-established service network with robust support for various operational needs. These engines, available in a wide range of sizes and configurations, are versatile to tolerate harsh conditions, such as dusty environments and heavy vibrations. Mechanical familiarity, established infrastructure, and the proven reliability of ICE technology offer the promise of fast practical implementation. In the context of the present topic, discussions are restricted to hydrogen engines and their applications. It will be clear from the following discussions that apart from major continuous development carried out in important industries, some significant achievements have been made by individual researchers and academic/research institutes on pure hydrogen and hydrogen in blends with other fuels.

Hydrogen engines have a long and rich history spanning more than two centuries, with successive innovations at each stage. Chapter 3 highlights the research efforts of numerous scholars, at different stages, bringing out both the advantages and limitations of hydrogen engines. Sporadic efforts have also been made to develop H_2ICE vehicles. It has been reported that François Isaac de Rivaz, in 1808, invented a hydrogen-fueled engine with an electric ignition system [2] and put it in an existing vehicle. Étienne Lenoir's contribution to the development of one of the earliest internal combustion engines using coal gas as a fuel is well known. Later in 1863, he operated the engine by igniting a hydrogen–air mixture through electrical ignition. Huysken et al. [3] installed a hydrogen port fuel-injection system on an Opel Combo bi-fuel (gasoline and CNG) and developed a flex-fuel vehicle operating on hydrogen and also on gasoline and/or compressed natural gas. As expected, it was not possible to achieve close to zero emission as the engine used CNG, which is a carbonaceous fuel.

BMW, Ford, and Toyota [4–9] have developed dedicated hydrogen vehicles built on ICEs. BMW's Hydrogen 7 was a revolutionary development in sustainable automotive technology. BMW's Hydrogen 7 vehicles are designed to operate in a bi-fuel mode. They have a 12-cylinder engine, with each cylinder having 500 cm^3 displacement. In the gasoline mode, the engine uses direct injection, and it switches when running on hydrogen. In this engine external mixture formations were adopted, where hydrogen was mixed with air before entering the combustion chamber. The engine exhibited a significant reduction in greenhouse gas emissions. This limited production vehicle was based on the overall vehicle concept carried over from the BMW 760i and BMW 760Li and featured a modified 6 L V12 engine. The dual-mode 12-cylinder engine could be operated on both hydrogen and gasoline with the maximum speed limited to 230 km/h. Maximum output was 191 kW. In 2004, BMW unveiled the BMW H2R, a modified 6.0 L V12 engine from the BMW 760i that was designed to run on liquid hydrogen fuel. With its lightweight design built on a carbon fiber body and aluminum frame, the H2R produced 232 hp, reaching a speed of 302.4 km/h and accelerating from 0 to 100 km/h in just 6.1 seconds. Safety features include an air-cooling process for each cylinder and a specialized fuel injection system for hydrogen. Liquid hydrogen was stored in a cryogenic tank at −253°C within a carbon fiber-reinforced plastic (CFRP) safety cell. At the 2001 Internationale Automobil-Ausstellung (IAA) in Frankfurt, the MINI Hydrogen Concept Car was demonstrated at the motor show as part of BMW Group's commitment to hydrogen power. The car featured a four-cylinder engine utilizing cryogenic liquid hydrogen. Ultra-cold liquid hydrogen was injected into the intake duct. Hydrogen tanks were placed under the rear seats so that the cabin and luggage space were not disturbed. A six-city global tour of these BMW 7 series hydrogen-powered vehicles established the feasibility of practical and scalable hydrogen integration into the automotive sector.

A 1992 Ford Ranger truck at the University of California, Riverside (UCR) was converted to run on hydrogen produced via water electrolysis. The modified 2.3 L engine used a lean-burn hydrogen fuel system with constant volume injection (CVI) and closed-loop control. The vehicle demonstrated excellent start, idle, and shutdown, with a range over 161 km and without any significant symptoms of abnormal combustion. At speeds above 64 km/h, performance was comparable to gasoline vehicles, but operation at low ranges of speed was not very satisfactory. Emissions of CO, CO_2, and hydrocarbons were negligible, and the very insignificant amount observed was caused by oil seeping from the piston rings. The NOx emission rate, measured by the Federal Test Procedure (FTP-77), was 0.23 g/km. On-road and chassis dynamometer tests indicated that a precise fuel injection system was essential for optimal performance, driveability, and low NOx emissions. Operating at an equivalence ratio below 0.5 was vital for reducing NOx, but it significantly impaired drivability [10]. In 2000, a gasoline-powered 427 Shelby Cobra was converted to run on hydrogen at the University of California, Riverside (CE-CERT) with the objective of assessing its capability to perform at higher speed ranges.

The modified 427 Ford FE engine used electronic fuel injection and operated on a quality control, lean operation mode to achieve near-zero (less than 10 ppm) NOx without any specific pollution control device [11]. In late 1997, Ford Motor Company began developing a low-cost hydrogen-fueled internal combustion engine (H_2ICE) vehicle to promote hydrogen infrastructure ahead of fuel cell vehicles. The P2000 H_2ICE concept vehicle, featuring a modified 2.0 L Zetec gasoline engine, significantly reduced emissions and improved efficiency by 25 to 30% compared to gasoline engines. The engine dynamometer development program and the unique triple-redundant vehicle safety system were implemented into the Ford P2000 experimental vehicle.

In its continued effort to demonstrate the feasibility of hydrogen-powered ICE vehicles, Ford developed a fleet of 30 E-450 shuttle buses with 6.8 L Triton engines running on hydrogen. The design, development, and engineering of the hydrogen fuel storage and delivery system, hydrogen monitoring, telematics, and the vehicle engineering validation program were considered. This study showed the relevance of the telematics system in monitoring data that can provide significant information on future decisions for improved design and development of new hydrogen vehicles [5]. The Ford 250 Super Chief is a versatile, multi-fuel vehicle equipped with a V10 turbocharged engine, producing nearly 542 Nm of torque. Its Tri-Flex system had the unique flexibility of running on gasoline, E85, or hydrogen. Fuel switching was done effectively via a dash-mounted switch while driving, except in the situation of operating in the turbocharged-hydrogen mode, which requires idling. At a speed range of 800 km in the hydrogen mode it produced 99% lower CO_2 emission. The safety innovation Belt Minder was used for fastening the seat belts, and the Blocker Beam improved crash compatibility with smaller vehicles.

Toyota put its GR Yaris H2—a special hydrogen-engine version of the Toyota GR Yaris—on a demonstration run during the ninth round of the World Rally Championship (WRC) in Ypres in August 2022. That was the first time Toyota drove an under-development hydrogen engine vehicle on public roads outside Japan. The GR Yaris H2 took to the rally course and covered 15 km along extremely narrow roads. The hydrogen fuel, fuel tanks, and refueling process of the experimental vehicle were the same as the Mirai, Toyota's commercially available fuel cell electric vehicle. The hydrogen-powered experimental GR Yaris and Corolla Sport both feature the same G16E-GTS, 1.6L, in-line three-cylinder and turbocharged engine with a modified fuel injection system to use hydrogen. The hydrogen-powered Corolla Sport demonstrated high performance at motorsport events in Japan with almost zero tailpipe emissions [9]. In April 22, Toyota announced its development of a hydrogen-fueled engine, integrating it into a Corolla Sport model for entry into the Super TEC 24 Hour Race [12]. The vehicle combined the Corolla road car chassis with running gear, including the four-wheel drive system from the Yaris GR. Minimum modifications were made to convert to hydrogen fuel. Initial tests used a blend of 50% gasoline and 50% hydrogen before switching over to 100% hydrogen. The major challenge was designing injection to achieve optimum performance with stable and efficient combustion over the entire range of operation.

Rolls-Royce's Phoenix project—a three-year collaboration with five companies and research institutes—aims at developing a highly efficient hydrogen combustion engine for combined heat and power (CHP). The consortium, including the Technical University of Munich, MAHLE Konzern, Fuchs Lubricants, BAM, and Robert Bosch AG, have concentrated on overcoming the technical challenges for making H_2ICE climate-friendly. Funded by the German Federal Ministry for Economic Affairs and Climate Protection, the project has a nearly five million euro investment. Cosworth, the British automotive company, in 2022 began developing hydrogen engines, marking a significant milestone in their journey toward eco-friendly automotive solutions.

Researchers at several academic and research institutes have developed hydrogen vehicles based on internal combustion engines, which also have been tested on the road. Research focused on conducting extensive tests on the performance, emissions, combustion, storage, distribution, and safety of hydrogen. Although most initial research was restricted to laboratory settings, the experience gathered from extensive tests offered crucial guidelines for adopting hydrogen engine technology in vehicles in subsequent phases. The optimized parameters generated through lab tests were transferred to field operations to maximize thermal efficiency, reduce fuel consumption, minimize NOx emissions, and ensure smooth combustion without backfire and knocking.

Research activities at Mushashi Institute of Technology (now Tokyo City University) were consistently followed from 1974 to 1997. Furuhama and his team worked on several engine configurations and integrated the optimized technology generated during engine tests into actual vehicles such as Mushashi 1, Mushashi 2, and Mushashi 10 that were developed [13–15].

In 1975, a group of dedicated students at the University of California, Los Angeles (UCLA), as part of their project developed a hydrogen-powered car. This efficient low-emission hydrogen vehicle demonstrated safe operation under the dynamic and complex traffic of Los Angeles, where the vehicle was regularly driven and tested [16].

As part of a UNIDO-sponsored project, Indian Institute of Technology (IIT), Delhi along with the well-known automobile Indian manufacturer Mahindra and Mahindra and Air Products (USA) collaborated to develop a hydrogen-powered three-wheeler. The base CNG operated vehicular engine was modified for hydrogen operation using an electronically controlled timed manifold injection system. IIT Delhi conducted engine tests in the lab under various operating conditions, followed by Mahindra and Mahindra's field tests on passenger and cargo vehicles using optimized settings. Ten of the fifteen vehicles were for passenger transport and five for carrying loads. After UNIDO's phase, the Government of India's Ministry of New and Renewable Energy (MNRE) funded continued field tests. These field tests confirmed the improved efficiency, almost zero-emission characteristics and safe operation of hydrogen-fueled vehicles [17,18].

Aston Martin, in collaboration with Alset GmbH, developed the hydrogen Rapide S engine for road tests to serve as a proof of concept, demonstrating the viability of hydrogen as an alternative to battery hybrid systems. The Rapide S was selected for its ample space to accommodate four hydrogen tanks: two located over the rear axle and two replacing the front passenger seat. These tanks store 3.3 kg of hydrogen at 350 bar pressure. Hydrogen fuel was delivered to Aston Martin's 6.0 L V12 engine through hydrogen injectors positioned alongside existing gasoline injectors. The engine itself required no additional hardware to switch between gasoline, hydrogen, or a combination of both fuels. Major modifications included a reduced compression ratio, a redesigned intake manifold, and an updated cylinder head gasket. Due to hydrogen's lower energy density than gasoline, turbochargers were added to the V12 engine to compensate for power loss. However, these turbochargers did not provide any power benefit in the gasoline mode but significantly improved midrange torque. Using hydrogen alone, the engine's power output was reduced by 50–60 BHP. The innovative design and field operation proved that appropriate modification of existing engines to hydrogen operation is a prospective technology [19].

Kohler Engines has launched a direct injection hydrogen (KDH) engine, built upon the KDI 2504 TCR engine by replacing diesel with hydrogen. However, the original engine's 74 hp, dimensions, installation, power take-offs (PTOs) were left undisturbed. This novel engine aligned well within the same space in diesel with no changes on critical components like cooling circuit, power take-offs, flywheel, engine mounts, and intake line. In addition, it eliminated the need for any after-treatment under Stage V emissions regulations [20]. The KDH engine with direct injection technology could successfully avoid backfires and preserve diesel-like performance, including robust transient response.

Several studies have been conducted on hydrogen use in heavy-duty vehicular engines and trucks. The Ministry of Land, Infrastructure, Transport and Tourism (MLIT), Japan initiated a multi-year project on development of next-generation environmentally friendly vehicles (EFVs). The implementation of this project was entrusted to the National Traffic Safety and Environment Laboratory (NTSEL) with the objective of developing heavy-duty direct injection multi-cylinder hydrogen engines with appropriate combustion strategies to meet NOx emission standards. The H_2ICE truck development project in Japan adopted direct-injection H_2ICE and a built-in liquid hydrogen pump within the liquid hydrogen tank system. A NOx storage-reduction (NSR) catalyst proved to be very effective in conversion rate with negligible impact on fuel economy [21]. Naganuma et al. studied HPDI strategy for high efficiency and low NOx emissions in heavy-duty hydrogen ICE trucks. An indicated thermal efficiency of over 45% was achieved at IMEP of 0.95 MPa. The researchers found that a hydrogen engine with a high-pressure direct-injection system has a high potential for improving thermal efficiency and reducing NOx emissions. Different engine management strategies involving the injection pressure, injection timing, ignition timing, lambda, and the EGR rate were also evaluated under a Japanese emissions test cycle. This study showed that with an optimized engine management strategy, the hydrogen engine could achieve high power output and energy efficiency comparable to diesel engines while maintaining low NOx emissions of 0.5 g/kWh [22].

10.3 Heavy-Duty Engines, Trucks, and Generating Sets

Many researchers in several parts of the world have been working on a hydrogen engine for high-capacity long-haul, heavy-duty trucks. Development work on a hydrogen truck in Japan has already been briefly described [21]. Kawamura et al. developed the appropriate combustion control strategy for low NOx emissions with high power output using HPDI for transient driving of multi-cylinder trucks [23]. A NOx storage-reduction (NSR) catalyst system demonstrated a high NOx conversion rate without affecting fuel economy. JE05 mode transient emission tests showed emissions of the direct injection hydrogen engine to be lower than Japan's 2009 regulations using an NSR catalyst and a diesel oxidation catalyst [23].

Volvo Trucks is making significant strides in developing hydrogen-fueled ICE trucks and plans to commercially launch these trucks by end of this decade after on-road tests scheduled for 2026. These trucks would have high-pressure direct injection (HPDI) technology with a small amount of fuel injected at high pressure to facilitate compression ignition before adding hydrogen. Volvo Group is establishing a joint venture in collaboration with Westport Fuel Systems to implement the HPDI technology in the near future. This approach improves energy efficiency and reduces fuel consumption and also increases engine power. Hydrogen trucks are particularly advantageous for long-distance heavy-duty transport and in areas with sparse charging infrastructure or limited-time battery recharging [24].

Sinnz et al. [25] operated a Volkswagen Polo 1.4 in a bi-fuel mode running on both hydrogen and gasoline by adding a 200 bar compressed hydrogen storage system placed in the car trunk. They developed an innovative electronic control unit for hydrogen operation and electrical junction box for seamless fuel switch-over without stalling the car. Successful road tests at a top speed of 125 km/h on hydrogen and a consumption rate of 1 kg per 100 km at an average speed of 90 km/h established the feasibility of bi-fuel operation. Figure 10.1 shows the machined intake manifold with low-pressure hydrogen accumulator where hydrogen injectors are assembled. As part of EU-funded Flemish hydrogen Ports Valley Project, trucks and ships were identified as end users [26]. Based on successful outcomes, the fleet was expanded to 75 trucks. This large-scale implementation provided crucial insights into designing an effective technology with its operational efficiency in a logistics environment. To support this, Colruyt also installed indoor hydrogen dispensers in 2016 to facilitate refueling speed and operational convenience for these forklifts [27].

Commercial vehicle manufacturer MAN Truck & Bus is set to launch hydrogen combustion engines and deliver about 200 units in Germany, the Netherlands, Norway, Iceland, and selected non-European countries in 2025. Powered by the innovative H45 hydrogen combustion engine, the hTGX (as the vehicle is called) is based on MAN's well-known D38 diesel engine modified for hydrogen operation. It adopts direct cylinder injection of compressed hydrogen gas stored at 700 bar, with tank capacity of 56 kg, refueling in under 15 minutes. The engine delivers 383 kW (520 hp) and 2500 N·m of torque between 900 and 1300 rpm. Emissions from the hTGX would be less than 1 g of CO_2 per km, meeting stringent criteria for zero-emission vehicles under upcoming EU CO_2 legislation. Designed for heavy goods transport, construction, tank transport, and timber hauling, the hTGX offers a solution in areas with insufficient charging infrastructure or abundant hydrogen supply [28].

The EU-funded HyICE Project [29] carried out extensive studies to optimize hydrogen combustion engines and generated lots of data applicable to vehicles. The project focused on two key mixture formation principles: direct injection (DI) and cryogenic port injection (CPI). CPI targets near-term economic drivetrains, while DI offers a wider scope beyond the CPI limits. High-pressure injection

Figure 10.1 Low-pressure hydrogen accumulator with injectors [25] / with permission of Elsevier.

allowed variable timing and facilitated controlled fuel stratification. Extensive experiments at Graz University of Technology (TUG) with a single-cylinder engine in a modern car configuration and a transparent engine provided extremely valuable information on mixture formation and combustion. A better understanding on stratification for optimal efficiency and strategies of injection from these studies were developed to improve noise behavior in passenger cars. A 12.8 L six-cylinder engine was designed for hydrogen, combining 3D-CFD simulation at MAN and optical measurements at University of Armed Forces Munich (UBW) and TUG. Various injector nozzles were tested and homogenized. With a turbocharger, the engine quickly reached a maximum power output of 200 kW and an efficiency of 42%. Lean burn operation further improved the emission characteristics. Detailed experiments and CFD data were validated for engine operation at high compression ratios with an indicated efficiency of 44% and stable combustion. Turbocharged operation reached 100 kW/L displacement. 3D calculation models for hydrogen mixture formation and combustion were validated by IFP and UBW through literature and experiments. UBW designed a mixing chamber test bed using schlieren imaging and high-speed cameras for detailed turbulent mixture data. Laser-induced fluorescence (LIF) and Raman spectroscopy were used in an optical engine by TUG to visualize hydrogen mixture formation and combustion processes.

In May 2022, Cummins debuted its 15 L hydrogen engine at the ACT Expo in Long Beach, California. Built on Cummins' versatile platform, the engine was basically designed to operate on multiple fuels with slight alteration in parts above the head gasket to accommodate different components for different fuels. Almost identical components were used below the head gasket for various types of fuel. Tests on hydrogen engines in July 2021 were successful in achieving production power suitable for medium-duty engines. Cummins plans to produce hydrogen internal combustion engines in both 15 L and 6.7 L displacements at lower costs than fuel cell or battery electric vehicles. The rapid penetration of hydrogen-powered vehicles in the marketplace is expected because of their high technology maturity, lower initial costs, extended range, quick refueling times, compatibility with existing powertrain installations, and end-user familiarity [30].

Tata Motors began integrating the B6.7H hydrogen internal combustion engines produced at their Jamshedpur (India) facility consequent to the formation of TCPL Green Energy Solutions (GES), a 50:50 joint venture between Tata Motors and Cummins Inc. In March 2024, the first B6.7H rolled off the production line and was integrated into Tata Motors' trucks. The goal is to decarbonize medium and heavy-duty commercial vehicles in India. The B6.7H engine, now in Tata Motors' trucks, brings robust technical capabilities. This 6.7 L in-line six-cylinder engine delivers 290 to 300 hp and over 1000 Nm of torque, using advanced high pressure direct injection (HPDI) technology for efficient hydrogen combustion. It shares many components with diesel and natural gas engines, installs easily into existing engine spaces, and offers quick refueling comparable to traditional diesel systems [31].

Hydroville is a hydrogen-powered passenger ferry developed in 2017 by Compagnie Maritime Belge-CMB Technologies for operations on the Scheldt River in Antwerp's port. It is a combination of a hydrogen fuel and diesel engine designed for agile and efficient river navigation. It cruises at an average speed of 22 kn and can reach up to 27 kn. The emission level is much lower than that of conventional fossil fuel–powered ships. Safety was of great concern during the development stage, with multiple risk analyses performed. Certification by Lloyd's Register Group has ensured its compliance with stringent maritime standards. Weighing 12 t with a full load displacement of 14 t, Hydroville represents a significant step toward CMB's vision of an emission-free passenger vessel.

Hydrobingo is another hydrogen-powered ferry with hydrogen–diesel dual-fuel technology launched on August 10, 2021. The vessel, operating in the Japanese inland sea, is the outcome of a collaborative effort between CMB.TECH and Japan's Tsuneishi Facilities & Craft (TFC).

Hydrobingo's marine D13 diesel engine has been converted to integrate a hydrogen injection system and thus operate in hydrogen–diesel dual-fuel mode. Rigorous safety measures have been implemented in collaboration with Japanese authorities to ensure the vessel's compliance with hydrogen fuel standards at all stages of storage, piping, and engine retrofitting. Hydrobingo's improved efficiency and reduced emissions mark a significant leap in maritime technology with the prospect of hydrogen–diesel dual fuel combustion in larger vessels like tugboats and barges.

Hydrotug, developed by CMB.TECH, is a tugboat powered by a dual-fuel (hydrogen and diesel) engine that arrived in Ostend, Belgium, in 2022. This tugboat, operated at the Port of Antwerp, is part of the port's comprehensive integrated greening program aimed at climate neutrality by 2050. The infrastructure of hydrogen refueling stations was also developed at the port [32]. Ronn Motors introduced a twin-turbocharged V6 hydrogen-operated Scorpion™ at the Specialty Equipment Market Association (SEMA) show in 2009 in Las Vegas. The base engine was derived from Acura's V6. The Scorpion was available with both 450 and 650 hp options, each equipped with the company's proprietary H2GO hydrogen injection system [33].

Hydrogen-fueled engine generators present a versatile and eco-friendly solution for portable power generation. At the Indian Institute of Technology, Delhi, both spark ignition (SI) and compression ignition (CI) engine gensets were developed. The SI engine genset (Figure 10.2) runs

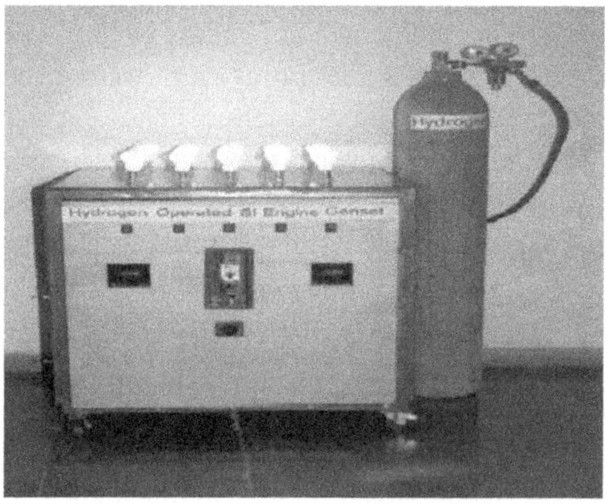

Figure 10.2 Hydrogen SI engine genset.

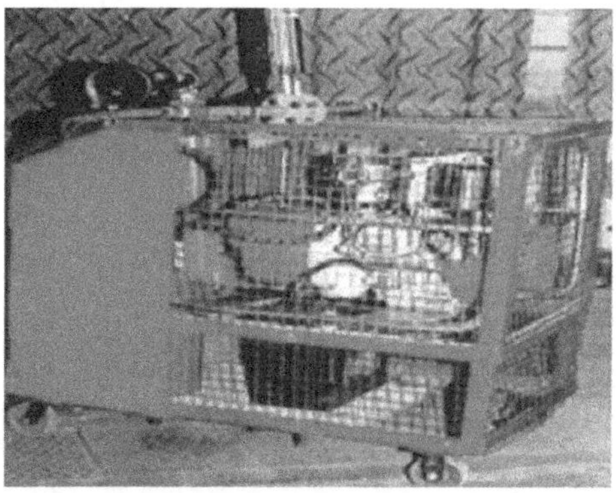

Figure 10.3 Hydrogen–diesel dual-fuel engine genset.

entirely on hydrogen, and diesel gensets operate in a dual-fuel mode with diesel and hydrogen (Figure 10.3). These are mounted on trolleys for easy transport and can be used at various locations in need of power. These engines can also be used in pump sets and irrigation.

10.4 The Rotary Engine

Mazda's pioneering journey of hydrogen technology in rotary engines started with the introduction of the Mazda HR-X at the 1991 Tokyo Motor Show with the Wankel rotary engine. Subsequently, in 1995 the Mazda Capella Hydrogen RE became the first vehicle to test a hydrogen rotary engine on public streets, demonstrating the feasibility of hydrogen cars. The journey continued with the 2003 Mazda RX-8 Hydrogen RE, a bi-fuel version (Figure 10.4) of the RX-8 sports car, which could switch between hydrogen and gasoline. In 2005, Mazda introduced the Premacy Hydrogen RE Hybrid, later known as the Mazda5 Hydrogen RE Hybrid, in which a hydrogen rotary engine was integrated with a hybrid electric drive. This unit showed wider application of hydrogen with improved efficiency and environmental sustainability. In collaboration with the Norwegian national hydrogen project, HyNor, the Mazda RX-8 Hydrogen RE (Rotary Engine) was introduced to Norway's public roads in 2008.

Ricardo [35] has developed hydrogen engines that offer promising solutions in situations where it is difficult to develop adequate infrastructure, such as for off-highway construction equipment, long-haul heavy-duty trucks, and marine propulsion. Existing designs of the Euro VI heavy-duty diesel and natural gas engine have been tested. The cylinder head for tests was a production CNG cylinder head modified for the direct injection of hydrogen, which precludes the possibility of back-fire. Specifications for tests with a single cylinder and a multi-cylinder engine are shown in Table 10.1.

Tests were carried out over varying speed and load, usually in the ranges applicable to heavy-duty engines. The effect of significant parameters such as injection timing, ignition timing, and equivalence ratio were studied. Smooth combustion without misfire could be achieved at an EGR rate of more than 45% [35]. At the American Trucking Association's annual management conference in Austin, Texas, Peterbilt demonstrated the DAF XF Euro-style high cab-over truck equipped with a modified Paccar MX-13 engine operated on hydrogen [36]. The engine produced 295 hp and emitted a much reduced amount of nitrogen oxides (NOx).

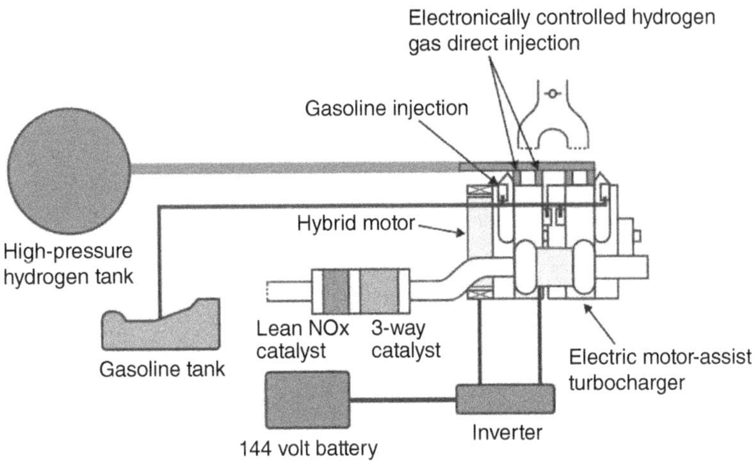

Figure 10.4 Hydrogen/gasoline bi-fuel rotary engine [34].

Table 10.1 Tests for single cylinder and multi-cylinder engine with hydrogen injection [35].

Parameters/Specifications	Single cylinder engine	Multi-cylinder engine (six cylinders)
Compression ratio	12.6:1	11.9:1
Rated power speed (rpm)	1800	1800
Piston and bottom end	Diesel piston with modified bowl	CNG engine piston
Fuel injection system	Direct injection	Direct injection twin port injection
Fuel pressure (bar)	35	35 (direct injection) 12 (port fuel injection)

Concluding Remarks

There have been efforts to phase out ICE-powered cars on which there have been conflicting views and pushbacks. Transport energy's future is dependent on several technical and economic challenges such as technology, infrastructure, economics, and regional differences. The influence of these drivers is currently at a very low base. Thus, a global ban on internal combustion engines from the transport sector is not possible in the foreseeable future. Alternatives such as EVs and FCEVs have been adopted in several countries, while many countries face economic and infrastructure hurdles to ban conventional IC engines. At such a critical juncture, a balanced strategy would be to move away from fossil fuels and switch over to hydrogen to ensure the safe survival of internal combustion engines. Mature technology has already evolved which shows that no substantial hardware modification is necessary to operate an existing engine design with hydrogen. Optimizing internal combustion engines by installing an appropriate fuel injection system and optimizing operating parameters such as compression ratio (CR), ignition timing, and equivalence ratio, etc. can significantly improve engine efficiency and reduce fuel consumption. It is possible to achieve precise control of the combustion process leading to near-zero emissions and improved overall performance.

Recent developments in hydrogen engines are encouraging for its mainstream adoption for sustainable transportation. Air Products has announced its plan to build a network of commercial-scale multi-modal hydrogen refueling stations, capable of fueling up to 200 heavy-duty trucks or 2000 light-duty vehicles daily in California. Alberta Motor Transport Association's (AMTA) research carried out the long-distance use of hydrogen trucks in Alberta's extreme cold weather conditions. Field trials were conducted to assess the effect of extreme cold temperatures on the 20 carriers using Hydra Energy dual-fuel trucks and Hyzon and Nikola FCEVs. The automotive industry, with BMW, Ford, and Mazda at the forefront, is embracing hydrogen as a fuel for internal combustion engines (ICE) and not restricted just to fuel cells.

References

1 Khan, A. BMW says goodbye to electric cars; it has now solved the problem of hydrogen engines. https://myelectricsparks.com/bmw-hydrogen-engines/.

2 Eckermann, E. (2001). *World History of the Automobile, Volume 272, Premiere Series Books.* Warrendale, PA: SAE International.

3 Huyskens, P., Van Oost, S., Goemaere, P. et al. (2011). The technical implementation of a retrofit hydrogen PFI system on a passenger car. SAE International. SAE Paper No. 2011-01-2004.

4 Kiesgen, G., Klüting, M., Bock, C. et al. (2006). The new 12-cylinder hydrogen engine in the 7 series. The H2 ICE age has begun. SAE Paper No. 2006-01-0431.

5 Gopalakrishnan, R., Throop, M., Richardson, A. et al. (2007). Engineering the ford H2 IC engine powered E-450 shuttle bus. SAE Technical Paper 2007-01-4095.

6 Lapetz, J. Natkin, R., Zanardelli, V. et al. (2006). The design, development, validation and delivery of the Ford H2 ICE E-450 shuttle bus. In: *1st (International) Symposium on Hydrogen Internal Combustion Engines*, pp. 20–33.

7 Wallner, T., Lohse-Busch, H., Gurski, S. et al. (2008). Fuel economy and emission evaluation of a BMW hydrogen 7 monofuel demonstration vehicle. *International Journal of Hydrogen Energy* 33 (24): 7607–7618.

8 BMW. (2006). Hydrogen 7 Rescue Guidelines.

9 Toyota Motor Corporation. "Morizo" Puts Hydrogen-Engine GR Yaris through Its Paces on Belgian Roads | Toyota | Global Newsroom. Toyota Motor Corporation Official Global Website. https://global.toyota/en/newsroom/toyota/37888377.html.

10 Heffel, J.W., McClanahan, M.N., Norbeck, J.M. et al. (1998). Turbocharged hydrogen fueled vehicle using constant volume injection (CVI). SAE Technical Paper Series. Vol. 1. https://doi.org/10.4271/981922.

11 Heffel, J.W., Johnson, D.C., and Shelby, C. (2001). Hydrogen powered Shelby Cobra: vehicle conversion. SAE Technical Paper No. 2001-01-2530. https://doi.org/ 10.4271/2001-01-2530.

12 Butcher, L. (2021). Toyota successfully completes Fuji 24-hour race with hydrogen engine. *Professional Motorsport World*. Automotive Powertrain Technology International.

13 Furuhama, S. (1981). State of the art and future trend of hydrogen-fueled engines. *JSAE Review* 4. Society of Automotive Engineers of Japan.

14 Furuhama, S., Hiruma, M., and Enomoto, Y. (1978). Development of a liquid hydrogen car. *International Journal of Hydrogen Energy* 8: 61–81.

15 Furuhama, S. and Fukuma, T. (1986). High output power hydrogen engine with high pressure fuel injection, hot surface ignition and turbocharging. *International Journal of Hydrogen Energy* 11: 399–407.

16 Finegold, J., Lynch, F., Baker, N. et al. (1973). The UCLA hydrogen car: design, construction, and performance. SAE Technical Paper 730507. https://doi.org/10.4271/730507.

17 Das, L.M. (2016). Hydrogen fuelled internal combustion engine (Chapter 7). In: *Compendium of Hydrogen Energy*, Volume 3. Sawston, UK: Woodhead Publishing.

18 Natarajan, S., Abraham, M., Rajesh, M. et al. (2013). DelHy 3W -- hydrogen fuelled Hy-Alfa three wheeler. SAE Technical Paper 2013-01-0224.

19 Tisshaw, M. (25 May 20). Hydrogen-powered Aston Martin–the story behind the racer. www.autocar.co.uk › car-news › motorsport.

20 Dokso, A. (15 Nov. 2023). Kohler's hydrogen-powered engine redefines clean combustion. energynews.biz/kohlers-hydrogen-powered-engine-redefines-clean-combustion.

21 Kawamura, A., Yanai, T., Sato, Y. et al. (2009). Summary and progress of the hydrogen ICE truck development project. SAE International. SAE Paper No. 2009-01-1922.

22 Naganuma, K., Honda, T., Yamane, K. et al. (2009). Efficiency and emissions-optimized operating strategy of a high-pressure direct injection hydrogen engine for heavy-duty trucks. SAE International. SAE Paper No. 2009-01-2683.

23 Kawamura, A., Sato, Y., Naganuma, K. et al. (2010). Development project of a multi-cylinder DISI hydrogen ICE system for heavy duty vehicles. SAE International. SAE Paper No. 2010-01-2175.

24 Heavy Duty Trucking. (24 May 2024). HDT Trucking info: Volvo plans trucks with hydrogen-powered engines by end of decade.

25 Sainz, D., Dieguez, P., Sopena, C. et al. (2012). Conversion of a commercial gasoline vehicle to run bi-fuel (hydrogen gasoline). *International Journal of Hydrogen Energy* 37: 1781–1789.

26 Flemish Hydrogen Ports Valley. https://h2v.eu/hydrogen-valleys/flemish-hydrogen-ports-valley.

27 Demonstration of hydrogen forklift trucks. Hydrogen Region 2, Flanders, Netherlands. https://www.waterstofnet.eu/en/projects/vehicles/demonstratie-heftrucks-op-waterstof-waterstofregio-2-0.

28 Green Car Congress (09 April 2024). MAN plans truck series with hydrogen combustion for 2025; initially around 200 vehicles for selected markets. https://www.greencarcongress.com/2024/04/20240409-man.html.

29 Final Report Summary–HYICE (Optimisation of hydrogen powered internal combustion engines). https://cordis.europa.eu/docs/results/506/506604/pre-070213-hyice-publishable-summary-hs-fg-v11-final.pdf.

30 Cummins. (2022). Cummins Inc. debuts 15-litre hydrogen engine at ACT Expo, 09 May 2022, Columbus, Indiana. https://www.cummins.com/news/releases/2022/05/09/cummins-inc-debuts-15-liter-hydrogen-engine-act-expo.

31 Cummins. (2024). Cummins produces hydrogen internal combustion engines for Tata Motors at new manufacturing facility, 01 May 2024, Jamshedpur. https://www.cummins.com/news/releases/2024/05/01/cummins-produces-hydrogen-internal-combustion-engines-tata-motors-new.

32 cmb.tech. (2019). Revolve Technologies Limited joins the CMB group.

33 Hardigree, M. (05 November 2008). Ronn Motors Scorpion hydrogen internal combustion supercar debuts at SEMA. https://jalopnik.com/ronn-motors-scorpion-hydrogen-internal-combustion-super-50775742.

34 Cogan, R. (27 August 2021). Mazda's hydrogen RX-8 RE. greencarjournal.com › dont-miss › mazdas-hydrogen-rx.

35 Osborne, R. Heavy-duty hydrogen engines for truck, marine and non-road applications Decarbonising propulsion. www.ricardo.com › media › ifpnbc2p.

36 *Tank Transport*. (21 December 2023). Peterbilt hydrogen engine: a sustainable revolution in trucking. https://tanktransport.com/2023/12/peterbilt-hydrogen-engine-innovation.

Index

Note: Page numbers in *italics* and **bold** refers to figures and tables respectively.

Hydrogen Engines: Design, Performance Evaluation, Combustion Analysis, and Exhaust Emissions, First Edition.
Lalit Mohan Das.

b

c

f

g

h

n

q

r

t